Chemically-Induced DNA Damage, Mutagenesis, and Cancer

Chemically-Induced DNA Damage, Mutagenesis, and Cancer

Special Issue Editors

Ashis K. Basu
Takehiko Nohmi

MDPI • Basel • Beijing • Wuhan • Barcelona • Belgrade

MDPI

Special Issue Editors

Ashis K. Basu
University of Connecticut
USA

Takehiko Nohmi
National Institute of Health Sciences
Japan

Editorial Office
MDPI
St. Alban-Anlage 66
Basel, Switzerland

This is a reprint of articles from the Special Issue published online in the open access journal *International Journal of Molecular Sciences* (ISSN 1422-0067) from 2017 to 2018 (available at: http://www.mdpi.com/journal/ijms/special_issues/chemical_induced_mutagenesis)

For citation purposes, cite each article independently as indicated on the article page online and as indicated below:

LastName, A.A.; LastName, B.B.; LastName, C.C. Article Title. *Journal Name* **Year**, *Article Number*, Page Range.

ISBN 978-3-03897-129-0 (Pbk)
ISBN 978-3-03897-130-6 (PDF)

Contents

About the Special Issue Editors

Ashis Basu, Ph.D., is Professor of Chemistry at the University of Connecticut. The research focus of the Basu laboratory is the determination of the consequences of DNA damaged by anti-tumor drugs, chemical carcinogens, oxidation, or radiation. This research at the interface of chemistry and biology involves the introduction of specific lesions in DNA by organic synthesis, investigation of the structural effects of the lesions using biophysical methods, and studying their repair and replication.

Takehiko Nohmi, Ph.D., became a staff scientist at the National Institute of Health Sciences (NIHS), Tokyo in 1978. He was a postdoctoral fellow/associate at the Massachusetts Institute of Technology (MIT) (1986 to 1988). After returning to Japan, he became a section chief in 1989 and the Head of the Division of Genetics and Mutagenesis, NIHS, in 2008. He developed Salmonella typhimurium YG1021 and YG1024, strains which are highly sensitive to environmental mutagens, and gpt delta transgenic mice/rats for in vivo mutagenesis. He was involved in the discovery of novel DNA polymerase DinB, a prototype of Y-family DNA polymerase. He received the JEMS award and annual award from the Japanese Environmental Mutagen Society (JEMS) in 2006 and 1992, respectively. He was the President of the International Association of Environmental Mutagenesis and Genomics Societies (IAEMGS, 2013–2017). He retired from NIHS (Tokyo) in 2012 and is currently a scientist emeritus and a visiting scientist at NIHS (Tokyo).

Preface to "Chemically-Induced DNA Damage, Mutagenesis, and Cancer"

This volume provides a snapshot of the current advances in chemical carcinogenesis. It promotes the view that DNA damage is an important first step in the process of carcinogenesis. DNA repair processes are active in every cell to correct these damages, yet the replication of damaged DNA may occur prior to repair, resulting in gene mutations and the generation of altered proteins. Mutations in an oncogene, a tumor-suppressor gene, or a gene that controls the cell cycle give rise to a clonal cell population with an advantage in proliferation. The complex process of carcinogenesis includes many such events, but has generally been considered to comprise the three main stages known as initiation, promotion, and progression, which ultimately give rise to the induction of human cancer.

Fifteen articles published in a Special Issue of IJMS entitled "Chemically-Induced DNA Damage, Mutagenesis, and Cancer" provide an overview on the topic of the "consequence of DNA damage" in the context of human cancer with their challenges and highlights.

The topics of the published articles cover a wide range, which include study of DNA adducts, DNA polymerase, DNA repair and repair defects, nanoparticle-induced DNA damage and cell death, metal toxicity, apoptosis, risk of melanoma, chemoprevention, a radiosensitizer for primary culture tumor cells, the signaling of a tumor suppressor gene, and an adjuvant for chemotherapy.

The variety of articles published in this Special Issue underscores the diversity of this field of research. Evidently, this is only a glimpse of this very large area of research that includes thousands of active researchers working in many complementary directions to better understand the etiology of cancer and develop superior strategies for chemoprevention and cancer treatment.

Ashis K. Basu, Takehiko Nohmi
Special Issue Editors

International Journal of
Molecular Sciences

MDPI

Review

DNA Damage, Mutagenesis and Cancer

Ashis K. Basu

Department of Chemistry, University of Connecticut, Storrs, CT 06269-3060, USA; ashis.basu@uconn.edu;
Tel.: +1-860-486-3965

Received: 29 January 2018; Accepted: 20 March 2018; Published: 23 March 2018

Abstract: A large number of chemicals and several physical agents, such as UV light and γ-radiation, have been associated with the etiology of human cancer. Generation of DNA damage (also known as DNA adducts or lesions) induced by these agents is an important first step in the process of carcinogenesis. Evolutionary processes gave rise to DNA repair tools that are efficient in repairing damaged DNA; yet replication of damaged DNA may take place prior to repair, particularly when they are induced at a high frequency. Damaged DNA replication may lead to gene mutations, which in turn may give rise to altered proteins. Mutations in an oncogene, a tumor-suppressor gene, or a gene that controls the cell cycle can generate a clonal cell population with a distinct advantage in proliferation. Many such events, broadly divided into the stages of initiation, promotion, and progression, which may occur over a long period of time and transpire in the context of chronic exposure to carcinogens, can lead to the induction of human cancer. This is exemplified in the long-term use of tobacco being responsible for an increased risk of lung cancer. This mini-review attempts to summarize this wide area that centers on DNA damage as it relates to the development of human cancer.

Keywords: carcinogenesis; carcinogen; mutagen; metabolism; DNA adduct; tumor; chronic exposure; somatic mutation

1. History

In 1761, after use of tobacco for recreation became popular in London, physician John Hill wrote a book entitled "Cautions Against the Immoderate Use of Snuff". Hill's observations that tobacco snuff can cause "polypus" (i.e., small vascular growth on the surface of a mucous membrane) led to epidemiology studies nearly 200 years later in 1950s and 1960s, which convincingly established that tobacco smoking causes lung cancer. A few years after Hill's book was available, in 1775, Sir Percivall Pott of Saint Bartholomew's Hospital in London published a groundbreaking essay showing that exposure to soot leads to high incidence of scrotal cancer in young men worked as chimney sweeps, which he named the chimney-sweepers' cancer [1]. This was the first occupational link to cancer. This also was the first association to materials such as soot (a complex mixture of chemicals) to the etiology of cancer. He further hypothesized that the scrotum of young men working as chimney sweeps were particularly susceptible for scrotal cancer later in life, due to their chronic exposure to soot. Sir Percival Pott's remarkable insight notwithstanding, it took nearly seventy years to pass a law in the UK to protect children from working as chimney sweeps. Perhaps more remarkably, almost 150 years passed when additional studies were attempted on chemical carcinogens, even though an association between certain chemicals and cancer has been reported from time to time. For instance, in 1895 Rehn described the first cases of bladder cancer in German fuchsin dye manufacturing workers.

An important advance was made in the early 20th century, when Yamagiwa and Ichikawa, two Japanese investigators, developed the first animal assay for carcinogens [2,3]. They repeatedly applied the test compound(s), such as coal tar, on the skin of rabbit ears. Tumors were developed in the experimental animals after a few weeks. Later, rats and mice were found to be better suited for this type of assays [4]. Even though these assays are slow, arduous, and expensive, it continues to be the experimental approach to determine if a compound or a mixture of compounds cause tumorigenesis in mammals. In the 1930s Cook, Kennaway and coworkers were able to isolate and identify benzo[*a*]pyrene (B[*a*]P), a polycyclic aromatic hydrocarbon (PAH), as a potent carcinogen present in soot and coal tar [5,6]. Subsequently, other PAHs were isolated from coal tar and synthetic methods to prepare them were also developed. Over the years, many other groups of compounds and mixtures have been recognized as human carcinogens. Specifically in the 1930s and 1940s, reports of bladder cancers from DuPont and other American dye manufacturers were documented [7,8]. In addition to PAHs (in soot and coal tar) and aromatic amines (present in dyes) [9], numerous other classes of compounds including nitroaromatics [10], asbestos [11], chromium, nickel, and arsenic compounds [12], vinyl chloride [13], aflatoxins [14], diesel exhaust [15], and most notably, tobacco smoke [16], were found to cause cancer. Physical agents like UV light [17] and gamma radiation [18,19] also turned out to be carcinogenic.

2. Metabolic Activation and DNA Damage

In 1950, Boyland proposed that arene oxides are the major metabolites of PAHs that give rise to the phenols, dihydrodiols, and other oxidation products [20]. But the mechanism of the in vivo effects of these carcinogens was little understood until DNA was shown to be the genetic material responsible for coding for all biological processes [21], and the structure of DNA was elucidated by Watson and Crick [22] on the basis of Rosalind Franklin's unpublished crystal structure of DNA. It became gradually clear that many of the carcinogenic chemicals are metabolically activated to electrophilic species that bind to DNA or cause DNA damage [23–25]. Extensive investigations were performed to establish how each carcinogenic agent, either directly or following metabolic changes in their structures, damage DNA or form DNA adducts. As for example, B[*a*]P is converted to 7*S*,8*R*-B[*a*]P oxide by cytochrome P-450 (CYP) 1A1/1B1, which is hydrolyzed by microsomal epoxide hydrolase to form the (−)-7*R*,8*R*-dihydroxydihydro-B[*a*]P [26,27] (Figure 1). This *trans* dihydrodiol is then oxidized again by the same CYP 1A1/1B1 enzymes to form predominantly (+)-*anti*-B[*a*]P-7,8-dihydrodiol-9,10-epoxide, the most mutagenic and tumorigenic metabolite of B[*a*]P. The major adduct formed by this B[*a*]P metabolite is the (+)-*trans-anti*-B[*a*]PDE (Figure 1).

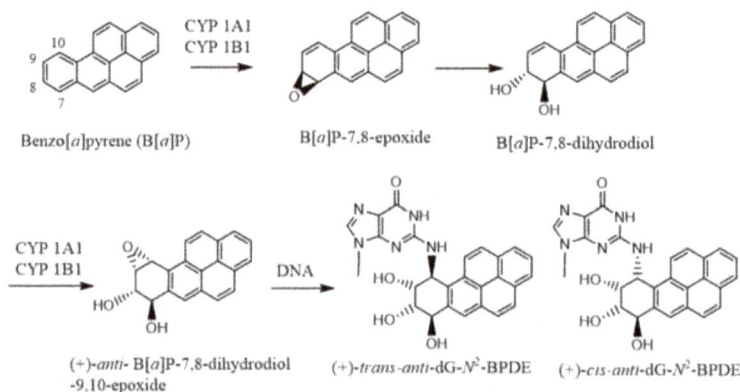

Figure 1. Microsomal metabolic activation of benzo[*a*]pyrene to its most reactive (+)-*anti*-B[*a*]P-7,8-dihydrodiol-9,10-epoxide, which reacts with DNA to form the dG adducts.

UV light, on the other hand, is an example of a direct acting agent that damages DNA [28–31], although it also damages DNA indirectly via reactive oxygen species. UV light is considered to be responsible for most skin cancers. UVB (280–320 nm) and UVC (240–280 nm) irradiation form *cis-syn* cyclobutane dimer and pyrimidine(6-4)pyrimidone photoproducts (Figure 2) as the main products in duplex DNA. The chemically stable (6-4) photoproduct may undergo conversion to its Dewar isomer by UVA or UVB light.

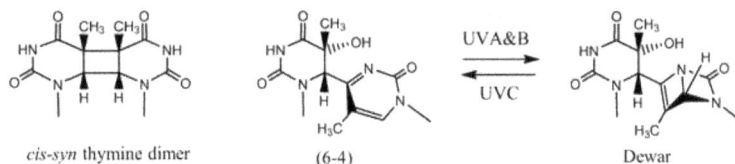

Figure 2. The chemical structures of UV light induced *cis-syn* thymine dimer, pyrimidine(6-4)pyrimidone and Dewar photoproducts formed by two adjacent thymines.

Similar to B[*a*]P, metabolism and DNA binding by a large number of chemical carcinogens have been reported. Figure 3 shows the chemical structures of a few of these carcinogenic compounds, which include PAHs, nitroaromatic compounds, aromatic amines, natural products, industrial chemicals, and a chemotherapeutic agent that also induces secondary tumor.

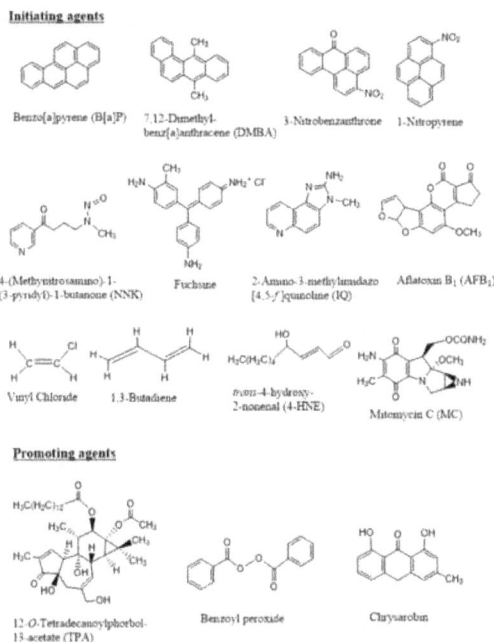

Figure 3. Chemical structures of a few initiating and promoting agents. The initiating agents shown here include polycyclic aromatic hydrocarbons (PAHs) (B[*a*]P and DMBA, present in soot, coal tar, and many environmental mixtures), nitroaromatic compounds (3-nitrobenzanthrone and 1-nitropyrene, present in diesel exhaust), tobacco-specific nitrosamine (NNK, present in tobacco smoke), an amine salt and a magenta dye (fuchsine), aromatic amine (IQ, formed during cooking of meat), a naturally occurring molecule produced by *Aspergillus flavus* (AFB₁, a food contaminant), industrial chemicals (vinyl chloride and 1,3-butadiene to make the polymer PVC and synthetic rubber, respectively), lipid peroxidation product (4-HNE, produced in cells and tissues of living organisms or in foods during processing or storage), and a chemotherapeutic agent (MC, a toxic drug used to treat upper gastrointestinal cancers). The promoting agents include the phorbol ester (TPA), benzoyl peroxide, and chrysarobin.

3. Multi-Step Process of Cancer

As early as in the 1940s, it became apparent that the process of carcinogenesis involves at least two distinct steps. In 1944, Mottram showed that a single application of a carcinogen, such as B[*a*]P, followed by multiple applications of an "irritant", such as croton oil, induce tumors in animals [32]. Berenblum and Shuvik followed up this study with application of either B[*a*]P or 7,12-dimethylbenz[*a*]anthracene (DMBA) and croton oil, and demonstrated that croton oil, a non-carcinogen, had no effect alone, but when applied after even a single dose of either B[*a*]P or DMBA on mouse skin, tumors were developed [32]. These results led to the hypothesis of "initiation" (result of application of the carcinogen like B[*a*]P) followed by "promotion" (caused by croton oil). Later, croton oil was shown to contain the phorbol ester, 12-*O*-tetradecanoylphorbol-13-acetate (TPA), as the active ingredient that is responsible for the promotion phase in carcinogenesis [33]. Additional tumor promoters, including benzoyl peroxide, okadaic acid, chrysarobin, have been identified (Figure 3).

There are several fundamental differences between these two stages (and the agents that trigger these processes) [34–37]. An initiating agent is also a "complete carcinogen", since either repeated exposure in small dosage or a single large exposure to such agents lead to carcinogenesis, whereas a promoting agent is not carcinogenic alone. The effect of an initiating agent, in addition, is irreversible and additive, in contrast to the reversible action of a promoting agent at the early stages. The initiating agents furthermore become electrophilic after metabolic activation, and bind to cellular macromolecules such as DNA, while there is no evidence of covalent binding by the promoting agents. The initiating agents are mutagenic and, as a result, quite a few short-term assays have been developed [38–41], whereas the promoting agents are not mutagenic. Experiments in rodents on the two-stage model, however, showed that mainly benign tumors were developed by tumor promoters [42]. It became gradually accepted that carcinogenesis involves multi-stages, which include initiation, promotion, and malignant progression, when benign neoplasms become malignant and invasive lesions [43] (Figure 4).

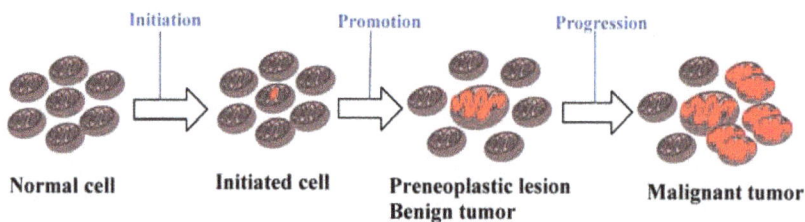

Figure 4. A brief depiction of initiation, promotion, and progression in the process of carcinogenesis.

Discovery of oncogenes and tumor suppressor genes added to the concept that carcinogenesis is a multi-step process [43,44]. Notably, continuous oxidative stress and chronic inflammation sustain each other, leading to neoplasm, and promote tumor progression. Inflammation has been associated with the development of cancer, and inflammatory mediators, like cytokines, chemokines, and eicosanoids, have been shown to stimulate the proliferation of both untransformed and tumor cells [45]. Certain initiating agents, such as UV light and tobacco smoke also exhibit strong tumor promoting activity.

Most of our understanding of tumor promotion comes from experiments performed on mouse skin [46]. The promotion stage in carcinogenesis induces a number of epigenetic changes, including proliferation of epidermal cells and activation of ornithine decarboxylase that leads to synthesis of polyamines [47–49]. Overall, the promotion stage is characterized by hyperplasia, that leads the initiated cells to form papillomas. Strong tumor promoters, such as the phorbol esters, activate membrane receptors like protein kinase C [50]. Activation of protein kinase C phosphorylation of critical proteins is considered an important event in skin tumor promotion. Several other tumor promotors, including benzoyl peroxide, appear to involve free-radical mechanisms, which indirectly lead to phosphorylation of certain proteins [51]. Tumor promotion is also characterized by clastogenic effect and genetic instability, resulting in chromosomal alterations. Consequently, tumor promotion includes a series of complicated epigenetic steps leading to formation of papillomas. Tumor promotion can also be induced by tumor necrosis factor-α (TNF-α) and TNF-α-inducing protein (Tipα) of *Helicobacter pylori* stimulates progression phase [52]. Recent studies on human cancer development includes upregulation of TNF-α and activation of NF-κB, an important transcription factor [52].

4. DNA Damage and DNA Repair

DNA damage occurs continuously in all organisms via a number of endogenous and exogenous factors, and it seems to play a central role in many biological processes, ultimately leading to cancer (Figure 5). Hence, robust DNA repair systems, which repair this damage, have evolved to maintain genomic integrity. The importance of DNA repair was underscored by conferring the Nobel Prize in Chemistry in 2015 to Tomas Lindahl, Paul Modrich, and Aziz Sancar for mapping, at a molecular level, how cells repair damaged DNA and protect the genetic information. There are a number excellent reviews on DNA repair, which summarize this rapidly evolving field [53–57].

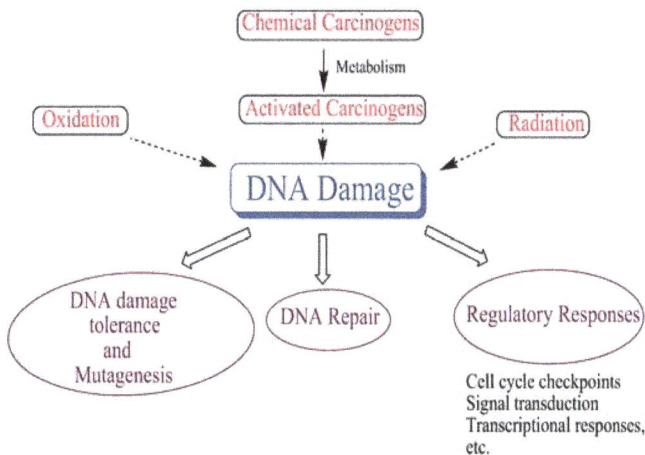

Figure 5. DNA damage plays a central role in many biological processes linked to cancer.

DNA replication occurs during the S (synthetic) phase of cell cycle, which is preceded by the G1 (Gap 1) phase. The nuclear division occurs in the M (mitosis) phase, which takes place after the G2 phase. The differentiated cells at the G0 phase do not proliferate, whereas the G1, S, and G2 phases of a proliferating cell constitute the time lapse between two consecutive mitoses. The progression of a cell during cell cycle is regulated by cyclin dependent kinase in order to avoid the initiation of a cell cycle before the preceding one is completed. DNA damage interferes with the cell cycle, and therefore, there are checkpoint proteins that delay cell cycle progression providing the necessary time for DNA repair. If the DNA damage exceeds the capability of repair, pathways to trigger cell death are activated by apoptosis. The checkpoint pathways accordingly play an integral role in DNA damage response, and dysfunction of these pathways are important for the pathogenesis of malignant cells [58].

5. Relationship between DNA Adducts and Tumor Incidence

Carcinogens and mutagens usually generate multiple DNA adducts, and it was shown that certain adducts are biologically more relevant than others. Many diseases in humans are the result of specific genetic mutations. Therefore, DNA adducts or lesions that lead to mutations became the focus of many studies. As for example, the predominant mutation induced by most methylating and ethylating agents are G:C→A:T transitions induced by O^6-alkylguanine, even though the major adduct is formed at the N7-position of guanine [59].

Characterization of a quantitative relationship between DNA adduct levels and tumor incidences in rats and mice was attempted by Ottender and Lutz [60]. Of the 27 different chemicals investigated, the range of carcinogenic potency of structurally different DNA adducts is typically within 2 orders of magnitude. In the rat, for instance, 53 adducts per 10^8 nucleotides for the aflatoxin B1 to 2082 adducts per 10^8 nucleotides for dimethylnitrosamine relate to the normalized 50% level of liver tumor incidences, suggesting that the aflatoxin–DNA adducts are 40 times more potent than the adducts formed by dimethylnitrosamine for inducing hepatocellular carcinoma.

6. Damaged DNA Replication

DNA replication causes mutations and DNA damage, or DNA adducts increases the rate of error-prone replication [61]. However, each DNA damage or adduct has a unique mutational signature, which is directly related to the identity of the DNA polymerase that bypass it and the mechanism of its nucleotide insertion and extension [61].

A human cell contains at least 17 different DNA polymerases. The DNA polymerases belong to seven families (A, B, C, D, X, Y, and RT) [62,63], of which the C family enzymes were only found in prokaryotes. In eukaryotes, the B-family enzymes pol ε and pol δ carry out a large fraction of nuclear DNA replication, whereas pol α of the same family performs initiation and priming. These three polymerases are essential for DNA replication in eukaryotes. In the current model of DNA replication, pol ε carries out majority of leading strand DNA replication of the undamaged genome, whereas pol δ primarily replicates the lagging strand. But this model has recently been challenged, and data supporting involvement of pol δ in both leading and lagging strand replication have been presented [64–66]. It is noteworthy that these important DNA polymerases are inefficient in bypassing most bulky or distorting DNA damages, such as the ones induced by PAHs and UV light.

The discovery of translesion synthesis (TLS) DNA polymerases in the 1990s and the study of their catalytic and non-catalytic roles in damaged DNA replication provided much of our current understanding of DNA adduct or lesion bypass [63]. Lesion bypass is carried out primarily by the Y-family polymerases. But X- and B-family polymerases are also involved in many cases.

TLS of various types of DNA damage have been conducted by genetic studies in repair and replication competent cells, by in vitro experiments using purified DNA polymerases and accessory proteins, and by structural and computational studies. The mechanistic information gathered from these studies is critical to understand the mechanism of mutagenesis, the underlying process for the development of cancer. These fundamental studies are now allowing therapeutic application, as inhibiting the activity of some of the TLS polymerases may enhance the effect of an antitumor agent.

7. Epidemiology

At the international cancer congress held in Tokyo in 1966, Sir Alexander Haddow, the President of international union against cancer, pronounced: "We are impressed by the probability that a much higher proportion of human cancer than we had ever recently suspected—perhaps amounting to as much as 80 percent—may be due to environmental causes" [67]. These remarks from an eminent cancer researcher are significant, because it suggests that most human cancers are preventable. The most common preventable risk factors for cancer are tobacco smoking, diet (low in fruits and vegetable and high in fatty foods, red meats, etc.), obesity, and alcohol [68–73]. Cancer rate also increases with age, but age-related cancer patterns are fairly complex.

Epidemiology showing the definitive link between tobacco smoke and cancer was a noteworthy achievement in the United States, and the Surgeon General's Report in 1964 had a significant positive effect on public health in this country. The smoking prevalence in males decreased by about 60%, while prevalence in females diminished by about 50% [74]. As a result, lung cancer mortality and other tobacco-related diseases continue to decrease. These facts reiterate the importance of tobacco control in prevention of cancer and other diseases [16,75–77].

Epidemiology of skin cancer has also been enlightening [78]. One in every three cancers diagnosed is a skin cancer, and one in every five Americans will develop skin cancer in their lifetime. Melanoma and nonmelanoma skin cancer (NMSC) are the most common types of cancer mainly in the white populations. Both types of tumors show an increasing incidence rate worldwide, but a stable or decreasing mortality rate, presumably due to earlier diagnosis and better treatments. NMSC is the most common cancer in fair-skinned individuals, which causes significant morbidity. The rising incidence rates of NMSC are believed to be triggered by a combination of increased exposure to direct UV rays or UV in sunlight, increased longevity, ozone depletion, genetics, and in a limited number of cases, immune suppression.

8. Mutation and Cancer

Several types of cancers are the result of at least a few mutations in critical genes [79,80]. The somatic mutation theory (SMT) of cancer, the most prevalent model, proposes that cancer is caused by mutation(s) in the body cells (as opposed to germ cells), especially nonlethal mutations associated with increased proliferation of the mutant cells. The SMT hypothesis originated from Theodore Boveri's postulate in 1914 that a combination of chromosomal defects could result in cancer [81]. After Watson and Crick's discovery of the structure of DNA that also implied that DNA contains the genetic information, in 1953, Carl O. Nordling proposed that several mutated genes may lead to cancer [82]. Ashley suggested that cancer may occur as a result of three to seven mutations [83]. Alfred Knudson modified Ashley's proposal, based on his observations of a number of retinoblastoma cases, proposing that cancer is the result of accumulated mutations to a cell's DNA, which could be as little as two hits [84]. The two-hit model proposes that dominantly inherited predisposition to cancer requires a germline mutation, while tumorigenesis necessitates a second somatic mutation. For colorectal carcinoma, Fearon and Vogelstein suggested that four to five gene mutations are necessary for the development of malignant tumor, and the accumulation of the mutations, rather than their specific order, is the critical determinant of tumorigenesis [85]. More recently, these mutations have been referred to as "driver" mutations conferring growth advantage to the cells [79]. In humans, more than 350 mutated genes that are implicated in the development of cancer have been identified. A large-scale sequencing study has shown that most somatic mutations in cancer cells are "passengers" that do not cause tumorigenesis, whereas 120 of the 518 genes screened (~23%) carry a "driver" mutation, which can function as cancer genes. Similar conclusions have been reached in other studies [79,86]. The basic premise of SMT, however, has been challenged from time to time [87].

Genes that contribute to cancer include oncogenes and tumor suppressor genes. Oncogenes change a normal healthy cell into a cancerous cell. Examples include the *ras* family of genes and *HER2*. The *ras* genes produce proteins engaged in cell communication pathways, cell growth, and cell death, whereas *HER2* makes specialized proteins controlling cell growth, and spread notably in breast and ovarian cancer cells. DNA adduct-induced mutations in the *ras* gene, at the activating codons 12, 13, 59, and 61, are considered to be of note. Aflatoxin B_1 (AFB_1) causes $G \cdot C \rightarrow A \cdot T$ or $G \cdot C \rightarrow T \cdot A$ substitutions at codon 12 in experimental animals [88]. Analyses of lung tumors in A/J mice by the tobacco-specific nitrosamine 4-(methylnitrosamino)-1-(3-pyridyl)-1-butanone (NNK) and related compounds showed high frequency of $G \rightarrow A$ mutations (GGT to GAT) in codon 12 [89]. By contrast, tumor suppressor genes protect a cell from becoming cancerous. The tumor suppressor proteins control cell growth by monitoring cell division, repairing base mismatches in DNA, and controlling cell death (apoptosis). Examples of tumor suppressor genes include *p53*, *BRCA1*, and *BRCA2*. More than 50% of human cancers are characterized by mutations in the *p53* gene, and most *p53* gene mutations are not hereditary. Germline mutations in *BRCA1* or *BRCA2* gene increases a woman's risk of hereditary breast and ovarian cancer. A convincing relationship between a chemical and *p53* mutation in human cancer has been shown in geographical areas where AFB_1-derived liver cancers accompanied unusually high frequency of G·C→T·A mutations at the third base of codon 249 of the *p53* gene [90]. Also, a human liver cell line following exposure to AFB₁ showed the same mutation at the third base of *p53* codon 249 [91]. Likewise, for lung cancer cases of smokers, ~40% of the mutations involved G→T transversions, and more than 90% of them are on a guanine on the non-transcribed strand [90]. Major hotspots are observed at codons 157, 248, and 273. Even though codon 157 is unique to lung cancer, the other two are hotspots for mutations in many other cancers, usually detected as transitions at these CpG sequences, whereas in lung cancers, G→T transversions are the most common mutations [92]. Pfeifer and colleagues have claimed that sequence specificity of G→T transversions in lung tumors is consistent with a direct mutagenic action of PAH compounds, such as B[*a*]P present in cigarette smoke [93]. In addition to the cancers induced by exogenous agents, few hereditary cancers, which include retinoblastoma and Li-Fraumeni syndrome, involve germline mutations in tumor suppressor genes [94,95].

Human tumors are largely heterogeneous. Loeb and coworkers suggest that this heterogeneity results from a mutator phenotype. They hypothesized that increased mutation rates are essential to account for the large number of mutations observed in cancer cells [96,97]. Consequently, an initial mutator mutation triggers additional mutations, including mutations in genes that maintain genetic stability, starting a cascade of mutations throughout the genome. Several types of cancers exhibit mutator phenotype resulting from mutations at loci responsible for DNA mismatch repair [98]. It was also proposed that *p53* mutations might give rise to mutator phenotype, because p53 is a gatekeeper of DNA damage responses [99]. However, others believe that a mutator phenotype is not necessary for tumor initiation and progression, in spite of the fact that some tumors may acquire it during tumorigenesis [100].

In one of the most highly cited articles, entitled "Hallmarks of Cancer", in the year 2000, Hanahan and Weinberg suggested that the complexity of cancer can be summarized in six hallmarks that enable normal cells to turn tumorigenic and ultimately malignant [101]. These hallmarks are as follows: (1) self-sufficiency in growth signals, implying the ability of tumor cells to grow in the absence of the signals that allow them to grow, (2) insensitivity to anti-growth signals, i.e., they resist the signals to stop growth, (3) evading apoptosis, i.e., they resist their programmed death, (4) limitless replicative potential, so that they can multiply indefinitely, (5) sustained angiogenesis, i.e., they stimulate the blood vessel growth in order to supply nutrients to the tumor cells, and (6) tissue invasion and metastases, i.e., they invade surrounding tissues and spread to distant sites. However, Lazebnik pointed out that hallmarks 1–5 are also the characteristics of benign tumors [102]. In an update of the *Hallmark* paper, in 2011, Hanahan and Weinberg proposed four additional hallmarks: (1) abnormal metabolic pathways, (2) evading the immune system, (3) genome instability, and (4) inflammation [103]. In principle, the cancer phenotypes proposed as hallmarks are based on the SMT and its cell-centered variants. Others, though, argued that cancer is a tissue-level disease and cataloguing such cellular-level hallmarks are misleading [104].

9. Conclusions

A detailed understanding of multi-stage carcinogenesis is important for both the treatment and prevention of cancer. This area of research, for the last fifty years, has provided us a great deal of mechanistic information on initiation, promotion, and progression, the three main steps leading to cancer. Consequently, many types of cancer deaths have been reduced in the USA over the last two decades, to an overall reduction of 23%, and more than 1.7 million cancer deaths were averted [105]. In spite of this progress, cancer is still the leading cause of death for much of the US population. Likewise, there has been significant reduction in several European countries. Unfortunately, progress has been limited in many other countries, due to the lack of adequate cancer diagnosis and limited medical treatment capabilities [106,107]. In fact, more than 60% of the world's new cancer cases take place in Africa, Asia, and Central and South America, and 70% of the world's cancer deaths occur in these continents. Therefore, it is imperative to continue further studies on the mechanism of carcinogenesis with the objective of prevention, treatment, as well as developing new strategies to combat this deadly disease.

Acknowledgments: Research in the AKB laboratory has been supported by the NIH (NIEHS grants ES09127, ES027558, and ES023350).

Conflicts of Interest: The author declares no conflict of interest.

Abbreviations

PAH   polycyclic aromatic hydrocarbon
B[*a*]P   benzo[*a*]pyrene
CYP   cytochrome P-450
UV   ultraviolet
TPA   12-*O*-tetradecanoylphorbol-13-acetate
TNF-α   tumor necrosis factor-α
TLS   translesion synthesis
NMSC   nonmelanoma skin cancer
SMT   somatic mutation theory

References

1. Brown, J.R.; Thornton, J.L. Percivall Pott (1714–1788) and chimney sweepers' cancer of the scrotum. *Br. J. Ind. Med.* **1957**, *14*, 68–70. [CrossRef] [PubMed]
2. Yamagiwa, K.; Ichikawa, K.J. Experimental Study of the Pathogenesis of Carcinoma. *Cancer Res.* **1918**, *3*, 1–29. [CrossRef]
3. Yamagiwa, K.; Ichikawa, K. Experimental study of the pathogenesis of carcinoma. *CA Cancer J. Clin.* **1977**, *27*, 174–181. [CrossRef] [PubMed]
4. Tsutsui, H. Uber das kustlich erzeugte cancroid bei der maus. *Gann* **1918**, *12*, 17–21.
5. Cook, J.W.; Hieger, I.; Kennaway, E.L.; Mayneord, W.V. The production of cancer by pure hydrocarbons. *R. Soc. Proc.* **1932**, *111*, 455–484. [CrossRef]
6. Cook, J.W.; Hewett, C.L.; Hieger, I. The isolation of a cancer-producing hydrocarbon from coal tar. Parts I, II, and III. *J. Chem. Soc.* **1933**, *0*, 395–405. [CrossRef]
7. Stern, F.B.; Murthy, L.I.; Beaumont, J.J.; Schulte, P.A.; Halperin, W.E. Notification and risk assessment for bladder cancer of a cohort exposed to aromatic amines. III. Mortality among workers exposed to aromatic amines in the last beta-naphthylamine manufacturing facility in the United States. *J. Occup. Med.* **1985**, *27*, 495–500. [PubMed]
8. Michaels, D. When Science Isn't Enough: Wilhelm Hueper, Robert A. M. Case, and the Limits of Scientific Evidence in Preventing Occupational Bladder Cancer. *Int. J. Occup. Environ. Health* **1995**, *1*, 278–288. [CrossRef] [PubMed]
9. Clayson, D.B. Specific aromatic amines as occupational bladder carcinogens. *Natl. Cancer Inst. Monogr.* **1981**, *58*, 15–19.
10. Purohit, V.; Basu, A.K. Mutagenicity of nitroaromatic compounds. *Chem. Res. Toxicol.* **2000**, *13*, 673–692. [CrossRef] [PubMed]
11. Nicholson, W.J. The carcinogenicity of chrysotile asbestos—A review. *Ind. Health* **2001**, *39*, 57–64. [CrossRef] [PubMed]
12. Hayes, R.B. The carcinogenicity of metals in humans. *Cancer Causes Control* **1997**, *8*, 371–385. [CrossRef] [PubMed]
13. Brady, J.; Liberatore, F.; Harper, P.; Greenwald, P.; Burnett, W.; Davies, J.N.; Bishop, M.; Polan, A.; Vianna, N. Angiosarcoma of the liver: An epidemiologic survey. *J. Natl. Cancer Inst.* **1977**, *59*, 1383–1385. [CrossRef] [PubMed]
14. Wogan, G.N. Aflatoxin as a human carcinogen. *Hepatology* **1999**, *30*, 573–575. [CrossRef] [PubMed]
15. Taxell, P.; Santonen, T. Diesel Engine Exhaust: Basis for Occupational Exposure Limit Value. *Toxicol. Sci.* **2017**, *158*, 243–251. [CrossRef] [PubMed]

16. Hecht, S.S. Tobacco smoke carcinogens and lung cancer. *J. Natl. Cancer Inst.* **1999**, *91*, 1194–1210. [CrossRef] [PubMed]

17. Soehnge, H.; Ouhtit, A.; Ananthaswamy, O.N. Mechanisms of induction of skin cancer by UV radiation. *Front. Biosci.* **1997**, *2*, d538–d551. [PubMed]

18. Robock, A.; Toon, O.B. Local nuclear war, global suffering. *Sci. Am.* **2010**, *302*, 74–81. [CrossRef] [PubMed]

19. IARC. *Ionizing Radiation: Part 1: X- and Gamma (γ)-Radiation, and Neutrons (IARC Monographs on the Evaluation of the Carcinogenic Risks to Humans) (Pt. 1)*, 1st ed.; IARC: Lyon, France, 2000.

20. Boyland, E.; Wolf, G. Metabolism of polycyclic compounds. 6. Conversion of phenanthrene into dihydroxydihydrophenanthrenes. *Biochem. J.* **1950**, *47*, 64–69. [CrossRef] [PubMed]

21. Avery, O.T.; Macleod, C.M.; McCarty, M. Studies on the Chemical Nature of the Substance Inducing Transformation of Pneumococcal Types: Induction of Transformation by a Desoxyribonucleic Acid Fraction Isolated from Pneumococcus Type III. *J. Exp. Med.* **1944**, *79*, 137–158. [CrossRef] [PubMed]

22. Watson, J.D.; Crick, F.H. The structure of DNA. *Cold Spring Harb. Symp. Quant. Biol.* **1953**, *18*, 123–131. [CrossRef] [PubMed]

23. Miller, E.C. Some current perspectives on chemical carcinogenesis in humans and experimental animals: Presidential Address. *Cancer Res.* **1978**, *38*, 1479–1496. [PubMed]

24. Wogan, G.N.; Hecht, S.S.; Felton, J.S.; Conney, A.H.; Loeb, L.A. Environmental and chemical carcinogenesis. *Semin. Cancer Biol.* **2004**, *14*, 473–486. [CrossRef] [PubMed]

25. Pullman, A.; Pullman, B. Electronic structure and carcinogenic activity of aromatic molecules; new developments. *Adv. Cancer Res.* **1955**, *3*, 117–169. [PubMed]

26. Harvey, R.G. Historical Overview of Chemical Carcinogenesis. *Curr. Cancer Res.* **2011**, *6*, 1–26.

27. Singer, B.; Grunberger, D. *Molecular Biology of Mutagens and Carcinogens*; Plenum Press: New York, NY, USA, 1983.

28. Varghese, A.J.; Wang, S.Y. Ultraviolet irradiation of DNA in vitro and in vivo produces a 3d thymine-derived product. *Science* **1967**, *156*, 955–957. [CrossRef] [PubMed]

29. Wang, S.Y.; Varghese, A.J. Cytosine-thymine addition product from DNA irradiated with ultraviolet light. *Biochem. Biophys. Res. Commun.* **1967**, *29*, 543–549. [CrossRef]

30. Wang, S.Y.; Patrick, M.H.; Varghese, A.J.; Rupert, C.S. Concerning the mechanism of formation of UV-induced thymine photoproducts in DNA. *Proc. Natl. Acad. Sci. USA* **1967**, *57*, 465–472. [CrossRef] [PubMed]

31. Varghese, A.J.; Wang, S.Y. *Cis-syn* thymine homodimer from ultra-violet irradiated calf thymus DNA. *Nature* **1967**, *213*, 909–910. [CrossRef] [PubMed]

32. Berenblum, I.; Shuvik, P. The persistence of latent tumor cells induced in the mouse's skin by a single application of 9,10-dimethyl-1,2-benzanthracene. *Br. J. Cancer* **1949**, *3*, 384–386. [CrossRef] [PubMed]

33. Hecker, E. Phorbol esters from croton oil. Chemical nature and biological activities. *Naturwissenschaften* **1967**, *54*, 282–284. [CrossRef] [PubMed]

34. Argyris, T.S.; Slaga, T.J. Promotion of carcinomas by repeated abrasion in initiated skin of mice. *Cancer Res.* **1981**, *41*, 5193–5195. [PubMed]

35. Reiners, J.J., Jr.; Nesnow, S.; Slaga, T.J. Murine susceptibility to two-stage skin carcinogenesis is influenced by the agent used for promotion. *Carcinogenesis* **1984**, *5*, 301–307. [CrossRef] [PubMed]

36. Naito, M.; Naito, Y.; DiGiovanni, J. Comparison of the histological changes in the skin of DBA/2 and C57BL/6 mice following exposure to various promoting agents. *Carcinogenesis* **1987**, *8*, 1807–1815. [CrossRef] [PubMed]

37. Walborg, E.F., Jr.; DiGiovanni, J.; Conti, C.J.; Slaga, T.J.; Freeman, J.J.; Steup, D.R.; Skisak, C.M. Short-term biomarkers of tumor promotion in mouse skin treated with petroleum middle distillates. *Toxicol. Sci.* **1998**, *45*, 137–145. [PubMed]

38. McCann, J.; Choi, E.; Yamasaki, E.; Ames, B.N. Detection of carcinogens as mutagens in the Salmonella/microsome test: Assay of 300 chemicals. *Proc. Natl. Acad. Sci. USA* **1975**, *72*, 5135–5139. [CrossRef] [PubMed]

39. Cupples, C.G.; Miller, J.H. A set of *lacZ* mutations in *Escherichia coli* that allow rapid detection of each of the six base substitutions. *Proc. Natl. Acad. Sci. USA* **1989**, *86*, 5345–5349. [CrossRef] [PubMed]

40. Clive, D.; Johnson, K.O.; Spector, J.F.; Batson, A.G.; Brown, M.M. Validation and characterization of the L5178Y/TK$^{+/-}$ mouse lymphoma mutagen assay system. *Mutat. Res.* **1979**, *59*, 61–108. [CrossRef]

41. Kraemer, K.H.; Seidman, M.M. Use of *supF*, an *Escherichia coli* tyrosine suppressor tRNA gene, as a mutagenic target in shuttle-vector plasmids. *Mutat. Res.* **1989**, *220*, 61–72. [CrossRef]

42. Ohkawa, Y.; Iwata, K.; Shibuya, H.; Fujiki, H.; Inui, N. A rapid, simple screening method for skin-tumor promoters using mouse peritoneal macrophages in vitro. *Cancer Lett.* **1984**, *21*, 253–260. [CrossRef]

43. Vogelstein, B.; Fearon, E.R.; Hamilton, S.R.; Kern, S.E.; Preisinger, A.C.; Leppert, M.; Nakamura, Y.; White, R.; Smits, A.M.; Bos, J.L. Genetic alterations during colorectal-tumor development. *N. Engl. J. Med.* **1988**, *319*, 525–532. [CrossRef] [PubMed]

44. Kinzler, K.W.; Vogelstein, B. Lessons from hereditary colorectal cancer. *Cell* **1996**, *87*, 159–170. [CrossRef]

45. Rakoff-Nahoum, S. Why cancer and inflammation? *Yale J. Biol. Med.* **2006**, *79*, 123–130. [PubMed]

46. Kemp, C.J. Animal Models of Chemical Carcinogenesis: Driving Breakthroughs in Cancer Research for 100 Years. *Cold Spring Harb. Protoc.* **2015**, 865–874. [CrossRef] [PubMed]

47. Fujiki, H.; Mori, M.; Nakayasu, M.; Terada, M.; Sugimura, T.; Moore, R.E. Indole alkaloids: Dihydroteleocidin B, teleocidin, and lyngbyatoxin A as members of a new class of tumor promoters. *Proc. Natl. Acad. Sci. USA* **1981**, *78*, 3872–3876. [CrossRef] [PubMed]

48. Fujiki, H.; Sugimura, T. New classes of tumor promoters: Teleocidin, aplysiatoxin, and palytoxin. *Adv. Cancer Res.* **1987**, *49*, 223–264. [PubMed]

49. DiGiovanni, J. Multistage carcinogenesis in mouse skin. *Pharmacol. Ther.* **1992**, *54*, 63–128. [CrossRef]

50. Castagna, M.; Takai, Y.; Kaibuchi, K.; Sano, K.; Kikkawa, U.; Nishizuka, Y. Direct activation of calcium-activated, phospholipid-dependent protein kinase by tumor-promoting phorbol esters. *J. Biol. Chem.* **1982**, *257*, 7847–7851. [PubMed]

51. Slaga, T.J.; Klein-Szanto, A.J.; Triplett, L.L.; Yotti, L.P.; Trosko, K.E. Skin tumor-promoting activity of benzoyl peroxide, a widely used free radical-generating compound. *Science* **1981**, *213*, 1023–1025. [CrossRef] [PubMed]

52. Fujiki, H.; Sueoka, E.; Suganuma, M. Tumor promoters: From chemicals to inflammatory proteins. *J. Cancer Res. Clin. Oncol.* **2013**, *139*, 1603–1614. [CrossRef] [PubMed]

53. Lindahl, T. Suppression of spontaneous mutagenesis in human cells by DNA base excision-repair. *Mutat. Res.* **2000**, *462*, 129–135. [CrossRef]

54. Friedberg, E.C.; Aguilera, A.; Gellert, M.; Hanawalt, P.C.; Hays, J.B.; Lehmann, A.R.; Lindahl, T.; Lowndes, N.; Sarasin, A.; Wood, R.D. DNA repair: From molecular mechanism to human disease. *DNA Repair (Amst.)* **2006**, *5*, 986–996. [CrossRef] [PubMed]

55. Arczewska, K.D.; Michalickova, K.; Donaldson, I.M.; Nilsen, H. The contribution of DNA base damage to human cancer is modulated by the base excision repair interaction network. *Crit. Rev. Oncog.* **2008**, *14*, 217–273. [CrossRef] [PubMed]

56. Spivak, G. Nucleotide excision repair in humans. *DNA Repair (Amst.)* **2015**, *36*, 13–18. [CrossRef] [PubMed]

57. Modrich, P. Mechanisms in *E. coli* and Human Mismatch Repair (Nobel Lecture). *Angew. Chem. Int. Ed. Engl.* **2016**, *55*, 8490–8501. [CrossRef] [PubMed]

58. Kastan, M.B.; Bartek, J. Cell-cycle checkpoints and cancer. *Nature* **2004**, *432*, 316–323. [CrossRef] [PubMed]

59. Basu, A.K.; Essigmann, J.M. Site-specifically alkylated oligodeoxynucleotides: Probes for mutagenesis, DNA repair and the structural effects of DNA damage. *Mutat. Res.* **1990**, *233*, 189–201. [CrossRef]

60. Otteneder, M.; Lutz, W.K. Correlation of DNA adduct levels with tumor incidence: Carcinogenic potency of DNA adducts. *Mutat. Res.* **1999**, *424*, 237–247. [CrossRef]

61. Basu, A.K. Mutagenesis: The Outcome of Faulty Replication of DNA. In *Chemical Carcinogenesis*; Penning, T.M., Ed.; Humana Press: New York, NY, USA, 2011; pp. 375–399.

62. Burgers, P.M.; Koonin, E.V.; Bruford, E.; Blanco, L.; Burtis, K.C.; Christman, M.F.; Copeland, W.C.; Friedberg, E.C.; Hanaoka, F.; Hinkle, D.C.; et al. Eukaryotic DNA polymerases: Proposal for a revised nomenclature. *J. Biol. Chem.* **2001**, *276*, 43487–43490. [CrossRef] [PubMed]

63. Ohmori, H.; Friedberg, E.C.; Fuchs, R.P.; Goodman, M.F.; Hanaoka, F.; Hinkle, D.; Kunkel, T.A.; Lawrence, C.W.; Livneh, Z.; Nohmi, T.; et al. The Y-family of DNA polymerases. *Mol. Cell* **2001**, *8*, 7–8. [CrossRef]

64. Burgers, P.M.; Gordenin, D.; Kunkel, T.A. Who Is Leading the Replication Fork, Pol epsilon or Pol delta? *Mol. Cell* **2016**, *61*, 492–493. [CrossRef] [PubMed]

65. Johnson, R.E.; Klassen, R.; Prakash, L.; Prakash, S. Response to Burgers et al. *Mol. Cell* **2016**, *61*, 494–495. [CrossRef] [PubMed]

66. Johnson, R.E.; Klassen, R.; Prakash, L.; Prakash, S. A Major Role of DNA Polymerase delta in Replication of Both the Leading and Lagging DNA Strands. *Mol. Cell* **2015**, *59*, 163–175. [CrossRef] [PubMed]

67. Phillips, A.J. The epidemiology of cancer. *Can. Fam. Physician* **1969**, *15*, 44–47. [PubMed]

68. Levin, M.L.; Goldstein, H.; Gerhardt, P.R. Cancer and tobacco smoking: A preliminary report. *J. Am. Med. Assoc.* **1950**, *143*, 336–338. [CrossRef] [PubMed]

69. Doll, R. On the etiology of lung cancer. *J. Natl. Cancer Inst.* **1950**, *11*, 638–640. [PubMed]

70. Doll, R.; Hill, A.B. Smoking and carcinoma of the lung; preliminary report. *Br. Med. J.* **1950**, *2*, 739–748. [CrossRef] [PubMed]

71. Willett, W.C. Diet and cancer: One view at the start of the millennium. *Cancer Epidemiol. Biomark. Prev.* **2001**, *10*, 3–8.

72. Willett, W.C. Diet and breast cancer. *J. Intern. Med.* **2001**, *249*, 395–411. [CrossRef] [PubMed]

73. Verhoeven, D.T.; Goldbohm, R.A.; van Poppel, G.; Verhagen, H.; van den Brandt, P.A. Epidemiological studies on brassica vegetables and cancer risk. *Cancer Epidemiol. Biomark. Prev.* **1996**, *5*, 733–748.

74. Surgeon General. *Reducing the Health Consequences of Smoking: 25 Years of Progress*; U.S. Government Publishing Office: Washington, DC, USA, 1989.

75. Hecht, S.S.; Carmella, S.G.; Chen, M.; Dor Koch, J.F.; Miller, A.T.; Murphy, S.E.; Jensen, J.A.; Zimmerman, C.L.; Hatsukami, D.K. Quantitation of urinary metabolites of a tobacco-specific lung carcinogen after smoking cessation. *Cancer Res.* **1999**, *59*, 590–596. [PubMed]

76. Mercincavage, M.; Souprountchouk, V.; Tang, K.Z.; Dumont, R.L.; Wileyto, E.P.; Carmella, S.G.; Hecht, S.S.; Strasser, A.A. A Randomized Controlled Trial of Progressively Reduced Nicotine Content Cigarettes on Smoking Behaviors, Biomarkers of Exposure, and Subjective Ratings. *Cancer Epidemiol. Biomark. Prev.* **2016**, *25*, 1125–1133. [CrossRef] [PubMed]

77. Chen, L.H.; Quinn, V.; Xu, L.; Gould, M.K.; Jacobsen, S.J.; Koebnick, C.; Reynolds, K.; Hechter, R.C.; Chao, C.R. The accuracy and trends of smoking history documentation in electronic medical records in a large managed care organization. *Subst. Use Misuse* **2013**, *48*, 731–742. [CrossRef] [PubMed]

78. Diepgen, T.L.; Mahler, V. The epidemiology of skin cancer. *Br. J. Dermatol.* **2002**, *146* (Suppl. 61), 1–6. [CrossRef] [PubMed]

79. Greenman, C.; Stephens, P.; Smith, R.; Dalgliesh, G.L.; Hunter, C.; Bignell, G.; Davies, H.; Teague, J.; Butler, A.; Stevens, C.; et al. Patterns of somatic mutation in human cancer genomes. *Nature* **2007**, *446*, 153–158. [CrossRef] [PubMed]

80. Lea, I.A.; Jackson, M.A.; Li, X.; Bailey, S.; Peddada, S.D.; Dunnick, J.K. Genetic pathways and mutation profiles of human cancers: Site- and exposure-specific patterns. *Carcinogenesis* **2007**, *28*, 1851–1858. [CrossRef] [PubMed]

81. Boveri, T. Concerning the origin of malignant tumours by Theodor Boveri. Translated and annotated by Henry Harris. *J. Cell Sci.* **2008**, *121* (Suppl. 1), 1–84. [CrossRef] [PubMed]

82. Nordling, C.O. A new theory on cancer-inducing mechanism. *Br. J. Cancer* **1953**, *7*, 68–72. [CrossRef] [PubMed]

83. Ashley, D.J. The two "hit" and multiple "hit" theories of carcinogenesis. *Br. J. Cancer* **1969**, *23*, 313–328. [CrossRef] [PubMed]

84. Knudson, A.G. Hereditary cancer: Two hits revisited. *J. Cancer Res. Clin. Oncol.* **1996**, *122*, 135–140. [CrossRef] [PubMed]

85. Fearon, E.R.; Vogelstein, B. A genetic model for colorectal tumorigenesis. *Cell* **1990**, *61*, 759–767. [CrossRef]

86. Bignell, G.R.; Santarius, T.; Pole, J.C.; Butler, A.P.; Perry, J.; Pleasance, E.; Greenman, C.; Menzies, A.; Taylor, S.; Edkins, S.; et al. Architectures of somatic genomic rearrangement in human cancer amplicons at sequence-level resolution. *Genome Res.* **2007**, *17*, 1296–1303. [CrossRef] [PubMed]

87. Brucher, B.L.; Jamall, I.S. Somatic Mutation Theory—Why it's Wrong for Most Cancers. *Cell. Physiol. Biochem.* **2016**, *38*, 1663–1680. [CrossRef] [PubMed]

88. Shen, H.M.; Ong, C.N. Mutations of the *p53* tumor suppressor gene and *ras* oncogenes in aflatoxin hepatocarcinogenesis. *Mutat. Res.* **1996**, *366*, 23–44. [CrossRef]

89. Ronai, Z.A.; Gradia, S.; Peterson, L.A.; Hecht, S.S. G to A transitions and G to T transversions in codon 12 of the Ki-*ras* oncogene isolated from mouse lung tumors induced by 4-(methylnitrosamino)-1-(3-pyridyl)-1-butanone (NNK) and related DNA methylating and pyridyloxobutylating agents. *Carcinogenesis* **1993**, *14*, 2419–2422. [CrossRef] [PubMed]

90. Greenblatt, M.S.; Bennett, W.P.; Hollstein, M.; Harris, C.C. Mutations in the *p53* tumor suppressor gene: Clues to cancer etiology and molecular pathogenesis. *Cancer Res.* **1994**, *54*, 4855–4878. [PubMed]
91. Aguilar, F.; Hussain, S.P.; Cerutti, P. Aflatoxin B_1 induces the transversion of G–>T in codon 249 of the *p53* tumor suppressor gene in human hepatocytes. *Proc. Natl. Acad. Sci. USA* **1993**, *90*, 8586–8590. [CrossRef] [PubMed]
92. Hainaut, P.; Soussi, T.; Shomer, B.; Hollstein, M.; Greenblatt, M.; Hovig, E.; Harris, C.C.; Montesano, R. Database of *p53* gene somatic mutations in human tumors and cell lines: Updated compilation and future prospects. *Nucleic Acids Res.* **1997**, *25*, 151–157. [CrossRef] [PubMed]
93. Pfeifer, G.P.; Hainaut, P. On the origin of G–> T transversions in lung cancer. *Mutat. Res.* **2003**, *526*, 39–43. [CrossRef]
94. Harris, C.C.; Hirohashi, S.; Ito, N.; Pitot, H.C.; Sugimura, T.; Terada, M.; Yokota, J. Multistage carcinogenesis: The Twenty-Second International Symposium of the Princess Takamatsu Cancer Research Fund. *Cancer Res.* **1992**, *52*, 4837–4840. [PubMed]
95. Harris, C.C. Tumour suppressor genes, multistage carcinogenesis and molecular epidemiology. *IARC Sci. Publ.* **1992**, 67–85.
96. Loeb, L.A.; Springgate, C.F.; Battula, N. Errors in DNA replication as a basis of malignant changes. *Cancer Res.* **1974**, *34*, 2311–2321. [PubMed]
97. Beckman, R.A.; Loeb, L.A. Efficiency of carcinogenesis with and without a mutator mutation. *Proc. Natl. Acad. Sci. USA* **2006**, *103*, 14140–14145. [CrossRef] [PubMed]
98. Branch, P.; Hampson, R.; Karran, P. DNA mismatch binding defects, DNA damage tolerance, and mutator phenotypes in human colorectal carcinoma cell lines. *Cancer Res.* **1995**, *55*, 2304–2309. [PubMed]
99. Strickler, J.G.; Zheng, J.; Shu, Q.; Burgart, L.J.; Alberts, S.R.; Shibata, D. *p53* mutations and microsatellite instability in sporadic gastric cancer: When guardians fail. *Cancer Res.* **1994**, *54*, 4750–4755. [PubMed]
100. Tomlinson, I.P.; Novelli, M.R.; Bodmer, W.F. The mutation rate and cancer. *Proc. Natl. Acad. Sci. USA* **1996**, *93*, 14800–14803. [CrossRef] [PubMed]
101. Hanahan, D.; Weinberg, R.A. The hallmarks of cancer. *Cell* **2000**, *100*, 57–70. [CrossRef]
102. Lazebnik, Y. What are the hallmarks of cancer? *Nat. Rev. Cancer* **2010**, *10*, 232–233. [CrossRef] [PubMed]
103. Hanahan, D.; Weinberg, R.A. Hallmarks of cancer: The next generation. *Cell* **2011**, *144*, 646–674. [CrossRef] [PubMed]
104. Sonnenschein, C.; Soto, A.M. The aging of the 2000 and 2011 Hallmarks of Cancer reviews: A critique. *J. Biosci.* **2013**, *38*, 651–663. [CrossRef] [PubMed]
105. Siegel, R.L.; Miller, K.D.; Jemal, A. Cancer statistics, 2016. *CA Cancer J. Clin.* **2016**, *66*, 7–30. [CrossRef] [PubMed]
106. Chen, W.; Zheng, R.; Baade, P.D.; Zhang, S.; Zeng, H.; Bray, F.; Jemal, A.; Yu, X.Q.; He, J. Cancer statistics in China, 2015. *CA Cancer J. Clin.* **2016**, *66*, 115–132. [CrossRef] [PubMed]
107. Jemal, A.; Bray, F.; Center, M.M.; Ferlay, J.; Ward, E.; Forman, D. Global cancer statistics. *CA Cancer J. Clin.* **2011**, *61*, 69–90. [CrossRef] [PubMed]

International Journal of
Molecular Sciences

MDPI

Perspective

The Future of DNA Adductomic Analysis

Peter W. Villalta [1,*] and Silvia Balbo [1,2]

1   Masonic Cancer Center, University of Minnesota, Minneapolis, MN 55455, USA; balbo006@umn.edu
2   Division of Environmental Health Sciences, University of Minnesota, Minneapolis, MN 55455, USA
*   Correspondence: villa001@umn.edu; Tel.: +1-612-626-8165

Received: 13 July 2017; Accepted: 22 August 2017; Published: 29 August 2017

Abstract: Covalent modification of DNA, resulting in the formation of DNA adducts, plays a central role in chemical carcinogenesis. Investigating these modifications is of fundamental importance in assessing the mutagenicity potential of specific exposures and understanding their mechanisms of action. Methods for assessing the covalent modification of DNA, which is one of the initiating steps for mutagenesis, include immunohistochemistry, ^{32}P-postlabeling, and mass spectrometry-based techniques. However, a tool to comprehensively characterize the covalent modification of DNA, screening for all DNA adducts and gaining information on their chemical structures, was lacking until the recent development of "DNA adductomics". Advances in the field of mass spectrometry have allowed for the development of this methodology. In this perspective, we discuss the current state of the field, highlight the latest developments, and consider the path forward for DNA adductomics to become a standard method to investigate covalent modification of DNA. We specifically advocate for the need to take full advantage of this new era of mass spectrometry to acquire the highest quality and most reliable data possible, as we believe this is the only way for DNA adductomics to gain its place next to the other "-omics" methodologies as a powerful bioanalytical tool.

Keywords: DNA adducts; DNA adductomics; DNA damage; genotoxicity; chemical carcinogenesis; high resolution accurate mass (HRAM) mass spectrometry; constant neutral loss

1. Introduction

Covalent modification of DNA plays a key role in the initiation phase of chemically induced carcinogenesis [1,2]. Modifications, typically referred to as DNA adducts, if not repaired can lead to genomic instability that may result in mutations, which can translate into altered gene expression, abnormal cell growth, and disruption of normal cell function [1,3,4]. Therefore, the measurement of DNA adducts is of fundamental importance in assessing the potential carcinogenic effects of specific exposures and understanding their mechanisms of action. Additionally, the characterization of this type of DNA damage is extremely valuable for the investigation of the safety of exposure to substances used in the industrial and manufacturing processes, pharmaceuticals, environmental pollutants, as well as life-style factors associated with increased cancer risk.

The identification and structural elucidation of DNA adducts in human tissues can be used to either identify specific exposures which resulted in genotoxicity or confirm that suspected exposures have occurred and led to DNA modification. Additionally, the identification and/or quantitation of DNA adducts can reveal important mechanistic aspects of cancer etiology, by elucidating the sequence of events occurring from human chemical exposure to DNA modification, and ultimately to the occurrence of a tumor [3,5]. Therefore, methods to detect and recognize these specific alterations can provide insight into the type of DNA damage resulting from the exposure studied and can provide opportunities for the design of more efficient intervention and prevention approaches.

There are several established methods for assessing the genotoxicity and mutagenicity induced by exposure to various compounds. The in vitro metaphase chromosome aberration assay, the in vitro micronucleus assay, and the mouse lymphoma gene mutation assay (MLA) are widely used and can be considered sufficiently validated. These three assays are currently considered equally appropriate for measurement of chromosomal damage when used together with other genotoxicity tests in a standard battery for testing for example pharmaceuticals. In vivo tests are included to account for absorption, distribution, metabolism, and excretion, with the analysis either of micronuclei in erythrocytes, or of chromosome aberrations in metaphase cells in bone marrow, currently being the most frequently used [6]. In vitro and in vivo tests that measure chromosomal aberrations in metaphase cells can detect a wide spectrum of changes in chromosomal integrity. These methods give a general overview of the DNA damage resulting from an exposure, however they do not provide specific information on the chemical structure of the modifications the damage may result from, or on the mechanism through which the damage may have occurred. DNA adducts analysis has the ability to provide this critical information.

Methods to directly detect and quantify DNA adducts in humans have been developed in the past 30 years, with immunohistochemistry, ^{32}P-postlabeling, and mass spectrometry-based techniques being the most common [7,8]. Among these, only ^{32}P-postlabeling has been used for DNA adduct screening with varying degrees of comprehensiveness, but lacking the ability to provide information on the specific chemical nature of the DNA adducts detected. Immunohistochemistry methods [3,9] rely on specific antibodies for detection of a particular adduct or type of adduct. Examples include polycyclic aromatic hydrocarbon-DNA adducts assayed using the monoclonal 5D11 antibody [10] and cisplatin-DNA adducts using rabbit antiserum NKI-A59 against cisplatin-modified calf thymus DNA [11]. Liquid chromatography-tandem mass spectrometry (LC-MS^2) has become the preferred technique for targeted DNA adduct analysis [12–16]. The popularity of this approach is due to its highly selective nature, sensitivity rivaling and at times surpassing that of ^{32}P-postlabeling, and the ability to perform accurate quantitation using stable isotope dilution [12–16]. Both endogenous adducts, including those related to epigenetic modifications, and exogenous adducts resulting from nucleobase alkylation, oxidation, deamination, and cross-linking due to various exposures, have been measured (see Section 3.2, Figure 3 for representative examples) [15,16]. However, traditionally LC-MS^2 approaches have focused on the analysis of a limited number of DNA adducts at a time, which does not allow them to provide a global picture of the DNA modifications resulting from an exposure or a combination of exposures. The ideal assessment of the potential DNA modification induced by the combination of various exposures requires a methodology which is capable of screening for adducts in a global and comprehensive fashion with as much structural information as possible.

2. Conventional Approach for DNA Adduct Screening: ^{32}P-Postlabeling

The ^{32}P-postlabeling methodology is well-suited to broad-based DNA adduct screening because of its ability to monitor many adducted nucleotides simultaneously in a given sample [17] and its high sensitivity with certain DNA adducts detectable at levels approaching 1 adduct per 10^{10} nucleotides. Adducts are identified either as spots on thin layer chromatography plates observed by autoradiographic detection or as peaks using high-performance liquid chromatography (HPLC) separation with radioactive detection. This method has been successfully employed to screen for DNA adducts in a variety of human tissues and white blood cells [18–23], in exfoliated epithelial cells in urine of smokers [24], in breast milk of lactating mothers [25], and the sputum of lung cancer patients [26,27]. These studies have revealed that human DNA is modified by many different electrophiles, including those formed endogenously as well as by both environmental and dietary genotoxicants. It has also been shown that the level of DNA modification can be influenced by lifestyle and host factors [20,28]. The ^{32}P-postlabeling methodology, however, does have some significant limitations [29], including being labor-intensive, needing significant amounts of radioactive phosphorus, and having potentially highly variable labeling efficiency [30]. Additionally, the most significant limitations are the lack of information regarding the structure of the DNA adduct detected and, at times, the presence of co-migrating adducts on the thin layer chromatography plate [19,31], both of which make the chemical structural determination of adducts very difficult.

3. New Approach for DNA Adduct Screening: DNA Adductomics Using Liquid Chromatography-Mass Spectrometry (LC-MS)

Ideally, what is required to comprehensively assess covalent modification of DNA in realistic scenarios, with various exposures, associated metabolism, and downstream endogenous effects, is an approach which combines the screening capability of ^{32}P-postlabeling and the structural information provided by targeted LC-MS^2 analysis, with little or no a priori assumptions regarding the nature of the adducts formed. An effort to address this need has led to the establishment of the field of LC-MS^n-based DNA adductomics intended to comprehensively screen for DNA modifications, including both known DNA adducts and those which have not been previously detected and/or identified. The first example of this basic approach that we are aware of was performed by Claereboudt and coworkers in 1990 [32], but its further development in subsequent years was limited by the sensitivity and selectivity of the available instrumentation. The rapidly improving instrumentation and technology of the past 10 years has paved the way for the development of more robust DNA adductomics approaches, able to perform a comprehensive characterization of the chemical nature of DNA modification. The field of DNA adductomics [29], while still in its infancy, has now become significantly more powerful with new approaches [33–38] taking advantage of modern mass spectrometry and the wide spread use of high resolution mass spectrometers, allowing for the elucidation of the chemical formula of adducts and their fragments. Details on the advantages of using high resolution mass spectrometry are described further in Section 3.4.2. As with any "-omics" based screening technique, DNA adductomics presents new analytical challenges and therefore requires development work aimed at optimizing chromatography, sample preparation, and data collection and analysis. The approaches this field is pursuing and the features and challenges they each present are described here, with special emphasis on the need for the development of robust, reliable and effective DNA adductomic methods.

3.1. Typical DNA Adductomics Workflow

The sample preparation for the basic LC-MS^n-based DNA adductomics workflow (Figure 1) is similar to that typically performed for targeted LC-MS^2 DNA adduct quantitation, with some modifications to make it more general so as to avoid the potential loss of unknown adducts during sample preparation [15,16]. First, DNA is isolated from the sample, typically tissue, cells, or blood, and usually hydrolyzed to nucleosides using a cocktail of enzymes or to nucleobases by mild acid treatment. When enzymes are used, they are often removed through protein precipitation using organic solvent or through the use of a molecular weight filter cartridge. Salts and other hydrophilic substances are commonly removed by solid phase extraction or fraction collection off of an HPLC column. The resulting samples are typically concentrated through drying and reconstitution to a small volume. The resulting samples are usually analyzed with LC-MS^2, utilizing a key feature of the fragmentation behavior of modified nucleosides, which is discussed in some detail below.

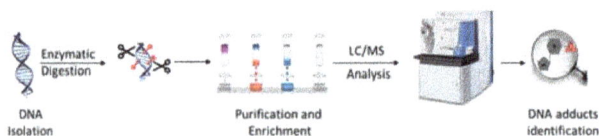

| DNA | Purification and | DNA adducts |
| Isolation | Enrichment | identification |

Figure 1. Summary of a typical sample preparation and analysis workflow for DNA adductomics analysis.

3.2. Key Feature of the Positive Ion LC-MS^n DNA Adductomics Methodology

The enzymatic hydrolysis of modified DNA results in the liberation of modified nucleoside adducts (DNA adducts) which share the same basic chemical structure, the modified nucleobase linked to a deoxyribose (dR) group. The primary and critical feature for DNA adductomic screening of nucleoside adducts is the nearly universal neutral loss (m/z 116 amu) of the dR moiety, upon fragmentation (MS/MS) of the positive ion of the precursor, as shown in Figure 2A [29]. This feature allows for the identification of a given trace level adduct from the multitude of more abundant chemical noise ions present in the LC-MS chromatogram of a given sample.

A recent study [37] expanded the DNA adductomic approach by combining the neutral loss of the bases (Figure 2B), a common ion fragmentation pathway of base adducts, with the conventional neutral loss of dR, allowing for the simultaneous screening of nucleoside adducts and aglycone base adducts. Aglycone base adducts can result upon loss of the deoxyribose from unstable nucleoside adducts upon enzymatic or thermal hydrolysis of the DNA, i.e., N7 position of guanine, N7/N3 of adenine, and the O^2 positions for both cytosine and thymine [39]. This ion fragmentation pathway (Figure 2B) can be very useful to broaden the basic DNA adductomic approach [37].

Another DNA adductomic analysis [40], which allows for the detection of guanine adducts, takes advantage of the fact that aglycone guanine adducts often fragment to form m/z 152 (guanine + H^+) and 135 (guanine-NH_3 + H^+) ions. This observation suggest that a similar detection scheme could be used for adenine adducts with characteristic fragments of m/z 136 (adenine + H^+) and 119 (adenine-NH_3 + H^+). It seems likely that this approach could be broadened to include all four bases (Figure 2C) and would be complimentary to the neutral loss of bases (Figure 2B), allowing for the detection of the majority of aglycone base adducts.

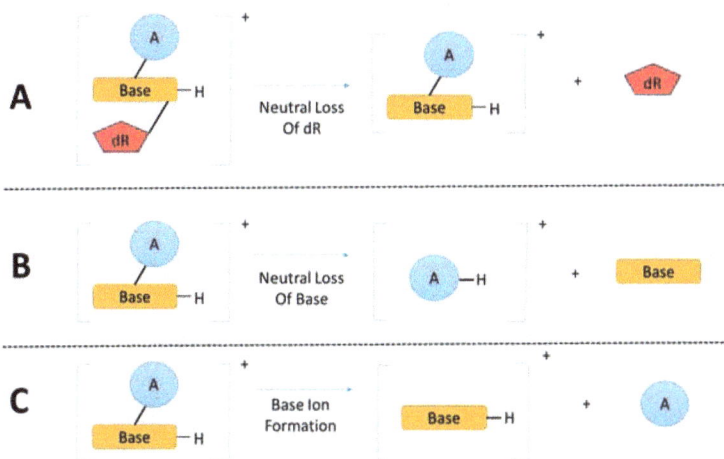

Figure 2. (**A**) The dominant fragmentation pathway of nucleoside adducts is the neutral loss of the 2′-deoxyribose moiety; (**B**) and (**C**) Common fragmentation pathways of nucleobase adducts. (Base = nucleobase, A = modification, and dR = 2′-deoxyribose).

The three fragmentation pathways outlined in Figure 2 could in theory be combined for a nearly comprehensive DNA adductomics methodology for enzymatic hydrolyzed DNA, allowing for the detection of both nucleoside DNA adducts as well as any base DNA adducts resulting from the loss of the deoxyribose group from unstable nucleoside adducts. It would be possible to combine all three fragmentation pathways into a data dependent MS^3 or MS^2 approach or a data independent MS^2 approach. These scanning modes are discussed below. Also, ion fragmentation pathways of the aglycone base adducts (Figure 2B,C) could be used for a DNA adductomic method screening for adducts formed upon DNA acid hydrolysis where the deoxyribose group is cleaved and only the modified base is present. A better understanding of the fragmentation of aglycone base adducts would be very useful in understanding the comprehensiveness of this approach and in confirming the identity of the features resulting from the analysis. The most widely used MS-based "omics" methodologies, proteomics and metabolomics, rely on vast databases for data analysis, which allows for an organized and automated workflow for data analysis and interpretation. DNA adductomics is lacking similar automated data analysis and bioinformatic tools and, therefore, we envision that efforts devoted to create a database containing fragmentation spectra of aglycone base and nucleoside adducts will be extremely helpful in advancing the field of DNA adductomics.

A list of DNA adducts which are representative of those which could be screened for using DNA adductomics, and which have been extensively studied using mass spectrometry based approaches in a variety of experimental settings, are shown in Figure 3.

Deoxyguanosine adducts

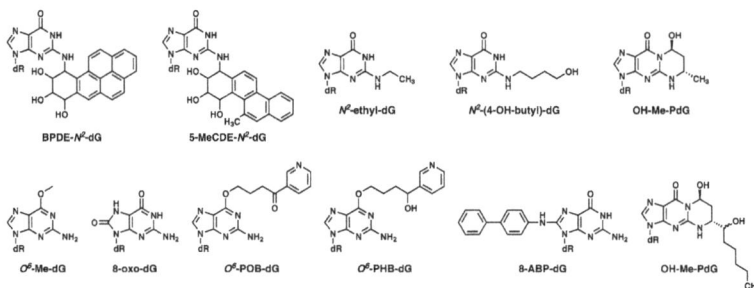

Deoxyadenosine adducts

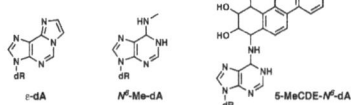

Thymidine adducts

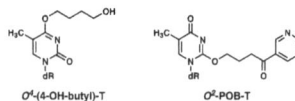

Deoxycytidine adducts

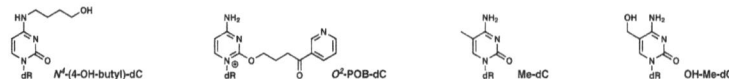

Figure 3. Examples of the type of DNA modifications that can be analyzed with mass spectrometry-based DNA adductomic approaches. The deoxyribose moiety of the structure is abbreviated as dR.

3.3. A Sensitive and Selective LC-MS^n Screening

Probing for DNA adduct formation in human samples requires maximum sensitivity due to the trace levels of these analytes (1 adduct in 10^6–10^{10} nucleotides) in often limited amounts of available DNA (typically 1–100 µg DNA); likewise, in cells or animal models, sensitivity is critical due to the need to keep dose levels low to approximate human exposure. High selectivity in adduct identification is also needed to differentiate DNA adducts from the significant background signal present in biological samples. The need for optimal sensitivity and selectivity is even greater than what is required for trace level targeted DNA adduct quantitation [15] due to the need to screen for multiple adducts, often of unknown identity, across large mass ranges and lacking isotopically labeled internal standards and the well characterized fragmentation patterns of the targeted DNA adduct analytes. This requirement for greater sensitivity and selectivity means that successful analysis is only possible when taking advantage of the technological advancements and scanning modes available with the latest generation of instrumentation. This is particularly true for the use of high resolution accurate mass (HRAM) MS^n detection, which is the acquisition of spectral data with typical mass resolving power sufficient to differentiate masses within 0.01–0.001 amu of each other and accuracy of mass measurement on the order of 0.001 amu, often sufficient to determine the molecular formula of the ion. This type of data acquisition greatly increases both the specificity of the analysis, allowing for precise characterization of the detected adducts, as well as increased sensitivity due to the ability to differentiate the adduct ion signals from isobaric background ions signals. The acquisition of HRAM DNA adductomics data can be performed by MS^1 mode consisting of full scan data, or MS^2 mode consisting of full scan and MS/MS mass spectral data, or MS^3 mode consisting of full scan, MS/MS and an additional fragmentation level (MS/MS/MS). Figure 4 illustrates this variety of data acquisition modes in the context of adduct screening.

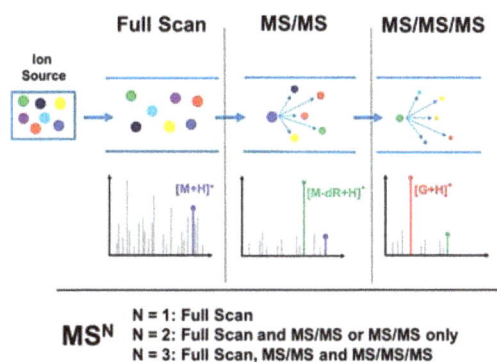

Figure 4. Illustration of different types of high resolution accurate mass (HRAM) DNA adductomic (MS^n) detection where N = 1 represents Full Scan, N = 2 represents Full Scan and MS/MS or MS/MS only, and N = 3 represents Full Scan, MS/MS, and MS/MS/MS. In this example, M is a nucleoside adduct with the general formula of M = X-G-dR where G is guanine, dR is the deoxyribose moiety, and X is the modification. In the Full Scan panel, the accurate mass of the DNA adduct ($[M + H]^+$) can be extracted to generate a chromatogram and provide molecular formula information (this step is common to both DDA and DIA approaches). In the second panel, the MS/MS signal can either be extracted to generate a chromatogram (as in the case of a DIA approach) or provide MS/MS spectral data for the adduct (as in the case of a DDA approach). Finally, in the third panel, the MS/MS/MS fragmentation data can be used to indicate presence of a DNA adduct as well as provide structural confirmation/information (this is the final step for the DDA approach, while it is done in a separate injection in a DIA approach, focusing on candidate adducts identified in the first analysis).

3.4. Rapidly Evolving Technology

The technological capabilities of mass spectrometers, propelled by the development of electrospray and MALDI (matrix-assisted laser desorption/ionization), have been improving rapidly for more than two decades. Further driving the improvements, over the past 15 years or so, has been the development and promise of proteomics. These two factors have resulted in the development of powerful, targeted small molecule and macromolecule quantitative and qualitative analytical capabilities as well as new -omic analyses, including metabolomics and lipidomics. DNA adductomics has now joined the list of -omic methodologies which are used to investigate biological systems. The improvement in technology is continuing unabated with steady advances occurring yearly (e.g., improved ion trap, quadrupole-trap, and quadrupole-TOF instrumentation) with the occasional quantum leap forwards such as the introduction of Orbitrap technology (Thermo Scientific), rapid scanning Q-TOF technology (AB Sciex Triple-TOF), powerful ion mobility capabilities (Waters Synapt technology), and advanced hybrid instruments (Thermo Scientific Fusion instrumentation).

3.4.1. Nanospray Ionization

Electrospray ionization and sampling efficiency increases dramatically as the flow rate is decreased. Proteomics takes advantage of this phenomenon by operating in "nanoflow" ionization mode with flow rates in the 100 s of nanoliters per minute. Nanospray operation has evolved from exotic—requiring flow splitting, handmade emitters, and self-packed columns, and the need to master using delicate low flow fittings and tubing—to truly routine, with the use of commercially produced ultra performance liquid chromatographs (UPLCs) designed for nanoflow operation, easy to use nanospray sources, and pre-made nanoflow columns. It is now possible with minimal training for new analysts to easily work in this mode, and has been used in the field of LC-MS[n] DNA adduct analysis, including DNA adductomics [36–38,41]. Due to the trace levels of DNA adducts and the need to screen for multiple and often unknown adducts, maximizing sensitivity is critical to successful DNA adductomic analysis.

3.4.2. High Resolution Accurate Mass (HRAM) Data

The ability to measure the mass of adducts with accuracies [16] sufficient to provide the selectivity necessary to discriminate them from chemical noise, as well as provide significant information regarding identity of the adducts, has become increasingly possible with recent advances in Orbitrap and Q-TOF instrumentation. The power of HRAM data acquisition comes from the combination of high resolution, which allows ions of similar mass to be resolved from each other, and the subsequent accurate mass measurement allowing for the precise measurement of the ion masses. Without sufficient resolution, only ions which dominate in intensity over adjacent unresolved ions can be measured accurately. With trace level analysis, the analyte ions of interest often have much lower intensity relative to background ions with similar m/z values, and this explains the need for higher levels of resolution. This is especially true in the case of MS[1] data where the number of background ions is dramatically larger than present in MS[n] fragmentation spectra, where the initial parent ion isolation dramatically reduces the number of ions which need to be resolved in the acquired spectra. When sufficient mass resolution is used for the selective detection of trace level DNA adducts, the accurate mass measurements can often provide sufficient information to determine the molecular formula of the analyte and fragment ions, especially when using internal lock masses for maximum accuracy and accounting for the abundances of their isotopic peaks.

3.4.3. Scanning Modes for HRAM MS[n] Data Acquisition

New instrumentation has made new operational modes possible. Early DNA adductomics [29] primarily utilized triple quadrupole instrumentation to perform neutral loss and pseudo-neutral loss screening, whereas more recent analyses have taken advantage of HRAM instrumentation for analyses. For example, Orbitrap detection with data dependent acquisition (DDA) of HRAM MS[1], MS/MS, and MS/MS/MS data has been performed [36–38]. DDA analysis has been a mainstay of LC-MS[2] proteomic analysis, however recently a new scanning mode, data independent acquisition (DIA), has become popular and made possible initially by faster scanning Q-TOF instruments, and more recently by faster scanning Orbitrap detectors. This scanning mode has recently [35] been utilized for DNA adductomics and will be discussed below along with a brief description of DDA, and their merits with regard to DNA adductomics will be addressed.

3.4.4. Data Dependent Acquisition (MS[n])

Data dependent acquisition (DDA) is a continuous scanning mode in which each full scan spectrum acquired is followed with multiple subsequent MS/MS fragmentation events with rapid, on-the-fly precursor ion selection by the instrument software. The detection, and possibly identification, of adducts is done analyzing the product ion spectra using the DNA adduct fragmentation types discussed above. This data acquisition mode was developed for shotgun proteomic analysis and is the conventional approach for this type of analysis. There are many features which are available for tailoring this scanning mode to the specific analysis, including those typically used for proteomics such as dynamic exclusion, exclusion lists, charge state selection, monoisotopic precursor selection, etc. as well as others more likely to be used for small molecule analysis such as inclusion lists, neutral loss triggering, product ion triggering, etc. The current sophistication of most LC-MS instrumentation makes programming of methods using this scanning mode straightforward, although there are typically many parameters which need to be optimized for a particular sample type and experiment. In contrast to proteomics, there are no software tools available for the automated analysis of the resulting data and it must be analyzed manually.

3.4.5. Data Independent Acquisition (MS^2)

Data independent acquisition (DIA) in its simplest terms is the acquisition of fragmentation spectra for all ions across a broad mass range rapidly enough to acquire multiple data points across the chromatographic peak shape of the analytes of interest. There are various forms of instrument scanning modes for acquisition of DIA data, typically with concurrent acquisition of the corresponding full scan data [42]. This approach [43–45] was developed as an alternative to the conventional proteomics approach of DDA, and has gained popularity; more recently researchers have started to implement it in the acquisition of metabolomics data [46,47]. The aim of DIA is to comprehensively fragment all analytes of interest present, thereby providing for a complete data set, in contrast to DDA, which uses ion intensity as a criterion for fragmentation and is prone to missing the detection of lower level analytes. This makes DIA amenable to comprehensive detection/quantification of adducts, either in a targeted fashion by extraction of parent and product ions from the full scan and MS/MS data, respectively, or in an untargeted fashion by relying on peak picking software to identify chromatographic peaks with the correct fragmentation characteristics. Both the targeted and untargeted analysis require co-elution of full scan and MS/MS chromatographic peaks as a criteria for adduct detection. Software and bioinformatics tools are required to handle the challenging amount of data produced if the promise of DIA is to be fulfilled. The DIA approach is rapidly evolving with significant progress being made, primarily in the field of proteomics [42,44,48], but adapting the approach for DNA adductomics will take significant development work to take advantage of the possibilities of the approach.

3.4.6. DDA and DIA for DNA Adductomics

The DDA and DIA scanning modes, which we feel take full advantage of the available instrumentation for DNA adductomics, are the constant neutral loss with triggering of MS/MS/MS fragmentation (CNL/MS^3) mode and wide range selected ion monitoring with corresponding MS/MS fragmentation (Wide SIM/MS^2) mode, respectively. The features characterizing these modes of operation are summarized in Table 1. Briefly, the DDA-CNL/MS^3 approach utilizes the triggering of MS/MS/MS fragmentation upon observation of the neutral loss of the mass used to identify adducts (typically deoxyribose, m/z 116.0473). This analysis provides the advantage of relatively straightforward data analysis, although software tools providing automated analysis are still lacking, as well as a rich set of fragmentation data (both MS/MS and MS/MS/MS) for adduct verification and/or identification in a single injection. The primary negative with this approach is the potential for incomplete sampling due to insufficient speed of analysis, resulting in low level ions not being fragmented. The advantages of the DIA-WideSIM/MS^2 approach is the completeness of the analysis and the archival nature providing for re-analysis of data to probe for newly found or expected adducts in previously analyzed samples. A negative of the DIA analysis is that fragmentation data is limited to MS/MS and principally only the adduct-identifying fragment ion is considered. There is the potential for generation of a pseudo-MS^2 fragmentation spectrum either by peak picking in the MS/MS spectra, and co-alignment with the identified chromatographic peak of the adduct in the full scan data, or manual interrogation of the data. In addition, while DIA has the potential to provide more thorough coverage than the DDA approach, it requires advanced data analysis [47–50], which is currently not available for DNA adductomics analysis, as well as mastery of the data acquisition parameters necessary for development of the data acquisition methodologies.

Table 1. Summary of MS^{2-3} data dependent acquisition (DDA) and MS^2 data independent acquisition (DIA) scanning modes.

Approach	Method	Scan Events	Frequency	Adduct Detection
DDA CNL/MS^3	Targeted	Full Scan	Continuous	MS/MS/MS Triggered Event
		MS/MS	Ions included in a list	
		MS/MS/MS	MS/MS ions selected by loss of 116.0474	
	Untargeted	Full Scan	Continuous	MS/MS/MS Triggered Event
		MS/MS	Most abundant ions	
		MS/MS/MS	MS/MS ions selected by loss of 116.0473	
DIA Wide SIM/MS^2	Targeted	Full Scan	Continuous	Post-run data analysis on ions from a list (characterized by co-eluters with NL = 116.0473)
		MS/MS	Continuous	
	Untargeted	Full Scan	Continuous	Post-run data analysis (any co-eluters with NL = 116.0473)
		MS/MS	Continuous	

4. Adductomic Studies

The field of DNA adductomics has been recently reviewed and [29,51,52] Table 2 summarizes the LC-MS based DNA adductomics analyses performed to date. Three recent studies [53–55] have used the conventional low resolution, nominal mass MS^2 approach for DNA adductomics, but the trend since our previous review is to use HRAM data acquisition for DNA adductomic experiments. Two studies [33,34] have relied on full scan HRAM data acquisition and a self-generated DNA adduct database searching for DNA adductomic analysis, whereas another recent study [35] used HRAM DIA data acquisition using the simultaneous acquisition of fragment ions resulting from high and low collision energy (MS^E). Our approach [36–38] takes advantage of DDA HRAM data acquisition with MS/MS/MS fragmentation upon observation of neutral loss of dR (Figure 2A) or base (Figure 2B). The new studies using HRAM detection are discussed briefly in Section 5.

Need for Methodology Comparisons

Comparisons of the various methodologies would be very useful for deciding upon an optimal DNA adductomic approach for a given experiment. It seems unlikely that one single approach would be best in all contexts, and considerations such as differences in analytical goals (e.g., targeted or untargeted), DNA amounts available, DNA adduct levels, hydrophobic vs. hydrophilic adducts, instrument availability, etc. will need to considered to determine the best approach to use. Comparisons of previously published studies are difficult if not impossible. For example, comparing analyses performed by DDA-CNL/MS^3 analysis using low resolution/nominal mass detection with an ion trap instrument [56] and high resolution/accurate mass detection with an Orbitrap instrument [36] would be informative, especially in the context of our advocacy for HRAM data acquisition for DNA adductomics, but these two studies used different sources of DNA to perform their proof-of-principle investigations. Ideally, direct comparisons of various DNA adductomic methodologies using identical samples would provide a true measure of their relative analytical power. Turesky and coworkers have recently performed a comparison of targeted DNA adductomics methodologies of Orbitrap-based DIA-WideSIM/MS^2 and DDA-CNL/MS^3 analyses and triple quadrupole-based CNL and pseudo-CNL analyses, all four of which were performed on the same samples with identical chromatography and ion source conditions [57]. Levels of synthetic DNA adduct standards were spiked in calf thymus DNA and analyzed with the different methods. The complete results of this study are beyond the scope of this paper but the performance of the four approaches were DIA-WideSIM/MS^2 > DDA-CNL/MS^3 > pseudo-CNL > CNL, where the number of adducts detected were 12, 7, 2, 0, respectively, out of 15 at the lowest level of spiking (4–8 adducts per 10^9 nucleotides).

Table 2. Summary of published LC-MS-based DNA adductomics studies.

Approach	Instrument	Sample Type	Adduct Type/Origin	Strengths	Weaknesses	Details	Reference
	DF-EB/Q	Reaction with nucleosides	PGE [c] (industrial chemical)	High resolution, First example of DNA adductomic analysis	Simplistic model	Nucleoside reacted with chemical of interest	Claereboudt et al., 1990 [32]
		Synthetic standards	Arylamine (industrial chemical)	Early report of DNA adductomics		Analysis of synthetic standards only	Bryant et al., 1992 [58]
		In vitro reaction	PhIP [a] (food)	Comparison made with ^{32}P-postlabeling		-	Vouros et al., 1995 [59]
		In vitro reaction and Animal tissues	IQ [b] (food)	First example of nanospray ionization		-	Vouros et al., 1999 [41]
CNL		Irradiated cells (human monocyte)	Radiation-induced	Only example of analysis of adducts due to exposure to radiation			Ravanat et al., 2004 [60]
		In vitro reaction	PAH (environmental/industrial exposure)	Automated data analysis		Small mass range (500–650 Da)	Singh et al., 2010 [61]
		Reaction with oligonucleotide	PGE [c], SO [d] (industrial chemicals)	-		Limited to oligonucleotides	Feng et al., 2016 [53]
	Triple Quad	Treated cells (from ovarian follicles)	PAH [e] (environmental/industrial exposure)	-	Nominal mass measurement and lack of fragmentation data		Feng et al., 2016 [54]
		Human lung tissue	Screening for all DNA modifications	Adductome map data analysis			Matsuda et al., 2006 [62]
		Human lung and esophagus tissue	Screening for all DNA modifications	Seven adducts unambiguously detected			Matsuda et al., 2007 [63]
		Various human tissues	LPO-induced (endogenous)	Reported lipid peroxidation-derived adducts in humans			Matsuda et al., 2010 [64]
Pseudo-CNL		Quorn, button mushrooms, brewer's yeast	Food	-		Only 7 SRM transitions per injection	Berdal et al., 2010 [65]
		Treated cells (Chinese hamster)	Micronucleus test-positive compounds	First comparison to micronucleus test		-	Yagi et al., 2011 [66]
		Human gastric mucosa	LPO (endogenous)	-		-	Matsuda et al., 2013 [67]
		Soil Bacterium	Screening for all DNA modifications	First DNA adductomic study of bacterial DNA		-	Kanaly et al., 2015 [55]

23

Table 2. *Cont.*

Approach	Instrument	Sample Type	Adduct Type/Origin	Strengths	Weaknesses	Details	Reference
DD-MS²	Q-TOF	Treated cells (immortalized human T lymphocyte)	Melphalan (chemotherapy drug)	First example of MS² spectral data acquisition	No MS³ fragmentation data, accurate mass data not reported	-	Esmans et al., 2004 [68]
MSᴱ (HRAM)		Mouse lung tissue	Magnetic nanoparticles	First application of MSᴱ	No MS² or MS³ data, reported accurate mass data limited to 10 mmu	-	Totsuka et al., 2015 [57]
Full Scan (HRAM)	Orbitrap	Human colon tumor tissue	Diet-related	-		Diet-related DNA adduct database, acid hydrolysis resulting in nucleobase adducts	Vanhaecke et al., 2015 [34]
		In vitro microbiota meat digests	Diet-related	-	-	Utilized methodology developed in [34]	Vanhaecke et al., 21016 [33]
DD-CNL-MS³	Ion Trap	Treated cells (human hepatocytes) Rat liver Human buccal cells	4-ABP f, MeIQx g Tobacco constituents	Human samples examined, First example of MS³ data acquisition	No accurate mass measurements		Turesky et al., 2009 [56]
		Treated cells (human colon adenocarcinoma)	Illudin S (chemotherapeutic natural product)			Used similar method to Turesky [56]	Sturla et al., 2013 [69]
DD-CNL-MS³ (HRAM)	Orbitrap	Mouse liver tissue	Tobacco constituents	Combination of HRAM, MS³ and nanospray	Extensive sample purification and multiple injections	-	Balbo et al., 2014 [36]
		Treated cells (human colon adenocarcinoma)	DNA alkylating drug	First targeted approach	-	-	Balbo et al., 2015 [37]
		Mouse lung tissue	Endogenous adducts	HRAM MS³ data acquisition	-	-	Balbo et al., 2017 [38]

[a] 2-Amino-1-methyl-6-phenylimidazo[4,5-b]pyridine (PhIP); [b] 2-Amino-3-methylimidazo[4,5-f]quinolone (IQ); [c] Phenyl glycidyl ether (PGE); [d] Styrene-7,8-oxide (SO); [e] Polycyclic aromatic hydrocarbons (PAH); [f] 4-Aminobiphenyl (4-ABP); [g] 2-Amino-3,8-dimethylimidazo[4,5-f]quinoxaline (MeIQx).

5. New HRAM DNA Adductomic Studies

5.1. Untargeted and Targeted Nanospray HRAM CNL-MS^3 Analysis

The DNA adductomic methodology using HRAM CNL-MS^3 detection, a relatively new approach for broad-based screening of DNA adducts, has emerged [36] and efforts are underway to improve the basic methodology, tailor the approach to specific applications, and demonstrate its capabilities [37,38]. The basic method involves HRAM full scan detection followed by DDA MS^2 fragmentation and subsequent MS^3 fragmentation of MS^2 events for which the neutral loss of the deoxyribose moiety was observed. The presence of the MS^3 fragmentation event serves as an indicator of probable adduct detection. The initial proof-of-principle analysis, expanding upon the work of Turesky and coworkers with ion trap detection [56], was performed with incorporation of HRAM detection (5 ppm) and nanospray ionization (300 nL/min) to further empower the CNL-MS^3 approach [36]. The method was optimized using a mix of 18 synthetic DNA adduct standards which included adducts of all 4 bases. Liver tissue from mice exposed to nitrosamine 4-(methylnitrosamino)-1-(3-pyridyl)-1-butanone (NNK) was analyzed with detection of both previously characterized and putative DNA adducts. The methodology was refined [37] for screening of anticipated and unknown adducts induced in cells treated with a chemotherapeutic DNA alkylating agent (PR104A, an experimental nitrogen mustard prodrug under investigation for treatment of leukemia) by incorporating neutral loss triggering of the four DNA bases (Figure 2B) into the methodology. In addition, an extensive ion mass list including all suspected ions from the alkylating agent and metabolites along with all four bases, including cross-link adducts, was utilized for data dependent triggering leading to the detection of many mono- and cross-linked adducts which had not been observed previously. Most recently, the method was used to successfully identify and semi-quantify endogenous and exogenous DNA adducts in the lung of mice exposed to NNK and the proinflammatory agent LPS to observe an adductomic profile [38]. This methodology utilized an extensive list of parent ions from previously observed endogenous adducts as well as suspected adducts resulting from exposure to NNK, and took advantage of an advanced hybrid Orbitrap instrumentation (Fusion).

5.2. Untargeted HRAM MS^E Analysis

Totsuka and coworkers developed a comprehensive DNA adductomic analysis for DNA samples derived from the lungs of mice exposed to nanosized-magnetite (MGT) using an MS^E approach [35] to identify DNA adducts resulting from inflammation. Briefly, the MS^E approach is data independent acquisition (DIA) scanning mode where all ions of interest undergo low and high energy fragmentation and the subsequent ion signal undergoes data analysis to reconstitute fragmentation spectra for individual ions with subsequent identification of the corresponding analytes. Data was acquired with a Waters Xevo QTOF mass spectrometer with a mass range of m/z 50–1000 and a scan duration of 0.5 s (1.0 total duty cycle). The resolution is not reported, however the data analysis was performed with a mass tolerance of 0.05 Da. Reversed phase UPLC separation was performed using a 1.0 mm ID × 150 mm C18 column with 1.7 μm particles and a flow rate of 25 μL/min. In total they detected 30 and 42 types of DNA adducts in the vehicle control and MGT-treated groups, respectively. They performed principal component analysis (PCA) against a subset of DNA adducts and several adducts, which are deduced to be formed by inflammation or oxidative stress (e.g., etheno-deoxycytidine (εdC)), revealed higher contributions to covalent DNA modification resulting from MGT exposure. The levels of εdC were quantified by LC-MS/MS and found to be significantly higher in MGT-treated mice than those of the vehicle control. This analysis is the first example of DIA data acquisition for DNA adductomics analysis.

5.3. Targeted HRAM Full Scan Analysis

An alternative approach to relying upon MS/MS and MS/MS/MS data as confirmation of adduct identity is to develop a DNA adduct database in a targeted DNA adductomics approach and rely upon the accurate mass of the adduct parent ion (MS^1) as an indication of adduct identity. This is the approach developed [33,34] by Vanhaecke and coworkers, in which they created a database of 123 diet-related DNA adducts. This approach used acid-hydrolysis of DNA such that the deoxyribose moiety, which is commonly used in DNA adductomic analysis as an indicator of DNA adduct identity, is not present. The exact mass (10 ppm) and $^{12}C/^{13}C$ ratio of the parent ion was used as confirmation of putatively identified adducts identity with full scan data collected using an Orbitrap detector at a resolution of 100,000 and 3 microscans per spectrum. The methodology was used [33] to analyze in vitro beef digests using fecal microbiota from human subjects and found various DNA adduct profiles consisting of adducts formed from DNA alkylation, oxidation, and reaction of DNA nucleobases by lipid peroxidation products.

5.4. Adduct-Tagging MALDI Ionization Approach

A DNA adductomic approach differing significantly from the conventional methodology described above has been performed, whereby nucleotides are derivatized with benzoylhistamine. Analysis is performed using MALDI-TOF and MALDI-TOF/TOF detection with adduct identification based upon phosphate-specificity of the tagging, detection of adducts as a pair of ions, and measurement of fragment ions characteristic of the presence of deoxyribose or ribose [70,71].

6. Challenges

In vitro DNA adductomics analysis can provide useful information regarding a given biological system, however we feel the ultimate goal should be to analyze in vivo systems and ultimately human samples. There are several challenges to making in vivo DNA adductomics analysis a robust and powerful approach to screening for DNA modification, and they are discussed below.

6.1. Selectivity

We feel that the path forward to fully realizing the promise of DNA adductomics, namely the comprehensive assessment of DNA modification in a variety of exposure contexts at trace levels in biological matrices, is to utilize the analytical power of HRAM and MS^n data acquisition available with the ever-improving modern LC-MS instrumentation. Targeted DNA adduct analysis, while typically performed with triple quadrupole instrumentation (low resolution nominal mass detection), relies upon stable isotopically labeled internal standards, not only for quantitation but just as importantly for confirmation of identity of the analyte being measured. In our experience, there are frequently peaks in these MS/MS chromatograms which are either adjacent or co-eluting with the analyte of interest, especially at the lower levels found in in vivo samples, and could and probably would be attributed to the analyte to be measured if not for the internal standard. While the triple quadrupole MS^2 approaches or MS^1 HRAM strategies may be useful for in vitro applications, where exposures are well defined and usually at higher levels, we feel that for in vivo applications, the power of HRAM MS^n is needed to provide the certainty required for analysis of DNA adducts in the absence of internal standards.

6.2. Sensitivity

One of the main factors limiting the sensitivity for screening DNA modifications is the amount of DNA available for analysis, which is especially true in human blood or biopsy samples, but can also be case for tumor and tumor adjacent tissue where the samples are precious and only a small amount acquired may be made available for DNA adductomic analysis.

The sample cleanliness and chemical complexity of the samples affects not only the selectivity but also the sensitivity. For traditional MS^2 based analysis, the background signal is directly related to the chemical complexity and limits the ability to detect low level adducts. This is less of an issue for HRAM analyses because of the discrimination provided by accurate mass measurements. Chemical noise can limit the sensitivity of trap-based instrumentation due to the finite capacity of the trap, which limits the ability to see trace level ions in the presence of abundant background ions. In addition, for methods using DDA for untargeted detection, the chemical complexity of the sample limits the ability to detect unknown trace level adducts since the scanning speed of the instrumentation is insufficient to sample all ions present as the chromatogram. The presence of matrix material can also impact sensitivity by suppression of the ion signal, a chronic problem with electrospray ionization of LC-MS analysis. Therefore, sample preparation needs to be thoroughly optimized and sensitivity maximized through the use of nanospray ionization.

Finally, nanospray ionization is a powerful option for increasing the inherent sensitivity of ESI-LC-MS analysis. The field of proteomics uses this approach nearly universally and this has, over time, made this a routine mode of operation. It is now possible with minimal experience or training to easily operate in nanospray mode. We feel nanospray should be the default mode of operation for DNA adductomics, due to the trace levels of adducts and the often limited amount of DNA available for analysis.

6.3. Quantitation

Accurate absolute quantitation by LC-MS requires the use of stable isotope-labeled internal standards of the analytes of interest. This is either not possible in the case of untargeted DNA adductomics or impractical in the case of targeted DNA adductomics monitoring for hundreds of adducts at a time. Fortunately, typically absolute quantitation is not necessary to draw the conclusions needed to answer the scientific questions that DNA adductomics is designed to answer. Namely, which DNA adducts are formed at measureable levels and what are the relative amounts of the individual adducts across the samples analyzed. The use of internal standards accounts for several issues, including ionization efficiency variation across analytes, possible losses during sample preparation (recovery), and ion suppression/enhancement due to sample matrix components [15,16]. Relative quantitation of individual analytes across samples is possible if either there is no ion suppression/enhancement and 100% recovery or the ion suppression/enhancement and recovery are consistent across samples. The probability of this being the case is most likely related to the complexity of the matrix. In the case of DNA adduct analysis, the matrix is much simpler than other common biological matrices commonly analyzed by LC-MS such as urine or plasma. The complexity of the DNA samples should consist of the unmodified bases, which are very hydrophilic and therefore elute much earlier than many adducts, hydrolysis enzymes, and any impurities in the enzymes and unpolymerized constituents entering the sample solution when using plastic components, such as solid-phase extraction cartridges and molecular weight filters for sample preparation [29,36,72]. The impurities due to the use of hydrolysis enzymes can be nearly eliminated, or at least greatly reduced, by careful enzyme source and vendor selection, cleaning of the enzymes prior to use, and determining the minimal enzyme necessary for the analysis [72,73]. Impurities due to the use of plastics can be greatly reduced or eliminated by avoiding the use plastics either entirely or as much is practically possible and by careful type/vendor selection of consumables used for the experiments.

A relative quantitation strategy, which will account for variable ion suppression/enhancement and recovery, has recently been demonstrated with the DNA adductomic analysis of a cell-based system [37,74]. It involves the generation of a mix of DNA adducts by the treatment of cells with an isotopically labeled version of the genotoxic substance of interest. This mix of isotopically labeled adducts was used for relative quantitation of many adducts in subsequent experiments by spiking the labeled adduct internal standard mixture into cell, animal, or human samples.

Additionally, relative quantitation is also possible, using a peak-area-based labeled free quantitation method. This approach is based on integrating the adduct precursor ion chromatogram peak areas in the full scan and on normalizing the signal intensity to the amount of DNA used for the analysis and to a quantitative reference, a labeled internal standard added in constant amounts to each sample.

Lastly, if it is determined that absolute quantitation or more precise relative quantitation is necessary, the targeted quantitation of those adducts of interest can be performed via the traditional quantitation approach after synthesis of the labeled internal standards.

6.4. Ease of Data Analysis

Software development for advanced HRAM data analysis has strived to keep pace with advances in mass spectrometry instrumentation to take full advantage of the technology. Unfortunately, the advances have focused upon proteomics, metabolomics, metabolite analysis, lipidomics, etc. and have not been geared toward the type of analysis required for DNA adductomics. Therefore, further development of software tools for the data analysis is necessary to interpret the results coming from DNA adductomics experiments. Software solutions are needed for both DDA and DIA data. In the case of DNA adductomics DDA data, software tools for the recognition, tabulation, and display of fragmentation data which corresponds to the neutral loss of deoxyribose and bases (see Figure 2A,B) would help with the automated and high throughput analysis of the data. For example, ideally for our DDA-CNL/MS^3 methodology [36–38], for each putative adduct identification, an output displaying the MS/MS and MS/MS/MS spectra as well as the integrated extracted ion chromatogram would be very useful, and tabulation of this data in a searchable format would be ideal. In the case of the DIA-Wide SIM/MS^2 methodology we recommend, software tools are needed which can perform a variety of tasks in an automated fashion, such as perform peak picking, extract ion chromatograms for the SIM and MS/MS data, integrate the resulting peak areas, compare retention times and mass differences, and tabulate and display the results.

7. Summary

The identification and structural characterization of DNA adducts in human tissues can be used to either identify specific genotoxic exposures or confirm that suspected exposures have occurred and led to DNA modification. Quantitation of these DNA adducts can be used to assess the extent of this damage. Recent advances in the field of mass spectrometry have led to the development of DNA adductomic methods which can be used to comprehensively identify, characterize, and semi-quantify the DNA adducts produced in an in vivo system or in human samples. We feel that the most advanced, sophisticated aspects of this new era of mass spectrometry should be harnessed to make this type of analysis a powerful tool for screening for DNA modification characterization in biologically relevant contexts. In addition, careful use of negative controls and scrutiny of HRAM data is necessary to assure signal is rightfully attributed to putative DNA adducts. As in the other "-omics" methodologies, we envision different basic approaches being used depending upon the needs of the specific experiments. For example, HRAM DIA-Wide SIM/MS^2 analysis might be better suited for screening for large numbers of known adducts, whereas HRAM DDA-CNL/MS^3 analysis may be better suited for identifying unknown DNA adducts.

The ultimate goal of DNA adductomics [29] is to characterize the modifications of DNA as a profile of specific adducts, rather than focusing only on a few adducts at a time. There are many applications of DNA adductomic analysis, including investigating the genotoxic effect of exposures from the environment [33–36,41,51,53–56,59–63,65,66], as well as endogenous adduct formation [38,64,67]. It can be used to investigate mechanisms of actions of genotoxic chemotherapeutic drugs [37,68,69], and the mutagenicity potential of pharmaceuticals or supplements in the context of cancer risk. It can be used in drug design for development of DNA alkylating chemotherapeutic agents, both in terms of maximizing the genotoxicity to cancer cells as well as minimizing the genotoxicity to healthy cells [37]. The ability to broadly screen for DNA adducts can also be used for a "precision medicine" approach to chemotherapeutic drug treatment [74]. Screening for DNA modification by non-cancer therapeutic drugs, as well as minimizing toxicity during drug development, are also possible applications. Lastly, screening of epigenetic changes [16] is also a possible use for DNA adductomics, whereby all known epigenetic modifications to DNA bases could be monitored while simultaneously screening for unknown modifications.

Currently, DNA adductomics offers the potential to fully characterize the chemical modification of DNA by detection and relative quantitation of known and unknown DNA adducts, providing information regarding the exposures which have occurred, resulting genotoxic effects, and, more importantly, elucidating mechanisms of interaction between chemicals and DNA. Overall this approach provides crucial complementary information to that acquired from mutagenicity assays.

The critical DNA modifications resulting from exposure to a particular compound may be on specific DNA sequences or chromatin structures. DNA adductomics requires the hydrolysis of DNA to allow for the analysis of the modified nucleosides, and therefore any information regarding the sites of the modifications is lost. For now, DNA adductomics should be combined with genomic-wide sequencing to correlate DNA adduct formation with biologically important mutations. However, the promising trend of improving instrumentation and molecular biology techniques leads us to believe that in the future we will be able to perform this analysis on specific sequences and targeting specific genes.

Acknowledgments: This work was supported by University of Minnesota start-up funds for Silvia Balbo and by the National Cancer Institute of the National Institute of Health under Award Number R50CA211256 (NCI Research Specialist Award) for Peter W. Villalta.

Conflicts of Interest: The authors declare no conflict of interest.

Abbreviations

LC-MS	Liquid chromatography—mass spectrometry
LC-MS^2	Liquid chromatography—tandem mass spectrometry
LC-MS^n	Liquid chromatography—multistage fragmentation mass spectrometry
HRAM	High resolution/accurate mass
MALDI	Matrix assisted laser desorption ionization
TOF	Time-of-flight
UPLC	Ultra high pressure liquid chromatography
Q-TOF	Quadrupole-Time-of-flight
DDA	Data dependent acquisition
DIA	Data independent acquisition
CNL/MS^3	Constant neutral loss/triple stage mass spectrometry
SIM/MS^2	Selected ion monitoring/tandem mass spectrometry
NNK	Nitrosamine 4-(methylnitrosamino)-1-(3-pyridyl)-1-butanone
LPS	Lipopolysaccharide
PCA	Principal component analysis
MALDI-TOF	Matrix assisted laser desorption ionization—time-of-flight
MALDI-TOF/TOF	Matrix assisted laser desorption ionization—time-of-flight/time-of-flight

References

1. Wiencke, J.K. DNA adduct burden and tobacco carcinogenesis. *Oncogene* **2002**, *21*, 7376–7391. [CrossRef] [PubMed]
2. Dipple, A. DNA adducts of chemical carcinogens. *Carcinogenesis* **1995**, *16*, 437–441. [CrossRef] [PubMed]
3. Poirier, M.C. Linking DNA adduct formation and human cancer risk in chemical carcinogenesis. *Environ. Mol. Mutagen.* **2016**, *57*, 499–507. [CrossRef] [PubMed]
4. Poirier, M.C.; Beland, F.A. DNA adduct measurements and tumor incidence during chronic carcinogen exposure in rodents. *Environ. Health. Perspect.* **1994**, *102* (Suppl. S6), 161–165. [CrossRef] [PubMed]
5. Hemminki, K.; Koskinen, M.; Rajaniemi, H.; Zhao, C. DNA adducts, mutations, and cancer 2000. *Regul. Toxicol. Pharmacol.* **2000**, *32*, 264–275. [CrossRef] [PubMed]
6. Food and Drug Administration, HHS. International conference on harmonisation; guidance on s2(r1) genotoxicity testing and data interpretation for pharmaceuticals intended for human use; availability. Notice. *Fed. Regist.* **2012**, *77*, 33748–33749.

7. Klaene, J.J.; Sharma, V.K.; Glick, J.; Vouros, P. The analysis of DNA adducts: The transition from ^{32}P-postlabeling to mass spectrometry. *Cancer Lett.* **2013**, *334*, 10–19. [CrossRef] [PubMed]

8. Farmer, P.B.; Singh, R. Use of DNA adducts to identify human health risk from exposure to hazardous environmental pollutants: The increasing role of mass spectrometry in assessing biologically effective doses of genotoxic carcinogens. *Mutat. Res.* **2008**, *659*, 68–76. [CrossRef] [PubMed]

9. Parry, J.M. *Genetic Toxicol. Principles Methods*; Humana: Totowa, NJ, USA, 2012; p. 433.

10. Rybicki, B.A.; Rundle, A.; Savera, A.T.; Sankey, S.S.; Tang, D. Polycyclic aromatic hydrocarbon-DNA adducts in prostate cancer. *Cancer Res.* **2004**, *64*, 8854–8859. [CrossRef] [PubMed]

11. Blommaert, F.A.; Michael, C.; Terheggen, P.M.; Muggia, F.M.; Kortes, V.; Schornagel, J.H.; Hart, A.A.; den Engelse, L. Drug-induced DNA modification in buccal cells of cancer patients receiving carboplatin and cisplatin combination chemotherapy, as determined by an immunocytochemical method: Interindividual variation and correlation with disease response. *Cancer Res.* **1993**, *53*, 5669–5675. [PubMed]

12. Andrews, C.L.; Vouros, P.; Harsch, A. Analysis of DNA adducts using high-performance separation techniques coupled to electrospray ionization mass spectrometry. *J. Chromatogr.* **1999**, *856*, 515–526. [CrossRef]

13. Koc, H.; Swenberg, J.A. Applications of mass spectrometry for quantitation of DNA adducts. *J. Chromatogr. B* **2002**, *778*, 323–343. [CrossRef]

14. Singh, R.; Farmer, P.B. Liquid chromatography-electrospray ionization-mass spectrometry: The future of DNA adduct detection. *Carcinogenesis* **2006**, *27*, 178–196. [CrossRef] [PubMed]

15. Tretyakova, N.; Goggin, M.; Sangaraju, D.; Janis, G. Quantitation of DNA adducts by stable isotope dilution mass spectrometry. *Chem. Res. Toxicol.* **2012**, *25*, 2007–2035. [CrossRef] [PubMed]

16. Tretyakova, N.; Villalta, P.W.; Kotapati, S. Mass spectrometry of structurally modified DNA. *Chem. Rev.* **2013**, *113*, 2395–2436. [CrossRef] [PubMed]

17. Beach, A.C.; Gupta, R.C. Human biomonitoring and the ^{32}P-postlabeling assay. *Carcinogenesis* **1992**, *13*, 1053–1074. [CrossRef] [PubMed]

18. Phillips, D.H. DNA adducts as markers of exposure and risk. *Mutat. Res.* **2005**, *577*, 284–292. [CrossRef] [PubMed]

19. Jones, N.J.; McGregor, A.D.; Waters, R. Detection of DNA adducts in human oral tissue: Correlation of adduct levels with tobacco smoking and differential enhancement of adducts using the butanol extraction and nuclease p1 versons of [32]p postlabeling. *Cancer Res.* **1993**, *53*, 1522–1528. [PubMed]

20. Phillips, D.H. Smoking-related DNA and protein adducts in human tissues. *Carcinogenesis* **2002**, *23*, 1979–2004. [CrossRef] [PubMed]

21. Di Paolo, O.A.; Teitel, C.H.; Nowell, S.; Coles, B.F.; Kadlubar, F.F. Expression of cytochromes P450 and glutathione s-transferases in human prostate, and the potential for activation of heterocyclic amine carcinogens via acetyl-coa-, paps- and atp-dependent pathways. *Int. J. Cancer* **2005**, *117*, 8–13. [CrossRef] [PubMed]

22. Nath, R.G.; Ocando, J.E.; Guttenplan, J.B.; Chung, F.L. 1,n^2-propanodeoxyguanosine adducts: Potential new biomarkers of smoking-induced DNA damage in human oral tissue. *Cancer Res.* **1998**, *58*, 581–584. [PubMed]

23. Arif, J.M.; Dresler, C.; Clapper, M.L.; Gairola, C.G.; Srinivasan, C.; Lubet, R.A.; Gupta, R.C. Lung DNA adducts detected in human smokers are unrelated to typical polyaromatic carcinogens. *Chem. Res. Toxicol.* **2006**, *19*, 295–299. [CrossRef] [PubMed]

24. Talaska, G.; Schamer, M.; Skipper, P.; Tannenbaum, S.; Caporaso, N.; Unruh, L.; Kadlubar, F.F.; Bartsch, H.; Malaveille, C.; Vineis, P. Detection of carcinogen-DNA adducts in exfoliated urothelial cells of cigarette smokers: Association with smoking, hemoglobin adducts, and urinary mutagenicity. *Cancer Epidemiol. Biomark. Prev.* **1991**, *1*, 61–66. [PubMed]

25. Gorlewska-Roberts, K.; Green, B.; Fares, M.; Ambrosone, C.B.; Kadlubar, F.F. Carcinogen-DNA adducts in human breast epithelial cells. *Environ. Mol. Mutagen.* **2002**, *39*, 184–192. [CrossRef] [PubMed]

26. Nia, A.B.; Maas, L.M.; van Breda, S.G.; Curfs, D.M.; Kleinjans, J.C.; Wouters, E.F.; van Schooten, F.J. Applicability of induced sputum for molecular dosimetry of exposure to inhalatory carcinogens: ^{32}P-postlabeling of lipophilic DNA adducts in smokers and nonsmokers. *Cancer Epidemiol. Biomark. Prev.* **2000**, *9*, 367–372.

27. Nia, A.B.; Maas, L.M.; Brouwer, E.M.; Kleinjans, J.C.; van Schooten, F.J. Comparison between smoking-related DNA adduct analysis in induced sputum and peripheral blood lymphocytes. *Carcinogenesis* **2000**, *21*, 1335–1340. [CrossRef]

28. Phillips, D.H. On the origins and development of the [32]p-postlabelling assay for carcinogen-DNA adducts. *Cancer Lett.* **2013**, *334*, 5–9. [CrossRef] [PubMed]

29. Balbo, S.; Turesky, R.J.; Villalta, P.W. DNA adductomics. *Chem. Res. Toxicol.* **2014**, *27*, 356–366. [CrossRef] [PubMed]

30. Totsuka, Y.; Fukutome, K.; Takahashi, M.; Takahashi, S.; Tada, A.; Sugimura, T.; Wakabayashi, K. Presence of N^2-(deoxyguanosin-8-yl)-2-amino-3,8-dimethylimidazo[4,5-*f*]quinoxaline (dG-C8-MeIQx) in human tissues. *Carcinogenesis* **1996**, *17*, 1029–1034. [CrossRef] [PubMed]

31. Phillips, D.H.; Arlt, V.M. The ^{32}P-postlabeling assay for DNA adducts. *Nat. Protocols* **2007**, *2*, 2772–2781. [CrossRef] [PubMed]

32. Claereboudt, J.; Esmans, E.L.; Vandeneeckhout, E.G.; Claeys, M. Fast-atom-bombardment and tandem mass-spectrometry for the identification of nucleoside adducts with phenyl glycidyl ether. *Nucleos. Nucleot.* **1990**, *9*, 333–344. [CrossRef]

33. Hemeryck, L.Y.; Rombouts, C.; van Hecke, T.; van Meulebroek, L.; Vanden Bussche, J.; de Smet, S.; Vanhaecke, L. In vitro DNA adduct profiling to mechanistically link red meat consumption to colon cancer promotion. *Toxicol. Res.* **2016**, *5*, 1346–1358. [CrossRef]

34. Hemeryck, L.Y.; Decloedt, A.I.; Vanden Bussche, J.; Geboes, K.P.; Vanhaecke, L. High resolution mass spectrometry based profiling of diet-related deoxyribonucleic acid adducts. *Anal. Chim. Acta* **2015**, *892*, 123–131. [CrossRef] [PubMed]

35. Ishino, K.; Kato, T.; Kato, M.; Shibata, T.; Watanabe, M.; Wakabayashi, K.; Nakagama, H.; Totsuka, Y. Comprehensive DNA adduct analysis reveals pulmonary inflammatory response contributes to genotoxic action of magnetite nanoparticles. *Int. J. Mol. Sci.* **2015**, *16*, 3474–3492. [CrossRef] [PubMed]

36. Balbo, S.; Hecht, S.S.; Upadhyaya, P.; Villalta, P.W. Application of a high-resolution mass-spectrometry-based DNA adductomics approach for identification of DNA adducts in complex mixtures. *Anal. Chem.* **2014**, *86*, 1744–1752. [CrossRef] [PubMed]

37. Stornetta, A.; Villalta, P.W.; Hecht, S.S.; Sturla, S.; Balbo, S. Screening for DNA alkylation mono and cross-linked adducts with a comprehensive LC-MS^3 adductomic approach. *Anal. Chem.* **2015**, *87*, 11706–11713. [CrossRef] [PubMed]

38. Carra', A.; Villalta, P.W.; Dator, R.P.; Balbo, S. Screening for inflammation-induced DNA adducts with a comprehensive high resolution LC-MS^3 adductomic approach. In Proceedings of the 64th ASMS Annual Meeting, San Antonio, TX, USA, 4–8 June 2016.

39. Gates, K.S.; Nooner, T.; Dutta, S. Biologically relevant chemical reactions of n7-alkylguanine residues in DNA. *Chem. Res. Toxicol* **2004**, *17*, 839–856. [CrossRef] [PubMed]

40. Inagaki, S.; Hirashima, H.; Esaka, Y.; Higashi, T.; Min, J.Z.; Toyo'oka, T. Screening DNA adducts by lc–esi–ms–ms: Application to screening new adducts formed from acrylamide. *Chromatographia* **2010**, *72*, 1043–1048. [CrossRef]

41. Gangl, E.T.; Turesky, R.J.; Vouros, P. Determination of in vitro- and in vivo-formed DNA adducts of 2-amino-3-methylimidazo[4,5-*f*]quinoline by Capillary Liquid Chromatography/Microelectrospray Mass Spectrometry. *Chem. Res. Toxicol.* **1999**, *12*, 1019–1027. [CrossRef] [PubMed]

42. Shi, T.; Song, E.; Nie, S.; Rodland, K.D.; Liu, T.; Qian, W.J.; Smith, R.D. Advances in targeted proteomics and applications to biomedical research. *Proteomics* **2016**, *16*, 2160–2182. [CrossRef] [PubMed]

43. Gillet, L.C.; Navarro, P.; Tate, S.; Rost, H.; Selevsek, N.; Reiter, L.; Bonner, R.; Aebersold, R. Targeted data extraction of the MS/MS spectra generated by data-independent acquisition: A new concept for consistent and accurate proteome analysis. *Mol. Cell Proteom.* **2012**, *11*, O111-016717. [CrossRef] [PubMed]

44. Gillet, L.C.; Leitner, A.; Aebersold, R. Mass spectrometry applied to bottom-up proteomics: Entering the high-throughput era for hypothesis testing. *Annu Rev. Anal. Chem.* **2016**, *9*, 449–472. [CrossRef] [PubMed]

45. Venable, J.D.; Dong, M.Q.; Wohlschlegel, J.; Dillin, A.; Yates, J.R. Automated approach for quantitative analysis of complex peptide mixtures from tandem mass spectra. *Nat. Methods* **2004**, *1*, 39–45. [CrossRef] [PubMed]

46. Xiao, J.F.; Zhou, B.; Ressom, H.W. Metabolite identification and quantitation in LC-MS/MS-based metabolomics. *Trends Anal. Chem.* **2012**, *32*, 1–14. [CrossRef] [PubMed]

47. Tsugawa, H.; Cajka, T.; Kind, T.; Ma, Y.; Higgins, B.; Ikeda, K.; Kanazawa, M.; VanderGheynst, J.; Fiehn, O.; Arita, M. MS-DIAL: Data-independent MS/MS deconvolution for comprehensive metabolome analysis. *Nat. Methods* **2015**, *12*, 523–526. [CrossRef] [PubMed]

48. Tsou, C.C.; Avtonomov, D.; Larsen, B.; Tucholska, M.; Choi, H.; Gingras, A.C.; Nesvizhskii, A.I. DIA-Umpire: comprehensive computational framework for data-independent acquisition proteomics. *Nat. Methods* **2015**, *12*, 258–264. [CrossRef] [PubMed]

49. Bilbao, A.; Varesio, E.; Luban, J.; Strambio-De-Castillia, C.; Hopfgartner, G.; Muller, M.; Lisacek, F. Processing strategies and software solutions for data-independent acquisition in mass spectrometry. *Proteomics* **2015**, *15*, 964–980. [CrossRef] [PubMed]

50. Li, H.; Cai, Y.; Guo, Y.; Chen, F.; Zhu, Z.J. MetDIA: Targeted metabolite extraction of multiplexed MS/MS spectra generated by data-independent acquisition. *Anal. Chem.* **2016**, *88*, 8757–8764. [CrossRef] [PubMed]

51. Hemeryck, L.Y.; Moore, S.A.; Vanhaecke, L. Mass spectrometric mapping of the DNA adductome as a means to study genotoxin exposure, metabolism and effect. *Anal. Chem.* **2016**, *88*, 7436–7446. [CrossRef] [PubMed]

52. Hemeryck, L.Y.; Vanhaecke, L. Diet-related DNA adduct formation in relation to carcinogenesis. *Nutr. Rev.* **2016**, *74*, 15. [CrossRef] [PubMed]

53. Yao, C.; Feng, Y.-L. A nontargeted screening method for covalent DNA adducts and DNA modification selectivity using liquid chromatography-tandem mass spectrometry. *Talanta* **2016**, *159*, 10. [CrossRef] [PubMed]

54. Yao, C.; Foster, W.G.; Sadeu, J.C.; Siddique, S.; Zhu, J.; Feng, Y.L. Screening for DNA adducts in ovarian follicles exposed to benzo [a] pyrene and cigarette smoke condensate using liquid chromatography-tandem mass spectrometry. *Sci. Total Environ.* **2016**, *575*, 742–749. [CrossRef] [PubMed]

55. Kanaly, R.A.; Micheletto, R.; Matsuda, T.; Utsuno, Y.; Ozeki, Y.; Hamamura, N. Application of DNA adductomics to soil bacterium Sphingobium sp strain KK22. *Microbiologyopen* **2015**, *4*, 841–856. [CrossRef] [PubMed]

56. Bessette, E.E.; Goodenough, A.K.; Langouet, S.; Yasa, I.; Kozekov, I.D.; Spivack, S.D.; Turesky, R.J. Screening for DNA adducts by data-dependent constant neutral loss-triple stage mass spectrometry with a linear quadrupole ion trap mass spectrometer. *Anal. Chem.* **2009**, *81*, 809–819. [CrossRef] [PubMed]

57. Guo, J.; Villalta, P.W.; Turesky, R.J. A data-independent mass spectrometry approach for screening and identification of DNA adducts. *Anal. Chem.*. submitted.

58. Bryant, M.S.; Lay, J.O.; Chiarelli, M.P. Development of fast atom bombardment mass spectral methods for the identification of carcinogen-nucleoside adducts. *J. Am. Soc. Mass Spectrom.* **1992**, *3*, 360–371. [CrossRef]

59. Rindgen, D.; Turesky, R.J.; Vouros, P. Determination of in vitro formed DNA adducts of 2-amino-1-methyl-6-phenylimidazo [4,5-b] pyridine using capillary liquid chromatography/electrospray ionization/tandem mass spectrometry. *Chem. Res. Toxicol.* **1995**, *8*, 1005–1013. [CrossRef] [PubMed]

60. Regulus, P.; Spessotto, S.; Gateau, M.; Cadet, J.; Favier, A.; Ravanat, J.L. Detection of new radiation-induced DNA lesions by liquid chromatography coupled to tandem mass spectrometry. *Rapid Commun. Mass Spectrom.* **2004**, *18*, 2223–2228. [CrossRef] [PubMed]

61. Singh, R.; Teichert, F.; Seidel, A.; Roach, J.; Cordell, R.; Cheng, M.K.; Frank, H.; Steward, W.P.; Manson, M.M.; Farmer, P.B. Development of a targeted adductomic method for the determination of polycyclic aromatic hydrocarbon DNA adducts using online column-switching liquid chromatography/tandem mass spectrometry. *Rapid Commun. Mass Spectrom.* **2010**, *24*, 2329–2340. [CrossRef] [PubMed]

62. Kanaly, R.A.; Hanaoka, T.; Sugimura, H.; Toda, H.; Matsui, S.; Matsuda, T. Development of the adductome approach to detect DNA damage in humans. *Antioxid. Redox. Signal* **2006**, *8*, 993–1001. [CrossRef] [PubMed]

63. Kanaly, R.A.; Matsui, S.; Hanaoka, T.; Matsuda, T. Application of the adductome approach to assess intertissue DNA damage variations in human lung and esophagus. *Mutat. Res.* **2007**, *625*, 83–93. [CrossRef] [PubMed]

64. Chou, P.H.; Kageyama, S.; Matsuda, S.; Kanemoto, K.; Sasada, Y.; Oka, M.; Shinmura, K.; Mori, H.; Kawai, K.; Kasai, H.; et al. Detection of lipid peroxidation-induced DNA adducts caused by 4-oxo-2 (E)-nonenal and 4-oxo-2 (E)-hexenal in human autopsy tissues. *Chem. Res. Toxicol.* **2010**, *23*, 1442–1448. [CrossRef] [PubMed]

65. Spilsberg, B.; Rundberget, T.; Johannessen, L.E.; Kristoffersen, A.B.; Holst-Jensen, A.; Berdal, K.G. Detection of food-derived damaged nucleosides with possible adverse effects on human health using a global adductomics approach. *J. Agric. Food Chem.* **2010**, *58*, 6370–6375. [CrossRef] [PubMed]

66. Kato, K.; Yamamura, E.; Kawanishi, M.; Yagi, T.; Matsuda, T.; Sugiyama, A.; Uno, Y. Application of the DNA adductome approach to assess the DNA-damaging capability of in vitro micronucleus test-positive compounds. *Mutat. Res.* **2011**, *721*, 21–26. [CrossRef] [PubMed]

67. Matsuda, T.; Tao, H.; Goto, M.; Yamada, H.; Suzuki, M.; Wu, Y.; Xiao, N.; He, Q.; Guo, W.; Cai, Z.; et al. Lipid peroxidation-induced DNA adducts in human gastric mucosa. *Carcinogenesis* **2013**, *34*, 121–127. [CrossRef] [PubMed]

68. Van den Driessche, B.; van Dongen, W.; Lemiere, F.; Esmans, E.L. Implementation of data-dependent acquisitions in the study of melphalan DNA adducts by miniaturized liquid chromatography coupled to electrospray tandem mass spectrometry. *Rapid Commun. Mass Spectrom.* **2004**, *18*, 2001–2007. [CrossRef] [PubMed]

69. Pietsch, K.E.; van Midwoud, P.M.; Villalta, P.W.; Sturla, S.J. Quantification of acylfulvene- and illudin s-DNA adducts in cells with variable bioactivation capacities. *Chem. Res. Toxicol.* **2013**, *26*, 146–155. [CrossRef] [PubMed]

70. Wang, P.; Gao, J.; Li, G.; Shimelis, O.; Giese, R.W. Nontargeted analysis of DNA adducts by mass-tag ms: Reaction of p-benzoquinone with DNA. *Chem. Res. Toxicol.* **2012**, *25*, 2737–2743. [CrossRef] [PubMed]

71. Wang, P.; Fisher, D.; Rao, A.; Giese, R.W. Nontargeted nucleotide analysis based on benzoylhistamine labeling-MALDI-TOF/TOF-MS: discovery of putative 6-oxo-thymine in DNA. *Anal. Chem.* **2012**, *84*, 3811–3819. [CrossRef] [PubMed]

72. Klaene, J.J.; Flarakos, C.; Glick, J.; Barret, J.T.; Zarbl, H.; Vouros, P. Tracking matrix effects in the analysis of DNA adducts of polycyclic aromatic hydrocarbons. *J. Chromatogr. A* **2016**, *1439*, 112–123. [CrossRef] [PubMed]

73. Villalta, P.W.H.; Hochalter, J.B.; Hecht, S.S. Ultra-sensitive high resolution mass spectrometric analysis of a DNA adduct of the carcinogen benzo [a] pyrene in human lung. *Anal. Chem.*. submitted.

74. Stornetta, A.; Villalta, P.W.; Gossner, F.; Wilson, W.R.; Balbo, S.; Sturla, S.J. DNA adduct profiles predict in vitro cell viability after treatment with the experimental anticancer prodrug pr104a. *Chem. Res. Toxicol.* **2017**, *30*, 830–839. [CrossRef] [PubMed]

International Journal of
Molecular Sciences

MDPI

Review

Every OGT Is Illuminated ... by Fluorescent and Synchrotron Lights

Riccardo Miggiano [1], Anna Valenti [2], Franca Rossi [1], Menico Rizzi [1], Giuseppe Perugino [2,*] and Maria Ciaramella [2,*]

[1] DSF-Dipartimento di Scienze del Farmaco, University of Piemonte Orientale, Via Bovio 6, 28100 Novara, Italy; riccardo.miggiano@uniupo.it (R.M.); franca.rossi@uniupo.it (F.R.); menico.rizzi@uniupo.it (M.R.)

[2] Institute of Biosciences and BioResources, National Research Council of Italy, Via Pietro Castellino 111, 80131 Naples, Italy; anna.valenti@ibbr.cnr.it

* Correspondence: giuseppe.perugino@ibbr.cnr.it (G.P.); maria.ciaramella@ibbr.cnr.it (M.C.); Tel.: +39-081-6132-496 (G.P.); +39-081-6132-274 (M.C.); Fax: +39-081-6132-646 (M.C.)

Received: 15 November 2017; Accepted: 30 November 2017; Published: 5 December 2017

Abstract: O^6-DNA-alkyl-guanine-DNA-alkyl-transferases (OGTs) are evolutionarily conserved, unique proteins that repair alkylation lesions in DNA in a single step reaction. Alkylating agents are environmental pollutants as well as by-products of cellular reactions, but are also very effective chemotherapeutic drugs. OGTs are major players in counteracting the effects of such agents, thus their action in turn affects genome integrity, survival of organisms under challenging conditions and response to chemotherapy. Numerous studies on OGTs from eukaryotes, bacteria and archaea have been reported, highlighting amazing features that make OGTs unique proteins in their reaction mechanism as well as post-reaction fate. This review reports recent functional and structural data on two prokaryotic OGTs, from the pathogenic bacterium *Mycobacterium tuberculosis* and the hyperthermophilic archaeon *Sulfolobus solfataricus*, respectively. These studies provided insight in the role of OGTs in the biology of these microorganisms, but also important hints useful to understand the general properties of this class of proteins.

Keywords: DNA repair; alkylation damage; conformational changes; protein structure; protein stability; protein-tag

1. Introduction

O^6-DNA-alkyl-guanine-DNA-alkyl-transferases (EC: 2.1.1.63) are small (17–22 kDa) proteins that catalyze repair of DNA lesions induced by alkylating agents, a class of mutagenic and carcinogenic agents present in the environment. These proteins are evolutionarily conserved across the three domains of life, and are known by different acronyms (AGT, for alkyl-guanine-DNA-alkyl-transferase; OGT, for O^6-alkyl-guanine-DNA-alkyl-transferase, or MGMT, for O^6-methylguanine-DNA methyltransferase) in different organisms [1–5] and hereafter will be collectively indicated as OGTs. Alkylating agents can be divided into two subgroups, namely the SN_1 and SN_2 types respectively, which act with different mechanisms. Those belonging to the SN_1 type (such as N-methyl-N-nitrosourea and N-methyl-N'-nitro-N-nitrosoguanidine) act by a monomolecular mechanism and can induce both N- and O-methylation, whereas the SN_2 type drugs (such as methyl methanesulfonate (MMS) and methyl iodide) mainly induce N-methylation by bimolecular mechanisms [6]. Alkylation occurs preferentially at position N^7 or O^6 of guanine and N^3 of adenine. Alkylation damage is particularly harmful for genomes because, if unrepaired, it can result in base mismatch, replication fork stalling, single or double strand breaks [6–10]. These lesions can be repaired by different enzymes. DNA glycosylases, such as AlkA and Tag remove N^3-methyladenine (3meA) and create abasic sites in DNA, which are subsequently repaired by the base excision repair system

(BER). Dioxygenases, such as AlkB, repair N^1-methyladenine (1meA) and N^3-methylcytosine (3meC) directly restoring the correct DNA sequence [6]. Finally, OGTs directly remove methyl groups from O^6-methylguanine (O^6meG) and O^4-methylthymine (O^4meT).

OGTs are unique in their structure, mechanism and post-reaction outcome. They are able to perform whole repair reactions without the assistance of other factors or the need of energy source. OGTs are able to catalyze a trans-alkylation reaction, in which an alkyl group in DNA, mainly at position O^6 of guanines or O^4 of tymines, is transferred to a cysteine residue in the protein active site [3,4,11,12] (Figure 1a). This reaction is irreversible and determines the protein inactivation, thus each OGT molecule works only once [13]. The lesion recognition and repair on DNA occurs by a peculiar flipping-out mechanism, in which the protein extrudes the alkylated base form the double helix [4,12,14].

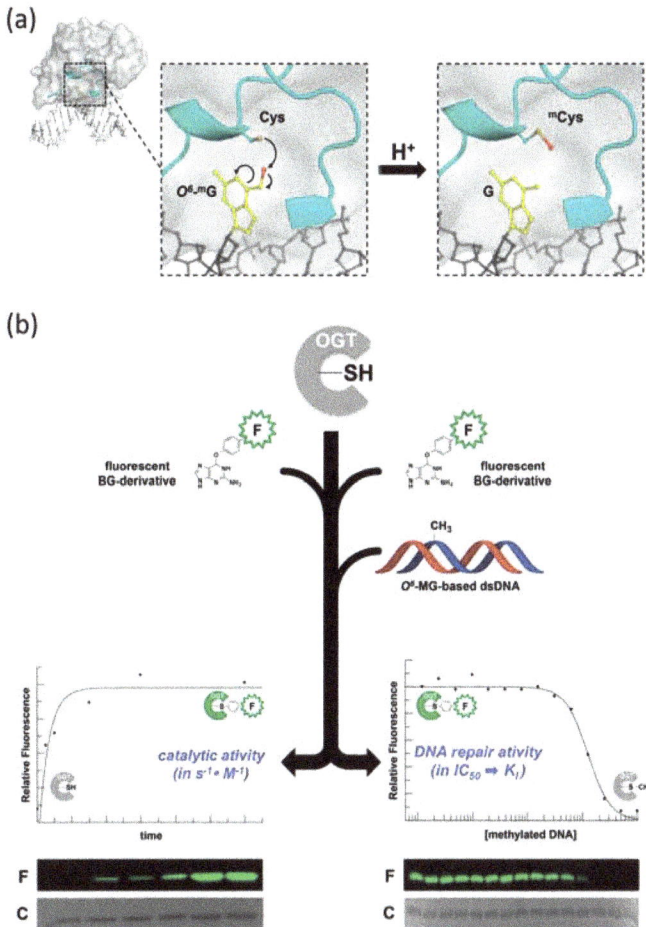

Figure 1. The DNA repair reaction by O^6-DNA-alkyl-guanine-DNA-alkyl-transferases (OGTs). (**a**) Upon DNA binding, OGT flips-out the damaged guanine from the DNA backbone and irreversibly transfers the alkyl group to its own catalytic cysteine. (**b**) The covalent bond between the alkyl group and the cysteine allowed developing a new assay, by using O^6-BG fluorescent derivatives alone (**left panel**), or in combination with natural non-fluorescent alkylated-DNA substrates (**right panel**).

Historically, knowledge of OGTs was provided by classic studies on two proteins encoded by *Escherichia coli*, called Ada (adaptive response) and OGT [15–17]. In recent years, the human DNA-alkyl-transferase protein, called hAGT, raised a lot of interest because its overexpression in cancer cells has been associated with the onset of resistance to alkylating chemotherapy agents, making hAGT a potential therapeutic target [18–20]. In addition, hAGT has recently been the starting point for a new protein labelling strategy, called SNAP-tag® technology [21–23].

In this review, we will focus on recent results reported on two prokaryotic OGTs, from the pathogenic bacterium *M. tuberculosis* (*Mt*OGT) and the hyperthermophilic archaeon *S. solfataricus* (*Ss*OGT), respectively. Among OGTs, these two proteins raised particular interest because of the potential involvement in virulence of the former and the peculiar thermal stability of the latter. In-deep biochemical characterization of both proteins has been performed, thanks to the development of novel assays and site-directed mutagenesis; moreover, a structural characterization of wild type and mutant versions of these two proteins allowed us to highlight important features of all OGTs. Finally, the peculiar thermostable nature of *Ss*OGT has allowed the development of a novel protein-tag for thermophilic organisms.

2. Development of Novel Alkyl-Transferase Assays

Although they are biocatalysts, OGTs are not conventional enzymes, as their reaction is irreversible and protein molecules are not recycled (Figure 1a). A number of assays have been used to measure OGTs activity; most of them rely on laborious and time-consuming procedures, implying the use of radiolabelled isotopes and/or chromatographic techniques, requiring high amount of purified proteins [24–27]. In order to overcome these limitations, novel DNA alkyl-transferase assays have been recently developed. These assays are based on the use of fluorescent derivatives of O^6-benzyl-guanine (O^6-BG), a competitive inhibitor of most OGTs. This molecule reacts with the catalytic cysteine of OGTs with the same mechanism as the natural substrate, resulting in covalent transfer of the benzyl group to the protein. If O^6-BG is conjugated with a fluorogenic group, the latter is irreversibly transferred to the protein, with a one to one stoichiometry; hence the fluorescence intensity is a direct measure of the protein activity. The method can be applied to all OGTs sensitive to O^6-BG inhibition, such as the human AGT and most bacterial and archaeal orthologs, and can be used under both native and denaturing conditions (Figure 1b, left side) [28]. The assay allows the determination of kinetics parameters for the trans-alkylation reaction [28,29]. In addition, if alkylated DNA is included in competition assays along with the fluorescent molecule, it is possible to measure the protein affinity for the natural substrate (Figure 1b, right side). The latter approach allows rapid determination of the half maximal inhibitory concentration (IC_{50}), which can be converted to the dissociation constant for DNA (K_{DNA}), giving an indirect measure of the efficiency of O^6-MG repair by OGTs [28–30].

3. DNA Alkylation Damage and OGT-Mediated DNA Repair Response in *M. tuberculosis*

M. tuberculosis is an extremely well adapted human pathogen that spends the majority of its life inside the host in a non replicative state, confined to granulomas, where it inhabits the most inhospitable cells of the body: the infected macrophages. During its life cycle, *M. tuberculosis* is continuously exposed to DNA damaging stresses that could compromise the bacillary fitness during the different phases of the infection [31]. DNA damaging agents are mainly represented by potent DNA-alkylating chemical species, which originate by the action of reactive oxygen species (ROS), reactive nitrogen species (RNS) and other antimicrobial factors generated by the host immune system [32,33]. Similar to most bacteria, *M. tuberculosis* counteracts the deleterious effect of alkylating agents by mounting multi-enzymatic DNA repair response as well as by the expression of inducible genes of Ada response [34,35].

Although adaptive response is conserved among many bacterial species, the domains of Ada (namely AdaA and AdaB), AlkA and AlkB proteins exist in different combinations in different prokaryotes. In particular, *M. tuberculosis* shows a gene fusion of *adaA* with *alkA* (*adaA-alkA*), and an independent *adaB* gene (Tuberculist code: Rv1316c) also annotated as *ogt*, that encodes for the *Mt*OGT protein.

Gene silencing experiments demonstrated that *ogt* is not essential for infectivity and survival either in vitro or in the mouse model of *M. tuberculosis* infection [33–36]. However, several studies support the importance of OGT protein in protecting the mycobacterial GC-rich DNA from the promutagenic potential of O^6-alkylguanine, occurring at different stage of the infection along *M. tuberculosis* bacilli exposure to different DNA-alkylating species. It was observed that the *ogt* gene expression undergoes a fine-tuning regulation during the infection and in response to alkylating agents [34,37,38]; moreover, the key role of *Mt*OGT protein in preserving mycobacterial genome from deleterious effects of alkylation damage is supported by trans-complementation experiment. Indeed, the heterologous expression of *Mt*OGT in the *E. coli* KT233 *ada-ogt*-defective strain suppresses its MNNG (*N*-methyl-*N'*-nitro-*N*-nitroso-guanidine) sensitivity [34] and rescues the hypermutator phenotype.

3.1. Biochemical and Structural Studies of M. tuberculosis OGT

*Mt*OGT, as its orthologs from other organisms, invariably performs the removal of alkyl adduct on modified guanines through a suicidal mechanism (Figure 1a), by performing the stoichiometric transfer of the O^6-alkyl group to the strictly conserved cysteine residue in the protein active site, which is hosted in the C-terminal (C_{ter}) domain along the -PCHR- signature protein sequence [29]. *Mt*OGT repairs methyl adducts on the O^6-position of a guanine base in double-stranded DNA (dsDNA), as well as bulky adducts on isolated bases, as demonstrated by in vitro experiment using the fluorescent derivative of O^6-BG (SNAP-Vista Green®) with a dissociation constant (K_{VG}) in the low micromolar range [29]. On the contrary, the *E. coli* Ada and OGT are inactive on O^6-BG adducts in dsDNA [39,40]. *Mt*OGT recognizes a double-stranded DNA fragment bearing an internal O^6-methylated guanine with affinity ranking in the micromolar range; moreover, it binds dsDNA in a cooperative manner with a dissociation constant value (K_D) of approximately 7.0 μM, a value comparable to that determined for hAGT by using the same method [41]. Recognition of a damaged site is the molecular determinant for the thermodynamic-driven alkyltransferase reaction with the protein that specifically recognizes damaged site upon a normal base that is transiently placed in the active site. Indeed, *Mt*OGT recognizes a modified dsDNA substrate with a 30-fold higher affinity with respect to unmodified DNA [29].

Similar to structures of other prokaryotic OGTs, as well as of hAGT in different ligand-bound states [30,42–44], *Mt*OGT folds into a roughly globular molecular architecture built up by two domains connected by a long loop and ending in a 10-residue-long tail (Figure 2). The *Mt*OGT N-terminal domain (N_{ter}) consists of an anti-parallel three-stranded β-sheet and connecting loops constrained between a mainly randomly coiled region, containing a single helical turn at its middle, on one side and a structurally conserved α-helix on the opposite one. The C_{ter} adopts the typical all-α-fold and houses the highly conserved elements that are functional to a proper reaction performing: (*i*) the helix-turn-helix (HTH) motif, which is involved in DNA binding at its minor groove and bears the arginine residue (R109) acting as a temporary steric substitute for the modified base upon its flipping-out from the regular base stacking; (*ii*) the "Asn hinge" contributing to the formation of the ligand binding pocket, that accepts the modified base; (*iii*) the catalytic cysteine (C126); (*iv*) the active-site loop that participates in the correct positioning of the alkylated base inside the ligand binding pocket; and (*v*) the structurally conserved H6 helix that, by building the ligand binding cavity on the opposite side of the Asn hinge, contributes essential residues for completing the modified-base bonding network [29,45].

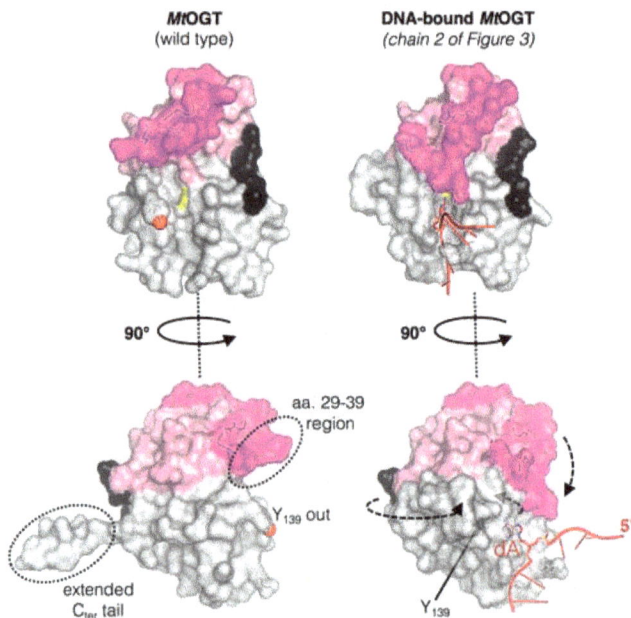

Figure 2. Structural rearrangements occurring on *M. tuberculosis* OGT (*Mt*OGT) in its ligand-free form (Protein Data Base, PDB ID: 4BHB) along DNA binding process (PDB ID: 4WX9); the regions affected by the conformational changes are indicated with dotted circle in ligand-free structure while the arrows indicate the direction of the movements of the N-terminal flap, the active site loop and the C-terminal tail of the protein as consequence of DNA binding (adapted from [45]).

3.2. The Active-Site Loop of MtOGT Is Intrinsically Flexible

The active site loop and the C$_{ter}$ tail of *Mt*OGT adopt different conformations, depending on the association of the protein with the DNA substrate (Figure 2). The C-side region of the active site loop (residues 136–141) of the ligand-free structures of *Mt*OGT and of all the mutated variants characterized until now is invariably oriented towards the bulk solvent, in a so-called "unbound/out" conformation [29,45]. This unprecedented orientation of the active site loop is reminiscent of that observed for the same region in the *Methanococcus jannaschii* OGT structure in solution [43], but strongly differs from what was observed in other OGTs for which the crystal structure has been solved [14,30,42,46–48]. On the contrary, the C-side of the active site loop of each protein chain that builds up the *Mt*OGT::dsDNA complex structure is oriented inwards the active site pocket, where it participates in fitting the ligand-binding cavity to the flipped-out base [45]. These observations raise the possibility that the active site of *Mt*OGT could exist in two alternative conformations: "active site loop out" in the ligand free state or "active site loop in" in the DNA-bound form (Figure 2). These observations highlight the high degree of structural flexibility of this region in the mycobacterial protein with respect to the equivalent region of the well-characterized human ortholog.

Site-directed mutagenesis and structural studies demonstrated that the conserved Y139 residue of the active site loop of *Mt*OGT could play a role not only in properly fixing the base inside the protein active site [29], as proposed for hAGT Y158 [25,26,40–42], but also in making *Mt*OGT able to discriminate between intact and alkylated dsDNA molecules, albeit through a molecular mechanism that is still clear. In fact, the Y139F mutated variant appears less affected in its ability to bind unmodified dsDNA, while its affinity for the modified dsDNA is clearly negatively affected. In principle, the substitution of Y139 with a phenylalanine should have little effect on catalysis, since the capability of narrowing the ligand-binding pocket by providing an aromatic environment for the alkyl adduct remains unaltered [14,48].

The high-resolution crystal structure of the recombinant *Mt*OGT R37L mutated variant, together with the biochemical analysis of highly homogeneous recombinant versions of *Mt*OGT T15S and *Mt*OGT R37L were reported [29]. *Mt*OGT R37L displayed a ten-fold reduced affinity towards methylated dsDNA, as compared with the wild-type protein, although the mutation did not affect the reaction rate. Moreover, the R37L aminoacidic substitution also affects the direct protein-DNA association, as it was observed in the electrophoretic mobility shift assay (EMSA)-based analysis using unmodified double-stranded oligonucleotides. The R37 residue is involved in the coordination of a peculiar network of bonds established between the core β-sheets and the facing random coil of the N$_{ter}$ and its substitution could affect the conformational stability of such a region. Taking into account the defective DNA repair activity of the R37L mutated protein, it was suggested that the conformational stability of peculiar regions at the N$_{ter}$ of the protein is of relevant importance for the catalysis taking place at the far C-terminal domain [29,45]. The analysis of the crystal structure of *Mt*OGT in complex with a modified dsDNA molecule, N^1-O^6-ethano-2′-deoxyxanthosine-containing dsDNA (*Mt*OGT::E1X-dsDNA, Figure 3; [45]) seems to exclude direct participation of R37 in DNA binding, because the protein residue and the sugar–phosphate backbone of the dsDNA substrate were at a distance of >16 Å. Instead, R37 could function as a hinge, limiting the conformational plasticity at the C-side of the flap of the N$_{ter}$, by participating to keep it in contact with the bulky core of the N-terminal domain, upon the protein-DNA complex formation. In principle, the absence of such an anchoring site could affect the capability of the flap to undergo discrete movements and the resulting unconstrained flexibility of the N$_{ter}$ random coil could hamper the correct assembly of *Mt*OGT cooperative clusters at the damaged DNA sites.

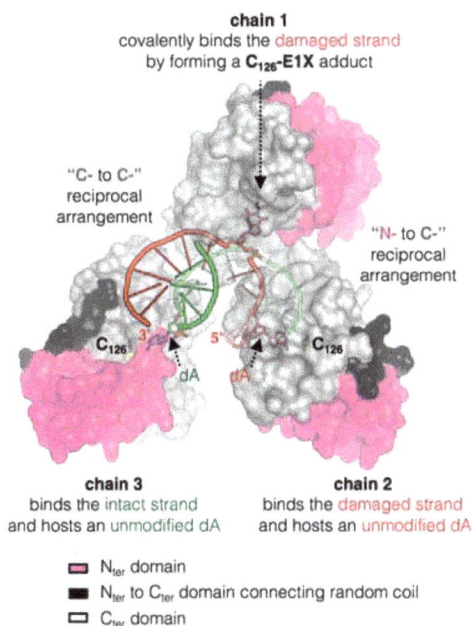

Figure 3. The peculiar cluster of *Mt*OGT. Surface representation of three *Mt*OGT units on an E1X-dsDNA which behaves as a mechanistic inhibitor of the protein (PDB ID: 4WX9); the two protein domains are depicted with different colors as indicated in the figure, the dsDNA molecule is rendered as cartoon and the extra-helical bases are rendered as sticks (adapted from [45]).

3.3. The N-Terminal Domain of MtOGT Attends DNA Cooperative Binding Process

A number of geographically distributed *M. tuberculosis* strains (like the W-Beijing strain and multi-drug resistant isolates) have been identified which carry non-synonymous SNPs in their ogt genes [49,50]. These point mutations result in aminoacid substitution at position 15 (*Mt*OGT T15S) or position 37 (*Mt*OGT R37L), mapping at the *Mt*OGT N-terminal domain. A limited number of studies put the focus on the OGTs N-terminal domain, which could play a role in the coordination of the catalytic cycle [11,46] and/or in mediating protein assembly at the site of damage upon DNA binding [11,14,51–53].

The high-resolution crystal structure of the recombinant *Mt*OGT R37L mutated variant, together with the biochemical analysis of highly homogeneous recombinant versions of *Mt*OGT T15S and *Mt*OGT R37L were reported [29]. *Mt*OGT R37L displayed a ten-fold reduced affinity towards methylated dsDNA, as compared with the wild-type protein, although the mutation did not affect the reaction rate. Moreover, the R37L substitution also affects the direct protein-DNA association, as it was observed in EMSA-based analysis using unmodified ds-oligonucleotides. The R37 residue is involved in the coordination of a peculiar network of bonds established between the core β-sheets and the facing random coil of the N-terminal domain; its substitution could affect the conformational stability of this region. Taking into account the defective DNA repair activity of the R37L mutated protein, it was suggested that the conformational stability of peculiar regions at the N-terminal domain of the protein is of relevant importance for the catalysis taking place at the far C-terminal domain [29,45]. The analysis of the crystal structure of *Mt*OGT in complex with a modified dsDNA molecule, N^1-O^6-ethano-2'-deoxyxanthosine-containing dsDNA (*Mt*OGT::dsDNA; [45]) (Figure 2) seems to exclude a direct participation of R37 in DNA binding, because the protein residue and the sugar–phosphate backbone of the dsDNA substrate were at a distance of >16 Å. Instead, R37 could function as a hinge, limiting the conformational plasticity at the C-side of the flap of the N-terminal domain, by participating to keep it in contact with the bulky core of the N-terminal domain, upon the protein-DNA complex formation. In principle, the absence of such an anchoring site could affect the capability of the flap to undergo discrete movements and the resulting unconstrained flexibility of the N-terminal domain random coil could hamper the correct assembly of *Mt*OGT cooperative clusters at the damaged DNA sites.

3.4. The Structure of MtOGT in Complex with Modified DNA Sheds Light on Cooperative Substrate Binding

The crystal structure of the *Mt*OGT::dsDNA complex revealed an unprecedented possible mode of assemblage of three adjacent protein chains onto the same damaged DNA duplex and could explain the cooperative DNA-binding mechanism of *Mt*OGT, which was demonstrated by EMSA-based analyses [29,45]. The association of *Mt*OGT with the dsDNA substrate induces the repositioning of three solvent-exposed protein regions (Figure 2): a random coiled segment (residues 29–39) of the N-terminal domain, part of the active site loop (residues 135–142) and the C-terminal tail (residues 156–165). Therefore, each protein monomer in the *Mt*OGT::dsDNA complex appears more compact than the ligand-free protein.

The architecture of *Mt*OGT as it was observed in the protein-DNA complex structure is a snapshot of a potential reaction step at which the modified base has been irreversibly bound by one protein unit, whereas two other subunits occlude available binding sites on both strands by flipping and housing in their active site the unmodified nucleobases (Figure 3). The *Mt*OGT protein cluster was more compact, if compared with the one proposed for hAGT [11,51–54], possibly due to the structural plasticity of *Mt*OGT, which could allow more crowded protein assembling onto DNA. Interestingly, the model built on further DNA-bound *Mt*OGT monomers towards the 5'-end of the modified strand, revealed that the association of a *Mt*OGT monomer with the region of the intact DNA strand facing the alkylated base hampers recruitment of additional protein subunits at the 5'-side of the damaged base. The analysis of a model built by omitting the protein chain bound to the undamaged strain revealed that the DNA binding-associated repositioning of the N_{ter} flap enables additional contacts between adjacent subunits. Finally, both short- and long-range steric hindrance phenomena could play a role in regulating *Mt*OGT-DNA association and dissociation, resulting in protein clusters that are capable of self-limiting their own size, similar to what has been experimentally determined by direct atomic force microscopy studies on human AGT (Figure 4) [55].

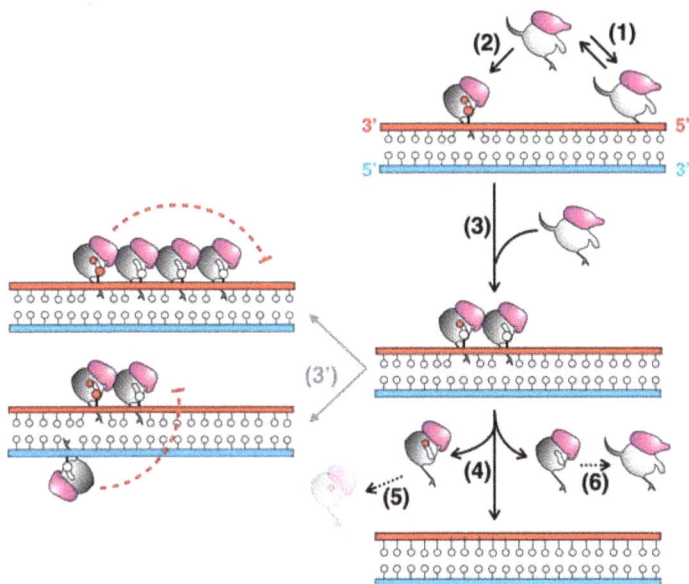

Figure 4. Proposed DNA repair mechanism involving *Mt*OGT. Sub-optimal OGT binding to undamaged dsDNA could result in protein-DNA complex dissociation (1); when a lesion is encountered, conformational modifications could take place (2); leading to the recruitment of additional OGT subunits tightly packed at a few bases apart, where they check the DNA for further modified bases (on both strands) (3); However, the formation of a continuous protein coat onto dsDNA could be hampered by long- and short-range steric hindrance phenomena (red dashed T-bars) (3′), until the protein-DNA complex disassembles upon repair (4); The final fate of alkylated MtOGT (5) and the capability of the released, un-reacted protein monomers to undergo further conformational changes (6) are still unknown (black dotted arrows) (adapted from [45]).

The cooperative assembly of protein-DNA complexes might contribute to the efficiency of lesion search and removal, due to the higher density of repair activity at the site of damage than that occurring in a non-cooperative DNA-binding mechanism [51]. Furthermore, the systematic scanning of DNA could be facilitated by the small and self-limiting *Mt*OGT protein clusters, allowing the alkylation damage search to be coupled with transcription and replication processes, as has been suggested for the human protein [55]. Moreover, an inherent capability of the protein to limit its own distribution on DNA could influence the rates of association to and dissociation from the target, and hence the kinetics of the lesion search; in fact, repositioning of a subunit placed in the middle of a single long protein cluster should probably be slower than repositioning of subunits mapping at the ends of many short clusters [51,55].

4. The *S. solfataricus* DNA Alkyl-Transferase, *Ss*OGT

Alkylating agents are widely present in the environment and are also produced by endogenous reactions, thus they pose a treat to the genome integrity of all cells. For thermophilic organisms, alkylation lesions are particularly harmful because high temperatures accelerate conversion of alkylated bases into DNA breaks, ultimately leading to DNA degradation [56]. Whereas OGTs are encoded by several thermophilic bacteria and archaea, limited information is available on these proteins [57–59]. Over the last few years, detailed biochemical, physiological and structural studies have been reported for the OGT protein from the hyperthermophilic archaeon *S. solfataricus* (called *Ss*OGT), a microorganism living in hot springs, at optimal temperature of 85 °C and pH 3.0. In order to maintain its genome stable under these highly challenging conditions, these organisms have evolved a number of very efficient repair and protection systems [28,30,60,61]. *Ss*OGT is a highly thermostable protein, which works under a variety of harsh conditions. Studies on this protein have been useful to understand its role in DNA damage response, elucidate structure-function relationships and describe conformational modifications occurring during the different steps of the reaction. The results have been useful to propose a general paradigm to correlate the structure and stability of OGTs with the active site status, which could be applied to many, if not all OGTs. Moreover, for its peculiar features, *Ss*OGT has been used to develop a useful protein-tag for protein imaging in thermophilic organisms [62,63].

4.1. Biochemical Characterization of SsOGT

Using a combination of the assays exploiting derivatives of O^6-BG described in paragraph 2, the properties of *Ss*OGT heterologously expressed in *E. coli* were carefully characterized (Table 1). The protein showed activity over a wide range of conditions: in agreement with the thermophilic nature of *S. solfataricus*, *Ss*OGT optimal catalytic activity was at 80 °C, but it was also relatively active at temperatures as low as 25 °C; in addition, the protein showed significant activity in the pH 5.0–8.0 interval, was tolerant to a number of different reaction conditions, such as ionic strength, low concentrations of detergents, organic solvents and ethylenediaminetetraacetic acid (EDTA) [28], and was strikingly resistant to proteases [62].

Table 1. Biochemical properties of *S. solfataricus* OGT (*Ss*OGT), using SNAP-Vista Green® as substrate (data are from [28,30]).

		80 °C
T_{opt}		
Activity	25 °C	25%
	37 °C	45%
	80 °C	100%
pH_{opt}		6.0
catalytic activity (M^{-1} s^{-1})	25 °C	$0.28 \pm 0.03 \times 10^4$
	70 °C	$5.33 \pm 1.49 \times 10^4$
thermal stability ($t_{\frac{1}{2}}$, min)	60 °C	257.2 ± 10.3
	70 °C	165.1 ± 16.5
	80 °C	18.7 ± 2.0
NaCl > 1.0 M		100% activity
EDTA > 10.0 mM		100% activity
Dithiotreithol (DTT)		not required
O^6-MG-DNA (K_i, μM)	close to 5′ end	4.29 ± 0.39
	central	0.83 ± 0.02
	close to 3′ end	56.6 ± 23.8

*Ss*OGT binds DNA in a cooperative manner, and efficient binding depends on the arginine 102 residue and the HTH domain [28]. Indeed, mutation of the R102 residue reduces DNA binding efficiency, whereas mutation of up to five residues in the HTH motif completely abolishes the protein ability to form stable complexes with DNA, although the mutant is normally folded and able to perform the trans-alkylation reaction if O^6-BG derivatives are used [28]. *Ss*OGT binds methylated oligonucleotides independently from the position of the lesion, but repair is position-dependent [29]:

efficient repair requires the presence of at least 2 bases from the 5′ end and 4 bases from the 3′ end of the DNA molecule. This is likely due to the asymmetric interactions formed by the two protein sides with the double helix, as confirmed by structural models [30].

4.2. SsOGT and the S. solfataricus Response to Alkylating Agents

In line with the notion that combination of high temperature with alkylating agents exacerbates DNA damage, *S. solfataricus* is highly sensitive to drugs such as MMS [28,56,64]. Treatment of *S. solfataricus* cultures with relatively low concentrations of this agent induces an apoptotic-like phenomenon characterized by degradation of a number of proteins involved in DNA damage response, DNA fragmentation and cell death [28,56,64]. At lower MMS concentrations, cell growth is restored after a transient arrest. The single *ogt* gene of this species is dispensable for growth; indeed, a knock-out strain obtained by CRISPR (Clustered Regularly Interspaced Short Palindromic Repeats)-Cas9 (CRISPR associated protein 9) technology is viable and does not show significant growth impairment [63]. The *ogt* steady-state RNA level was increased after treatment with MMS, suggesting that the protein takes part in an inducible response to DNA damage [28]. Interestingly, despite the transcriptional induction, the *Ss*OGT protein is degraded in *S. solfataricus* cells in response to treatment with MMS, and degradation is triggered by alkylation of *Ss*OGT and some pathway activated in vivo in response to alkylation damage. Since a similar phenomenon occurs in human and yeast cells [13,65], these results suggest a striking evolutionary conservation, from archaea to higher eukaryotes, of an important repair system, as shown for all DNA metabolic processes [65]. In eukaryotes, OGT degradation is triggered by ubiquitination, which targets the alkylated protein to the proteasome [13,65]. Whereas ubiquitin is not present in archaea, ubiquitin-like small proteins have been found, along with a proteasome devoted to degradation of damaged proteins. It is thus possible that some post-translational modification might target the protein to degradation pathways, either directly or after interaction with other factors [66].

4.3. Crystal Structure of Free and DNA-Bound SsOGT

The crystal structure of *Ss*OGT, solved at 1.8 Å resolution showed a very well conserved folding with respect to the other proteins of the family, with the typical two domains joined by a connecting loop [30]. As for the latter protein, the N-terminal domain was less conserved, whereas higher conservation is found in the C-terminal domain, which contains all amino acid residues and structures important for both DNA binding and repair, including the active site loop with the catalytic C119 residue, the HTH motif for DNA binding and the so called "arginine finger" R102. An interesting feature, not present in other AGTs, is the C29-C31 S-S bridge of the N-terminal domain, which is an important structural element, contributing to the impressive thermal stability of *Ss*OGT (Table 2; [30]). Although disulfide bonds are rare in intracellular mesophilic proteins, recent genomic and biochemical data show that they are present in intracellular proteins of hyperthermophilic archaea, thus suggesting a role for disulfide bonding in stabilizing at least some thermostable proteins [67,68].

The structure of *Ss*OGT bound to methylated DNA was obtained thanks to the availability of an inactive C119A mutant, which is able to form a stable complex, but not to repair the alkylated base [30]. The crystal structure of the complex showed that each *Ss*OGT molecule occupies 4 base-pairs (bp) on dsDNA substrate. The protein forms several interactions at the 3′ side of the methylated base, which are important to stabilize the complex, confirming the results obtained in binding experiments, showing that at least 2 bases downstream to the O^6-MG are needed to establish correct interactions, whereas at the 5′ end of the O^6-MG two bases are sufficient for efficient repair.

Table 2. Analysis of the thermal stability of *Ss*OGT and related mutants, by using Differential Scanning Fluorimetry. *Ss*OGT C119^m is the *Ss*OGT protein methylated by incubation in the presence of O^6-methyl-guanine; *Ss*OGT H^5 is a mutant carrying five mutations in the HTH domain (data from [30,61]).

Type	Mutation	ΔT_m (°C)
-	*Ss*OGT	a
disulphide bond	*Ss*OGT C29A	−17.1
catalytic cysteine	*Ss*OGT C119^m	−17.3
	*Ss*OGT C119L	−35.4
	*Ss*OGT C119F	−35.1
N$_{ter}$-C$_{ter}$ ion pair	*Ss*OGT D27A	−8.0
	*Ss*OGT D27K	−35.3
N$_{ter}$-loop network	*Ss*OGT E44L	+0.7
	*Ss*OGT K48A	−8.0
	*Ss*OGT K48L	−9.6
DNA binding	*Ss*OGT H^5	−5.0

a Wild type *Ss*OGT exhibited a T_m of 80.0 °C.

The crystal structure of *Ss*OGT::DNA complex showed that substrate binding does not affect significantly the protein overall structure [30]; however, structural changes in specific protein regions occur, in order to accommodate the double helix as well as the methylated base. Indeed, in silico mapping, performed by using difference distance matrix plots (DDMPs), of all interactions occurring in the molecule showed changes in the distance of a number of residues when the protein contacts DNA and the O^6-MG is extruded into its active site [61]. These changes suggest that the molecule undergoes subtle movements, which might result in lost and gained interactions among residues forming intra-domain interactions. Moreover, important changes at the interface of the two domains are also observed upon DNA binding, involving a complex network of interacting aminoacid residues (K48, N59, R61 and E62 residues, called the K48-network). In the free *Ss*OGT, these residues are at a distance compatible with formation of an ionic/hydrogen bond network. Upon DNA binding, the E44 and K48 residues flip out, moving away from the protein core, leading to strong perturbation of the interaction network (Figure 5; [61]). As discussed in the next paragraph, these conformational changes are maintained in the post reaction form of the protein, when the active site is alkylated and DNA is released [61].

4.4. Alkylation of the Active Site Induces Dramatic Conformational Changes and Destabilization of SsOGT

Once the repair reaction has been completed, trans-alkylated OGTs undergo inactivation and dramatic destabilization. Thus, the efficiency of repair of alkylation lesions in vivo greatly depends on the OGT neosynthesis; this phenomenon has important consequences for the cell response to DNA damage as well as cancer treatments [3,20]. The structural mechanism of alkylation-induced destabilization is difficult to investigate exactly because once alkylated, OGTs unfold, aggregate and/or undergo degradation. Indeed, alkylated forms of the human protein could not be subjected to extensive biochemical analysis and were not stable enough to be crystallized. Structural information on alkylated hAGT could be obtained by incubating hAGT crystals in solutions containing alkylating agents; the presence of a methyl group could be accommodated in crystals, allowing resolution of the structure, which showed limited conformational rearrangements [14,46]. Bulkier alkyl adducts (such as a benzyl group) determined rapid crystal dissolution [46].

In contrast, although alkylation destabilized *Ss*OGT at its physiological temperature (>70 °C), at 25 °C the protein was enough stable to allow structural and biochemical analyses. In particular, resolution of the three-dimensional structure was obtained for the methylated form of *Ss*OGT, (*Ss*OGT^m) obtained after incubation in solutions containing O^6-MG, as well as of mutant proteins carrying substitution of the catalytic cysteine with either a leucine or a phenylalanine (C119L and

C119F), which mimicked the presence of larger adducts in the active site (an isopropyl and a benzyl group, respectively). The structures of these three proteins showed extensive remodelling of interactions between aminoacid residues upon alkylation and large movements in the backbone structure [30,61]. These movements are determined by an active site expansion, as a consequence of binding of the alkyl group to the active site pocket, resulting in the increase of the distance between the active site loop and the recognition helix H4 of the HTH motif, which in turn affects the position of the adjacent structural elements (Figure 5).

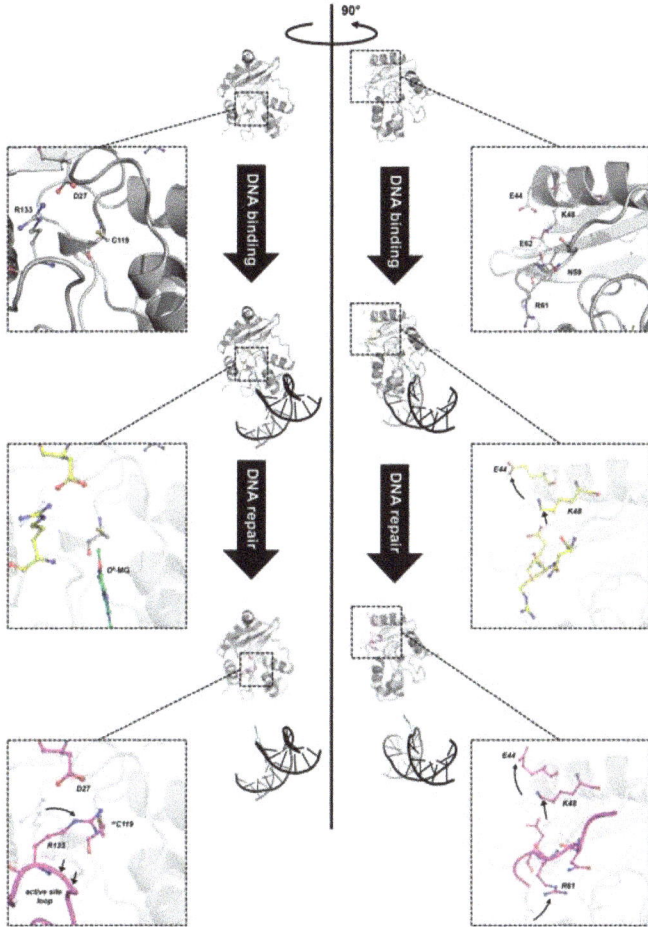

Figure 5. Different stages of DNA repair reaction of *Ss*OGT. The free form of *Ss*OGT (PDB ID: 4ZYE, in grey) recognizes and binds the alkylated DNA, leading to reversible conformational changes in its own structure (PDB ID: 4ZYD, in yellow), mainly involving the K48-network. Upon the alkyl transfer, the protein undergoes dramatic changes (PDB ID: 4ZYG, in magenta), which cause its inactivation and degradation. Insets: zoom-in of regions involved in the main conformational changes (adapted from [30,61]). Green, methylated guanine. Amino acid atoms are coloured according the CPK convention (carbon, in the corresponding colour of each 3D structure; oxygen, in red; nitrogen, in blue; sulphur, in yellow).

In addition, structural and in silico analyses showed that significant changes in specific interdomain interactions occur at the interface between the N_{ter} and C_{ter} domains upon alkylation. One main change is observed at the level of D27 residue of the N_{ter} domain and R133 residue of the C_{ter} domain. Before reaction, these residues are at a distance compatible with ionic interaction; after alkylation, rotation of R133 could weaken or impair the interaction (Figure 5, left panel). Similarly, the H2 helix moves away from the protein core. Both modifications occur only upon alkylation, but not in the DNA-bound protein, thus suggesting that they are a direct consequence of the active site alkylation and not just substrate binding. In addition, the conformational changes observed in the protein-DNA complex at the interface between the N_{ter} domain and the connecting loop at the level of the K48-network are maintained after alkylation [61]. Thus, lesion recognition and alkylation trigger distinct modifications of intramolecular interactions in *Ss*OGT: whereas the changes at the level of the D27-R133 ion pair occur only upon irreversible trans-alkylation of the catalytic cysteine, those observed in the K48-network might be a consequence of lesion recognition/steric hindrance of the active site, since they are already found in protein in complex with the methylated DNA and are retained in the post-repair protein structure devoid of DNA (Figure 5, right panel) [61].

Biochemical and mutational analyses showed that alkylation of the active site leads to dramatic loss of stability of *Ss*OGT, (Table 2; [30,61]); whereas the Tm of wild type protein is 80 °C, a reduction of 17–35 °C is observed in variants were an adduct is present in the active site, and the extent of destabilization is linear with the hindrance of the adduct. Interestingly, a remarkable extent of destabilization is observed also in mutants where inter-domain interactions are perturbed, (Table 2) thus suggesting that the latter play important role in protein stability. On the basis of structure and biochemical data, it was suggested that the K48-network and D27-R133 ion pair are "locks", which contribute to the correct folding of the protein in its free state. Active site/recognition helix conformational changes, which occur after binding to DNA and lesion recognition, determine opening of one "lock" (the K48-network). However, this modification is reversible until the C119 trans-alkylation occurs; once this reaction is completed, a second set of conformational changes determines the opening of the second "lock", thus resulting in irreversible protein destabilization (Figure 5; [61]).

The specific inter-domain interactions which affect *Ss*OGT stability are not conserved in the structures of other OGTs. However, it is likely that the connection between the two domains play an important role in all OGTs. In the case of the protein from *Thermococcus kodakaraensis*, an inter domain ion pair network is important for the protein stability [42,69]. In hAGT, perturbation of interactions at the interface between the two protein domains is observed after alkylation, which resulted in dramatic hAGT destabilization [70]. These observations suggested a possible general model for OGTs, in which alkylation of the active site determines conformational changes at the level of the active site as well as of the interactions between the two domains, which trigger protein destabilization [30,61].

5. Development of OGT-Based Novel *Protein-Tags*

In recent years several *protein-tags*, which can be fused to proteins of interest to allow their detection and analysis, have been used in a number of organisms, enabling a wide variety of biological studies. These include the popular Green Fluorescent Protein (GFP) and its derivatives, which are intrinsically fluorescent [71,72], as well as proteins that can be labelled by an external substrate. In this context, a modified version of the hAGT, called SNAP-tag®, which is impaired in DNA binding and of higher operational stability, has been developed [21–23]. The advantage of the system consists in the possibility of labelling the chimeric protein with virtually any chemical group (such as a fluorophore, biotin, and so on), provided the latter is linked to the O^6-BG molecule. This approach proved to be a highly specific and versatile tool for in vivo and in vitro specific labelling of proteins [21–23].

Thanks to its thermostable nature, *Ss*OGT was used to develop a version of the SNAP-tag® protein suitable for thermophilic microorganisms. To this aim, a modified version of *Ss*OGT was used, in which DNA binding was abolished by the introduction of five mutations in the HTH domain [28]. This protein, called H^5, was successfully used as a protein-tag in both mesophilic (*E. coli*) and thermophilic microorganisms (the bacterium *Thermus thermophilus* and the archaeon *Sulfolobus islandicus*). Plasmids expressing fusions of the H^5-tag with two thermostable proteins were constructed, namely the *S. solfataricus* β-glycosidase and the thermophile-specific DNA topoisomerase reverse gyrase [62,73–76]. Both proteins were correctly expressed, folded, functional and stable, when the expression plasmids were introduced by transformation in *T. thermophilus* and *S. islandicus*, respectively. The presence and the activity of the H^5-tag could be imaged in living cells as well as in cell-free protein extracts [62,63]. In addition, the H^5-tag did not interfere with the enzyme activity of target proteins and could be fused to thermostable proteins overexpressed in *E. coli*, allowing to perform purification protocols including thermal treatment to precipitate aggregated host proteins [62].

The H^5-tag might be used as a thermostable version of the SNAP-tag® protein in other (hyper)thermophilic archaea and bacteria allowing detection and sub-cellular localization of proteins and protein interactions. In addition, the use of different fluorescent ligands could be used to label different proteins, including in pulse-chase analysis, and follow their movements and fate in the cell in real time. Both experiments were performed in the absence of endogenous OGT activity; indeed, *T. thermophilus* is a natural *ogt*$^{(-)}$ species [62,77], whereas a *S. islandicus* mutant strain deleted for the *ogt* gene obtained by a CRISPR-based genome-editing method was used as a host [63]. Thus, it remains to be determined whether possible alkyl-transferase background activity might interfere with the efficacy of the system.

6. Conclusions and Perspective

Despite their completely different lifestyle, both *M. tuberculosis* and *S. solfataricus* face challenging external conditions, where alkylation damage is a serious treat for their genome integrity. Thus, studies on *Mt*OGT and *Ss*OGT help in understanding the function of these proteins in the biology of both microorganisms and their protection mechanisms. In addition, these proteins are useful models to elucidate the role of each structural element in the different OGTs activities, the details of DNA binding and lesion recognition, as well as the conformational changes associated with each step of the reaction. These results could have a wide impact, providing a general picture of how OGTs work, which is the foundation for structure-based design of novel OGTs inhibitors to be used in cancer and possibly other pathological conditions.

Acknowledgments: The authors would like to thank Giovanni del Monaco for technical assistance. This work was supported by the European Community (project SysteMTb HEALTH-F4-2010-241587), Fondazione Cariplo (Ricerca biomedica condotta da giovani ricercatori, project 2016-0604) and FIRB-Futuro in Ricerca Nematic Project No. BFR12OO1G_002.

Conflicts of Interest: The authors declare no conflict of interest.

References

1. Pegg, A.E. Repair of O^6-alkylguanine by alkyltransferases. *Mutat. Res.* **2000**, *462*, 83–100. [CrossRef]
2. Drabløs, F.; Feyzi, E.; Aas, P.A.; Vaagbø, C.B.; Kavli, B.; Bratlie, M.S.; Peña-Diaz, J.; Otterlei, M.; Slupphaug, G.; Krokan, H.E. Alkylation damage in DNA and RNA-repair mechanisms and medical significance. *DNA Repair* **2004**, *11*, 1389–1407. [CrossRef] [PubMed]
3. Pegg, A.E. Multifaceted roles of alkyltransferase and related proteins in DNA repair, DNA damage, resistance to chemotherapy, and research tools. *Chem. Res. Toxicol.* **2011**, *24*, 618–639. [CrossRef] [PubMed]
4. Tubbs, J.L.; Pegg, A.E.; Tainer, J.A. DNA binding, nucleotide flipping, and the helix-turn-helix motif in base repair by O^6-alkylguanine-DNA-alkyltransferase and its implications for cancer chemotherapy. *DNA Repair* **2007**, *6*, 1100–1115. [CrossRef] [PubMed]
5. Hoeijmakers, J.H. Genome maintenance mechanisms for preventing cancer. *Nature* **2001**, *411*, 366–374. [CrossRef] [PubMed]
6. Mielecki, D.; Wrzesiński, M.; Grzesiuk, E. Inducible repair of alkylated DNA in microorganisms. *Mutat. Res. Rev. Mutat. Res.* **2015**, *763*, 294–305. [CrossRef] [PubMed]

7. Soll, J.M.; Sobol, R.W.; Mosammaparast, N. Regulation of DNA Alkylation Damage Repair: Lessons and Therapeutic Opportunities. *Trends Biochem. Sci.* **2017**, *3*, 206–218. [CrossRef] [PubMed]

8. Mishina, Y.; Duguid, E.M.; Chuan, H. Direct reversal of DNA alkylation damage. *Chem. Rev.* **2006**, *106*, 215–232. [CrossRef] [PubMed]

9. Debiak, M.; Nikolova, T.; Kaina, B. Loss of ATM sensitizes against O^6-methylguanine triggered apoptosis, SCEs and chromosomal aberrations. *DNA Repair* **2004**, *3*, 359–368. [CrossRef] [PubMed]

10. Gillingham, D.; Sauter, B. Genomic Studies Reveal New Aspects of the Biology of DNA Damaging Agents. *Chembiochem* **2017**. [CrossRef] [PubMed]

11. Fang, Q.; Kanugula, S.; Pegg, A.E. Function of domains of human O^6-alkyl-guanine-DNA alkyltransferase. *Biochemistry* **2005**, *44*, 15396–15405. [CrossRef] [PubMed]

12. Jena, N.R.; Shukla, P.K.; Jena, H.S.; Mishra, P.C.; Suhai, S. O^6-methylguanine repair by O^6-alkylguanine-DNA alkyltransferase. *J. Phys. Chem. B* **2009**, *113*, 16285–16290. [CrossRef] [PubMed]

13. Xu-Welliver, M.; Pegg, A.E. Degradation of the alkylated form of the DNA repair protein, O^6- alkylguanine-DNA alkyltransferase. *Carcinogenesis* **2002**, *23*, 823–830. [CrossRef] [PubMed]

14. Daniels, D.S.; Woo, T.T.; Luu, K.X.; Noll, D.M.; Clarke, N.D.; Pegg, A.E.; Tainer, J.A. DNA binding and nucleotide flipping by the human DNA repair protein AGT. *Nat. Struct. Mol. Biol.* **2004**, *11*, 714–720. [CrossRef] [PubMed]

15. Mielecki, D.; Grzesiuk, E. Ada response—A strategy for repair of alkylated DNA in bacteria. *FEMS Microbiol. Lett.* **2014**, *355*, 1–11. [CrossRef] [PubMed]

16. Sedgwick, B.; Lindahl, T. Recent progress on the Ada response for inducible repair of DNA alkylation damage. *Oncogene* **2002**, *58*, 8886–8894. [CrossRef] [PubMed]

17. Takano, K.; Nakamura, T.; Sekiguchi, M. Roles of two types of O^6-methylguanine-DNA methyltransferases in DNA repair. *Mutat. Res.* **1991**, *254*, 37–44. [CrossRef]

18. Khan, O.; Middleton, M.R. The therapeutic potential of O^6-alkylguanine DNA alkyltransferase inhibitors. *Expert Opin. Investig. Drugs* **2007**, *10*, 1573–1584. [CrossRef] [PubMed]

19. Ishiguro, K.; Zhu, Y.L.; Shyam, K.; Penketh, P.G.; Baumann, R.P.; Sartorelli, A.C. Quantitative relationship between guanine O^6-alkyl lesions produced by Onrigin™ and tumor resistance by O^6-alkyl-guanine-DNA alkyltransferase. *Biochem. Pharmacol.* **2010**, *80*, 1317–1325. [CrossRef] [PubMed]

20. Philip, S.; Swaminathan, S.G.; Kuznetsov, S.; Kanugula, K.; Biswas, S.; Chang, N.A.; Loktionova, D.C.; Haines, P.; Kaldis, A.E.; Pegg, S.K.; et al. Degradation of BRCA2 in alkyltransferase-mediated DNA repair and its clinical implications. *Cancer Res.* **2008**, *68*, 9973–9981. [CrossRef] [PubMed]

21. Keppler, A.; Pick, H.; Arrivoli, C.; Vogel, H.; Johnsson, K. Labeling of fusion proteins with synthetic fluorophores in live cells. *Proc. Natl. Acad. Sci. USA* **2004**, *10*, 9955–9959. [CrossRef] [PubMed]

22. Keppler, A.; Gendreizig, S.; Gronemeyer, T.; Pick, H.; Vogel, H.; Johnsson, K. A general method for the covalent labeling of fusion proteins with small molecules in vivo. *Nat. Biotechnol.* **2003**, *21*, 86–89. [CrossRef] [PubMed]

23. Gautier, A.; Juillerat, A.; Heinis, C.; Corrêa, I.R., Jr.; Kindermann, M.; Beaufils, F.; Johnsson, K. An engineered protein-tag for multi-protein labeling in living cells. *Chem. Biol.* **2008**, *15*, 128–136. [CrossRef] [PubMed]

24. Ishiguro, K.; Shyam, K.; Penketh, P.G.; Sartorelli, A.C. Development of an O^6-alkyl-guanine-DNA alkyltransferase assay based on covalent transfer of the benzyl moiety from [benzene-^{3}H]-O^6-benzylguanine to the protein. *Anal. Biochem.* **2008**, *83*, 44–51. [CrossRef] [PubMed]

25. Tintoré, M.; Aviñó, A.; Ruiz, F.M.; Eritja, R.; Fábrega, C. Development of a novel fluorescence assay based on the use of the thrombin binding aptamer for the detection of O^6-alkyl-guanine-DNA alkyltransferase activity. *J. Nucleic Acids* **2010**, *2010*, 632041. [CrossRef] [PubMed]

26. Gronemeyer, T.; Chidley, C.; Juillerat, A.; Heinis, C.; Johnsson, K. Directed evolution of O^6-alkylguanine-DNA alkyltransferase for applications in protein labeling. *Protein Eng. Des. Sel.* **2006**, *19*, 309–316. [CrossRef] [PubMed]

27. Hinner, M.J.; Johnsson, K. How to obtain labeled proteins and what to do with them. *Curr. Opin. Biotechnol.* **2010**, *21*, 766–776. [CrossRef] [PubMed]

28. Perugino, G.; Vettone, A.; Illiano, G.; Valenti, A.; Ferrara, M.C.; Rossi, M.; Ciaramella, M. Activity and regulation of archaeal DNA alkyltransferase: Conserved protein involved in repair of DNA alkylation damage. *J. Biol. Chem.* **2012**, *287*, 4222–4231. [CrossRef] [PubMed]

29. Miggiano, R.; Casazza, V.; Garavaglia, S.; Ciaramella, M.; Perugino, G.; Rizzi, M.; Rossi, F. Biochemical and structural studies of the Mycobacterium tuberculosis O^6-methylguanine methyltransferase and mutated variants. *J. Bacteriol.* **2013**, *195*, 2728–2736. [CrossRef] [PubMed]

30. Perugino, G.; Miggiano, R.; Serpe, M.; Vettone, A.; Valenti, A.; Lahiri, S.; Rossi, F.; Rossi, M.; Rizzi, M.; Ciaramella, M. Structure-function relationships governing activity and stability of a DNA alkylation damage repair thermostable protein. *Nucleic Acids Res.* **2015**, *43*, 8801–8816. [CrossRef] [PubMed]

31. Gorna, A.E.; Bowater, R.P.; Dziadek, J. DNA repair systems and the pathogenesis of *Mycobacterium tuberculosis*: Varying activities at different stages of infection. *Clin. Sci.* **2010**, *119*, 187–202. [CrossRef] [PubMed]

32. Friedberg, E.C.; Walker, G.C.; Siede, W. *DNA Repair and Mutagenesis*; ASM Press: Washington, DC, USA, 1995.

33. Durbach, S.I.; Springer, B.; Machowski, E.E.; North, R.J.; Papavinasasundaram, K.G.; Colston, M.J.; Bottger, E.C.; Mizrahi, V. DNA alkylation damage as a sensor of nitrosative stress in *Mycobacterium tuberculosis*. *Infect. Immun.* **2003**, *71*, 997–1000. [CrossRef] [PubMed]

34. Yang, M.; Aamodt, R.M.; Dalhus, B.; Balasingham, S.; Helle, I.; Andersen, P.; Tonjum, T.; Alseth, I.; Rognes, T.; Bjoras, M. The ada operon of *Mycobacterium tuberculosis* encodes two DNA-methyltransferases for inducible repair of DNA alkylation damage. *DNA Repair* **2011**, *10*, 595–602. [CrossRef] [PubMed]

35. Shrivastav, N.; Li, D.; Essigmann, J.M. Chemical biology of mutagenesis and DNA repair: Cellular responses to DNA alkylation. *Carcinogenesis* **2010**, *3*, 59–70. [CrossRef] [PubMed]

36. Sassetti, C.M.; Boyd, D.H.; Rubin, E.J. Genes required for mycobacterial growth defined by high density mutagenesis. *Mol. Microbiol.* **2003**, *48*, 77–84. [CrossRef] [PubMed]

37. Boshoff, H.I.; Myers, T.G.; Copp, B.R.; McNeil, M.R.; Wilson, M.A.; Barry, C.E., III. The transcriptional responses of *Mycobacterium tuberculosis* to inhibitors of metabolism: Novel insights into drug mechanisms of action. *J. Biol. Chem.* **2004**, *279*, 40174–40184. [CrossRef] [PubMed]

38. Schnappinger, D.; Ehrt, S.; Voskuil, M.I.; Liu, Y.; Mangan, J.A.; Monahan, I.M.; Dolganov, G.; Efron, B.; Butcher, P.D.; Nathan, C.; et al. Transcriptional adaptation of *Mycobacterium tuberculosis* within macrophages: Insights into the phagosomal environment. *J. Exp. Med.* **2003**, *198*, 693–704. [CrossRef] [PubMed]

39. Wilkinson, M.C.; Potter, P.M.; Cawkwell, L.; Georgiadis, P.; Patel, D.; Swann, P.F.; Margison, G.P. Purification of the *E. coli ogt* gene product to homogeneity and its rate of action on O^6-methylguanine, O^6-ethylguanine and O^4-methylthymine in dodecadeoxyribonucleotides. *Nucleic Acids Res.* **1989**, *17*, 8475–8484. [CrossRef] [PubMed]

40. Goodtzova, K.; Kanugula, S.; Edara, S.; Pauly, G.T.; Moschel, R.C.; Pegg, A.E. Repair of O^6-benzylguanine by the *Escherichia coli* Ada and Ogt and the human O^6-alkylguanine-DNA alkyltransferase. *J. Biol. Chem.* **1997**, *272*, 8332–8339. [CrossRef] [PubMed]

41. Spratt, T.E.; Wu, J.D.; Levy, D.E.; Kanugula, S.; Pegg, A.E. Reaction and binding of oligodeoxynucleotides containing analogues of O^6-methylguanine with wild-type and mutant human O^6-alkylguanine-DNA alkyltransferase. *Biochemistry* **1999**, *38*, 6801–6806. [CrossRef] [PubMed]

42. Hashimoto, H.; Inoue, T.; Nishioka, M.; Fujiwara, S.; Takagi, M.; Imanaka, T.; Kai, Y. Hyperthermostable protein structure maintained by intra and inter-helix ion-pairs in archaeal O^6-methylguanine-DNA methyltransferase. *J. Mol. Biol.* **1999**, *292*, 707–716. [CrossRef] [PubMed]

43. Roberts, A.; Pelton, J.G.; Wemmer, D.E. Structural studies of MJ1529, an O^6-methylguanine-DNA methyltransferase. *Magn. Reson. Chem.* **2006**, *44*, S71–S82. [CrossRef] [PubMed]

44. Moore, M.H.; Gulbis, J.M.; Dodson, E.J.; Demple, B.; Moody, P.C. Crystal structure of a suicidal DNA repair protein: The Ada O^6-methylguanineDNA methyltransferase from *E. coli*. *EMBO J.* **1994**, *13*, 1495–1501. [PubMed]

45. Miggiano, R.; Perugino, G.; Ciaramella, M.; Serpe, M.; Rejman, D.; Páv, O.; Pohl, R.; Garavaglia, S.; Lahiri, S.; Rizzi, M.; et al. Crystal structure of *Mycobacterium tuberculosis* O^6-methylguanine-DNA methyltransferase protein clusters assembled on to damaged DNA. *Biochem. J.* **2016**, *473*, 123–133. [CrossRef] [PubMed]

46. Daniels, D.S.; Mol, C.D.; Arvai, A.S.; Kanugula, S.; Pegg, A.E.; Tainer, J.A. Active and alkylated human AGT structures: A novel zinc site, inhibitor and extrahelical base binding. *EMBO J.* **2000**, *19*, 1719–1730. [CrossRef] [PubMed]

47. Wibley, J.E.; Pegg, A.E.; Moody, PC. Crystal structure of the human O^6-alkylguanine-DNA alkyltransferase. *Nucleic Acids Res.* **2000**, *28*, 393–401. [CrossRef] [PubMed]

48. Duguid, E.M.; Rice, P.A.; He, C. The structure of the human AGT protein bound to DNA and its implications for damage detection. *J. Mol. Biol.* **2005**, *350*, 657–666. [CrossRef] [PubMed]

49. Olano, J.; Lopez, B.; Reyes, A.; Lemos, M.P.; Correa, N.; Del Portillo, P.; Barrera, L.; Robledo, J.; Ritacco, V.; Zambrano, M.M. Mutations in DNA repair genes are associated with the Haarlem lineage of *Mycobacterium tuberculosis* independently of their antibiotic resistance. *Tuberculosis* **2007**, *87*, 502–508. [CrossRef] [PubMed]

50. Rad, M.E.; Bifani, P.; Martin, C.; Kremer, K.; Samper, S.; Rauzier, J.; Kreiswirth, B.; Blazquez, J.; Jouan, M.; van Soolingen, D.; et al. Mutations in putative mutator genes of *Mycobacterium tuberculosis* strains of the W-Beijing family. *Emerg. Infect. Dis.* **2003**, *9*, 838.

51. Adams, C.A.; Melikishvili, M.; Rodgers, D.W.; Rasimas, J.J.; Pegg, A.E.; Fried, M.G. Topologies of complexes containing O^6-alkylguanine-DNA alkyltransferase and DNA. *J. Mol. Biol.* **2009**, *389*, 248–263. [CrossRef] [PubMed]

52. Adams, C.A.; Fried, M.G. Mutations that probe the cooperative assembly of O^6-alkylguanine-DNA alkyltransferase complexes. *Biochemistry* **2011**, *50*, 1590–1598. [CrossRef] [PubMed]

53. Melikishvili, M.; Fried, M.G. Quaternary interactions and supercoiling modulate the cooperative DNA binding of AGT. *Nucleic Acids Res.* **2017**, *45*, 7226–7236. [CrossRef] [PubMed]

54. Tessmer, I.; Fried, M.G. Insight into the cooperative DNA binding of the O^6-alkylguanine DNA alkyltransferase. *DNA Repair* **2014**, *20*, 14–22. [CrossRef] [PubMed]

55. Begley, T.J.; Samson, L.D. Reversing DNA damage with a directional bias. *Nat. Struct. Mol. Biol.* **2004**, *11*, 688–690. [CrossRef] [PubMed]

56. Valenti, A.; Napoli, A.; Ferrara, M.C.; Nadal, M.; Rossi, M.; Ciaramella, M. Selective degradation of reverse gyrase and DNA fragmentation induced by alkylating agent in the archaeon *Sulfolobus solfataricus*. *Nucleic Acids Res.* **2006**, *34*, 2098–2108. [CrossRef] [PubMed]

57. Leclere, M.M.; Nishioka, M.; Yuasa, T.; Fujiwara, S.; Takagi, M.; Imanaka, T. The O^6-methylguanine-DNA methyltransferase from the hyperthermophilic archaeon *Pyrococcus* sp. KOD1: A thermostable repair enzyme. *Mol. Gen. Genet.* **1998**, *258*, 69–77. [CrossRef] [PubMed]

58. Skorvaga, M.; Raven, N.D.; Margison, G.P. Thermostable archaeal O^6-alkylguanine-DNA alkyltransferases. *Proc. Natl. Acad. Sci. USA* **1998**, *95*, 6711–6715. [CrossRef] [PubMed]

59. Kanugula, S.; Pegg, A.E. Alkylation damage repair protein O^6-alkyl-guanine-DNA alkyltransferase from the hyperthermophiles *Aquifex aeolicus* and *Archaeoglobus fulgidus*. *Biochem. J.* **2003**, *375*, 449–455. [CrossRef] [PubMed]

60. Vettone, A.; Perugino, G.; Rossi, M.; Valenti, A.; Ciaramella, M. Genome stability: Recent insights in the topoisomerase reverse gyrase and thermophilic DNA alkyltransferase. *Extremophiles* **2014**, *18*, 895–904. [CrossRef] [PubMed]

61. Morrone, C.; Miggiano, R.; Serpe, M.; Massarotti, A.; Valenti, A.; Del Monaco, G.; Rossi, M.; Rossi, F.; Rizzi, M.; Perugino, G.; et al. Interdomain interactions rearrangements control the reaction steps of a thermostable DNA alkyltransferase. *Biochim. Biophys. Acta* **2017**, *1861*, 86–96. [CrossRef] [PubMed]

62. Vettone, A.; Serpe, M.; Hidalgo, A.; Berenguer, J.; del Monaco, G.; Valenti, A.; Ciaramella, M.; Perugino, G. A novel thermostable protein-tag: Optimization of the Sulfolobus solfataricus DNA-alkyl-transferase by protein engineering. *Extremophiles* **2016**, *20*, 1–13. [CrossRef] [PubMed]

63. Visone, V.; Han, W.; Perugino, G.; Del Monaco, G.; She, Q.; Rossi, M.; Valenti, A.; Ciaramella, M. In vivo and in vitro protein imaging in thermophilic archaea by exploiting a novel protein tag. *PLoS ONE* **2017**, *10*, e0185791. [CrossRef] [PubMed]

64. Valenti, A.; Perugino, G.; Nohmi, T.; Rossi, M.; Ciaramella, M. Inhibition of translesion DNA polymerase by archaeal reverse gyrase. *Nucleic Acids Res.* **2009**, *37*, 4287–4295. [CrossRef] [PubMed]

65. Hwang, C.S.; Shemorry, A.; Varshavsky, A. Two proteolytic pathways regulate DNA repair by cotargeting the Mgt1 alkyl-guanine transferase. *Proc. Natl. Acad. Sci. USA* **2009**, *106*, 2142–2147. [CrossRef] [PubMed]

66. Maupin-Furlow, J.A.; Humbard, M.A.; Kirkland, P.A.; Li, W.; Reuter, C.J.; Wright, A.J.; Zhou, G. Proteasomes from structure to function: Perspectives from archaea. *Curr. Top. Dev. Biol. Rev.* **2006**, *75*, 125–169.

67. Mallick, P.; Boutz, D.R.; Eisenberg, D.; Yeates, T.O. Genomic evidence that the intracellular proteins of archaeal microbes contain disulfide bonds. *Proc. Natl. Acad. Sci. USA* **2002**, *99*, 9679–9684. [CrossRef] [PubMed]

68. Porcelli, M.; De Leo, E.; Del Vecchio, P.; Fuccio, F.; Cacciapuoti, G. Thermal unfolding of nucleoside hydrolases from the hyperthermophilic archaeon *Sulfolobus solfataricus*: Role of disulfide bonds. *Protein Pept. Lett.* **2012**, *19*, 369–374. [CrossRef] [PubMed]

69. Nishikori, S.; Shiraki, K.; Yokota, K.; Izumikawa, N.; Fujiwara, S.; Hashimoto, H.; Imanaka, T.; Takagi, M. Mutational effects on O^6-methylguanine-DNA methyltransferase from hyperthermophile: Contribution of ion-pair network to protein thermostability. *J. Biochem.* **2004**, *135*, 525–532. [CrossRef] [PubMed]

70. Crone, T.M.; Goodtzova, K.; Pegg, A.E. Amino acid residues affecting the activity and stability of human O^6-alkylguanine-DNA alkyltransferase. *Mutat. Res.* **1996**, *63*, 15–25. [CrossRef]

71. Chalfie, M.; Tu, Y.; Euskirchen, G.; Ward, W.W.; Prasher, D.C. Green fluorescent protein as a marker for gene expression. *Science* **1994**, *263*, 802–805. [CrossRef] [PubMed]

72. Tsien, R.Y. The green fluorescent protein. *Annu. Rev. Biochem.* **1998**, *67*, 509–554. [CrossRef] [PubMed]

73. Perugino, G.; Valenti, A.; D'Amaro, A.; Rossi, M.; Ciaramella, M. Reverse gyrase and genome stability in hyperthermophilic organisms. *Biochem. Soc. Trans.* **2009**, *37*, 69–73. [CrossRef] [PubMed]

74. Jamroze, A.; Perugino, G.; Valenti, A.; Rashid, N.; Rossi, M.; Akhtar, M.; Ciaramella, M. The reverse gyrase from *Pyrobaculum calidifontis*, a novel extremely thermophilic DNA topoisomerase endowed with DNA unwinding and annealing activities. *J. Biol. Chem.* **2014**, *289*, 3231–3243. [CrossRef] [PubMed]

75. Valenti, A.; Perugino, G.; Rossi, M.; Ciaramella, M. Positive supercoiling in thermophiles and mesophiles: Of the good and evil. *Biochem. Soc. Trans.* **2011**, *39*, 58–63. [CrossRef] [PubMed]

76. Valenti, A.; Perugino, G.; D'Amaro, A.; Cacace, A.; Napoli, A.; Rossi, M.; Ciaramella, M. Dissection of reverse gyrase activities: Insight into the evolution of a thermostable molecular machine. *Nucleic Acids Res.* **2008**, *36*, 4587–4597. [CrossRef] [PubMed]

77. Morita, R.; Nakagawa, N.; Kuramitsu, S.; Masui, R. An O^6-methylguanine-DNA methyltransferase-like protein from *Thermus thermophilus* interacts with a nucleotide excision repair protein. *J. Biochem.* **2008**, *144*, 267–277. [CrossRef] [PubMed]

International Journal of
Molecular Sciences

MDPI

Article

Enhanced Susceptibility of Ogg1 Mutant Mice to Multiorgan Carcinogenesis

Anna Kakehashi * 🄳, Naomi Ishii, Takahiro Okuno, Masaki Fujioka, Min Gi and Hideki Wanibuchi

Department of Molecular Pathology, Osaka City University Graduate School of Medicine, Asahi-machi 1-4-3, Abeno-ku, Osaka 545-8585, Japan; m1159070@med.osaka-cu.ac.jp (N.I.); m2026860@med.osaka-cu.ac.jp (T.O.); m2066048@med.osaka-cu.ac.jp (M.F.); mwei@med.osaka-cu.ac.jp (M.G.); wani@med.osaka-cu.ac.jp (H.W.)
* Correspondence: anna@med.osaka-cu.ac.jp; Tel.: +81-6-6645-3737; Fax: +81-6-6646-3093

Received: 18 July 2017; Accepted: 15 August 2017; Published: 18 August 2017

Abstract: The role of deficiency of oxoguanine glycosylase 1 (*Ogg1*) *Mmh* homolog, a repair enzyme of the 8-hydroxy-2′-deoxyguanosine (8-OHdG) residue in DNA, was investigated using the multiorgan carcinogenesis bioassay in mice. A total of 80 male and female six-week-old mice of C57BL/6J background carrying a mutant *Mmh* allele of the *Mmh/Ogg1* gene ($Ogg1^{-/-}$) and wild type ($Ogg1^{+/+}$) mice were administered *N*-diethylnitrosamine (DEN), *N*-methyl-*N*-nitrosourea (MNU), *N*-butyl-*N*-(4-hydroxybutyl) nitrosamine (BBN), *N*-bis (2-hydroxypropyl) nitrosamine (DHPN) and 1,2-dimethylhydrazine dihydrochloride (DMH) (DMBDD) to induce carcinogenesis in multiple organs, and observed up to 34 weeks. Significant increase of lung adenocarcinomas incidence was observed in DMBDD-treated $Ogg1^{-/-}$ male mice, but not in DMBDD-administered $Ogg1^{+/+}$ animals. Furthermore, incidences of lung adenomas were significantly elevated in both $Ogg1^{-/-}$ males and females as compared with respective $Ogg1^{-/-}$ control and DMBDD-treated $Ogg1^{+/+}$ groups. Incidence of total liver tumors (hepatocellular adenomas, hemangiomas and hemangiosarcomas) was significantly higher in the DMBDD-administered $Ogg1^{-/-}$ males and females. In addition, in DMBDD-treated male $Ogg1^{-/-}$ mice, incidences of colon adenomas and total colon tumors showed a trend and a significant increase, respectively, along with significant rise in incidence of simple hyperplasia of the urinary bladder, and a trend to increase for renal tubules hyperplasia in the kidney. Furthermore, incidence of squamous cell hyperplasia in the forestomach of DMBDD-treated $Ogg1^{-/-}$ male mice was significantly higher than that of $Ogg1^{+/+}$ males. Incidence of small intestine adenomas in DMBDD $Ogg1^{-/-}$ groups showed a trend for increase, as compared to the wild type mice. The current results demonstrated increased susceptibility of *Ogg1* mutant mice to the multiorgan carcinogenesis induced by DMBDD. The present bioassay could become a useful tool to examine the influence of various targets on mouse carcinogenesis.

Keywords: oxoguanine glycosylase 1 (*Ogg1*); 8-hydroxy-2′-deoxyguanosine; DNA repair; multiorgan carcinogenesis bioassay

---

1. Introduction

DNA damage and disruption of DNA repair are considered key factors in the susceptibility of mammals to endogenous and exogenous carcinogens, as well as processes of aging and cancer development [1]. The oxidative DNA damage includes a variety of oxidative lesions in DNA and the main attack site of reactive oxygen species (ROS) is at the 8 position of guanine, producing strongly mutagenic base 8-hydroxy-2′-deoxyguanosine (8-OHdG) [2]. 8-OHdG is used as an oxidative DNA damage marker which mispairs with adenine (A) residues, thus resulting in increase of spontaneous G:C to T:A transversion mutations [3].

Three DNA repair enzymes from various bacteria and *Saccharomyces cerevisiae*, namely, the MutM (Fpg), MutY and MutT DNA glycosylase homologs are known to prevent spontaneous mutagenesis induced by 8-OHdG [4]. In mammalian cells, the MutM homolog (MMH; the glycosylase/apurinic, apyrimidinic (AP) lyase), MutY and MutT homolog enzymes have also been identified [5–7]. In both mammalian and yeast cells, cloned human and mouse cDNAs encode distinct nuclear and mitochondrial forms of the DNA glycosylase, the product of the *Ogg1* gene, which is generated by alternative RNA splicing [8–10]. MutY and MutM homologs prevent G:C to T:A transversions in DNA, while MutT protein hydrolyzes 8-oxo-dGTP to 8-oxo-dGMT and pyrophosphate, thus avoiding the occurrence of A:T to C:G transversion mutations during DNA replication [11,12]. Analysis of the mutation spectrum revealed that the frequency of G:C to T:A transversions increased five-fold in *Ogg1* mutant mice compared with wild-type animals [8].

Mmh/Ogg1 homozygous mutant ($Ogg1^{-/-}$) mice used in our studies have physically normal appearance but exhibit three- and seven-fold increased accumulation of 8-OHdG adduct at 9 and 14 weeks of age, respectively, in comparison with or heterozygous or wild-type animals [13]. We have previously demonstrated that treatment of $Ogg1^{-/-}$ mice with dimethylarsinic acid (DMA) and phenobarbital (PB) for 78 weeks resulted in enhancement of lung and liver carcinogenesis, respectively [14,15]. The tremendous increase of 8-OHdG levels with consequent G:C to T:A transversions and deletions in the kidney DNA of $Ogg1^{-/-}$ mice were reported following administration of potassium bromate ($KBrO_3$) [8]. Furthermore, a significant increase of mutation frequency in $Ogg1^{-/-}$ mice livers was observed during liver regeneration after partial hepatechtomy following $KBrO_3$ treatment [16]. In addition, Sakumi et al. and Xie et al. demonstrated spontaneous development of lung, ovary tumors and lymphomas in *Myh* and *Ogg1* knockout mice [5,17]. However, it is still unknown how the ablation of these enzymes affects the tumorigenicity of various chemical carcinogens.

Previously, several in vivo bioassay systems for carcinogenicity detection of test compounds have been developed. However, these bioassays usually predict carcinogenicity of test chemicals only in single organs with known strategies of carcinogenesis initiation. To develop the experimental approach for the determination of carcinogenicity in numerous target organs, multiorgan wide-spectrum initiation bioassay (namely, the multiorgan carcinogenicity bioassay: DMBDD model) has been established [18–21]. This bioassay was applied in rats and included treatment with five genotoxic carcinogens, *N*-diethylnitrosamine (DEN), *N*-methyl-*N*-nitrosourea (MNU), *N*-butyl-*N*-(4-hydroxybutyl) nitrosamine (BBN), *N*-bis (2-hydroxypropyl) nitrosamine (DHPN) and 1,2-dimethylhydrazine dihydrochloride (DMH) (DMBDD), as initiators of liver, lungs, kidneys, urinary bladder, stomach, small intestine, colon and thyroid gland carcinogenesis [19,22,23]. It has been demonstrated that DMBDD-induced organ-specific DNA damage could be attributed to free radicals, methylation, and accumulation of non-repaired DNA damage [23]. In rats, DEN is usually used as initiator of liver carcinogenesis, BBN as initiator of bladder carcinogenesis, DMH as initiator of intestine carcinogenesis, and MNU as initiator of stomach, bladder and liver carcinogenesis [24]. DHPN is a wide-spectrum carcinogen in rats, which induces lung, thyroid, kidney, bladder and liver cancers [23,25]. In previous studies, DMBDD treatment has been proposed to inactivate the tumor suppressor p53 in the bladder tumors of Zucker diabetic rats [26]. However, to our knowledge, the DMBDD model was never applied in mice and there is no information how the DMBDD treatment influences oncogenes and tumor suppressor genes.

The aim of the present study was to investigate the differences in susceptibility of *Ogg1* mutant and wild type mice of C57BL/6J background to the treatment with five types of genotoxic carcinogens (DEN, MNU, BBN, DHPN and DMH: DMBDD) by applying the multiorgan carcinogenesis bioassay in mice.

2. Results

2.1. General Observations

All control $Ogg1^{-/-}$ male or female mice were alive at the end of the study. They were healthy and long-lived as compared to the control $Ogg1^{+/+}$ mice. Three DMBDD-treated $Ogg1$ knockout male and three female mice were found moribund at Weeks 11, 12, and 17, and 10, 20 and 25, respectively. The causes of death of $Ogg1^{-/-}$ male mice were malignant lymphomas/leukemia, lung adenocarcinoma and fibrosarcoma, while $Ogg1^{-/-}$ female mice died due to the development of lymphoma/leukemia. Four DMBDD-administered $Ogg1^{+/+}$ male and one female mice died at Weeks 16, 29, 33, 35 and 32, respectively. The main causes of death in male and female wild type mice were malignant lymphoma/leukemia, T cell lymphoma and bladder transitional cell carcinoma (TCC). One non-treated control $Ogg1^{+/+}$ male mouse was found moribund at Week 27 due to a urinary tract infection.

As lung, liver and colon tumors were observed in the DMBDD-treated $Ogg1^{-/-}$ mice that were found moribund during the study, effective number of animals used for the histopathological analysis included all mice. Body weight and survival curves, final body weight and absolute and relative organ weights of mice are shown in Table 1 and Figure 1. Body weights of control $Ogg1$ mutant male and female mice were significantly lower than those of wild type mice all through the experiment. DMBDD treatment induced significant decrease of body weight of both $Ogg1$ mutant and wild type mice, However, at Experimental Week 14, mean body weight of $Ogg1^{-/-}$ mice became equal to that of the corresponding control animals of the same genotype, while the body weight of the DMBDD-administered $Ogg1^{+/+}$ mice continued to be significantly lower compared to the control $Ogg1^{+/+}$ until the end of the study (Figure 1A). Therefore, at Week 34, final body weight of the DMBDD-treated $Ogg1^{+/+}$ but not $Ogg1^{-/-}$ mice were significantly decreased as compared with the control mice of the same genotype.

DMBDD administration inhibited food intake of the $Ogg1^{+/+}$, but not the $Ogg1^{-/-}$ mice compared with the control mice of the same genotype (Figure S1A). Water intakes were similarly decreased in all DMBDD-treated $Ogg1^{-/-}$ and $Ogg1^{+/+}$ animals (Figure S1B). Thus, the body weight of the DMBDD-administered $Ogg1^{+/+}$ mice appeared to be significantly lower than the control $Ogg1^{+/+}$ due to the inhibited food intake. The absolute and relative liver, kidneys and spleen weights of the control $Ogg1^{-/-}$ male and female mice were significantly lower than those of the control $Ogg1^{+/+}$ mice (Table 1). DMBDD treatment induced significant increases of relative liver, kidneys and spleen weights of $Ogg1$ mutant but not wild type both male and female mice in comparison with corresponding $Ogg1^{-/-}$ controls. The absolute and relative weights of the lungs were significantly increased in both $Ogg1^{-/-}$ and $Ogg1^{+/+}$ male and female mice (Table 1).

Table 1. Final survival ratios, final body and relative organ weights of *Ogg1*^−/− and *Ogg1*^+/+ mice.

Final Body and Organ Weights	*Ogg1*^−/−				*Ogg1*^+/+			
Group	G1	G2	G3	G4	G5	G6	G7	G8
Gender	Male	Male	Female	Female	Male	Male	Female	Female
Treatment	DMBDD	Control	DMBDD	Control	DMBDD	Control	DMBDD	Control
Effective No. of mice	20	20	20	20	20	20	20	20
No. of surviving animals [a] (%)	17(85)	20(100)	17(85)	20(100)	16(80)	19(95)	19(95)	20(100)
Final body weight (g) [e]	29.0 ± 3.3	30.6 ± 1.9 [c]	24.7 ± 1.5	25.0 ± 2.6 [a]	26.7 ± 2.7 ****	33.8 ± 2.8	23.7 ± 1.6 ***	26.9 ± 2.5
Organ weights								
Liver (g)	1.48 ± 0.21 ***	1.26 ± 0.13 [d]	1.07 ± 0.11 **,[b]	0.95 ± 0.13 [d]	1.45±0.18 **	1.64 ± 0.22	0.94 ± 0.15 ****	1.13 ± 0.12
Liver (%)	5.14 ± 0.71 ****	4.11 ± 0.32 [d]	4.36 ± 0.40 ***,[a]	3.83 ± 0.38 [b]	5.35 ± 0.77	4.93 ± 0.68	4.02 ± 0.53	4.22 ± 0.31
Kidneys (g)	0.43 ± 0.05 *	0.39 ± 0.07 [d]	0.30 ± 0.02	0.28 ± 0.04 [b]	0.40 ± 0.05 ****	0.53 ± 0.11	0.28 ± 0.04 **	0.34 ± 0.06
Kidneys (%)	1.51 ± 0.22 **	1.28 ± 0.20 [c]	1.21 ± 0.07 *	1.14 ± 0.11 [b]	1.50 ± 0.22	1.60 ± 0.30	1.22 ± 0.15	1.27 ± 0.17
Spleen (g)	0.14 ± 0.15 *	0.06 ± 0.01 [d]	0.07 ± 0.02 **	0.05 ± 0.01 [d]	0.11 ± 0.08	0.09 ± 0.02	0.08 ± 0.02	0.08 ± 0.02
Spleen (%)	0.47 ± 0.49 *	0.20 ± 0.04 [b]	0.30 ± 0.08 ***	0.23 ± 0.03 [c]	0.39 ± 0.28	0.26 ± 0.08	0.35 ± 0.08	0.28 ± 0.05
Lungs (g)	0.25 ± 0.06 ****	0.17 ± 0.02	0.24 ± 0.04 ****	0.15 ± 0.02	0.27 ± 0.05 ****	0.19 ± 0.03	0.23 ± 0.04 ****	0.15 ± 0.03
Lungs (%)	0.90 ± 0.28 ****	0.56 ± 0.07	0.97 ± 0.16 ****	0.61 ± 0.08 [a]	1.00 ± 0.22 ****	0.56 ± 0.09	0.98 ± 0.18 ****	0.56 ± 0.08

Data are Mean ± SD for the surviving animals at the end of the study. Relative organ weights were calculated with the following equation: Absolute organ weight/final body weight × 100. * $p < 0.05$; ** $p < 0.01$; *** $p < 0.001$; **** $p < 0.0001$: significantly different vs. the respective control groups of the same genotype. [a] $p < 0.05$; [b] $p < 0.01$; [c] $p < 0.001$; [d] $p < 0.0001$ significantly different vs. the respective *Ogg1*^+/+ DMBDD-treated or control groups. [e] Final body weights of all survived mice at the termination of the experiment.

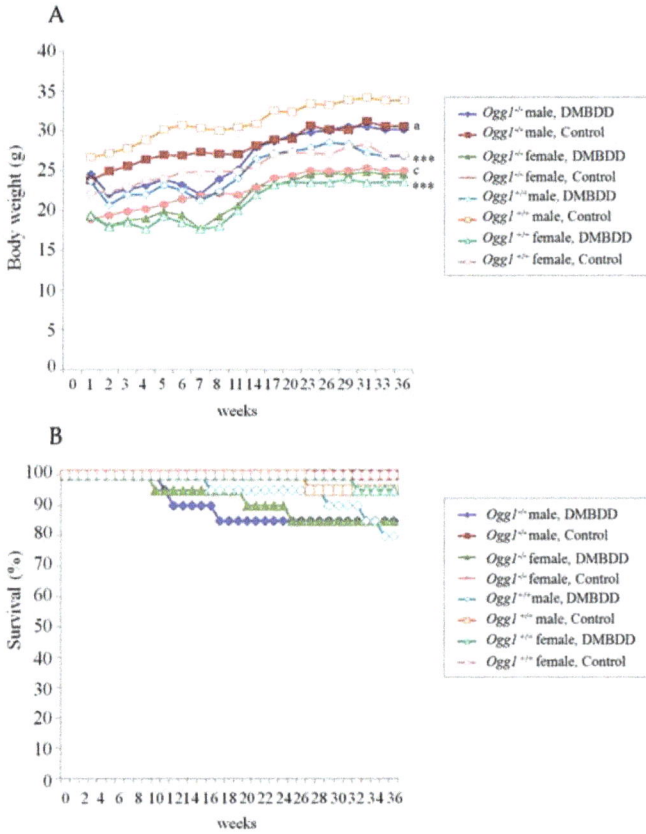

Figure 1. Body weight (**A**); and survival (**B**) curves for DMBDD-treated and control $Ogg1^{-/-}$ and $Ogg1^{+/+}$ male and female mice. *** $p < 0.001$ significantly different vs. respective control group of the same genotype; [a] $p < 0.05$ and [c] $p < 0.001$ significantly different vs. the respective $Ogg1^{+/+}$ control groups.

2.2. Survival Curves

Survival curves for the DMBDD-administered and control $Ogg1^{-/-}$ and $Ogg1^{+/+}$ mice are presented in Figure 1B. In the present model, there were no significant differences in survival between the control $Ogg1$ homozygous mutant and wild type mice. Trends for decrease in survival were observed in both $Ogg1^{-/-}$ and $Ogg1^{+/+}$ DMBDD-treated animals. However, in DMBDD-treated $Ogg1^{-/-}$ male and female mice earlier decrease in survival (males: Week 11; females: Week 10), respectively, as compared to the wild type mice (males: Week 16; females: Week 32) was found (Figure 1B). Importantly, the earlier development of tumors in DMBDD-administered $Ogg1^{-/-}$ males and females, mostly malignant lymphomas/leukemias, lung adenocarcinoma and subcutaneous tumors (fibrosarcomas) was the reason for their earlier mortality.

2.3. Results of Histopathological Examination

Table 2 summarizes the data on the incidence of preneoplastic, neoplastic and some non-neoplastic lesions and general distribution of tumors induced by DMBDD administration in $Ogg1$ knockout and wild type mice. Representative pictures of neoplastic lesions observed in the lungs, livers and colons of mice are presented in Figure 2. Neoplastic nodules induced in the DMBDD-treated group of

$Ogg1^{-/-}$ and $Ogg1^{+/+}$ mice were mainly lung, liver, colon, small intestine, urinary bladder tumors, malignant lymphomas/leukemias and subctaneous tumors (fibrosarcomas). In male $Ogg1^{+/+}$ mice higher number of tumors were induced by the DMBDD treatment as compared to the $Ogg1^{+/+}$ females likely due to the lower susceptibility to genotoxic carcinogens in females.

Table 2. Neoplastic and preneoplastic proliferative lesions in male and female $Ogg1^{-/-}$ and $Ogg1^{+/+}$ mice.

Incidence in Males (No. Mice (%))	$Ogg1^{-/-}$		$Ogg1^{+/+}$	
Group	G1	G2	G5	G6
Gender	Male	Male	Male	Male
Treatment	DMBDD	Control	DMBDD	Control
Effective No. mice	20	20	20	20
No. tumor-bearing mice (%)	20(100) ****	1(5)	17(85) ****	0
No. tumors/mouse	5.9 ± 3.5 ****,a	0.1 ± 0.2	3.5 ± 2.8 ****	0
Lung				
Adenoma	20(100) ****,a	1(5)	14(70) ****	0
Adenocarcinoma	7(35) **	0	2(10)	0
Total tumors	20(100) ****,a	1(5)	15(75) ****	0
HPL	20(100) ****	1(5)	20(100) ****	0
Liver				
HCA	4(20) (i)	0	1(5)	0
Hemangioma	2(10)	0	1(5)	0
Hemangiosarcoma	3(15)	0	0	0
Total tumors	5(25) *	0	1(5)	0
Basophilic PPFs	2(10)	0	3(15)	0
Eosinophilic PPFs	0	0	1(5)	0
Mixed type PPFs	0	0	1(5)	0
Kidneys				
Tubular cell HPL	5(25) (i)	1(5)	2(10)	1(5)
Urinary Bladder				
Papilloma	1(5)	0	0	0
TCC	1(5)	0	2(10)	0
Total tumors	2(10)	0	2(10)	0
Simple HPL	5(25) *	0	5(25)	1(5)
PN HPL	4(20) (i)	0	2(10)	0
Colon				
Adenoma	4(20) (i)	0	2(10)	0
Adenocarcinoma	1(5)	0	0	0
Total tumors	5(25) *	0	2(10)	0
Small Intestine				
Adenoma	3(15)	0	0	0
AdCa	1(5)	0	1(5)	0
Total tumors	4(20) (i)	0	1(5)	0
Forestomach				
Squamous cell HPL	14(70) ****,a	0	7(35) **	0
Glandular Stomach				
Adenoma	1(5)	0	0	0
Adenomatous cell HPL	1(5)	0	1(5)	0
Lymphoma/Leukemia	3(15)	0	1(5)	0
Skin/Subcutis				
Fibrosarcoma	3(15)	0	0	0

Table 2. *Cont.*

Incidence in Females (No. Mice (%))	Ogg1 $^{-/-}$		Ogg1 $^{+/+}$	
Group	G3	G4	G7	G8
Gender	Female	Female	Female	Female
Treatment	DMBDD	Control	DMBDD	Control
Effective No. mice	20	20	20	20
No. tumor-bearing mice (%)	20(100) ****,b	0	8(40) **	0
No. tumors/mouse	4.6 ± 2.1 ****,b	0	0.7 ± 0.9 **	0
Lung				
Adenoma	19(95) ****,b	0	12(60) ****	0
AdCa	2(10)	0	1(5)	0
Total tumors	20(100) ****,b	0	12(60) ****	0
HPL	20(100) ****	1(5)	17(85) ****	0
Liver				
HCA	4(20) (i)	0	0	0
Hemangioma	1(5)	0	1(5)	0
Hemangiosarcoma	1(5)	0	0	0
Total tumors	5(25) *	0	1(5)	0
Basophilic PPFs	1(5)	0	2(10)	0
Kidneys				
Renal cell adenoma	0	0	1(5)	0
Tubular cell HPL	2(10)	0	2(10)	0
Urinary Bladder				
TCC	1(5)	0	0	0
Simple HPL	6(30)	3(15)	2(10)	2(10)
PN HPL	3(15)	0	0	0
Colon				
Adenoma	2(10)	0	4(20) (i)	0
Small Intestine				
Adenoma	1(5)	0	2(10)	0
Forestomach				
Squamous cell HPL	11(55) ***	1(5)	8(40) **	0
Glandular Stomach				
AdCa	1(5)	0	0	0
Adenomatous cell HPL	1(5)	2(10)	0	0
Thyroid				
Follicular cell Adenoma	0	0	1(5)	0
Lymphoma/Leukemia	4(20) (i)	0	0	0
T Cell Lymphoma	0	0	1(5)	0
Adrenals				
Cortical HPL	1(5)	0	0	0

* $p < 0.05$; ** $p < 0.01$; *** $p < 0.001$; **** $p < 0.0001$ and (i) $p = 0.05$ vs. respective control mice of the same genotype. a $p < 0.05$; b $p < 0.01$; d $p < 0.0001$ vs. wild type control or DMBDD-treated mice. HPL, hyperplasia; HCA, hepatocellular adenoma; AdCa, adenocarcinoma; PN, papillary or nodular; TCC, transitional cell carcinoma, PPFs, putative preneoplastic foci.

Lung	Liver	Colon

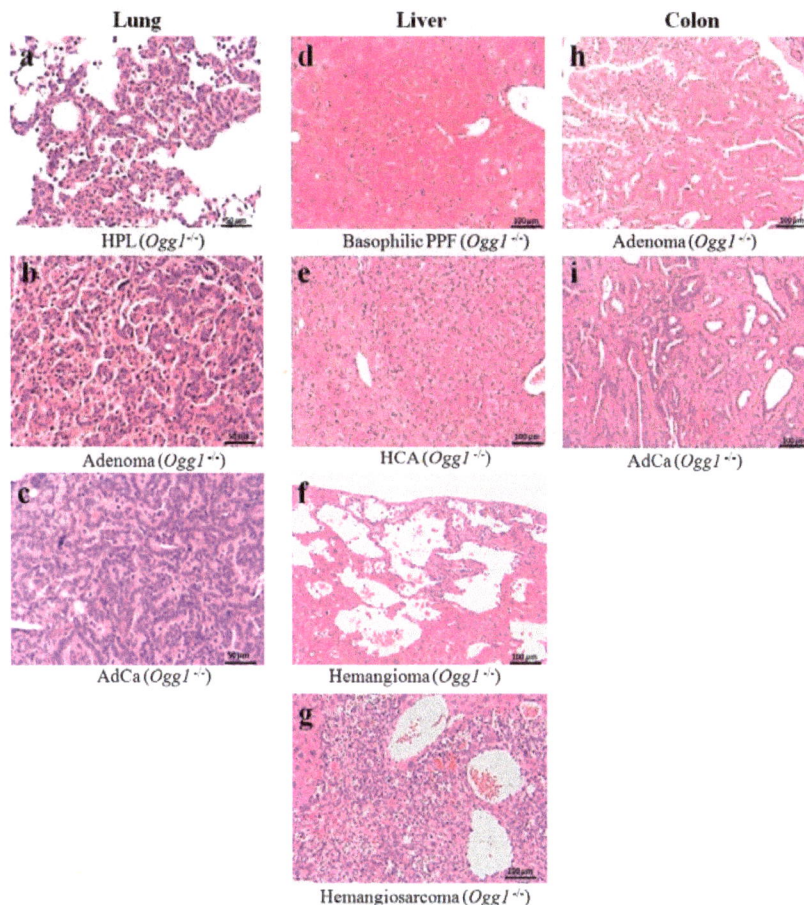

Figure 2. Representative histopathological pictures (H&E staining) of: lung hyperplasia (**a**); adenoma (**b**); adenocarcinoma (**c**); liver PPF (basophilic foci) (**d**); HCA (**e**); hemangioma (**f**); hemangiosarcoma (**g**); colon adenoma (**h**); and adenocarcinoma (**i**) developed in DMBDD-treated $Ogg1^{-/-}$ mice. HPL, hyperplasia; HCA, hepatocellular adenoma; AdCa, adenocarcinoma; PPFs, putative preneoplastic foci.

Macroscopically, no tumors were found in the non-treated control $Ogg1^{+/+}$ mice, however, spontaneous development of lung nodules (hyperplasia and adenoma) was detected in the control $Ogg1^{-/-}$ animals. Furthermore, DMBDD-treated $Ogg1^{-/-}$ mice were more susceptible to the induction of different tumors as compared to $Ogg1^{-/-}$ control and wild type mice. DMBDD treatment induced elevation of total tumor incidence and number of tumor bearing mice in mutant, predominantly $Ogg1^{-/-}$ female animals (males, 100%, 5.9 ± 3.5/mouse; females, 100%, 4.6 ± 2.1/mouse, $p < 0.0001$), as compared to the wild type mice (males, 85%, 3.5 ± 2.8/mouse; females, 40%, 0.7 ± 0.9/mouse). All DMBDD-treated male and female $Ogg1^{-/-}$ mice developed many nodules in the lungs, while incidences and multiplicities of lung nodules in DMBDD-treated $Ogg1^{+/+}$ was lower as compared to the $Ogg1^{-/-}$ DMBDD-administered animals. Furthermore, the incidence of liver lesions in $Ogg1^{-/-}$ mice, as well as their multiplicity was also increased after carcinogens treatment as compared to the corresponding controls. Moreover, in male, but not female DMBDD-treated $Ogg1^{-/-}$ mice, incidences of colon tumors and fibrosarcomas, were elevated.

Histopathological examination demonstrated significant elevation of incidences and multiplicities of lung adenoma and total lung tumors in DMBDD-treated $Ogg1^{-/-}$ male and female mice lungs (total tumors: males, 100%, 4.1 ± 2.7/mouse; females, 100%, 4.1 ± 2.2/mouse; adenoma: males, 100%, 4.0 ± 2.3/mouse; females, 95%, 2.7 ± 2.8/mouse) as compared to both respective $Ogg1^{-/-}$ controls (total tumors: males, 5%, 0.1 ± 0.2/mouse; females, 0%, 0/mouse; adenoma: males, 5%, 0.1 ± 0.2/mouse; females, 0%, 0/mouse) and DMBDD-treated $Ogg1^{+/+}$ mice (total tumors: males, 75%, 3.0 ± 2.7/mouse; females, 60%, 1.4 ± 2.2/mouse; adenoma: males, 70%, 2.7 ± 2.8/mouse; females, 60%, 1.4 ± 2.0/mouse) (Table 2). Interestingly, significant increase of lung adenocarcinoma incidence and multiplicity was found in DMBDD-administered $Ogg1^{-/-}$ male mice (35%, 0.5 ± 0.7/mouse, $p < 0.01$), but not in the wild type males (10%, 0.1 ± 0.3/mouse) as compared to corresponding controls of the same genotype. Furthermore, incidences and multiplicities of lung adenocarcinoma were higher in female $Ogg1^{-/-}$ mice of DMBDD group as compared to the DMBDD-treated wild type counterparts. In addition, increases of lung hyperplasia incidences due to the DMBDD application were observed in both $Ogg1$ homozygous mutant and wild type mice.

In the liver of $Ogg1$ knockout and wild type mice, DMBDD treatment caused development of putative preneoplastic foci of mostly basophilic phenotype, hepatocellular adenomas (HCAs), hemangiomas and hemangiosarcomas. Importantly, hemangiosarcomas were detected only in the $Ogg1^{-/-}$ mice (males, 15%, 0.6 ± 1.4/mouse; females, 5%, 0.1 ± 0.2/mouse). No hepatocellular carcinomas were apparent in DMBDD-treated $Ogg1$ knockout and wild type animals. Increases of HCA incidences (males, 20%; females, 20%; $p = 0.05$) and multiplicities (males: 0.3 ± 0.6/mouse, $p = 0.05$; females, 0.2 ± 0.4) were detected in livers of the DMBDD-initiated $Ogg1^{-/-}$ mice as compared to the $Ogg1^{-/-}$ controls (males, 0%, 0.1 ± 0.2/mouse; females, 0%, 0/mouse) and DMBDD-treated $Ogg1^{+/+}$ groups (males, 5%, 0.1 ± 0.2/mouse; females, 0%, 0/mouse). Significant elevations of total liver tumor incidences were observed in $Ogg1^{-/-}$ males (25%, $p < 0.05$) and females (25%, $p < 0.05$), but not $Ogg1^{+/+}$ mice as compared to the corresponding controls of the same genotype (0%). Furthermore, a trend for increase and a significant elevation of total liver tumors multiplicity were observed in DMBDD-treated $Ogg1^{-/-}$ males (0.9 ± 2.0/mouse, $p = 0.05$) and females (0.3 ± 0.6/mouse, $p < 0.05$), in respect of control $Ogg1^{-/-}$ group. In addition, DMBDD administration caused elevation of bile duct proliferation in the liver of $Ogg1^{-/-}$ mice as compared to the $Ogg1^{-/-}$ and $Ogg1^{+/+}$ counterparts. Significant increases of biliary cysts formation in the DMBDD groups were observed in the liver of both $Ogg1^{-/-}$ and $Ogg1^{+/+}$ animals in respect of corresponding controls (Table 2).

In kidneys, significant increase of renal tubular degeneration (80%, $p < 0.05$) and a trend for increase of tubular renal cell HPL was found in the DMBDD-treated $Ogg1^{-/-}$ male mice as compared to corresponding controls of the knockout and wild type genotypes (Table 2). Only one $Ogg1^{+/+}$ DMBDD-treated mouse developed renal adenoma.

In the urinary bladder, significant increase of simple hyperplasia (25%, $p < 0.05$) and a trend for increase of papillary and nodular (PN) hyperplasia (20%) incidences as compared to the corresponding controls of the same genotype was detected (Table 2).

The incidence of total colon tumors (25%, $p < 0.05$) was significantly increased in the DMBDD-treated $Ogg1^{-/-}$ male mice but not in the DMBDD-administered wild type males (10%). Development of adenocarcinoma (5%) was found in one male $Ogg1^{-/-}$ mouse of the DMBDD group. Furthermore, incidences of small intestine total tumors showed a trend for increase in the DMBDD-administered $Ogg1^{-/-}$ male mice (20%) as compared to the $Ogg1^{-/-}$ control group and DMBDD-treated $Ogg1^{+/+}$ animals (5%). In females, incidences and multiplicities of colon tumors induced by DMBDD were comparable with that of observed in wild type mice, pointing out the sex differences in susceptibility to colonic tumorigenesis.

In the forestomach, the DMBDD treatment resulted in significant elevation of the squamous cell HPL incidence in $Ogg1^{-/-}$ (male, 70%, $p < 0.0001$; female, 55%, $p < 0.001$) and $Ogg1^{+/+}$ (male, 35%, $p < 0.01$; female, 40%, $p < 0.01$) mice as compared to the corresponding controls of the same genotype.

Furthermore, in male mice, it was significantly increased in comparison to wild type DMBDD-treated males ($p < 0.05$).

Trends for increase of malignant lymphomas/leukemias were observed in *Ogg1* homozygous mutant males (15%) and females (20%) treated with DMBDD, as compared to wild type mice (Table 2). One *Ogg1*[+/+] female mouse in DMBDD group developed T cell lymphoma (5%).

2.4. Blood Biochemistry

The results of the blood biochemistry analysis are shown in Table 3. Aspartate aminotransferase (AST) and alanine aminotransferase (ALT) levels in the blood of both *Ogg1* null and wild type mice showed strong trends for increase, or were significantly elevated by the DMBDD treatment. Furthermore, this induction was higher in *Ogg1*[−/−] animals. Serum sodium (Na) levels were elevated by the DMBDD administration in both *Ogg1* mutant and wild type mice. Moreover, creatinine level was higher in the blood of DMBDD-treated *Ogg1*[−/−] mice as compared to the respective *Ogg1*[−/−] control groups. Alkaline phosphatase (ALP), T-cholesterol and chloride (Cl) levels were lowered in the wild type DMBDD-treated animals, but not altered in DMBDD *Ogg1*[−/−] mice. Serum calcium (Ca) level was significantly decreased in the DMBDD-treated *Ogg1* knockout male mice as compared to the wild type males administered DMBDD. In addition, inorganic phosphorus (IP) levels showed a trend for increase or the significant elevation in the blood of DMBDD-treated and control *Ogg1*[−/−] male and female mice, respectively, as compared to the wild type groups receiving the same treatment.

Table 3. Blood biochemistry data of DMBDD-treated and control *Ogg1* knockout and wild type mice.

Parameter	*Ogg1*[−/−]		*Ogg1*[+/+]		*Ogg1*[−/−]		*Ogg1*[+/+]	
Group	G1	G2	G3	G4	G5	G6	G7	G8
Gender	Male	Female	Male	Female	Male	Female	Male	Female
Treatment	DMBDD	Control	DMBDD	Control	DMBDD	Control	DMBDD	Control
Effective No. mice	7	9	9	9	8	9	9	9
AST (IU/L)	88.9 ± 39.7 [(i)]	52.3 ± 3.2	78.7 ± 14.6 **	58.3 ± 5.9 [a]	75.1 ± 11.2 **	57.3 ± 9.1	65.1 ± 12.1 *	51.9 ± 5.6
ALT (IU/L)	98.7 ± 82.9 *	26.9 ± 9.7	50.7 ± 15.4 **	27.2 ± 7.4	61.4 ± 22.8 **	32.8 ± 7.9	42.0 ± 23.0 *	22.1 ± 2.8
ALP (IU/L)	298.6 ± 105.5	245.0 ± 53.7	535.6 ± 138.3	452.4 ± 88.3	382.3 ± 93.7 **	224.0 ± 35.9	472.4 ± 82.2 *	375.0 ± 93.0
γ-GTP (IU/L)	1.0 ± 0.0	1.0 ± 0.0	1.3 ± 0.7	1.0 ± 0.0	1.1 ± 0.4	1.0 ± 0.0	1.0 ± 0.0	1.0 ± 0.0
T-protein (g/dl)	4.5 ± 0.9	5.2 ± 0.3	5.1 ± 0.3	5.2 ± 0.2	5.0 ± 0.4	5.3 ± 0.2	5.0 ± 0.2 *	5.3 ± 0.3
Albumin (g/dL)	2.0 ± 0.5 *	2.5 ± 0.1	2.4 ± 0.3	2.6 ± 0.2	2.1 ± 0.2 **	2.4 ± 0.1	2.4 ± 0.2	2.6 ± 0.1
A/G ratio	0.8 ± 0.1	0.9 ± 0.1	0.9 ± 0.1	1.0 ± 0.1	0.7 ± 0.1 **	0.9 ± 0.1	1.0 ± 0.1	1.0 ± 0.1
T-BiL (mg/dL)	0.2 ± 0.0	0.2 ± 0.0	0.2 ± 0.0	0.2 ± 0.0	0.2 ± 0.0	0.2 ± 0.0	0.2 ± 0.0	0.2 ± 0.0
Na (mEq/L)	154.1 ± 2.7 *	153.4 ± 1.2 [(i)]	152.2 ± 1.6 *	150.7 ± 1.9	154.9 ± 1.7 *	152.8 ± 1.2	153.0 ± 1.7 **	150.8 ± 1.4
K (mEq/L)	6.7 ± 1.3	6.1 ± 0.9	5.7 ± 0.9	6.2 ± 1.7	6.9 ± 0.7	6.6 ± 1.5	5.5 ± 0.9	5.3 ± 0.8
Cl (mEq/L)	109.7 ± 3.0	109.8 ± 1.8	111.0 ± 1.9	109.2 ± 2.3	111.3 ± 2.0 *	109.1 ± 1.5	112.7 ± 2.3 *	110.6 ± 2.2
Ca (mEq/L)	8.8 ± 0.5 [a]	8.9 ± 0.5	8.6 ± 0.6	8.7 ± 0.5	9.4 ± 0.6	9.1 ± 0.4	8.7 ± 0.4	8.7 ± 0.3
IP (mEq/L)	10.3 ± 1.1	9.8 ± 1.1	9.5 ± 1.8 [a]	9.4 ± 2.0 [a]	9.5 ± 0.8	9.8 ± 1.0	7.8 ± 1.4	7.4 ± 0.9
T-Cholesterol (mg/dL)	78.3 ± 21.4	87.8 ± 9.1	65.3 ± 6.5	74.1 ± 17.3	78.0 ± 20.5 *	103.0 ± 29.8	59.4 ± 14.6 **	100.3 ± 9.6
TG (mg/dL)	114.0 ± 211.2	80.4 ± 32.1	33.4 ± 24.7	32.3 ± 16.1	72.9 ± 25.4	68.2 ± 41.3	27.1 ± 14.9	33.8 ± 16.6
BUN (mg/dL)	35.6 ± 6.0	33.4 ± 5.4	35.1 ± 4.0	32.1 ± 4.3	35.3 ± 3.5	31.4 ± 4.3	31.6 ± 7.2	26.3 ± 8.5
Creatinine (mg/dL)	0.05 ± 0.02 *	0.03 ± 0.02	0.05 ± 0.03 *	0.03 ± 0.02 [b]	0.05 ± 0.03	0.04 ± 0.01	0.07 ± 0.02	0.07 ± 0.03

Values are means ± SD; * $p < 0.05$; ** $p < 0.01$; [(i)] $p = 0.05$ vs. respective control group of mice of same genotype. [a] $p < 0.05$; [b] $p < 0.01$ vs. respective wild type control or DMBDD group. TG, triglycerides; T-Bil, T-bilirubin; IP, inorganic phosphorus.

In the blood serum of *Ogg1* mutant and wild type mice, levels of total protein showed a trend (*Ogg1*[−/−] males, females and *Ogg1*[+/+] males) and a significant decrease (*Ogg1*[+/+] females) as compared to the non-treated respective control groups. Furthermore, albumin levels were lower in DMBDD-treated *Ogg1*[−/−] and *Ogg1*[+/+] groups, with significant differences observed for *Ogg1*[−/−] and *Ogg1*[+/+] DMBDD-administered males. Albumin/globulin (A/G) ratio was significantly lower in the *Ogg1*[+/+] male DMBDD group.

3. Discussion

The present study revealed that *Ogg1* mutant mice are more susceptible to the induction of tumors due to the treatment with DMBDD, than wild type C57Bl/6J mice. In the DMBDD-treated *Ogg1*[−/−]

mice, main causes of death besides malignant lymphoma/leukemia were lung adenocarcinoma and skin/subcutis fibrosarcoma, while $Ogg1^{+/+}$ animals died from malignant lymphoma/leukemia and urinary bladder carcinoma. Furthermore, the earlier mortality of DMBDD-administered $Ogg1^{-/-}$ mice appeared to be due to the earlier tumor development. Importantly, DMBDD caused significant increases of incidences and multiplicities of lung adenocarcinoma in $Ogg1^{-/-}$ males, liver tumors in $Ogg1^{-/-}$ males and females and colon tumors in $Ogg1^{-/-}$ male mice as compared to the $Ogg1^{-/-}$ controls. In the kidneys, urinary bladder, stomach, small intestine and subcutis of $Ogg1$ mutant mice, increases of carcinogenicity as compared to the DMBDD-treated wild type animals were obvious.

Lungs of $Ogg1$ null mice were strongly affected by DMBDD initiation, which could be concluded from significant increases of lung adenocarcinoma incidence in DMBDD-treated $Ogg1^{-/-}$ male mice and incidences and multiplicities of adenomas and total lung tumors in $Ogg1^{-/-}$ males and females. As lungs are strongly exposed to molecular oxygen, it is likely the most carcinogenicity sensitive organ in $Ogg1$ knockouts. Furthermore, in the lung of non-treated $Ogg1^{-/-}$ animals, spontaneously developed tumors were observed, possibly due to the accumulation of non-repaired oxidative DNA base modifications even in the absence of initiation. Several authors have reported an increase of spontaneous lung tumors in MutM, MutY and MutT-deficient mice [5,14,15,27]. Previously, lung tumors were also shown to be significantly induced in $Ogg1^{-/-}$ mice by the DMA treatment [14]. Furthermore, significant enhancement of spontaneous lung tumorigenesis was observed when the $Ogg1$ mutation was combined with a MutY homolog (MUTYH) or MSH2-deficient condition, and the G:C to T:A transversions in the *K-ras* gene were detected in the lung tumors [17]. In our previous study, genes related to cancer, cellular growth, proliferation and cell cycle (e.g., polymerase (DNA-directed), delta 4 (Pold4), cyclin C and mitogen activated protein kinase 8) and angiogenesis (e.g., matrix metalloproteinases 13, 14, and 17) were found to be up-regulated in non-treated $Ogg1^{-/-}$ mice lungs, but those involved in free radical scavenging, lipid metabolism, drug and endocrine system development and function were suppressed comparing to the $Ogg1^{+/+}$ case [14]. From the present and previous results, MutM, MutY and MutT homologs responsible for the repair of oxidative DNA modifications are extremely important for suppression of lung tumorigenesis in mammal.

DMBDD treatment induced significant increases of relative liver weights in $Ogg1$ homozygous mutant but not wild type mice. Furthermore, higher elevation of AST and ALT serum levels reflecting the pathological processes in the liver supported our histopathological findings in the DMBDD-treated $Ogg1$ knockout mice. The mechanism of DMBDD carcinogenicity in the liver of $Ogg1^{-/-}$ mice might be accumulation of non-repaired oxidative base modifications in DNA leading to increase of cell proliferation, occurrence of mutations and further elevation of cell proliferation, resulting in promotion and progression of liver carcinogenesis [14]. Interestingly, hemangiosarcomas were detected only in the DMBDD-treated $Ogg1^{-/-}$ mice. Hemangiomas and hemangiosarcomas are known to arise as primary vascular neoplasms in the liver and could be initiated in mice by DHPN [28]. They are usually not sharply demarcated from the surrounding parenchyma and the neoplastic cells are generally elongated or spindle-shaped and may form solid areas occupying dilated hepatic sinusoids and are typically locally invasive (Figure 3). Tsutsumi et al. previously demonstrated that incidences and multiplicities of hemangiomas and hemangiosarcomas in the liver were markedly higher in the poly(ADP-ribose) polymerase-1 (Parp-1)-null mice, while Parp-1 is one of the poly(ADP-ribose) polymerase family proteins taking part in genomic stability, DNA repair and cell death triggered by DNA damage [29]. Thus, the relationship between defective DNA repair and development of hemangiosarcomas may exist.

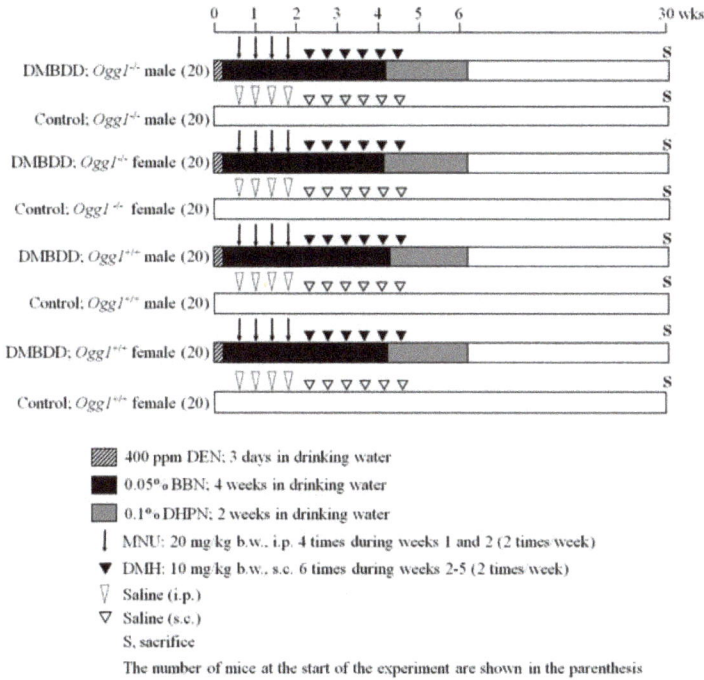

Figure 3. Experimental protocol of medium-term multiorgan carcinogenesis bioassay applied in $Ogg1^{-/-}$ and $Ogg1^{+/+}$ mice. wks: weeks.

Increase of DNA 8-OHdG levels has been previously reported by the DEN treatment in the livers of rats and mice [3,30]. Furthermore, *mutT*-deficient mice were also reported to be susceptible to liver carcinogenesis [27]. Moreover, the effect of potassium bromate, which has been reported to induce oxidative stress, was investigated in $Ogg1^{-/-}$ mouse liver after partial hepatectomy [16], and the results indicated a significant increase of mutation frequency and liver tumorigenicity being consistent with our present and previous data showing the promotion and progression of hepatocarcinogenesis in DMBDD and PB-treated $Ogg1^{-/-}$ mice [15]. Arai et al. suggested that high levels of cell proliferation are very important for the fixation of mutations induced by oxidative stress conditions in the liver [16]. Furthermore, in our previous study, it has been detected that cell proliferation and DNA 8-OHdG levels in the liver of $Ogg1^{-/-}$ mice treated with PB are much higher than that of wild type mice. Therefore, they are highly susceptible to the carcinogens treatment [15]. Thus, it could be suggested that accumulation of unrepaired 8-OHdG in the livers of DMBDD-treated $Ogg1^{-/-}$ animals might cause a significant increase of cellular proliferation, resulting in acceleration of hepatocarcinogenesis. With regard to specific elevation of cell proliferation in DMBDD target organs, elevation of cell proliferation has been previously shown in the lung, liver, colon, urinary bladder, thyroid and kidney by initiation with BBN, DEN, DMH, DHPN and MNU [31].

From our previous results, in contrast to the wild type mice, in the livers of *Ogg1*-deficient animals, Nrf2 phosphorylation, and likely, its transformation to the nuclear did not occur, resulting in increase of oxidative stress and DNA damage of liver cells [15]. The accumulation of reactive oxygen species and non-repaired DNA oxidative base modifications in the $Ogg1^{-/-}$ livers, thus, could become the reason of higher susceptibility to liver tumorigenesis. At present, Nrf2 is recognized as important protein involved in regulation of broad transcriptional response preventing DNA, proteins and lipids damage, recognition, repair and removal of macromolecular damage, and tissue renewal after application of toxic substance. Mice that lack the Nrf2 transcription factor were more sensitive to the genotoxic and cytotoxic and effects of foreign chemicals and oxidants than wild-type animals [32]. Multiple studies demonstrated enhanced tumorigenicity in Nrf2-disrupted mice compared to wild-type in models of lung disease and cancer, hepatocarcinogenesis, colon cancer, stomach cancer, bladder cancer, mammary cancer, skin cancer, and inflammation [33]. Furthermore, Nrf2 has been shown to upregulate the activity of multiple DNA repair, including the process of removal of oxidative stress-induced endogenous DNA interstrand cross-links [33,34].

The histological examination revealed a trend and a significant increase of renal tubular hyperplasia and degeneration, respectively, in DMBDD-treated $Ogg1^{-/-}$, predominantly male mice. The observation of an increased kidney weights and blood biochemistry data supported the finding concerning serious kidneys dysfunction in $Ogg1$ mutant mice administered DMBDD. It has been previously suggested that $Ogg1$ plays a major role in renal tumorigenesis [35], thus the observed increase of renal tubular hyperplasia could be related to the insufficient repair of 8-OHdG in kidneys. Furthermore, significantly elevated sodium (Na), creatinine and IP level and lowered calcium (Ca) in the blood serum of DMBDD-initiated Ogg1$^{-/-}$ mice signified about the impaired kidney function or kidney disease.

It has been reported that the incidence of bladder cancer induced by BBN is significantly higher in C57BL/6 mouse strains [36]. In this study, we observed development of simple, preneoplastic nodular (PN) hyperplasia and TCC in both DMBDD-administered $Ogg1$ homozygous mutant and wild type mice. However, a trend either significant increase for PN and simple urinary bladder hyperplasia incidences was observed, indicating increased susceptibility to bladder carcinogenesis in DMBDD-treated $Ogg1^{-/-}$ male and female mice.

In DMBDD-treated $Ogg1^{-/-}$ male mice, significantly enhanced incidence and multiplicity of colon tumors as compared to the $Ogg1^{-/-}$ control has been found. However, in females, inductions of colon tumors induced by DMBDD in $Ogg1^{-/-}$ and $Ogg1^{+/+}$ animals were comparable, pointing out the sex differences in susceptibility to colonic tumorigenesis. Furthermore, in our study, an increase of carcinogenicity in the small intestine of $Ogg1^{-/-}$ male mice was also observed. Previously, the MutY homolog (MUTYH)-null mice have been reported to have a higher susceptibility to intestinal adenoma and adenocarcinoma [37]. Thus, both the MutM and MutY deficiency leading to high levels of 8-OHdG in the colonic mucosa could be responsible for the tumorigenesis in the colon and small intestine.

In the study with *mutT* homolog-1 (MTH1)-deficient mice, 18 months after birth, increases of tumorigenicity were also detected in stomachs, as compared with wild type mice [27]. These data support our results on enhancement of forestomach squamous cell HPL in DMBDD- treated $Ogg1^{-/-}$ male mice, suggesting that *mutT* and $Ogg1$ deficiency may promote carcinogenesis in the forestomach.

$Ogg1$-null mice have been reported to show an increased susceptibility to UVB-induced skin tumorigenesis [38]. They developed more malignant tumors (squamous cell carcinomas and sarcomas) than did wild-type mice. In line with these results, in the present study, we observed increase of fibrosarcoma incidence in DMBDD-treated $Ogg1^{-/-}$ male mice. Furthermore, trends for increase of incidences of malignant lymphomas/leukemias induced by the DMBDD treatment in $Ogg1^{-/-}$ mice as compared to the $Ogg1^{-/-}$ controls and DMBDD-treated $Ogg1^{+/+}$ animals were found. One DMBDD-treated female $Ogg1^{+/+}$ mouse developed T cell lymphoma, likely due to the MNU treatment, as previously reported in C57Bl/6J mice [39], but in our study no such thymic lymphomas were observed in $Ogg1^{-/-}$ mice. It is necessary to mention, that no increased risk of thyroid cancer in DMBDD-treated $Ogg1$ knockout mice was found in this study.

It is important to note that mutations in the tumor genome induced by the Ogg1 deficiency could also cause tumors to express large number of mutant tumor specific proteins (neoantigens) which have been recently demonstrated to become one of key elements for efficacy of immuno-checkpoint inhibitors as anticancer therapeutics [40].

In conclusion, this study provides the experimental evidence for a strong relationship between repair of the oxidative base modifications and multiorgan carcinogenesis. The mechanism of DMBDD carcinogenicity in the tissues of $Ogg1^{-/-}$ mice could be related to the accumulation of non-repaired oxidative DNA modifications leading to mutations and elevation of cell proliferation what likely resulted in promotion and progression of carcinogenesis. The multiorgan carcinogenesis bioassay is concluded to become an important tool to examine the effects of different factors on carcinogenicity in mice.

4. Materials and Methods

4.1. Chemicals

DEN, BBN and DMH (purity $\geq$ 98%) were purchased from Tokyo Chemical Industry Co., Ltd. (Tokyo, Japan). DHPN and MNU were purchased from Nacalai Tesque Inc. (Kyoto, Japan) and Wako Pure Chemicals Industries (Osaka, Japan), respectively. Other chemicals were from Sigma or Wako Pure Chemical Industries (Osaka, Japan).

4.2. Animals

Mmh/Ogg1 homozygous mutant ($Ogg1^{-/-}$) generated previously [13] and wild type mice ($Ogg1^{+/+}$) of C57Bl/6J background were bred and placed in an environmentally controlled room maintained at a constant temperature of 22 $\pm$ 1 °C, relative humidity of 44 $\pm$ 5% and 12 h (7:00–19:00) light/dark cycle. During all the experimental period they were given free access to drinking water and food (Oriental CE-2 pellet diet, Oriental Yeast Co., Tokyo, Japan). Mice body weights, food and water consumptions were measured weekly for the first 12 weeks of the study and subsequently once every 4 weeks. The time when the animal should be euthanized was decided due to the specific signs, such as no response to stimuli or the comatose condition, loss of body weight loss and related changes in food and water consumption, hypothermia, heart rate and external physical appearance changes, dyspnea and prostration. The experiments were performed according to the Guidelines of the Public Health Service Policy on the Humane Use and Care of Laboratory Animals and approved by the Institutional Animal Care and Use Committee of Osaka City University Graduate School of Medicine (Approval No.597; 26 November 2015).

4.3. Experimental Design

We developed the new protocol for multiorgan carcinogenicity bioassay which could be applied in mice (Figure 3). In the present study, $Ogg1^{-/-}$ (80) and $Ogg1^{+/+}$ (80) six-week-old male and female mice were randomly divided into 4 groups each comprising of 20 mice. The treatment with five genotoxic carcinogens, including DEN, MNU, BBN, DMH, and DHPN, was performed as followers: DEN at a dose of 400 ppm was administered in a drinking water for 3 days from the very beginning of the experiment. We decided to perform DEN treatment in drinking water, as in the preliminary experiment too strong toxic effect was observed with $Ogg1^{-/-}$ mice after the DEN intraperitoneal (i.p.) injection. After finishing DEN administration, four i.p. injections of MNU (20 mg/kg b.w.) were done (2 times/week), following by six subcutaneous (s.c.) injections of DMH (10 mg/kg b.w.) during Weeks 3 and 4. BBN at a dose of 0.05% was administered in drinking water for 4 weeks starting immediately after finishing the DEN treatment, and 0.1% DHPN was applied for 2 weeks in drinking water during Weeks 4 and 5. Mice in the control groups were administered saline as injections (i.p. or s.c.) or the tap water for drinking. Animals were observed every day and euthanized in case of becoming moribund during the study, or at the end of the experiment at Week 34. All surviving mice were killed under the isofluorene treatment and the DMBDD target organs, including liver, lung, kidneys, urinary bladder, small intestine and colon and thyroid gland, were immediately excised and fixed in 10% phosphate-buffered formalin, Thereafter, tissues were embedded in paraffin, sections of 4 μm in thickness were prepared and stained with hematoxylin and eosin (H & E) for the routine histology. We assessed the incidences of hyperplasia (HPL), adenoma and adenocarcinomas in the lungs, putative preneoplastic foci (PPFs), tumors, bile duct proliferation and biliary cysts in the liver, incidences of adenoma and adenocarcinoma in the small intestine and colon. Intestines were excised and intraluminally injected and fixed with 10% phosphate-buffered formalin.

4.4. Blood Biochemical Analysis

Blood was collected via the abdominal aorta from 7–9 mice per group per sex at the end of the study period after overnight fasting. Automatic analyzer (Olympus AJ-5200, Tokyo, Japan) was employed for the blood biochemical analysis to detect total protein (T-protein, g/dL), albumin/globulin ratio (A/G ratio), albumin (g/dL), total bilirubin (T-bil, mg/dL), aspartate aminotransferase (AST, IU/L), alanine aminotransferase (ALT, IU/L), γ-glutamyl transpeptidase (γ-GTP, IU/L), alkaline phosphatase (ALP, IU/L), triglycerides (TG, mg/dL), total cholesterol (T-chol, mg/dL), blood urea nitrogen (BUN, mg/dL), creatinine (mg/dL), chloride (Cl), sodium (Na), potassium (K), calcium (Ca) and inorganic phosphorus (IP) (mEq/L).

4.5. Statistical Analysis

The statistical analysis of the significance of differences between mean values was performed with the StatLight-2000(C) program (Yukms corp, Tokyo, Japan). The inter group differences detected for the incidences of histopathological findings were analyzed with the by χ^2 test Fisher's exact probability test (two-sided). Kaplan–Meier analysis was used to examine the changes in survival rates of *Ogg1* knockout and wild type mice. Homogeneity of variance between of *Ogg1$^{-/-}$* and *Ogg1$^{+/+}$* groups was detected by the F test. Student's *t*-test (two-sided) was applied in the case the data were homogeneous; otherwise, Welch test was used. *p* Values less than 0.05 were considered significant.

Supplementary Materials: Supplementary materials can be found at www.mdpi.com/1422-0067/18/8/1801/s1.

Acknowledgments: We gratefully appreciate the expert technical assistance of Masayo Inoue, Kaori Nakakubo, Rie Onodera, Keiko Sakata and Yuko Hisabayashi and the assistance of Yukiko Iura in preparation of this manuscript. This work was supported in part by a Grant-in-Aid for Scientific Research from the Ministry of Health, Labour and Welfare of Japan.

Author Contributions: Conception and design of the experiments: Anna Kakehashi, and Hideki Wanibuchi. Performance of the experiments: Anna Kakehashi. Analysis of the data: Anna Kakehashi and Naomi Ishii. Writing of the paper: Anna Kakehashi, Naomi Ishii, Takahiro Okuno, Masaki Fujioka, Min Gi.

Conflicts of Interest: The authors declare no conflict of interest.

References

1. Anisimov, V.N.; Alimova, I.N. The use of mutagenic and transgenic mice for the study of aging mechanisms and age pathology. *Adv. Gerontol.* **2001**, *7*, 72–94. [PubMed]
2. Kasai, H. Analysis of a form of oxidative DNA damage, 8-hydroxy-2′-deoxyguanosine, as a marker of cellular oxidative stress during carcinogenesis. *Mutat. Res.* **1997**, *387*, 147–163. [CrossRef]
3. Nakae, D.; Kobayashi, Y.; Akai, H.; Andoh, N.; Satoh, H.; Ohashi, K.; Tsutsumi, M.; Konishi, Y. Involvement of 8-hydroxyguanine formation in the initiation of rat liver carcinogenesis by low dose levels of N-nitrosodiethylamine. *Cancer Res.* **1997**, *57*, 1281–1287. [PubMed]
4. Klungland, A.; Rosewell, I.; Hollenbach, S.; Larsen, E.; Daly, G.; Epe, B.; Seeberg, E.; Lindahl, T.; Barnes, D.E. Accumulation of premutagenic DNA lesions in mice defective in removal of oxidative base damage. *Proc. Natl. Acad. Sci. USA* **1999**, *96*, 13300–13305. [CrossRef] [PubMed]
5. Sakumi, K.; Tominaga, Y.; Furuichi, M.; Xu, P.; Tsuzuki, T.; Sekiguchi, M.; Nakabeppu, Y. *Ogg1* knockout-associated lung tumorigenesis and its suppression by Mth1 gene disruption. *Cancer Res.* **2003**, *63*, 902–905. [PubMed]
6. Slupska, M.M.; Baikalov, C.; Luther, W.M.; Chiang, J.H.; Wei, Y.F.; Miller, J.H. Cloning and sequencing a human homolog (hMYH) of the Escherichia coli *mutY* gene whose function is required for the repair of oxidative DNA damage. *J. Bacteriol.* **1996**, *178*, 3885–3892. [CrossRef] [PubMed]

7.   Aburatani, H.; Hippo, Y.; Ishida, T.; Takashima, R.; Matsuba, C.; Kodama, T.; Takao, M.; Yasui, A.; Yamamoto, K.; Asano, M. Cloning and characterization of mammalian 8-hydroxyguanine-specific DNA glycosylase/apurinic, apyrimidinic lyase, a functional mutM homologue. *Cancer Res.* **1997**, *57*, 2151–2156. [PubMed]

8.   Arai, T.; Kelly, V.P.; Minowa, O.; Noda, T.; Nishimura, S. High accumulation of oxidative DNA damage, 8-hydroxyguanine, in *Mmh/Ogg1* deficient mice by chronic oxidative stress. *Carcinogenesis* **2002**, *23*, 2005–2010. [CrossRef] [PubMed]

9.   Radicella, J.P.; Dherin, C.; Desmaze, C.; Fox, M.S.; Boiteux, S. Cloning and characterization of *hOGG1*, a human homolog of the *OGG1* gene of *Saccharomyces cerevisiae*. *Proc. Natl. Acad. Sci. USA* **1997**, *94*, 8010–8015. [CrossRef] [PubMed]

10. Rosenquist, T.A.; Zharkov, D.O.; Grollman, A.P. Cloning and characterization of a mammalian 8-oxoguanine DNA glycosylase. *Proc. Natl. Acad. Sci. USA* **1997**, *94*, 7429–7434. [CrossRef] [PubMed]

11. Michaels, M.L.; Tchou, J.; Grollman, A.P.; Miller, J.H. A repair system for 8-Oxo-7,8-dihydrodeoxyguanine. *Biochemistry* **1992**, *31*, 10964–10968. [CrossRef] [PubMed]

12. Michaels, M.L.; Cruz, C.; Grollman, A.P.; Miller, J.H. Evidence that MutY and MutM combine to prevent mutations by an oxidatively damaged form of guanine in DNA. *Proc. Natl. Acad. Sci. USA* **1992**, *89*, 7022–7075. [CrossRef] [PubMed]

13. Minowa, O.; Arai, T.; Hirano, M.; Monden, Y.; Nakai, S.; Fukuda, M.; Itoh, M.; Takano, H.; Hippou, Y.; Aburatani, H.; et al. *Mmh/Ogg1* gene inactivation results in accumulation of 8-hydroxyguanine in mice. *Proc. Natl. Acad. Sci. USA* **2000**, *97*, 4156–4161. [CrossRef] [PubMed]

14. Kinoshita, A.; Wanibuchi, H.; Morimura, K.; Wei, M.; Nakae, D.; Arai, T.; Minowa, O.; Noda, T.; Nishimura, S.; Fukushima, S. Carcinogenicity of dimethylarsinic acid in *Ogg1*-deficient mice. *Cancer Sci.* **2007**, *98*, 803–814. [CrossRef] [PubMed]

15. Kakehashi, A.; Ishii, N.; Okuno, T.; Fujioka, M.; Gi, M.; Fukushima, S.; Wanibuchi, H. Progression of hepatic adenoma to carcinoma in *Ogg1* mutant mice induced by phenobarbital. *Oxid. Med. Cell. Longev.* **2017**, *2017*, 1–16. [CrossRef] [PubMed]

16. Arai, T.; Kelly, V.P.; Komoro, K.; Minowa, O.; Noda, T.; Nishimura, S. Cell proliferation in liver of *Mmh/Ogg1*-deficient mice enhances mutation frequency because of the presence of 8-hydroxyguanine in DNA. *Cancer Res.* **2003**, *63*, 4287–4292. [PubMed]

17. Xie, Y.; Yang, H.; Cunanan, C.; Okamoto, K.; Shibata, D.; Pan, J.; Barnes, D.E.; Lindahl, T.; McIlhatton, M.; Fishel, R.; Miller, J.H. Deficiencies in mouse *Myh* and *Ogg1* result in tumor predisposition and G to T mutations in codon 12 of the K-*ras* oncogene in lung tumors. *Cancer Res.* **2004**, *64*, 3096–3102. [CrossRef] [PubMed]

18. Takahashi, S.; Hasegawa, R.; Masui, T.; Mizoguchi, M.; Fukushima, S.; Ito, N. Establishment of multiorgan carcinogenesis bioassay using rats treated with a combination of five carcinogens. *J. Toxicol. Pathol.* **1992**, *5*, 151–156. [CrossRef]

19. Ito, N.; Hasegawa, R.; Imaida, K.; Hirose, M.; Shirai, T. Medium-term liver and multi-organ carcinogenesis bioassays for carcinogens and chemopreventive agents. *Exp. Toxicol. Pathol.* **1996**, *48*, 113–119. [CrossRef]

20. Imaida, K.; Tamano, S.; Hagiwara, A.; Fukushima, S.; Shirai, T.; Ito, N. Application of rat medium-term bioassays for detecting carcino-genic and modifying potentials of endocrine active substances. *Pure Appl. Chem.* **2003**, *75*, 2491–2495. [CrossRef]

21. Fukushima, S.; Morimura, K.; Wanibuchi, H.; Kinoshita, A.; Salim, E.I. Current and emerging challenges in toxicopathology: Carcinogenic threshold of phenobarbital and proof of arsenic carcinogenicity using rat medium-term bioassays for carcinogens. *Toxicol. Appl. Pharmacol.* **2005**, *207*, 225–229. [CrossRef] [PubMed]

22. Fukushima, S.; Hagiwara, A.; Hirose, M.; Yamaguchi, S.; Tiwawech, D.; Ito, N. Modifying effects of various chemicals on preneoplastic and neoplastic lesion development in a wide-spectrum organ carcinogenesis model using F344 rats. *Jpn. J. Cancer Res.* **1991**, *82*, 642–649. [CrossRef] [PubMed]

23. Hasegawa, R.; Furukawa, F.; Toyoda, K.; Takahashi, M.; Hayashi, Y.; Hirose, M.; Ito, N. Inhibitory effects of antioxidants on N-bis(2-hydroxypropyl)nitrosamine-induced lung carcinogenesis in rats. *Jpn. J. Cancer Res.* **1990**, *81*, 871–877. [CrossRef] [PubMed]

24. Ito, N.; Hirose, M.; Fukushima, S.; Tsuda, H.; Shirai, T.; Tatematsu, M. Studies on antioxidants: Their carcinogenic and modifying effects on chemical carcinogenesis. *Food Chem. Toxicol.* **1986**, *24*, 1071–1082. [CrossRef]

25. Shirai, T.; Masuda, A.; Imaida, K.; Ogiso, T.; Ito, N. Effects of phenobarbital and carbazole on carcinogenesis of the lung, thyroid, kidney, and bladder of rats pretreated with N-bis(2-hydroxypropyl)nitrosamine. *Jpn. J. Cancer Res.* **1988**, *79*, 460–465. [CrossRef] [PubMed]

26. Ishii, N.; Wei, M.; Kakehashi, A.; Doi, K.; Yamano, S.; Inaba, M.; Wanibuchi, H. Enhanced Urinary Bladder, Liver and Colon Carcinogenesis in Zucker Diabetic Fatty Rats in a Multiorgan Carcinogenesis Bioassay: Evidence for Mechanisms Involving Activation of PI3K Signaling and Impairment of p53 on Urinary Bladder Carcinogenesis. *J. Toxicol. Pathol.* **2011**, *24*, 25–36. [CrossRef] [PubMed]

27. Tsuzuki, T.; Egashira, A.; Igarashi, H.; Iwakuma, T.; Nakatsuru, Y.; Tominaga, Y.; Kawate, H.; Nakao, K.; Nakamura, K.; Ide, F.; et al. Spontaneous tumorigenesis in mice defective in the *MTH1* gene encoding 8-oxo-dGTPase. *Proc. Natl. Acad. Sci. USA* **2001**, *98*, 11456–11461. [CrossRef] [PubMed]

28. Hirata, A.; Tsukamoto, T.; Yamamoto, M.; Sakai, H.; Yanai, T.; Masegi, T.; Donehower, L.A.; Tatematsu, M. Organ-specific susceptibility of p53 knockout mice to N-bis(2-hydroxypropyl)nitrosamine carcinogenesis. *Cancer Lett.* **2006**, *238*, 271–283. [CrossRef] [PubMed]

29. Tsutsumi, M.; Masutani, M.; Nozaki, T.; Kusuoka, O.; Tsujiuchi, T.; Nakagama, H.; Suzuki, H.; Konishi, Y.; Sugimura, T. Increased susceptibility of poly(ADP-ribose) polymerase-1 knockout mice to nitrosamine carcinogenicity. *Carcinogenesis* **2001**, *22*, 1–3. [CrossRef] [PubMed]

30. Yoshida, M.; Miyajima, K.; Shiraki, K.; Ando, J.; Kudoh, K.; Nakae, D.; Takahashi, M.; Maekawa, A. Hepatotoxicity and consequently increased cell proliferation are associated with flumequine hepatocarcinogenesis in mice. *Cancer Lett.* **1999**, *141*, 99–107. [CrossRef]

31. Yoshida, Y.; Tatematsu, M.; Takaba, K.; Iwasaki, S.; Ito, N. Target organ specificity of cell proliferation induced by various carcinogens. *Toxicol. Pathol.* **1993**, *21*, 436–442. [CrossRef] [PubMed]

32. Chanas, S.A.; Jiang, Q.; McMahon, M.; McWalter, G.K.; McLellan, L.I.; Elcombe, C.R.; Henderson, C.J.; Wolf, C.R.; Moffat, G.J.; Itoh, K.; et al. Loss of the Nrf2 transcription factor causes a marked reduction in constitutive and inducible expression of the glutathione S-transferase Gsta1, Gsta2, Gstm1, Gstm2, Gstm3 and Gstm4 genes in the livers of male and female mice. *Biochem. J.* **2002**, *365*, 405–416. [CrossRef] [PubMed]

33. Slocum, S.L.; Kensler, T.W. Nrf2: Control of sensitivity to carcinogens. *Arch. Toxicol.* **2011**, *85*, 273–284. [CrossRef] [PubMed]

34. Klaunig, J.E.; Kamendulis, L.M.; Hocevar, B.A. Oxidative stress and oxidative damage in carcinogenesis. *Toxicol. Pathol.* **2010**, *38*, 96–109. [CrossRef] [PubMed]

35. Habib, S.L. Tuberous sclerosis complex and DNA repair. *Adv. Exp. Med. Biol.* **2010**, *685*, 84–94. [PubMed]

36. Saito, B.; Ohashi, T.; Togashi, M.; Koyanagi, T. The study of BBN induced bladder cancer in mice. Influence of Freund complete adjuvant and associated immunological reactions in mice. *Nihon Hinyokika Gakkai Zasshi* **1990**, *81*, 993–996. [PubMed]

37. Tsuzuki, T.; Nakatsu, Y.; Nakabeppu, Y. Significance of error-avoiding mechanisms for oxidative DNA damage in carcinogenesis. *Cancer Sci.* **2007**, *98*, 465–470. [CrossRef] [PubMed]

38. Kunisada, M.; Sakumi, K.; Tominaga, Y.; Budiyanto, A.; Ueda, M.; Ichihashi, M.; Nakabeppu, Y.; Nishigori, C. 8-Oxoguanine formation induced by chronic UVB exposure makes Ogg1 knockout mice susceptible to skin carcinogenesis. *Cancer Res.* **2005**, *65*, 6006–6010. [CrossRef] [PubMed]

39. Liu, J.; Xiang, Z.; Ma, X. Role of IFN regulatory factor-1 and IL-12 in immunological resistance to pathogenesis of N-methyl-N-nitrosourea-induced T lymphoma. *J. Immunol.* **2004**, *173*, 1184–1193. [CrossRef] [PubMed]

40. Yarchoan, M.; Johnson, B.A., III; Lutz, E.R.; Laheru, D.A.; Jaffee, E.M. Targeting neoantigens to augment antitumour immunity. *Nat. Rev. Cancer* **2017**, *17*, 209–222. [CrossRef] [PubMed]

International Journal of
Molecular Sciences

MDPI

Review

DNA Damage Tolerance by Eukaryotic DNA Polymerase and Primase PrimPol

Elizaveta O. Boldinova [1,†], Paulina H. Wanrooij [2,†] ⬨, Evgeniy S. Shilkin [1], Sjoerd Wanrooij [2,*] and Alena V. Makarova [1,*] ⬨

[1] Institute of Molecular Genetics of Russian Academy of Sciences, Kurchatov sq. 2, 123182 Moscow, Russia; lizaboldinova@yandex.ru (E.O.B.); shilkinevgeniy.chem@gmail.com (E.S.S.)
[2] Department of Medical Biochemistry and Biophysics, Umeå University, 901 87 Umeå, Sweden; paulina.wanrooij@umu.se
* Correspondence: sjoerd.wanrooij@umu.se (S.W.); amakarova-img@yandex.ru (A.V.M.); Tel.: +46-72-246-03-09 (S.W.); +7-499-196-00-15 (A.V.M.)
† These authors contributed equally to this work.

Received: 26 June 2017; Accepted: 16 July 2017; Published: 21 July 2017

Abstract: PrimPol is a human deoxyribonucleic acid (DNA) polymerase that also possesses primase activity and is involved in DNA damage tolerance, the prevention of genome instability and mitochondrial DNA maintenance. In this review, we focus on recent advances in biochemical and crystallographic studies of PrimPol, as well as in identification of new protein-protein interaction partners. Furthermore, we discuss the possible functions of PrimPol in both the nucleus and the mitochondria.

Keywords: PrimPol; replication; DNA damage; mitochondria

1. Introduction

Human cells contain a variety of deoxyribonucleic acid (DNA) polymerases that differ in function and fidelity. Multisubunit replicative DNA polymerases Pol δ and Pol ε possess high fidelity and play a pivotal role in the replication of genomic DNA due to the stringent requirements of their active site [1], while Pol α, which forms a complex with primase, is responsible for the initiation of DNA replication during *de novo* ribonucleic acid (RNA)-primer synthesis at the origins of replication [2,3]. Pol γ is essential for mitochondrial DNA replication and repair [4–6].

In living organisms, however, DNA is subject to damage by various endogenous and exogenous chemical and physical factors such as reactive oxygen and nitrogen species, naturally occurring ultraviolet and ionizing radiation, and reactive chemicals from environmental, food and therapeutic sources [7–11]. The replication of damaged DNA (DNA translesion synthesis (TLS)) relies on specialized DNA polymerases, also called translesion DNA polymerases. Human translesion DNA polymerases include all members of the Y-family of DNA polymerases (Pol η, Pol ι, Pol κ, Rev1) as well as the B-family DNA polymerase Pol ζ [8,12,13]. They also include some A- and X-family DNA polymerases, such as Pol ν, Pol θ, Pol β, Pol λ, Pol μ [14–17]. Translesion DNA polymerases possess unique DNA damage bypass and fidelity profiles. Lesion bypass can be error-free or error-prone depending on the type of lesion and the particular translesion DNA polymerase that is involved in synthesis. Generally, translesion polymerases possess a wide and flexible active site and/or utilize non-canonical interactions during base-pairing and can therefore efficiently incorporate nucleotides opposite the site of damage. Because of the tolerance of the active site and the lack of 3′–5′ exonuclease activity, translesion DNA polymerases often demonstrate low accuracy of DNA synthesis; consequently, error-prone lesion bypass constitutes a leading mechanism of mutagenesis in eukaryotes.

In addition to DNA damage, non-B DNA structures and collisions between the replication and transcription machineries can also lead to replication fork stalling and cause replication stress and genome instability [18–20]. Our understanding of the processes and factors that help to resolve such collisions and the mechanisms of replication through natural DNA obstacles is still far from complete.

For a long time, a primase that forms a complex with Pol α was the only known eukaryotic primase. The Pol α-primase complex consists of the DNA polymerase catalytic subunit POLA1, the regulatory subunit POLA2 and the small catalytic and large regulatory primase subunits PriS (Prim1) and PriL (Prim2), respectively [21–23].

Prim1 belongs to the archaea-eukaryotic primase (AEP) superfamily. Many members of this superfamily possess both primase and DNA polymerase activities and play an essential role not only in initiation of DNA replication, but also undertake a wide variety of cellular roles in DNA replication, damage tolerance and repair, in addition to primer synthesis [24]. In 2005, Iyer L.M. et al. *in silico* predicted the existence of a new hypothetical single subunit human primase encoded by the gene *CCDC111* on chromosome 4q35.1 [25]. The protein encoded by *CCDC111* belongs to the NCLDV-herpesvirus clade of the AEP primases. In 2012 and 2013, this new enzyme was purified and characterized as a translesion DNA polymerase with low accuracy of DNA synthesis and primase activity. It was initially presented by the L. Blanco group at several meetings in 2012 and 2013 (Sevilla 2012, Banff 2013 and others). In 2013, three groups published research articles describing the new enzyme [26–28]. The protein shares the same active site for the DNA polymerase and primase activities and was named PrimPol ("Prim"—primase, "Pol"—polymerase).

Subcellular fractionation and immunodetection studies indicated that human PrimPol is present in both the nucleus and the mitochondrial matrix in human cells. In particular, in Hela cells PrimPol is distributed between the cytosol, mitochondria and nucleus with 47%, 34% and 19%, respectively, in each respective compartment [27]. With some exceptions, homologues of human PrimPol were found in many eukaryotic unicellular and multicellular organisms, including animals, plants, fungi and protists [27,29]. However, PrimPol-related proteins were not identified in such common model organisms as *Drosophila melanogaster*, *Caenorhabditis elegans* and *Saccharomyces cerevisiae*.

2. Activities and Fidelity of PrimPol

Since 2013, the biochemical activities of human PrimPol have been extensively studied. In vitro, human PrimPol possesses properties of a translesion DNA polymerase. Like other translesion DNA polymerases, PrimPol lacks $3'$–$5'$-exonuclease activity and exhibits low fidelity of DNA synthesis. In the presence of Mn^{2+} ions as a cofactor of DNA polymerization, PrimPol makes one error per 10^2–10^5 nucleotides on undamaged DNA templates [30–33], an error rate comparative with the fidelity of the error-prone Y-family, human Pol η, Pol ι and Pol κ [13]. However, the error specificity of PrimPol uniquely differs from other human DNA polymerases. In particular, PrimPol has a preference to generate base insertions and deletions (indels) over base misincorporations [30,34]. The high rate of indel mutations potentially leads to frame-shift mutagenesis and has a deleterious effect on cells. PrimPol also preferentially incorporates non-complementary nucleotides opposite the templating bases C and G and efficiently extends from primers with terminal mismatched base pairs contributing to mutation fixation [30]. In addition to low fidelity, another feature that PrimPol shares with translesion DNA polymerases is low processivity of DNA synthesis, as it incorporates only a few nucleotides per binding event [35]. The poor processivity can be explained by the lack of contacts with DNA (see below) and the low affinity of PrimPol to DNA [33,36].

It was shown that both DNA polymerase and DNA primase activities of PrimPol are significantly stimulated by Mn^{2+} ions [27,31,33]. The stimulation of the catalytic activity can be explained by the strong increase of PrimPol's affinity to DNA in the presence of Mn^{2+} ions [31–33]. However, Mn^{2+} ions decrease the fidelity of nucleotide incorporation by PrimPol [31–33]. Previous studies demonstrated that Mn^{2+} ions alter catalytic properties of many translesion DNA polymerases and, in particular, stimulate catalysis and DNA damage bypass by human Pol ι [37,38], Pol λ [39] and

Pol µ [40]. Mn^{2+}-dependent DNA synthesis may therefore play a role in the regulation of TLS in vivo. However, the direct evidence of the requirement of Mn^{2+} ions as a physiological cofactor for translesion DNA polymerases is lacking.

As previously mentioned, human PrimPol possesses properties of a translesion DNA polymerase in vitro, and bypasses several types of DNA lesions [26,27,33,41] (Table 1). PrimPol perform efficient and quite accurate synthesis through the most common type of oxidative damage—8-oxo-G (Table 1). Interestingly, Keen B.A. et al. observed the efficient and error-free bypass of *cis-syn* cyclobutane pyrimidine dimers by the PrimPol catalytic core (residues 1–354), but did not observe bypass by full-length PrimPol protein [35]. These data suggest that the TLS activity of PrimPol may be modulated by conformational changes.

Contrasting results were obtained on DNA templates with an abasic site (AP-site). Several studies did not observe AP-site bypass by PrimPol [26,42], whereas in the study by Garcia-Gomez S. et al., PrimPol efficiently bypassed an AP-site by the lesion "skipping" mechanism (also called pseudo-TLS and the template "scrunching" mechanism) [27] (Table 1). In this latter case, PrimPol does not insert a nucleotide opposite the AP site but skips the lesion, copying the next template base available. This scenario suggests that PrimPol re-anneals the primer to the nucleotide located downstream of the lesion and loops out the templating lesion. A similar mechanism was reported for T–T (6-4) photoproducts by Mouron S. and Martinez-Jimenez M. et al. [34,41]. It is likely that this mechanism is sequence-dependent and is facilitated by flanking microhomologies [27,34,41]. The preference of PrimPol to generate base deletions on undamaged DNA is in agreement with the observation of the lesion skipping mechanism. Along with different sequence context, the differences in reaction conditions could contribute to inconsistency in TLS activities of PrimPol in experiments of different groups.

Table 1. The translesion synthesis (TLS) activity of human PrimPol.

DNA Damage		PrimPol TLS In Vitro
oxidative lesions	8-oxo-G	- bypasses 8-oxo-G incorporating dATP and dCTP with equal efficiency [26,27]; - bypasses 8-oxo-G and preferentially incorporates dC [33,42]
	TG (thymidine glycol)	- does not bypass TG [26,35]; - PrimPol $_{1-354}$ incorporates a nucleotide opposite TG but cannot extend from the lesion [35]
photo-products	*cis-syn* T–T dimers	- bypasses CPD *cis-syn* T–T dimers [41]; - does not bypass *cis-syn* T–T dimers but extends a primer terminus with two dA residues annealed opposite the T–T CPD [26]; - PrimPol$_{1-354}$ bypasses *cis-syn* T–T dimers with high efficiency and fidelity [35]
	T–T (6-4) photoproducts	- bypasses T–T (6-4) photoproducts in error-prone manner incorporating dTTP opposite 3'T and dGTP/dCTP opposite 5'T [26] or by skipping mechanism [27,34,41]
abasic sites		- does not bypass lesion [26,35,42]; - bypasses lesion with high efficiency using skipping mechanism [27,34]; - bypasses with very weak efficiency and shows nearly equal preference for either skipping the abasic site or inserting dAMP [33]
deoxyuracil		- bypasses as T and incorporates dATP opposite the lesion [35]

PrimPol also possesses an unique primase activity and is the second primase found in human cells after the Pol α-primase complex. Strikingly, unlike the Pol α-primase, PrimPol can catalyze the incorporation of deoxyribonucleotides (dNTPs) to make primers during *de novo* DNA synthesis [26,27]. The incorporation of dNTPs into the newly synthesized DNA does not require the removal of an RNA primer after the initiation of DNA synthesis.

The primase activity of PrimPol is dependent on a template T and on ATP or dATP as a starting nucleotide. On homopolymeric single-stranded DNA, PrimPol possesses primase activity only on a poly(dT) template [26]. Accordingly, on a 3'-GTCC-5' template PrimPol preferentially forms initiating dinucleotides 5'-A-dG-3', and 5'-dA-dG-3' PrimPol incorporates deoxyribo- and ribonucleotides with similar efficiency for initiation of *de novo* synthesis but prefers deoxyribonucleotides at the second position [27]. In contrast, the initiation of the primer synthesis by the eukaryotic Pol α-primase complex requires ATP or GTP on poly(dT) and poly(dC) templates, with a slight preference for an ATP substrate [43–45]. Interestingly, PrimPol possesses higher affinity towards poly(dT) as well as towards poly(dG) and G-quadruplexes [46]. The template specificity during DNA binding and initiation of synthesis may reflect a dependence of PrimPol on certain natural initiation sites and likely plays a role in the regulation of PrimPol repriming activity in vivo. Indeed, it was shown that PrimPol initiates specific repriming almost immediately downstream of the G-quadruplex structures in vitro [46].

Finally, PrimPol has been reported to be able to connect two separate oligonucleotides by an oligonucleotide end bridging mechanism. To this end, PrimPol was shown to be able to induce alignment of non-complementary oligonucleotides based on microhomologies as short as one or two base pairs, and use these aligned oligonucleotides as a template for elongating the opposite strand. Such activity would open up the possibility of a role for PrimPol in non-homologous end joining [34]. Moreover, it was suggested that PrimPol possesses the terminal transferase activity (template-independent primer extension activity) which is stimulated by Mn^{2+} ions. In this case, PrimPol extends a primer to a homopolymeric strand, which is non-complementary to the template strand [35]. However, a conventional terminal transferase activity was not observed by Martínez-Jiménez M. et al. Alternatively, it was suggested that the transferase activity of PrimPol can be a result of the connecting activity. In this scenario, PrimPol extends DNA oligonucleotides incorporating the sequence homologous to the connected nucleotide [34]. Future studies are required to determine the role of methodological differences which may affect the activities of PrimPol in vitro, verify the terminal transferase and the connecting activities of PrimPol in vitro and in vivo and elucidate their biological roles in cells.

3. Structure of PrimPol

Human PrimPol is a 560 amino acid protein. It contains an N-terminal AEP-like catalytic domain and a C-terminal zinc finger (ZnF) domain that forms contacts to the DNA template [25,27,35]. The conserved I, II, III-motifs in the AEP-like domain are required for both the DNA polymerase and primase activities (Figure 1) [26,27].

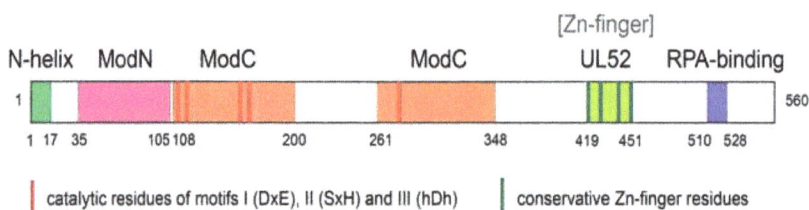

Figure 1. The schematic domain structure of human PrimPol. The N-helix, the Module N (ModN) and Module C (ModC) modules, the C-terminal zinc finger (ZnF) and replication protein A (RPA)-binding domains as well as conservative catalytic residues of I, II and III-motifs and Cys residues coordinating [Zn] are indicated.

The first crystal structure of the N-terminal catalytic core of human PrimPol (residues 1–354) in a ternary complex with a DNA template-primer, an incoming dATP and one Ca^{2+} ion was recently reported [36]. A comparison of human PrimPol with catalytic cores of other DNA polymerases and PriS provides some insight into the ability of PrimPol to function as both a polymerase and a primase.

Generally, the structures of most *DNA polymerases* resemble a *right hand*. *Y-family translesion* DNA polymerases contain the palm, fingers, thumb and little finger (or polymerase-associated domain), wherein the palm domain contains the active site, the finger domain interacts with the nascent base pair, the little finger and the thumb domains contact with the primer and the templating DNA [47].

However, the PrimPol catalytic core contains only the N-helix (an N-terminal helix, residues 1–17) and two modules called Module N (ModN) (residues 35 to 105) and Module C (ModC) (residues 108 to 200 and 261 to 348). The N-helix is connected to the ModN via a long flexible linker (residues 18–34) [36].

The ModC module encompasses functions of both the finger and palm domains and harbors key active site residues interacting with the nascent T-dATP base pair. The conserved ModC motifs I (DxE) and III (hDh) contain the acidic catalytic residues Asp114/Glu116 and Asp280, which are involved in coordination of Me^{2+} ions, while motif II (SxH) contains S167 and His169 participating in the incoming nucleotide binding. Mutations in these residues abrogate the polymerase and primase activities of human PrimPol [26–28,35]. Residues 201 to 260 inside ModC correspond to an unstructured region, which may have a regulatory role.

In PrimPol, ModN together with ModC functions as the finger domain and these modules are in contact with the template DNA strand and the templating base (T). The N-helix interacts with the template strand and resembles the little finger domain in Y-family polymerases but makes far fewer contacts in the major groove. Moreover, PrimPol does not have an analogue to the thumb domain to grip the template-primer. As a result, PrimPol demonstrates an almost complete lack of contacts to the DNA primer strand. This feature can play a key role in the primase activity of PrimPol, as the lack of contacts to the DNA primer strand eliminates the need for a pre-existing primer and leaves room for a dNTP at the initiation site during *de novo* DNA synthesis. The PrimPol catalytic core structure also differs from the structure of the primase subunit PriS because PriS has no equivalent of the N-helix. The contacts of the N-helix with the template DNA, likely, play an important role in the DNA polymerase activity of PrimPol [36].

The C-terminal domain of PrimPol contains a conserved ZnF motif (consisting of key residues C419, H426, C446, C451), which shares high sequence similarity with the viral UL52 primase domain. The Zn^{2+} ion is coordinated by the first conserved cysteine and histidine residues of the motif [26,35,41]. The presence of Zn^{2+} ions in a protein sample of the PrimPol C-terminal domain was confirmed by inductively coupled mass spectrometry [35]. The structure of the C-terminal domain of PrimPol has not been reported, but the C-terminal ZnF is indispensable for *de novo* synthesis by PrimPol while not being necessary for primer elongation [28,35]. In particular, mutations of residues C419 and H426 abrogate the primase activity of PrimPol but retain its DNA polymerase activity [28,35,41]. Nevertheless, the deletion and mutations of the ZnF modulate the processivity and fidelity of DNA synthesis by PrimPol as the presence of the ZnF reduces the processivity of the enzyme and allows a slower, higher-fidelity incorporation of complementary nucleotides [35]. The mutation of the ZnF also abrogates template-independent dNTP incorporation by PrimPol in the presence of Mn^{2+} ions [35].

The C-terminal domain is required for the binding to single-stranded DNA downstream of the primer-template junction [35] and likely is responsible for template recognition during repriming. It was also suggested that the ZnF may function as a "translocation" site to capture the initiating nucleotide triphosphate [36] analogous to the Fe–S domain in the PriS regulatory subunit PriL [21]. Moreover, the C-terminal domain was shown to be involved in protein interactions with replication protein A (RPA) (see below) [28,30,48]. Altogether, the data suggest that the C-terminal domain is a key regulator of PrimPol function. The structures of PrimPol with DNA lesions are yet to be determined. Human PrimPol replicates through photoproducts including the highly distorting T–T (6-4) lesion but, in contrast with Pol η, PrimPol has a constrained active-site cleft with respect to the templating base [36]. This suggests that PrimPol bypasses photoproducts and other bulky DNA lesions by looping the lesion out in the space between the ModN and the N-helix near the flexible linker and also explains the skipping of DNA lesions observed in some works [27,34,41].

4. Functions of PrimPol in Cells

4.1. The Role of PrimPol in Nuclear Replication and DNA Translesion Synthesis

Similar with other AEP enzymes, PrimPol possesses the versatile activities which provide a possibility to participate in different cellular processes including replication in unperturbed cells, DNA damage tolerance and, possibly, repair [24]. Studies demonstrated that PrimPol plays an important role in maintaining genome stability by protecting cells from replication stress derived from DNA damage, as well as by assisting fork progression on undamaged DNA. In accordance with these roles, PrimPol was shown to be recruited to the sites of DNA damage and stalled replication forks in the nucleus in vivo [26,28,41].

Mammalian and avian cells deficient in PrimPol display sensitivity to ultraviolet (UV) irradiation [26,28,35,41,49–51], indicating that PrimPol is important for recovery from UV damage. Because the loss of PrimPol in human xeroderma pigmentosum variant (XPV) cells lead to an increase in UV sensitivity, PrimPol's contribution to tolerance of UV photoproducts is likely to involve a pathway that is independent of Pol η [26]. Furthermore, chicken *PRIMPOL*$^{-/-}$ DT40 cells are also hypersensitive to cisplatin and methylmethane sulfonate; this effect was not epistatic to the Pol η- and Pol ζ-dependent pathways [52]. Finally, PrimPol appears not to play a major role in recovery from double-strand breaks as human *PRIMPOL*$^{-/-}$ cells showed little or no hypersensitivity to ionizing radiation [28].

Moreover, PrimPol is crucial for recovery of stalled replication forks in HeLa and DT40 cells after treatment with the dNTP depleting agent hydroxyurea and chain-terminating nucleoside analogues [28,41,52], and possibly plays an important role in replication even in unperturbed cells. In the absence of induced DNA damage, disruption of PrimPol function in mammalian cells slows down replication and induces replicative stress, thereby leading to the accumulation of DNA breaks and chromosome instability [26,41]. In particular, PrimPol may facilitate the replication across non-B DNA [46]. However, L. Wan et al. and B. Pilzecker et al. observed only modest or no effect of PrimPol defects on the replication speed during unperturbed replication in human cells [28,51].

PrimPol may restart a stalled replication fork by acting as either a translesion DNA polymerase or by repriming DNA synthesis downstream of the lesion. Biochemical studies have indicated that PrimPol is capable of TLS synthesis in vitro. However, recent in vivo studies with a zinc-finger (ZnF) primase-null PrimPol mutant in human and avian DT40 cells (see below) suggested that the primary function of PrimPol in nuclear replication is repriming at sites of DNA damage and at stalled replication forks on the leading strand. It was shown that PrimPol reprimes efficiently downstream of UV-induced DNA lesions, AP-sites and cisplatin lesions [35,41,51,52]. Moreover, it was suggested in Rad51-depleted and UV-treated cells PrimPol promotes dysregulated excessive elongation of nascent DNA by repriming after UV-induced lesions accumulated behind the replication fork. Rad51 recombinase protects the DNA synthesized before UV irradiation from degradation and prevents PrimPol-mediated repriming and excessive elongation of nascent DNA after UV irradiation [53].

It is likely that the primase activity of PrimPol also plays a pivotal role in the re-initiation of DNA synthesis after dNTP depletion by hydroxyurea and chain termination with nucleoside analogues [41,52]. It was also suggested that PrimPol contributes to replication across G-quadruplexes using close-coupled downstream repriming mechanisms on the leading DNA strand [46]. Taken together, the most likely role of PrimPol is to initiate *de novo* DNA synthesis downstream of not only DNA lesions, but also non-B DNA structures during normal chromosomal duplication.

The discovery of the repriming role of PrimPol in cells explains the previous evidence for chromosomal single-stranded gaps formed on the leading strand in S phase during replication and gap-filling during TLS in G2 phase of the cell cycle [54,55]. The mechanism of repriming is especially important on the leading strand which is replicated continuously. The remaining gaps can be filled by translesion DNA polymerases or restored by homology-directed repair. More information about repriming function of PrimPol in cells can be found in recent review [56].

However, a TLS function of PrimPol in cells cannot be completely ruled out. First, it is possible that the PrimPol ZnF mutants used as primase-deficient but DNA polymerase-proficient variants in the separation of function studies lack some posttranslational modifications or interactions with accessory proteins. Mutations of the ZnF may also affect PrimPol stability and folding. Any of these scenarios could impair the activity of the ZnF mutants and hence lead to an interpretation that overestimates the role of the primase activity. Second, several studies have evidenced a role for the polymerase activity of PrimPol in replication. In particular, Keen B.A. with co-authors suggested that in avian *PRIMPOL*$^{-/-}$ DT40 cells expressing human PrimPol, its primase activity is required to restore wild-type replication fork rates after UV irradiation, while the DNA polymerase activity of PrimPol is sufficient to maintain regular replisome progression in unperturbed cells [35]. Moreover, the African trypanosome *Trypanosoma brucei* encodes two forms of PrimPol-like proteins, PPL1 and PPL2. Both PPL1 and PPL2 proteins are translesion DNA polymerases and only the PPL1 form possesses primase activity. However, in the bloodstream form of trypanosome, PPL2 was suggested to be essential for the post-replication tolerance of DNA damage using its TLS activity, while PPL1 appears to be dispensable [29]. PPL2 was also required to complete genome replication in *T. brucei* even in the absence of external DNA damage. Therefore, it is possible that functions of PrimPol-related proteins and the regulation of their activity differ among species.

4.2. Functions of PrimPol in Mitochondria

As previously mentioned, a considerable fraction of PrimPol in cultured human cells localizes to mitochondria [27], whereby it may be available to assist Pol γ, the mitochondrial replicase, in the synthesis of mitochondrial DNA (mtDNA). In support of a mitochondrial function of PrimPol, its silencing in HEK293T cells causes a decrease in mtDNA copy number [27], and human or mouse cells lacking PrimPol show delayed recovery after transient drug-induced mtDNA depletion [27]. However, as *PRIMPOL*$^{-/-}$ knockout mice are viable, PrimPol is not essential for mtDNA maintenance [27].

A role for PrimPol in mtDNA maintenance is further supported by the fact that the DNA polymerization activity of PrimPol is affected by bona fide components of the mitochondrial replisome. Specifically, PrimPol interacts with and is inhibited by the mitochondrial single-stranded DNA-binding protein (mtSSB), most likely due to displacement of PrimPol from single-stranded DNA (ssDNA) [30]. Furthermore, PrimPol is stimulated by the mitochondrial replicative helicase Twinkle [42] and by polymerase delta-interacting protein 2 (PolDIP2) (Table 2) [50], a protein initially discovered as a Pol δ interactor [57] and that at least in some cell types appears to be localized to the mitochondrial matrix where also PrimPol is found [27,58].

Despite these functional interactions, the precise contribution of PrimPol to mtDNA maintenance has remained unclear. Intuitively, given its translesion synthesis abilities, PrimPol could be expected to synthesize past oxidative lesions created due to the high levels of reactive oxygen species that originate from the mitochondrial electron transport chain. However, PrimPol is unable to assist the mitochondrial replisome in bypassing the most common types of oxidative lesions, 8-oxo-G and abasic sites, both of which pose considerable blocks to the progression of the mitochondrial replication machinery [42]. Therefore, as has been suggested for the nucleus, the main contribution of PrimPol to mtDNA maintenance is likely to involve its ability to reprime replication. Indeed, our current unpublished data suggest PrimPol to be required for the repriming of stalled mtDNA replication after UV damage or treatment with the chain terminating nucleoside analog ddC in mouse cells, and to be able to provide primers for DNA replication by Pol γ both in vivo and in vitro [59].

Given that human mtDNA contains a number of sites that can form G-quadruplex structures [60–64], the repriming of mitochondrial replication may be required even in the absence of mtDNA damage. Because PrimPol has the ability to reprime replication immediately downstream of G4 structures in vitro [46], it may facilitate replication progression past these, and possibly other, mtDNA secondary structures. Depletion and/or deletions of mtDNA are implicated in rare genetic mitochondrial disorders, and causative gene defects include mutations in Pol γ, the Twinkle helicase and other factors involved in mtDNA maintenance [5,6]. The involvement of PrimPol in proper maintenance of mtDNA therefore opens up the possibility that PrimPol defects could give rise to mitochondrial pathologies.

5. Regulation of PrimPol Activity in Cells

To date, little is known about the mechanisms that regulate TLS and the re-initiation of DNA synthesis by PrimPol at the sites of DNA damage. Generally, translesion DNA polymerases are tightly regulated in the cell due to their high mutagenic potential. Numerous protein factors control the catalytic activity of DNA polymerases and their access to the replication fork. In nuclei, RPA and the trimeric protein-clamp proliferating cell nuclear antigen (PCNA) play a key role in the regulation of DNA polymerase activity. In mitochondria, mtSSB and helicase Twinkle are essential for DNA replication by Pol γ.

PCNA is involved in the replication process as a processivity factor. Functional interaction with PCNA facilitates access to replication fork and stimulates activity of both replicative and translesion DNA polymerases [65–68]. However, PrimPol does not interact with PCNA in vivo and is not stimulated by PCNA in vitro, suggesting that the regulation of PrimPol differs from other DNA polymerases and is independent of PCNA [30].

The main functions of RPA and mtSSB include stabilization of single-stranded DNA and positioning of proteins for the formation of DNA-protein and protein-protein complexes. RPA and mtSSB interact with PrimPol, and RPA has been shown to regulate PrimPol localization in cells in response to DNA damage and replication stress in vivo (Table 2) [28,30,48]. In particular, the deletion of the C-terminal RPA binding domain of PrimPol abrogates its interaction with RPA and nuclear foci formation after treatment of cells with *hydroxyurea*, *ionizing radiation* and UV irradiation [28,48]. Therefore, it is likely that RPA plays a key role in the recruitment of PrimPol to the stalled replication fork, and that mtSSB may play a similar role in mitochondria. Interestingly, while both RPA and mtSSB inhibit the DNA polymerase and primase activities of PrimPol on short ssDNA templates in vitro [30,42,69], RPA stimulates PrimPol activity on long M13 ssDNA templates [48,69]. The biochemical results are consistent with a model where full coating of ssDNA by RPA, such as is expected during normal replication, prevents PrimPol from accessing the DNA. In contrast, under conditions where the ssDNA is not fully coated with RPA and simultaneous binding of RPA and PrimPol is possible, RPA stimulates priming and polymerization by PrimPol. The latter scenario has been suggested to take place during replicative stress when the leading strand polymerase and the replicative helicase become uncoupled and longer stretches of ssDNA are exposed. The interaction between RPA and PrimPol is mediated by two acidic motifs in the C-terminal RPA binding domain (RBD) of PrimPol that have the ability to bind the basic cleft on the N-terminus of the RPA1 subunit in vitro [30,48]. However, in vivo the first of these RPA-binding motifs (amino acids 510–528 of PrimPol) has been suggested to be the primary mediator of the PrimPol-RPA interaction, and it is required for chromatin recruitment of PrimPol following UV irradiation [48].

Table 2. Proteins interacting with PrimPol.

• Protein-Partner	Localization of Protein in DNA Compartment	Effect on PrimPol Activity
• RPA (replication protein A)	nuclear	inhibits primase and polymerase activities on short DNA templates [30,42,69] but stimulates primase and polymerase activities on long DNA templates when non-saturating in vitro [48,69], targets PrimPol to DNA damage sites in nuclei in vivo [28,30,48]
• mtSSB (mitochondrial single-stranded DNA-binding protein)	mitochondrial	inhibits primase and polymerase activities on short DNA templates in vitro [42]
• PolDIP2 (polymerase delta-interacting protein 2)	mitochondrial and possibly nuclear	stimulates DNA polymerase activity in vitro [50]
• Twinkle	mitochondrial	stimulates DNA polymerase activity in vitro [42]

Recently, it was found that the DNA polymerase activity of PrimPol is stimulated by the mitochondrial helicase Twinkle (Table 2) [42]. *Twinkle stimulates synthesis of longer DNA replication products by PrimPol at high dNTP concentration. Interestingly, the helicase activity of Twinkle is not required for this stimulation. The stimulation by Twinkle was also not specific to damaged DNA as it was observed on undamaged DNA.*

Recently it was also shown that the DNA polymerase activity of human PrimPol is stimulated by interaction of its AEP domain with PolDIP2 (Table 2) [50]. PolDIP2 enhances the binding of PrimPol to DNA and stimulates the processivity of DNA synthesis but does not stimulate the primase activity of PrimPol [50]. The depletion of PolDIP2 in UV irradiated human cells causes a decrease in the replication fork rate, similar to that observed in *PrimPol$^{-/-}$* cells. Moreover, no further decrease in replication rate was observed when PolDIP2 was depleted in *PrimPol$^{-/-}$* cells. These data suggest that PrimPol and PolDIP2 work epistatically in the same *pathway to promote DNA replication in the presence of UV damage* [50].

Importantly, PolDIP2 also interacts with PCNA, the regulatory p50 subunit of replicative Pol δ and several translesion DNA polymerases [57,70,71]. It was shown that PolDIP2 stimulates Pol δ by increasing its affinity for PCNA binding [70]. Therefore, PolDIP2 may play a role in the coordination of replisome proteins providing docking sites for DNA polymerases and PCNA and facilitating their functional interactions. However, it was shown that PolDIP2 together with PCNA inhibits the DNA polymerase activity of PrimPol in vitro. These data argue against the "docking hypothesis" and suggests that PolDIP2 does not play a "bridging role" between PrimPol and PCNA [50].

PolDIP2 has been found both in the nucleus and mitochondria. While Klaile et al. showed that endogenous human and rat PolDIP2 localized primarily in the cytoplasmic and nuclear fractions, with lesser amounts entering the mitochondria [72], Cheng X. et al. and Xie B. et al. localized PolDIP2 almost exclusively in mitochondria [58,73]. Furthermore, it was shown that PolDIP2 is associated with the mitochondrial DNA nucleoid and it co-immunoprecipitated with mtSSB and mitochondrial transcription factor A (TFAM) [58]. Nevertheless, the role of PolDIP2 as a regulator of PrimPol function in mitochondria has been questioned by biochemical studies, because *the interaction between PolDIP2 and PrimPol is mediated by the N-terminal region of PolDIP2 that also contains the mitochondrial targeting signal that is likely cleaved off upon mitochondrial entry* [50]. The truncated form of PolDIP2 lacking the first 50 amino acids did not stimulate the DNA polymerase activity of PrimPol [50]. Therefore, it is possible that the stimulatory effects of PolDIP2 on PrimPol may be rather nuclear than mitochondrial.

Finally, to date, no proteins that act to stimulate the primase activity of PrimPol have been reported excepting for RPA [48,68]. Therefore, the mechanisms regulating repriming by PrimPol remain to be elucidated.

6. PrimPol Dysfunction and Disease

PRIMPOL$^{-/-}$ knockout mice are viable but *PRIMPOL$^{-/-}$* deficient cells show replication stress, genetic instability and defects in mitochondrial replication [27]. It is likely that PrimPol defects can be compensated in vivo by alternative mechanisms such as the use of (other) translesion DNA polymerases and helicases to bypass DNA damage and non-B structures, template switching mechanisms, homology-dependent repair and firing of dormant origins. Indeed, it was shown that avian DT40 cells deficient only in PrimPol or in Pol η/Pol ζ are viable and proliferate with nearly normal kinetics. However, cells deficient in PrimPol and Pol η/Pol ζ proliferated slowly and exhibited increased cell death. These data suggest that repriming by PrimPol and TLS by Pol η/Pol ζ are compensatory DNA damage tolerance mechanisms [52].

Nevertheless, the established roles of PrimPol in DNA damage tolerance and mitochondrial DNA replication suggest that PrimPol mutations could lead to some inherited diseases including cancer predisposition and mitochondriopathies. To date, however, little is known about the influence of PrimPol defects on human health. Recent analysis of gene expression data from The Cancer Genome Atlas (TCGA) demonstrated anti-mutagenic activity of PrimPol in genome maintenance and suggested a possible protective role of PrimPol in human breast cancer [51]. The study reported a high number of *PRIMPOL*-deficient tumors in breast cancer patients diagnosed with invasive lobular and ductal carcinoma, and the *PRIMPOL*$^{-/-}$ tumors were found to exhibit a mutation load that was nearly twice as high as in *PRIMPOL*-proficient tumors.

Only one missense *PRIMPOL* mutation, a naturally occurring minor PrimPol variant, has been reported to potentially associate with human disease [74]. This missense mutation (NM_152683.2: c.265T > G) results in the Y89D amino acid change in PrimPol and causes global alterations in polymerase domain structure and significantly decreased PrimPol affinity for both dNTPs and DNA. Being unable to interact firmly with DNA, the PrimPol^{Y89D} variant has dramatically lower processivity than wild-type PrimPol, which in turn causes a significant slowing of replication fork progression and increased UV-sensitivity in vivo [75]. Zhao et al. [74] hypothesized a link between the PrimPol^{Y89D} variant and the human ocular disease high myopia. This mutation was found in a single family and four additional sporadic patients with high myopia. Within the family, the mutation is inherited in an incomplete autosomal dominant manner, and when heterozygous, a lighter form of the disease results [74]. Nevertheless, further screening of genomes did not confirm this hypothesis as the PrimPol^{Y89D} variant was found among individuals with high myopia, other forms of genetic eye diseases and normal controls [76].

Many translesion DNA polymerases promote the resistance of cancer cells to chemotherapy [77,78]. As PrimPol is involved in tolerance to DNA damage, it can be assumed that PrimPol may play a role in the development of tolerance to some chemotherapeutic drugs inhibiting DNA replication and may therefore represent a promising drug target for the treatment of chemotherapy-resistant tumors.

7. Conclusions

Many human translesion DNA polymerases have been found and characterized in the last 15 years, and PrimPol is the most recent DNA polymerase to be identified. PrimPol is unique due to its dual activities as a translesion DNA polymerase and a primase. These activities provide extreme flexibility of DNA damage bypass as PrimPol can contribute to bypass through classical TLS, by skipping the lesion and/or by *de novo* priming of DNA synthesis downstream of the lesion. Since its discovery in 2013, extensive biochemical and structural studies have provided insight into the biochemical properties and functions of PrimPol and the mechanisms of its DNA polymerase and primase activities. Growing evidence suggests that the main biological function of human PrimPol during replication of chromosomal DNA is a repriming of stalled replication downstream of DNA damage or naturally occurring obstacles. However, the mechanisms that regulate the repriming by PrimPol in cells are yet to be understood. The functions of PrimPol in mitochondria and the association of PrimPol defects with human diseases are also largely unexplored and constitute important directions for future research.

Acknowledgments: This work was supported by the Russian Academy of Sciences Presidium Program "Molecular and Cellular Biology. New Groups" and the Russian Foundation for Basic Research (15-04-08398-a, 15-34-70002-$_{MOA}$-a-$_{MOC}$) to Alena V. Makarova, by the Wallenberg Foundation to Sjoerd Wanrooij and by postdoctoral fellowships from the Swedish Society for Medical Research and the Swedish Cancer Society to Paulina H. Wanrooij.

Conflicts of Interest: The authors declare no conflict of interest.

References

1. Lujan, S.A.; Williams, J.S.; Kunkel, T.A. DNA Polymerases Divide the Labor of Genome Replication. *Trends Cell Biol.* **2016**, *26*, 640–654. [CrossRef] [PubMed]
2. Baranovskiy, A.G.; Tahirov, T.H. Elaborated Action of the Human Primosome. *Genes* **2017**, *8*, 62. [CrossRef] [PubMed]
3. Harrington, C.; Perrino, F.W. Initiation of RNA-primed DNA synthesis in vitro by DNA polymerase α-primase. *Nucleic Acids Res.* **1995**, *23*, 1003–1009. [CrossRef] [PubMed]
4. Ropp, P.A.; Copeland, W.C. Cloning and characterization of the human mitochondrial DNA polymerase, DNA polymerase γ. *Genomics* **1996**, *36*, 449–458. [CrossRef] [PubMed]

5. Wanrooij, S.; Falkenberg, M. The human mitochondrial replication fork in health and disease. *Biochim. Biophys. Acta* **2010**, *1797*, 1378–1388. [CrossRef] [PubMed]

6. Young, M.J.; Copeland, W.C. Human mitochondrial DNA replication machinery and disease. *Curr. Opin. Genet. Dev.* **2016**, *38*, 52–62. [CrossRef] [PubMed]

7. Cadet, J.; Wagner, J.R. DNA base damage by reactive oxygen species, oxidizing agents, and UV radiation. *Cold Spring Harb. Perspect. Biol.* **2013**, *5*, a012559. [CrossRef] [PubMed]

8. Ignatov, A.V.; Bondarenko, K.A.; Makarova, A.V. Non-bulky lesions in human DNA: Ways of formations, repair and replication. *Acta Nat.* **2017**, in press.

9. Irigaray, P.; Belpomme, D. Basic properties and molecular mechanisms of exogenous chemical carcinogens. *Carcinogenesis* **2010**, *31*, 135–148. [CrossRef] [PubMed]

10. Mao, P.; Wyrick, J.J.; Roberts, S.A.; Smerdon, M.J. UV-Induced DNA Damage and Mutagenesis in Chromatin. *Photochem. Photobiol.* **2017**, *93*, 216–228. [CrossRef] [PubMed]

11. Tubbs, A.; Nussenzweig, A. Endogenous DNA Damage as a Source of Genomic Instability in Cancer. *Cell* **2017**, *168*, 644–656. [CrossRef] [PubMed]

12. Makarova, A.V.; Burgers, P.M. Eukaryotic DNA polymerase ζ. *DNA Repair* **2015**, *29*, 47–55. [CrossRef] [PubMed]

13. Vaisman, A.; Woodgate, R. Translesion DNA polymerases in eukaryotes: What makes them tick? *Crit. Rev. Biochem. Mol. Biol.* **2017**, *52*, 274–303. [CrossRef] [PubMed]

14. Beard, W.A.; Wilson, S.H. Structures of human DNA polymerases ν and θ expose their end game. *Nat. Struct. Mol. Biol.* **2015**, *22*, 273–275. [CrossRef] [PubMed]

15. Belousova, E.A.; Lavrik, O.I. DNA polymerases β and λ and their roles in cell. *DNA Repair* **2015**, *29*, 112–126. [CrossRef] [PubMed]

16. Yamtich, J.; Sweasy, J.B. DNA polymerase family X: Function, structure, and cellular roles. *Biochim. Biophys. Acta* **2010**, *1804*, 1136–1150. [CrossRef] [PubMed]

17. Yousefzadeh, M.J.; Wood, R.D. DNA polymerase POLQ and cellular defense against DNA damage. *DNA Repair* **2013**, *12*, 1–9. [CrossRef] [PubMed]

18. Boyer, A.S.; Grgurevic, S.; Cazaux, C.; Hoffmann, J.S. The human specialized DNA polymerases and non-B DNA: Vital relationships to preserve genome integrity. *J. Mol. Biol.* **2013**, *425*, 4767–4781. [CrossRef] [PubMed]

19. Helmrich, A.; Ballarino, M.; Nudler, E.; Tora, L. Transcription-replication encounters, consequences and genomic instability. *Nat. Struct. Mol. Biol.* **2013**, *20*, 412–418. [CrossRef] [PubMed]

20. Wang, G.; Vasquez, K.M. Impact of alternative DNA structures on DNA damage, DNA repair, and genetic instability. *DNA Repair* **2014**, *19*, 143–151. [CrossRef] [PubMed]

21. Baranovskiy, A.G.; Babayeva, N.D.; Zhang, Y.; Gu, J.; Suwa, Y.; Pavlov, Y.I.; Tahirov, T.H. Mechanism of Concerted RNA-DNA Primer Synthesis by the Human Primosome. *J. Biol. Chem.* **2016**, *291*, 10006–10020. [CrossRef] [PubMed]

22. Baranovskiy, A.G.; Zhang, Y.; Suwa, Y.; Babayeva, N.D.; Gu, J.; Pavlov, Y.I.; Tahirov, T.H. Crystal structure of the human primase. *J. Biol. Chem.* **2015**, *290*, 5635–5646. [CrossRef] [PubMed]

23. Kilkenny, M.L.; Longo, M.A.; Perera, R.L.; Pellegrini, L. Structures of human primase reveal design of nucleotide elongation site and mode of Pol α tethering. *Proc. Natl. Acad. Sci. USA* **2013**, *110*, 15961–15966. [CrossRef] [PubMed]

24. Guilliam, T.A.; Keen, B.A.; Brissett, N.C.; Doherty, A.J. Primase-polymerases are a functionally diverse superfamily of replication and repair enzymes. *Nucleic Acids Res.* **2015**, *43*, 6651–6664. [CrossRef] [PubMed]

25. Iyer, L.M.; Koonin, E.V.; Leipe, D.D.; Aravind, L. Origin and evolution of the archaeo-eukaryotic primase superfamily and related palm-domain proteins: Structural insights and new members. *Nucleic Acids Res.* **2005**, *33*, 3875–3896. [CrossRef] [PubMed]

26. Bianchi, J.; Rudd, S.G.; Jozwiakowski, S.K.; Bailey, L.J.; Soura, V.; Taylor, E.; Stevanovic, I.; Green, A.J.; Stracker, T.H.; Lindsay, H.D.; et al. PrimPol bypasses UV photoproducts during eukaryotic chromosomal DNA replication. *Mol. Cell* **2013**, *52*, 566–573. [CrossRef] [PubMed]

27. García-Gómez, S.; Reyes, A.; Martínez-Jiménez, M.I.; Chocrón, E.S.; Mourón, S.; Terrados, G.; Powell, C.; Salido, E.; Méndez, J.; Holt, I.J.; et al. PrimPol, an archaic primase/polymerase operating in human cells. *Mol. Cell* **2013**, *52*, 541–553. [CrossRef] [PubMed]

28. Wan, L.; Lou, J.; Xia, Y.; Su, B.; Liu, T.; Cui, J.; Sun, Y.; Lou, H.; Huang, J. hPrimpol1/CCDC111 is a human DNA primase-polymerase required for the maintenance of genome integrity. *EMBO Rep.* **2013**, *14*, 1104–11012. [CrossRef] [PubMed]

29. Rudd, S.G.; Glover, L.; Jozwiakowski, S.K.; Horn, D.; Doherty, A.J. PPL2 translesion polymerase is essential for the completion of chromosomal DNA replication in the African trypanosome. *Mol. Cell* **2013**, *52*, 554–565. [CrossRef] [PubMed]

30. Guilliam, T.A.; Jozwiakowski, S.K.; Ehlinger, A.; Barnes, R.P.; Rudd, S.G.; Bailey, L.J.; Skehel, J.M.; Eckert, K.A.; Chazin, W.J.; Doherty, A.J. Human PrimPol is a highly error-prone polymerase regulated by single-stranded DNA binding proteins. *Nucleic Acids Res.* **2015**, *43*, 1056–1068. [CrossRef] [PubMed]

31. Mislak, A.C.; Anderson, K.S. Insights into the Molecular Mechanism of Polymerization and Nucleoside Reverse Transcriptase Inhibitor Incorporation by Human PrimPol. *Antimicrob. Agents Chemother.* **2015**, *60*, 561–569. [CrossRef] [PubMed]

32. Tokarsky, E.J.; Wallenmeyer, P.C.; Phi, K.K.; Suo, Z. Significant impact of divalent metal ions on the fidelity, sugar selectivity, and drug incorporation efficiency of human PrimPol. *DNA Repair* **2017**, *49*, 51–59. [CrossRef] [PubMed]

33. Zafar, M.K.; Ketkar, A.; Lodeiro, M.F.; Cameron, C.E.; Eoff, R.L. Kinetic analysis of human PrimPol DNA polymerase activity reveals a generally error-prone enzyme capable of accurately bypassing 7,8-dihydro-8-oxo-2′-deoxyguanosine. *Biochemistry* **2014**, *53*, 6584–6594. [CrossRef] [PubMed]

34. Martínez-Jiménez, M.I.; García-Gómez, S.; Bebenek, K.; Sastre-Moreno, G.; Calvo, P.A.; Díaz-Talavera, A.; Kunkel, T.A.; Blanco, L. Alternative solutions and new scenarios for translesion DNA synthesis by human PrimPol. *DNA Repair* **2015**, *29*, 127–138. [CrossRef]

35. Keen, B.A.; Jozwiakowski, S.K.; Bailey, L.J.; Bianchi, J.; Doherty, A.J. Molecular dissection of the domain architecture and catalytic activities of human PrimPol. *Nucleic Acids Res.* **2014**, *42*, 5830–5845. [CrossRef] [PubMed]

36. Rechkoblit, O.; Gupta, Y.K.; Malik, R.; Rajashankar, K.R.; Johnson, R.E.; Prakash, L.; Prakash, S.; Aggarwal, A.K. Structure and mechanism of human PrimPol, a DNA polymerase with primase activity. *Sci. Adv.* **2016**, *2*, e1601317. [CrossRef] [PubMed]

37. Frank, E.G.; Woodgate, R. Increased catalytic activity and altered fidelity of human DNA polymerase iota in the presence of manganese. *J. Biol. Chem.* **2007**, *282*, 24689–24696. [CrossRef] [PubMed]

38. Makarova, A.V.; Ignatov, A.; Miropolskaya, N.; Kulbachinskiy, A. Roles of the active site residues and metal cofactors in noncanonical base-pairing during catalysis by human DNA polymerase iota. *DNA Repair* **2014**, *22*, 67–76. [CrossRef] [PubMed]

39. Blanca, G.; Shevelev, I.; Ramadan, K.; Villani, G.; Spadari, S.; Hubscher, U.; Maga, G. Human DNA polymerase lambda diverged in evolution from DNA polymerasebeta toward specific Mn(++) dependence: A kinetic and thermodynamic study. *Biochemistry* **2003**, *42*, 7467–7476. [CrossRef] [PubMed]

40. Martin, M.J.; Garcia-Ortiz, M.V.; Esteban, V.; Blanco, L. Ribonucleotides and manganese ions improve non-homologous end joining by human Pol μ. *Nucleic Acids Res.* **2013**, *41*, 2428–2436. [CrossRef] [PubMed]

41. Mourón, S.; Rodriguez-Acebes, S.; Martínez-Jiménez, M.I.; García-Gómez, S.; Chocrón, S.; Blanco, L.; Méndez, J. Repriming of DNA synthesis at stalled replication forks by human PrimPol. *Struct. Mol. Biol.* **2013**, *20*, 1383–1389. [CrossRef] [PubMed]

42. Stojkovič, G.; Makarova, A.V.; Wanrooij, P.H.; Forslund, J.; Burgers, P.M.; Wanrooij, S. Oxidative DNA damage stalls the human mitochondrial replisome. *Sci. Rep.* **2016**, *6*, a28942. [CrossRef] [PubMed]

43. Badaracco, G.; Valsasnini, P.; Foiani, M.; Benfante, R.; Lucchini, G.; Plevani, P. Mechanism of initiation of in vitro DNA synthesis by the immunopurified complex between yeast DNA polymerase I and DNA primase. *Eur. J. Biochem.* **1986**, *161*, 435–440. [CrossRef] [PubMed]

44. Gronostajski, R.M.; Field, J.; Hurwitz, J. Purification of a primase activity associated with DNA polymerase alpha from HeLa cells. *J. Biol. Chem.* **1984**, *259*, 9479–9486. [PubMed]

45. Tseng, B.Y.; Ahlem, C.N. DNA primase activity from human lymphocytes. Synthesis of oligoribonucleotides that prime DNA synthesis. *J. Biol. Chem.* **1982**, *257*, 7280–7283. [PubMed]

46. Schiavone, D.; Jozwiakowski, S.K.; Romanello, M.; Guilbaud, G.; Guilliam, T.A.; Bailey, L.J.; Sale, J.E.; Doherty, A.J. PrimPol is Required for Replicative Tolerance of G Quadruplexes in Vertebrate Cells. *Mol. Cell* **2016**, *61*, 16–169. [CrossRef] [PubMed]

47. Yang, W. An overview of Y-Family DNA polymerases and a case study of human DNA polymerase η. *Biochemistry* **2014**, *53*, 2793–2803. [CrossRef] [PubMed]

48. Guilliam, T.A.; Brissett, N.C.; Ehlinger, A.; Keen, B.A.; Kolesar, P.; Taylor, E.M.; Bailey, L.J.; Lindsay, H.D.; Chazin, W.J.; Doherty, A.J. Molecular basis for PrimPol recruitment to replication forks by RPA. *Nat. Commun.* **2017**, *8*, 15222. [CrossRef] [PubMed]

49. Bailey, L.J.; Bianchi, J.; Hégarat, N.; Hochegger, H.; Doherty, A.J. PrimPol-deficient cells exhibit a pronounced G2 checkpoint response following UV damage. *Cell Cycle* **2016**, *15*, 908–918. [CrossRef] [PubMed]

50. Guilliam, T.A.; Bailey, L.J.; Brissett, N.C.; Doherty, A.J. PolDIP2 interacts with human PrimPol and enhances its DNA polymerase activities. *Nucleic Acids Res.* **2016**, *44*, 3317–3329. [CrossRef] [PubMed]

51. Pilzecker, B.; Buoninfante, O.A.; Pritchard, C.; Blomberg, O.S.; Huijbers, I.J.; van den Berk, P.C.; Jacobs, H. PrimPol prevents APOBEC/AID family mediated DNA mutagenesis. *Nucleic Acids Res.* **2016**, *44*, 4734–4744. [CrossRef] [PubMed]

52. Kobayashi, K.; Guilliam, T.A.; Tsuda, M.; Yamamoto, J.; Bailey, L.J.; Iwai, S.; Takeda, S.; Doherty, A.J.; Hirota, K. Repriming by PrimPol is critical for DNA replication restart downstream of lesions and chain-terminating nucleosides. *Cell Cycle* **2016**, *15*, 1997–2008. [CrossRef] [PubMed]

53. Vallerga, M.B.; Mansilla, S.F.; Federico, M.B.; Bertolin, A.P.; Gottifredi, V. Rad51 recombinase prevents Mre11 nuclease-dependent degradation and excessive PrimPol-mediated elongation of nascent DNA after UV irradiation. *Proc. Natl. Acad. Sci. USA* **2015**, *112*, E6624–E6633. [CrossRef] [PubMed]

54. Diamant, N.; Hendel, A.; Vered, I.; Carell, T.; Reissner, T.; de Wind, N.; Geacinov, N.; Livneh, Z. DNA damage bypass operates in the S and G2 phases of the cell cycle and exhibits differential mutagenicity. *Nucleic Acids Res.* **2012**, *40*, 170–180. [CrossRef] [PubMed]

55. Elvers, I.; Johansson, F.; Groth, P.; Erixon, K.; Helleday, T. UV stalled replication forks restart by re-priming in human fibroblasts. *Nucleic Acids Res.* **2011**, *39*, 7049–7057. [CrossRef] [PubMed]

56. Guilliam, T.A.; Doherty, A.J. PrimPol-Prime Time to Reprime. *Genes* **2017**, *8*, 20. [CrossRef] [PubMed]

57. Liu, L.; Rodriguez-Belmonte, E.M.; Mazloum, N.; Xie, B.; Lee, M.Y. Identification of a novel protein, PDIP38, that interacts with the p50 subunit of DNA polymerase delta and proliferating cell nuclear antigen. *J. Biol. Chem.* **2003**, *278*, 10041–10047. [CrossRef] [PubMed]

58. Cheng, X.; Kanki, T.; Fukuoh, A.; Ohgaki, K.; Takeya, R.; Aoki, Y.; Hamasaki, N.; Kang, D. PDIP38 associates with proteins constituting the mitochondrial DNA nucleoid. *J. Biochem.* **2005**, *138*, 673–678. [CrossRef] [PubMed]

59. Torregrosa-Muñumer, R.; Forslund, J.M.E.; Goffart, S.; Stojkovic, G.; Pfeiffer, A.; Carvalho, G.; Blanco, L.; Wanrooij, S.; Pohjoismäki, J.L.O.; et al. PrimPol is required for replication re-initiation after mitochondrial DNA damage. *Proc. Natl. Acad. Sci. USA* **2017**. under review.

60. Bharti, S.K.; Sommers, J.A.; Zhou, J.; Kaplan, D.L.; Spelbrink, J.N.; Mergny, J.L.; Brosh, R.M., Jr. DNA sequences proximal to human mitochondrial DNA deletion breakpoints prevalent in human disease form G-quadruplexes, a class of DNA structures inefficiently unwound by the mitochondrial replicative Twinkle helicase. *J. Biol. Chem.* **2014**, *289*, 29975–29993. [CrossRef] [PubMed]

61. Dong, D.W.; Pereira, F.; Barrett, S.P.; Kolesar, J.E.; Cao, K.; Damas, J.; Yatsunyk, L.A.; Johnson, F.B.; Kaufman, B.A. Association of G-quadruplex forming sequences with human mtDNA deletion breakpoints. *BMC Genom.* **2014**, *15*, a677. [CrossRef] [PubMed]

62. Huang, W.C.; Tseng, T.Y.; Chen, Y.T.; Chang, C.C.; Wang, Z.F.; Wang, C.L.; Hsu, T.N.; Li, P.T.; Chen, C.T.; Lin, J.J.; et al. Direct evidence of mitochondrial G-quadruplex DNA by using fluorescent anti-cancer agents. *Nucleic Acids Res.* **2015**, *43*, 10102–10113. [CrossRef] [PubMed]

63. Wanrooij, P.H.; Uhler, J.P.; Shi, Y.; Westerlund, F.; Falkenberg, M.; Gustafsson, C.M. A hybrid G-quadruplex structure formed between RNA and DNA explains the extraordinary stability of the mitochondrial R-loop. *Nucleic Acids Res.* **2012**, *40*, 10334–10344. [CrossRef] [PubMed]

64. Wanrooij, P.H.; Uhler, J.P.; Simonsson, T.; Falkenberg, M.; Gustafsson, C.M. G-quadruplex structures in RNA stimulate mitochondrial transcription termination and primer formation. *Proc. Natl. Acad. Sci. USA* **2010**, *107*, 16072–16077. [CrossRef] [PubMed]

65. Kelman, Z. PCNA: Structure, functions and interactions. *Oncogene* **1997**, *14*, 629–640. [CrossRef] [PubMed]

66. Makarova, A.V.; Stodola, J.L.; Burgers, P.M. A four-subunit DNA polymerase ζ complex containing Pol δ accessory subunits is essential for PCNA-mediated mutagenesis. *Nucleic Acids Res.* **2012**, *40*, 11618–11626. [CrossRef] [PubMed]

67. Masuda, Y.; Kanao, R.; Kaji, K.; Ohmori, H.; Hanaoka, F.; Masutani, C. Different types of interaction between PCNA and PIP boxes contribute to distinct cellular functions of Y-family DNA polymerases. *Nucleic Acids Res.* **2015**, *43*, 7898–7910. [CrossRef] [PubMed]

68. Vidal, A.E.; Kannouche, P.; Podust, V.N.; Yang, W.; Lehmann, A.R.; Woodgate, R. Proliferating cell nuclear antigen-dependent coordination of the biological functions of human DNA polymerase iota. *J. Biol. Chem.* **2004**, *279*, 48360–48368. [CrossRef] [PubMed]

69. Martínez-Jiménez, M.I.; Lahera, A.; Blanco, L. Human PrimPol activity is enhanced by RPA. *Sci. Rep.* **2017**, *7*, a783. [CrossRef] [PubMed]

70. Maga, G.; Crespan, E.; Markkanen, E.; Imhof, R.; Furrer, A.; Villani, G.; Hübscher, U.; van Loon, B. DNA polymerase δ-interacting protein 2 is a processivity factor for DNA polymerase λ during 8-oxo-7,8-dihydroguanine bypass. *Proc. Natl. Acad. Sci. USA* **2013**, *110*, 18850–18855. [CrossRef] [PubMed]

71. Tissier, A.; Janel-Bintz, R.; Coulon, S.; Klaile, E.; Kannouche, P.; Fuchs, R.P.; Cordonnier, A.M. Crosstalk between replicative and translesional DNA polymerases: PDIP38 interacts directly with Pol eta. *DNA Repair* **2010**, *9*, 922–928. [CrossRef] [PubMed]

72. Klaile, E.; Müller, M.M.; Kannicht, C.; Otto, W.; Singer, B.B.; Reutter, W.; Obrink, B.; Lucka, L. The cell adhesion receptor carcinoembryonic antigen-related cell adhesion molecule 1 regulates nucleocytoplasmic trafficking of DNA polymerase δ-interacting protein 38. *J. Biol. Chem.* **2007**, *282*, 26629–26640. [CrossRef] [PubMed]

73. Xie, B.; Li, H.; Wang, Q.; Xie, S.; Rahmeh, A.; Dai, W.; Lee, M.Y. Further characterization of human DNA polymerase δ interacting protein 38. *J. Biol. Chem.* **2005**, *280*, 22375–22384. [CrossRef] [PubMed]

74. Zhao, F.; Wu, J.; Xue, A.; Su, Y.; Wang, X.; Lu, X.; Zhou, Z.; Qu, J.; Zhou, X. Exome sequencing reveals CCDC111 mutation associated with high myopia. *Hum. Genet.* **2013**, *132*, 913–921. [CrossRef] [PubMed]

75. Keen, B.A.; Bailey, L.J.; Jozwiakowski, S.K.; Doherty, A.J. Human PrimPol mutation associated with high myopia has a DNA replication defect. *Nucleic Acids Res.* **2014**, *42*, 12102–12111. [CrossRef] [PubMed]

76. Li, J.; Zhang, Q. PRIMPOL mutation: Functional study does not always reveal the truth. *Investig. Ophthalmol. Vis. Sci.* **2015**, *56*, 1181–1182. [CrossRef] [PubMed]

77. Roos, W.P.; Tsaalbi-Shtylik, A.; Tsaryk, R.; Güvercin, F.; de Wind, N.; Kaina, B. The translesion polymerase Rev3L in the tolerance of alkylating anticancer drugs. *Mol. Pharmacol.* **2009**, *76*, 927–934. [CrossRef] [PubMed]

78. Wu, F.; Lin, X.; Okuda, T.; Howell, S.B. DNA polymerase zeta regulates cisplatin cytotoxicity, mutagenicity, and the rate of development of cisplatin resistance. *Cancer Res.* **2004**, *64*, 8029–8035. [CrossRef] [PubMed]

International Journal of
Molecular Sciences

MDPI

Review

Exposure to Engineered Nanomaterials: Impact on DNA Repair Pathways

Neenu Singh [1,*], **Bryant C. Nelson** [2] (ID), **Leona D. Scanlan** [3,†], **Erdem Coskun** [3], **Pawel Jaruga** [3] **and Shareen H. Doak** [4,*]

[1] School of Allied Health Sciences, Faculty of Health & Life Sciences, De Montfort University, The Gateway, Leicester LE1 9BH, UK

[2] Material Measurement Laboratory, Biosystems and Biomaterials Division, National Institute of Standards and Technology, 100 Bureau Drive, Gaithersburg, MD 20899, USA; bryant.nelson@nist.gov

[3] Material Measurement Laboratory, Biomolecular Measurement Division, National Institute of Standards and Technology, 100 Bureau Drive, Gaithersburg, MD 20899, USA; scanlan.leona@gmail.com (L.D.S.); erdem.coskun@nist.gov (E.C.); pawel.jaruga@nist.gov (P.J.)

[4] Swansea University Medical School, Institute of Life Science, Centre for NanoHealth, Swansea University Medical School, Wales, SA2 8PP, UK

* Correspondence: neenu.singh@dmu.ac.uk (N.S.); s.h.doak@swansea.ac.uk (S.H.D.); Tel.: +44-116-250-6521 (N.S.); +44-179-229-5388 (S.H.D.)

† Current affiliation: Department of Pesticide Regulation, California Environmental Protection Agency, 1001 I Street, Sacramento, CA 95814, USA.

Received: 13 June 2017; Accepted: 4 July 2017; Published: 13 July 2017

Abstract: Some engineered nanomaterials (ENMs) may have the potential to cause damage to the genetic material in living systems. The mechanistic machinery functioning at the cellular/molecular level, in the form of DNA repair processes, has evolved to help circumvent DNA damage caused by exposure to a variety of foreign substances. Recent studies have contributed to our understanding of the various DNA damage repair pathways involved in the processing of DNA damage. However, the vast array of ENMs may present a relatively new challenge to the integrity of the human genome; therefore, the potential hazard posed by some ENMs necessitates the evaluation and understanding of ENM-induced DNA damage repair pathways. This review focuses on recent studies highlighting the differential regulation of DNA repair pathways, in response to a variety of ENMs, and discusses the various factors that dictate aberrant repair processes, including intracellular signalling, spatial interactions and ENM-specific responses.

Keywords: engineered nanomaterials; DNA damage; nanotoxicity; DNA repair proteins/genes; DNA repair pathways

---

1. Introduction

The unique properties of engineered nanomaterials (ENMs), intentionally manufactured particles or objects with at least one dimension in the size range of 1 nm to 100 nm (with 50% or more particles in the number size distribution), have contributed to the exponential growth and innovative advances in nanotechnology [1–3]. Monodisperse ENMs, as well as agglomerated and/or aggregated ENMs display a diverse range of magnetic, optical, electrical, catalytic and antibacterial properties which has led to their incorporation into a plethora of consumer and industrial products. Some of these products include personal care items (cosmetics, sun-creams, scratch resistant nail polishes, deodorants, toothpastes, etc.) medical devices (tissue scaffolds, drug delivery systems, orthopaedic implants, imaging modalities, biosensors, etc.) and nanoelectronics (field effect transistors, photonic crystals, field emission displays, etc.). The most commonly used ENMs include metal nanoparticles—NPs (silver, gold, cobalt, cobalt-chromium), metal oxide NPs (titanium dioxide, zinc oxide, silica, iron oxide), quantum dots (cadmium, tellurium, selenium) and carbon nanomaterials [3].

Ongoing research on evaluating the safety of ENMs has highlighted the potential of these nano-entities to cause perturbation of various cellular pathways and functional processes [2,3]. The alterations that may occur in the intracellular milieu in response to ENM-exposure can have unpredictable consequences on the functioning of the entire cellular system, as well as on the fidelity of DNA replication and cell division [4]. Cellular exposure to some ENMs has been linked to DNA damage resulting in wide ranging DNA lesions, which include genome rearrangements, single strand breaks (SSBs), double strand breaks (DSBs), intra/inter strand breaks (SBs) and the formation of modified bases (thymine glycol, 5-hydroxy-5-methylhydantoin, 8-hydroxyguanine) [5,6]. These different types of DNA lesions can lead to chromosomal aberrations, gene mutations, apoptosis, carcinogenesis or cellular senescence if left unrepaired [7] (Figure 1). Therefore, thorough studies on DNA damage response and repair related genes, pertaining to ENM exposure testing, are necessary for evaluating and characterizing the safety of ENMs.

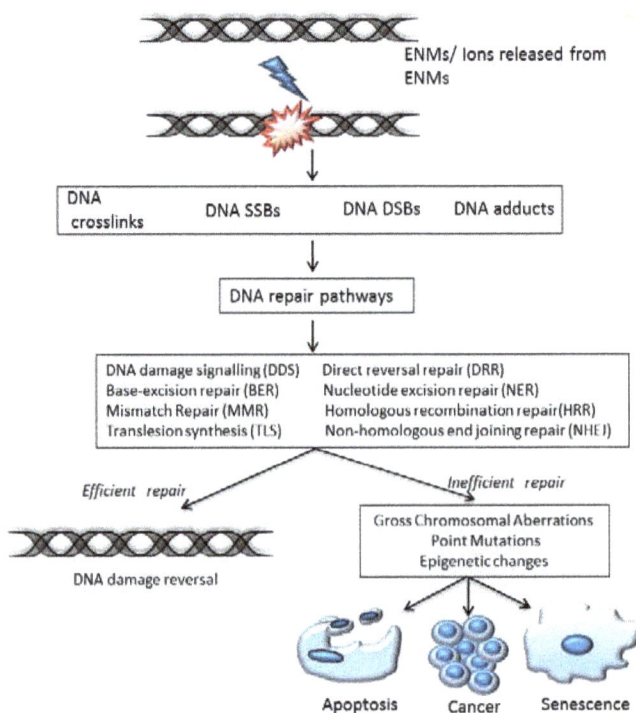

Figure 1. Schematic to illustrate various types of DNA damage caused by engineered nanomaterials (ENMs) that may result in efficient or inefficient repair activity, leading to either DNA damage reversal or progression to carcinogenesis, apoptosis and/or senescence, respectively. SSB: single strand breaks, DSB: double strand breaks.

The integrity of the genome is maintained at three levels. Level 1 involves phase I (involved in hydrolysis, oxidation etc.) and/or phase II (involved in methylation, conjugation with glutathione etc.) metabolizing enzymes that can process, inactivate or intercept the mechanistic processes that lead to ENM-induced DNA damage [8]. For example, a main antioxidant defence molecule, glutathione (along with antioxidant enzymes) binds and neutralizes reactive oxygen species (ROS) [9].

Although these defences at level 1 are generally effective, the generation of excess ROS subsequent to ENM exposure can tip the balance in favour of oxidative stress, which can lead to mutations and chromosomal aberrations [10,11]. Level 2 includes signal transduction pathways involving molecules that act as sensors for DNA damage and activate defence checkpoints. Level 3 mainly involves DNA repair processes, which play a pivotal role in maintaining the integrity of the genome by repairing DNA damage [8].

DNA repair is a complex process: it is comprised of >168 genes (that encode for proteins) involved in numerous, diverse processes encompassing intracellular signalling, cell cycle checkpoints, enzymatic reactions and chemical and structural modifications and transformations which eventually culminate in DNA repair [12]. Each pathway is represented by a set of proteins and enzymes with distinct functions and enzymatic activities. Many of the proteins implicated in DNA repair are well-defined in terms of kinetic activity, substrate specificity, mode of action and 3D structure. Therefore, knowledge about the DNA repair systems and their components is critical to our understanding of how cells control and repair the constantly occurring damage in their genomes.

There are multiple DNA repair pathways targeting various levels and extent of DNA damage. Researchers over the last few decades have progressively deciphered and revealed responses to DNA damage (including ENM-induced); currently, the DNA damage repair pathways can be divided into eight major categories:

- DNA damage signalling (DDS): this pathway is induced in response to DNA damage caused by various agents including environmental, ENM and endogenous. DDS pathways are programmed to induce several cellular responses including checkpoint activity, triggering of apoptotic pathways and DNA repair [13].
- Direct reversal repair (DRR): reverses/eliminates the DNA damage caused by chemical reversal or modification by restoring the original nucleotide. It is also known as direct DNA damage reversal.
- Base-excision repair (BER): this repair mechanism is initiated by the excision of modified bases from DNA by DNA glycosylases. The length of the DNA that needs to undergo re-synthesis can be variable; thus, the pathway can be subdivided into short-path or long-path BER. Although various pathways are involved in this repair process, one of the most widely studied mechanisms that triggers the BER pathway is oxidatively induced damage. Since oxidative stress is one of the most common mechanisms of ENM-induced DNA damage, oxidatively induced DNA lesions are predominantly repaired by the BER pathway (see Table 1). The key enzymes involved in the BER process are DNA glycosylases, which remove damaged bases by cleavage of the N-glycosylic bonds (between the bases and deoxyribose moieties) of the nucleotide residues. The DNA glycosylase action is followed by an incision step, DNA synthesis, an excision step, and DNA ligation. Various metal oxide based ENMs, quantum dots and carbon nanomaterials have been implicated in activating the BER pathway (Table 1).
- Nucleotide excision repair (NER): is involved in removing bulky DNA adducts. The damage from the active strand of transcribed DNA and DNA damage elsewhere in the genome is removed in this pathway by transcription-coupled repair and global genome repair, respectively. Silver and cadmium based ENMs have been shown to interfere with the NER pathway (Table 1).
- Mismatch repair (MMR): this pathway is involved in post-replicational DNA repair that removes errors including mismatched nucleotides, insertions, deletions, etc.
- Homologous recombination repair (HRR): this pathway involves repair of DSBs using the homologous DNA strand as a template for re-synthesis.
- Non-homologous end joining repair (NHEJ): helps to ligate the DNA ends resulting from DSBs.
- Translesion synthesis (TLS): this pathway employs specialized polymerases that use damaged DNA as templates, to finish replication across lesions. Although the mechanism is error-prone, and cell survival may be associated with an increased risk of mutagenesis/carcinogenesis, it helps to prevent a stalled replication fork.

Table 1. Summary of studies showing ENMs-induced changes in components of DNA repair pathway. The arrow indicates upregulation or downregulation of the specified molecule at either the gene/protein level or enzymatic activity, depending on the analytical technique used.

Study	Analysis Technique Applied	Cell/Tissue Used	NP	DNA Repair Pathway and Its Corresponding Component Involved					
				Homologous Recombination Repair (HRR)	Non-Homologous End Joining (NHEJ)	DNA Damage Signalling (DDS)	Base Excision Repair (BER)	Nucleotide Excision Repair (NER)	Mismatch Repair (MMR)
AshaRani et al., 2012 [14]	mRNA and array hybridisation RT-PCR	Human lung fibroblast, IMR 90	AgNPs	↓ BRCA1			↓ APEX1, MUTYH, MBD4, OGG1		↓ PMS1, MSH2
Korvuru et al., 2014 [15]	DNA repair RT2 Profiler PCR array	Liver	AgNPs	↓ RAD51/1 RAD51; ↑ RAD51C RAD52; ↑ RAD51			↓ APEX2 NEIL3 NEIL1 PARP1 NTHL1 MUTYH RPA1 XRCC1; ↑ TDG CCNO PARP2 UNG	↓ RAD23B ERCC8 XPC LIG1 RAD23A RPA1	
Asare et al., 2015 [16]	PCR	Lung tissue	AgNPs			↑ ATM			

Table 1. *Cont.*

Study	Analysis Technique Applied	Cell/Tissue Used	NP	DNA Repair Pathway and Its Corresponding Component Involved					
				Homologous Recombination Repair (HRR)	Non-Homologous End Joining (NHEJ)	DNA Damage Signalling (DDS)	Base Excision Repair (BER)	Nucleotide Excision Repair (NER)	Mismatch Repair (MMR)
Satapathy et al., 2014 [17]	In Vivo Base Excision Repair (BER) Assay	Oral squamous cell carcinoma	QAgNPs				↓ LIG1 FEN1 POLB POLD1 POLE		
Van Berlo et al., 2010 [18]	mRNA expression	Lung tissue	Carbon				↑ OGG1 APEX1		
Tang et al., 2013 [19]	RT-PCR	Daphnia pulex	CdSO4 or CdTeQDs		↑ Ku80		↑ OGG1	↑ XPC XPA	
Tang et al., 2015 [20]	RT-PCR	Daphnia pulex	CdTe/ZnS				↑ OGG1	↑ XPA XPC	
Ahamed et al., 2010 [21]	Western blotting	Human pulmonary epithelial cells (A549)	CuO	↑ RAD51					↑ MSH2
Khatri et al., 2013 [22]	RT-PCR	THP-1, Primary human nasal, Small airway epithelial	ENMs emitted from photocopiers	↑ RAD51	↑ Ku70				
Prasad et al., 2013 [13]	Western blot (phosphorylation)	Human dermal fibroblasts	TiO2			Activation of ATM/Chk2 DNA damage signalling pathway			
El-said et al., 2014 [23]	RT-PCR	HepG2	TiO2			↑ ATM	↑ APEX1 MBD4		
Hanot-Roy et al., 2014 [24]	Western blot (phosphorylation)	Alveolar macrophages (THP-1), Epithelial cells (A549), Human Pulmonary Endothelial Cells (HPMEC-ST1.6R cells)	TiO2			↑ ATM ATR			
Pati et al., 2016 [25]	Western blot	Macrophages	Zinc oxide nanoparticles (ZnO-NPs)				↓ POLB FEN1		

Int. J. Mol. Sci. **2017**, *18*, 1515

2. Activation/Up-Regulation of DNA Damage Signalling Pathways

DNA repair pathways/proteins seldom work in isolation in the cell, i.e., the repair pathways are interdependent and interconnected via shared proteins and components of the DNA repair system. More than one pathway may be up-regulated/down-regulated in response to cellular exposure by a given ENM. Moreover, repair pathways (genes/proteins/enzymes) induced because of DNA damage do not follow a similar trend with respect to being up-regulated or down-regulated; different studies on ENM exposure have shown varied DNA damage responses, i.e., the same gene/protein is up-regulated in one study, while being down-regulated in another ENM-exposure study. For example, several studies have shown apurinic/apyrimidinic endonuclease (APEX), involved in the BER pathway to be either up-regulated or down-regulated in response to exposure by ENMs (Table 2). This lack of a trend may be due to differences in physico-chemical characteristics (e.g., composition, size, structure, charge, morphology, coating, presence of impurities due to synthesis processes) in the tested ENMs or to differences in exposure concentrations, cell lines utilized and other experimental factors.

Table 2. Function of important enzymes/proteins involved in the major DNA repair pathways.

Enzyme/ Protein	Function
DDS Pathway	
ATM (ataxia-telangiectasia mutated)	Cell cycle checkpoint kinase protein, which belongs to the PI3/PI4-kinase family. Serves as a DNA damage sensor and regulator of a wide variety of downstream proteins, including, 1) Tumour suppressor protein p53 and 2) Serine/threonine protein kinase that activates checkpoint signalling upon double strand breaks (DSBs), apoptosis, and genotoxic stresses.
ATR Rad3-related kinase	PI3 kinase-related kinase family member (like ATM), which phosphorylates multiple substrates on serine/ threonine residues (that are followed by a glutamine) in response to DNA damage or replication blocks. Causes cell cycle delay, in part, by phosphorylating checkpoint kinase (CHK)1, CHK2, and p53.
CHK1 and CHK2 (Checkpoint kinase 1 and 2)	Downstream protein kinases of ATM/ATR, which play an important role in DNA damage checkpoint control.
BER Pathway	
APEX1 (Apurinic/apyrimidinic endonuclease 1)	Multifunctional DNA repair enzyme, apurinic/apyrimidinic endonuclease 1/redox factor-1 (APE1/Ref-1) responsible for abasic site cleavage activity. Plays a critical role in the DNA base excision repair (BER) pathway and in the redox regulation of transcriptional factors. Activated/ induced by oxidative DNA damage. Localisation signals, post-translational modifications and dynamic regulation determines the localisation of APE protein in the nucleus with subcellular localization in the mitochondria, endoplasmic reticulum and cytoplasm.
APEX2 (Apurinic/apyrimidinic endonuclease 2)	AP endonuclease 2 is characterized by a weak AP endonuclease activity, 3'-phosphodiesterase activity and 3'- to 5'-exonuclease activity. Involved in removal of mismatched 3'-nucleotides from DNA and ATR-Chk1 checkpoint signalling in response to oxidative stress.
(POLB) DNA polymerase β	Contributes to DNA synthesis and deoxyribose-phosphate removal.
(FEN1) Flap endonuclease 1	Possesses 5'–3' exonuclease activity and cleaves 5' overhanging "flap structures" in DNA replication and repair.
LIG1 (Ligase 1)	Seals SSB ends.
MBD4 (methyl-CpG binding domain protein 4)	Belongs to a family of nuclear proteins that possess a methyl-CpG binding domain (MBD). These proteins bind specifically to methylated DNA, possess DNA N-glycosylase activity and can remove uracil or 5-fluorouracil in G:U mismatches.

Table 2. *Cont.*

Enzyme/ Protein	Function
MUTYH (mutY DNA glycosylase)	Serves as DNA glycosylase (excises adenine mispaired with 8-oxoguanine). Maintains chromosome stability by inducing ATR-mediated checkpoint activation, cell cycle arrest and apoptosis.
NEIL1, NEIL3 (Nei-like 1; Nei-like 3)	Generate apurinic/apyrimidinic (AP) sites and/or SSBs with blocked ends.
NTHL1	Serve as oxidized base-specific DNA glycosylases that remove oxidized and/or mismatched DNA bases.
OGG1 (8-oxoguanine DNA glycosylase)	Excises and repairs oxidatively damaged guanine bases in DNA, which occur as a result of exposure to ROS.
PCNA (Proliferating cell nuclear antigen)	Co-factor for DNA polymerase and essential for DNA synthesis and repair.
PARP1 (Poly ADP ribose polymerase)	PARP1—serves as sensor of SSBs.
XRCC1 (X-ray repair cross-complementing protein 1)	XRCC1—serves as a scaffold for recruiting and activating BER proteins.
NER Pathway	
RPA1 (replication protein A1)	Largest subunit of the replication protein A (RPA), the heterotrimeric single-stranded DNA-binding protein involved in replication, repair, recombination and DNA damage check point activation.
XPC (xeroderma pigmentosum group C protein)	Recognizes bulky DNA adducts. Pairs up with RAD23 and helps in the assembly of the other core proteins involved in NER pathway progression.
XPA (xeroderma pigmentosum group A protein)	Attaches to damaged DNA, interacts along with other proteins in the NER pathway to unwind, excise and replace the damaged DNA.
HRR Pathway	
BRCA1/ BRCA2 (breast cancer type 1 and type 2 susceptibility proteins)	BRCA1 and BRCA2 are coded by human tumour suppressor genes that are involved in DNA damage repair, cell cycle progression, transcription, ubiquitination and apoptosis. Aberrant proteins coded by mutated genes are found in hereditary breast and ovarian cancers; activation of various kinases in response to DNA-damage have been shown to phosphorylate sites on BRC1 and BRC2 in a cell cycle-dependent manner.
RAD51	Involved in the homologous recombination and repair of double strand DNA breaks.
NHEJ Pathway	
Ku	Ku, a heterodimer of two related proteins, Ku70 and Ku80, is involved in DSB repair and V(D)J recombination.
LIG4 (Ligase 4)	LIG4 is the DNA ligase required for, and specific to, c-NHEJ. It catalyzes the same ATP-dependent transfer of phosphate bonds that results in strand ligation in all eukaryotic DNA repair. LIG4 is the only ligase with the mechanistic flexibility to ligate one strand independently of another or to ligate incompatible DSB ends as well as gaps of several nucleotides.
XRCC4 (X-ray repair cross-complementing protein 1)	XRCC4 is a non-enzymatic protein that is required for the conformational stability and functioning levels of LIG4. XRCC4 interacts with LIG4 facilitated by carboxy-terminal repeats at the LIG4 carboxyl terminus, resulting in a coiled-coil like conformation. Most of the enzymatic domain of LIG binds to and interacts with XRCC4, except for the small region implicated in DNA binding.

3. Up-Regulation of DNA Repair Genes

Prasad et al. recently showed that titanium dioxide NPs (TiO$_2$ NPs) induce the activation of serine/threonine kinase ATM/Chk2, involved in the DDS signalling pathway [13] (see Table 1). The study showed that TiO$_2$ NPs behave like ionizing radiation (IR), a well-known trigger for both the ATM/Chk2 pathway and the intra-S-phase DNA-damage checkpoint response [13,26]. TiO$_2$ NPs-induced increased expression (>1.5-fold) of ATM in hepatocellular carcinoma cells (HepG2), which is consistent with the induction of DSBs, chromatin condensation, nuclear fragmentation and apoptosis due to increased ROS production and subsequent DNA damage [23]. This enhanced ROS generation, which correlated with toll-like receptor 4 (TLR4) over-expression (vs. TLR3 over-expression, which protects against ROS-induced DNA damage) activated caspase-3 and oxidative stress-induced apoptosis. Similar increases in the expression of ATM (but decreased expression of ATR) with a corresponding increase in DNA damage (as indicated by micronucleus frequency), are attributed to increased ROS generation and oxidative stress; the experimental evidence suggested that DSBs were involved [27]. Therefore, up-regulation of the ATM protein is associated with its role as a DNA damage sensor that activates checkpoint signalling events subsequent to DSBs induced by TiO$_2$ NPs.

Apart from DDS, various other pathways have been shown to be activated by some ENMs. Tang et al., have demonstrated similar up-regulation of certain DNA repair genes in zebrafish liver cells including *Ku80* (NHEJ), *OGG1* (BER), *XPC* (NER) and *XPA* (NER) using cytotoxic concentrations of CdSO$_4$ salt or CdTe quantum dots (QDs) [19]. A QD exposure study by the same group tested cadmium selenide/zinc sulfide (CdTe/ZnS) core-shell QDs on the fresh water crustacean *Daphnia pulex*, and showed significant increases in OGG1 levels in response to CdTe QDs, but not for the CdTe/ZnS QD exposure. Genes involved in the NER pathway, namely *XPA* and *XPC*, showed significant up-regulation in response to treatment with both CdTe and CdSe/ZnS QDs. In addition to the BER and NER pathways, the MMR pathway was also affected in response to CdTe and CdSe/ZnS QDs [20]. Therefore, subsequent to exposure by a particular ENM, various genes of the repair pathways can be triggered, which work in conjunction with other proteins/enzymes/co-factors to eliminate the DNA damage. For example, copper oxide (CuO) NPs were shown to upregulate the expression of two DNA damage repair proteins RAD51 (HRR) and MSH2 (MMR) in lung epithelial cells, while up-regulation of *RAD51* along with increased levels of *Ku70* (implicated in NHEJ pathway) was observed in in THP-1 cells exposed to photocopier-emitted NPs [21,22].

Interestingly, some studies have also shown tissue-specific up-regulation of DNA damage genes/proteins in response to ENM exposure. Van Berlo et al. observed increased levels of DNA damage response genes *OGG1* and *APE1* in C57BL/6J mice exposed to carbon NPs, in a short-term inhalation study [18]. However, elevated mRNA levels of the two genes were seen in lung tissue, while the olfactory bulb cerebellum and other parts of the mice brain were not affected. Nevertheless, long-term studies are needed to evaluate any adverse effects on the brain, particularly with respect to other and perhaps more toxic ENMs, which may be released into the environment.

Similarly, a tissue-specific response was seen in a TiO$_2$ NP exposure study that used three model cell lines representing an alveolar-capillary barrier. The cell system consisted of alveolar macrophage-like THP-1 cells, alveolar epithelial A549 cells and human pulmonary microvascular endothelial, HPMEC-ST1.6R, cells [24]. Following exposure to the test ENM, significant levels of ROS were generated in all three cell lines. Differentiated THP-1 macrophages showed increased phosphorylation of ATR and ATM with increasing concentrations of TiO$_2$-NPs, (200 to 800 µg/mL). This correlated with increased phosphorylation of H2AX histone (γH2AX) revealing a link between deleterious DNA lesions and activation of the DNA damage repair pathway [24]. On the other hand, in HPMEC-ST1.6R cells, phosphorylation of H2AX histones did not correlate with activation of ATR or ATM proteins. However, an increased phosphorylation of p53 and checkpoint protein CHK1A was observed to correlate with cell cycle arrest. Interestingly, the A549 cell line showed no activation of signalling pathways related to DNA damage. This study thus sheds light on the differential profile of tissue specific DNA repair responses generated by the three cell lines under investigation, with only THP-1 and HPMEC-ST1.6R cells showing apoptosis, sensitivity to redox changes and concomitant activation of DNA damage and repair proteins.

4. Inactivation/Downregulation of DNA Repair Pathway Genes

Any causative agent that results in DNA damage may be anticipated to bear a positive correlation to an increased repair capacity, as described in studies in the previous section. This is possible by induction of DNA damage pathway genes that ensure high fidelity DNA synthesis to rectify the observed damage in order to maintain genome stability. However, a number of genes that participate in DNA damage repair processes and induction of cell cycle checkpoints are either up-regulated (e.g., *RAD9*, *PARP1*) or down-regulated/mutated (e.g., *BRCA1/2*, *ATM* and *TP53*); diminished repair capacity has also been associated with carcinogenesis [7]. Many ENM exposure studies have shown downregulation in the expression/activity of key candidate genes/proteins/enzymes involved in DNA repair pathways (Table 2).

Such downregulation of genes involved in the BER pathway has been shown by Kovvuru et al. following exposure to polyvinylpyrrolidone (PVP) coated AgNPs; the genes involved mediate and contribute to the observed oxidatively induced DNA damage, DSBs, and chromosomal damage in peripheral blood and bone marrow [15]. The study provided evidence that some of the BER pathway genes, which play a pivotal role in the repair of oxidatively induced DNA damage, were down-regulated—these genes include *NEIL1*, *NEIL3*, *NTHL1*, *MUTHY*, *APEX2*, *RPA1*, *XRCC1*, *PARP1* and *LIG1* (Table 2). The genes and proteins implicated in the BER pathway, are collectively involved in (1) recognition and base excision of ENM-induced DNA damage (2) repair, intermediate processing, synthesis and (3) nick sealing or ligation.

A correlation between chromosomal damage and impairment of repair pathways was also established using mice deficient in a BER pathway protein MutY homologue (MUTYH). These animal models were observed to be hypersensitive to PVP-coated AgNPs and resulted in increased micronuclei frequency indicating chromosomal damage [15]. MUTYH knock-down is also associated with decreased ATR, CHK1 and CHK2 phosphorylation induced by hydroxyurea, ultraviolet light and topoisomerase II inhibitor treatment [28,29]. This downregulation correlated with decreased apoptosis and reduced activation of ATR, which regulates cell cycle arrest and apoptosis [29]. MUTYH has been described as a trigger for cell death pathways in cells that have accumulated DNA lesions and SSBs [30], thus protecting the cells from permanent genome alterations in the form of damaged DNA. The study on AgNPs sheds light on the importance of DNA damage checkpoint activation as any defects in the execution of apoptosis may impact genomic integrity; it also highlights the interplay of different repair pathways that work in co-ordination to maintain stability in the genome. Other than downregulating key genes related to the BER pathway, genes implicated in other pathways were also down-regulated or up-regulated in response to treatment by PVP-coated AgNPs (Table 1).

The downregulation of DNA repair proteins/genes and its link to ROS induced DNA damage has been studied in depth by Pati et al. [25]. The authors observed a decreased expression of two DNA repair proteins that play an important role in the BER pathway, POLB (DNA polymerase β) and FEN1 (Flap endonuclease 1), in zinc oxide (ZnO) NP-treated macrophages. This correlated with ZnO NP-induced micronuclei frequency, chromosomal disintegration, cellular protrusions, cytotoxicity, reduced cell migration, phosphorylation of the H2A histone family and modulation in actin polymerization in peripheral blood macrophages and bone marrow cells. Interestingly, treatment with N-acetyl cysteine (NAC) after ZnO NP exposure alleviated the observed genotoxicity and clastogenicity and up-regulated the expression of DNA repair genes. This finding may indicate that ZnO NPs exhibit genotoxic, clastogenic, cytotoxic and actin depolymerization effects via generation of ROS in macrophages.

In a recent study comparing the responses of normal cells vs. cancer cells following exposure to AgNPs, both inhibition and activation of DNA repair responses were demonstrated. A comparison of normal lung fibroblasts (IMR-90) cells vs. U251 glioblastoma cancer cells revealed increased ATM and ATR levels, which activate downstream targets for DNA repair in response to DSB [14]. The gene expression results were associated with increased γ-H2AX foci in U251 cells as compared to IMR-90 cells, suggesting that the cancer cells were more prone to the accumulation of DSBs in response to AgNP exposure. The authors showed down-regulation of BER pathway genes (*MBD4*, *PCNA*, *APEX1*, *OGG1* and *MUTYH*) and MMR pathway genes (*PMS1* and *MSH2*) in the IMR-90 cell line, indicating differential gene expression in normal vs. cancer cells [14]. Additionally, the study showed downregulation of *BRCA1* and MMR-dependent *ABL1* gene expression in IMR-90 cells in comparison to the increased expression of *BRCA1* and *ABL1* in glioblastoma cells; indicating involvement of the HRR and MMR pathways. However, the molecular mechanisms of these alterations and regulation have yet to be elucidated.

A relatively new method (multiplexed excision/synthesis assay) developed by Millau et al., which evaluates repair capacity by both the NER and BER pathways was utilised to evaluate the repair ability of A549 cells exposed to TiO_2 NPs [31]. The TiO_2 NPs inactivated both the NER and BER pathways, thus impairing the cells' ability to repair damage to DNA. The induction of DNA damage coupled with the compromised repair ability suggest a synergistic response to TiO_2 NPs exposure that could potentially lead to mutagenesis or carcinogenesis. However, further detailed studies are necessary to corroborate the results.

Although inefficient/down-regulated repair systems could have unfavourable consequences on various cellular functions, unrepaired DNA damage can have positive implications in cancer therapy. An interesting study based on hybrid NPs composed of bioactive quinacrine (Q, an ideal antimalarial agent) and AgNPs (QAgNPs) demonstrated the capacity to inhibit the BER pathway and subsequently induce apoptosis in cancer cells, thus offering potential new directions in anticancer treatment for oral squamous cell carcinoma [17]. The inherent high DNA repair efficiency in cancer stem cells (CSCs) poses a challenge for chemotherapeutic drug treatment, as CSCs avoid DNA damage-mediated apoptosis during cancer therapy. However, treatment with QAgNPs showed significant reduction of LIG1, FEN1 and POLB, POLD1 (DNA polymerase δ) and POLE (DNA polymerases ε), showing the involvement of the BER pathway, while PRKDC (protein kinase, DNA-activated, catalytic polypeptide) and RAD51, components of NHEJ and HRR respectively, remained unaltered. These results corresponded to a >5-fold increase in the expression of γ-H2AX, indicating sustained DNA damage and formation of DSBs, which is associated with compromised BER activity within the cells and resultant apoptosis. Interestingly, no significant changes were observed when the cells were treated with either quinacrine (NQC) or AgNPs alone.

Steric hindrance is another important mechanism that governs the decreases in DNA repair protein activity. An interesting study on carbon ENMs (cENMs) revealed that these ENMs could physically perturb the incision activity of APEX1 [32]. The study showed that the cENMs were susceptible to adsorption onto custom synthesized DNA oligonucleotides with abasic sites. This study in a cell-free system effectively demonstrated that APE1's accessibility to the DNA lesions (abasic sites) could be sterically hindered, thereby inhibiting its enzymatic incision activity. This conformational alteration, resulting in stalled repair activity, could potentially lead to mutagenesis through impaired DNA repair processes in a cellular environment.

Release of ions governs repair capacity: co-exposure studies using Cd^{2+} or CdTe QDs and benzo[*a*]pyrene-7,8-diol-9,10-epoxide, (BPDE—a toxic compound that interacts covalently with DNA base pairs and forms adducts, thus leading to DNA damage and carcinogenesis) on zebrafish liver cells revealed that exposure to soluble Cd^{2+} (also released from CdTe QDs) significantly reduced NER repair capacity. This observation was attributed to Cd^{2+} interacting with DNA and hindering interactions, that lead to BPDE adduct formation, cellular uptake, intracellular distribution and possibly metabolism [19]. Indeed, Cd^{2+}-inhibition of DNA repair of BPDE-induced DNA lesions has been studied previously [33,34]. Therefore, in addition to affecting the BER pathway, Cd has also been implicated in the interference of the NER pathway and suggests that the process of adduct formation is influenced by Cd^{2+} in the nucleus [35].

Apart from Cd ions, studies have shown that Fe ions can inhibit APE1's enzymatic activity [36]. McNeill et al. demonstrated that Fe ions could prevent APE1 from excising a single, centrally located abasic site in a 26-mer oligonucleotide duplex [36]. To further assess the specificity of Fe dependent inhibition, and to better understand the effects in a physiological environment (that contains a whole array of different proteins), the study also investigated the impact of Fe ions on whole cell extracts. Similar results obtained on protein extracts provided substantiating evidence that Fe ions demonstrate specific inhibition of APE1 incision activity. Other studies have shown downregulation of APE1 in response to exposure by ENMs including AgNPs and TiO_2 NPs [13,14]. Moreover, Jugan et al. has also shown impaired cellular DNA repair in lung epithelial cells in response to TiO_2 NPs. The authors showed inhibition of both BER and NER pathway, as evaluated using a DNA repair microarray assay [37].

5. Conclusions & Future Considerations

The evolving development of ENMs has enabled innovative breakthroughs in a variety of sectors such as healthcare, manufacturing, agriculture and transportation [38,39]. The potential for human exposure to ENMs has increased what warrants a careful and thorough risk assessment of ENM interactions at the cellular level. Recent reports have shown that some ENMs may affect the fidelity of DNA repair processes/pathways. DNA damage occurs continuously as a result of aerobic respiration; thus, DNA repair pathways play an important role in maintaining genomic integrity. Potential perturbations in these pathways may result in biological alterations at the cellular/tissue/organism level.

Besides the adverse impact on genome integrity, it is worth mentioning that any defects in DNA repair genes involving aberrant regulation may impact a range of physiological functions and cellular processes. This is because the DNA repair pathway genes/proteins are also involved in apoptosis, transcriptional regulation, migration, telomere maintenance, chromatin remodelling and dNTP synthesis [7]. Altered DNA repair activity is further associated with tumour initiation and progression, and can result in increased resistance to DNA damage therapy.

Given the importance of DNA repair, it is vital to understand the role it plays in the genotoxic responses observed, and its specific impact on molecular interactions necessary for maintaining genomic stability. This review of the current literature demonstrates that there are several different factors that control DNA damage-induced mechanistic repair pathways. Important aspects to consider in the future, in order to more fully understand the potential impact of ENMs on DNA repair fidelity include, but are not limited to:

1. Characterisation of induced DNA damage lesions: a given ENM may have a primary mechanism for the induction of DNA damage, which triggers the initiation of a specific repair pathway. For example, metal and metal oxide based ENMs tend to cause oxidatively induced DNA damage, which is mainly repaired via the BER pathway. Therefore, characterising the type of DNA damage is critically important in future studies, as it will enable predictive models to be developed that can be used to predict which types of ENMs might affect specific DNA repair pathways.

2. Role of ions: inorganic NPs could via corrosion and dissolution release metal ions such as Cd^{2+}, Fe^{3+}, Zn^{2+}, and Ag^+ and hence influence the upregulation/downregulation (measured as excision activity) of various pathways. Additionally, metal ions released from ENMs have been shown to interact/bind with protein domains and amino acids of DNA repair proteins (e.g., zinc finger structures contained in the DNA repair protein, XPA) resulting in distorted protein structure and inefficient DNA repair activity [40]. Therefore, a thorough physicochemical characterisation of ENMs is imperative, to discriminate between the actual causative factor (ENM vs. metal ions), as the impact of ENMs on DNA repair pathway may be strongly associated with the presence of metal(s) either in their composition, or as undesirable impurities.

3. Dose-dependent DNA damage response: presently, the doses of ENMs administered in in vitro studies/test species to generate dose-response analysis may not mimic a potential human exposure level. This is because concentration-dependent activation of genes/pathways as well as transition in gene changes can be highly dose dependent. Therefore, dose ranges that are relevant

to true exposure levels of ENMs need to be included when studying DNA damage responses pertaining to repair pathways. However, ENM exposure assessment currently presents a technical challenge and more work is needed to evaluate emissions of ENMs into the environment [41]. For example, it will be necessary to perform more thorough background measurements at workplaces to determine accurate occupational exposure levels, to develop appropriate metrics for ENM exposure assessments and to validate personal air samplers.

4. Method/technique: various techniques and methods with different endpoints are utilized for evaluating DNA damage repair and/or DNA damage responses, e.g., Western blots for translational changes/modifications and/or phosphorylation events; RT-PCR for transcriptional alterations; excision or incision assays for DNA repair enzyme activity; mass spectrometry methods for measuring adduct or lesion formation and multiplexed excision/synthesis assays for DNA repair enzyme inhibition activity [31]. Each method has its own sensitivity, specificity and endpoints, which makes it challenging to compare results across different studies. Hence, to enable an appropriate intra-laboratory/interlaboratory comparison of DNA damage repair responses, statistically appropriate analysis on normalised data must be performed in order to identify reproducible upregulation or downregulation of ENM-induced DNA repair responses.

5. Tissue specific detection /expression: different tissues and cell types (including primary cells, normal/cancer cell lines) exhibit varying DNA repair responses, which may correlate with the degree of DNA damage and susceptibility following exposure to some ENMs. Hence, it is imperative to measure the levels and activity of DNA repair genes and proteins, respectively, in all relevant cells, tissues or organs of interest as their expression and responses are largely "site-specific".

6. Effect of acute vs. chronic exposure of ENMs: the type of exposure may affect the severity of the DNA damage and the resultant activation of specific DNA repair pathway(s). The human population may be exposed to natural, environmental or ENMs in a cumulative manner [42]. On the other hand, occupational, lifestyle or behaviour-related exposure to various nano-entities may induce acute responses [8,43]. Therefore, it is important to understand how various kinds of exposure scenarios dictate not only DNA damage, but also trigger specific repair pathways.

7. Effect of potential ENM artefacts on the interpretation of DNA damage repair or DNA damage response: as described in previous reports, the solution state physico-chemical properties of ENMs are not like the solution state physicochemical properties of chemicals [44]. Depending upon the category of ENMs under investigation, ENMs are prone to disparate rates of dissolution, aggregation/agglomeration phenomena, nutrient depletion and other behaviours that can potentially result in false-positive and/or false-negative responses in DNA damage repair and DNA damage response assays. These types of artefactual effects have been observed in many types of nanotoxicity [45] and nanoecotoxicity [44], but can be avoided by including appropriate experimental controls in the assays and having a thorough understanding of the assay variability parameters.

Although DNA repair process(es) are highly regulated and multifaceted involving cross-talk and overlapping functions, their evaluation in ENM-based studies contributes to an additional tier of complexity. Also, it is worth mentioning that it is extremely difficult at this stage to extrapolate from the in vitro/test species data discussed in this review, to actual ENMs exposure in vivo, e.g., outcomes on human health. The impact of ENM exposure on the fidelity of DNA repair processes is not fully understood at present and requires more in-depth investigation to elucidate the pathways of importance and the intrinsic or extrinsic ENM features that interfere with the maintenance of genomic stability.

Acknowledgments: One of the authors, Leona D. Scanlan, acknowledges funding and support from the National Academy of Sciences, National Research Council Postdoctoral Research Associateship Program. The authors thank Miral Dizdaroglu (NIST) for his helpful advice during the preparation of the manuscript.

Conflicts of Interest: The authors declare no conflicts of interest.

NIST Disclaimer: Certain commercial equipment, instruments and materials are identified in this paper to specify an experimental procedure as completely as possible. In no case does the identification of particular equipment or materials imply a recommendation or endorsement by the National Institute of Standards and Technology nor does it imply that the materials, instruments, or equipment are necessarily the best available for the purpose.

References

1. European Commission Recommendation on the Definition of Nanomaterial Text with EEA Relevance, 2011/696/EU. Available online: http://data.europa.eu/eli/reco/2011/696/oj (accessed on 30 June 2017).
2. Arora, S.; Rajwade, J.M.; Paknikar, K.M. Nanotoxicology and in vitro studies: The need of the hour. *Toxicol. Appl. Pharmacol.* **2012**, *258*, 151–165. [CrossRef] [PubMed]
3. Singh, N.; Manshian, B.; Jenkins, G.J.; Griffiths, S.M.; Williams, P.M.; Maffeis, T.G.; Wright, C.J.; Doak, S.H. NanoGenotoxicology: The DNA damaging potential of engineered nanomaterials. *Biomaterials* **2009**, *30*, 3891–3914. [CrossRef] [PubMed]
4. Evans, S.J.; Clift, M.J.; Singh, N.; de Mallia, J.O.; Burgum, M.; Wills, J.W.; Wilkinson, T.S.; Jenkins, G.J.; Doak, S.H. Critical review of the current and future challenges associated with advanced in vitro systems towards the study of nanoparticle (secondary) genotoxicity. *Mutagenesis* **2017**, *32*, 233–241. [CrossRef] [PubMed]
5. Biola-Clier, M.; Beal, D.; Caillat, S.; Libert, S.; Armand, L.; Herlin-Boime, N.; Sauvaigo, S.; Douki, T.; Carriere, M. Comparison of the DNA damage response in BEAS-2B and A549 cells exposed to titanium dioxide nanoparticles. *Mutagenesis* **2017**, *32*, 161–172. [CrossRef] [PubMed]
6. Singh, N.; Jenkins, G.J.; Nelson, B.C.; Marquis, B.J.; Maffeis, T.G.; Brown, A.P.; Williams, P.M.; Wright, C.J.; Doak, S.H. The role of iron redox state in the genotoxicity of ultrafine superparamagnetic iron oxide nanoparticles. *Biomaterials* **2012**, *33*, 163–170. [CrossRef] [PubMed]
7. Broustas, C.G.; Lieberman, H.B. DNA damage response genes and the development of cancer metastasis. *Radiat. Res.* **2014**, *181*, 111–130. [CrossRef] [PubMed]
8. Langie, S.A.; Koppen, G.; Desaulniers, D.; Al-Mulla, F.; Al-Temaimi, R.; Amedei, A.; Azqueta, A.; Bisson, W.H.; Brown, D.G.; Brunborg, G.; et al. Causes of genome instability: The effect of low dose chemical exposures in modern society. *Carcinogenesis* **2015**, *36* (Suppl. S1), S61–S88. [CrossRef] [PubMed]
9. Guengerich, F.P. Common and uncommon cytochrome P450 reactions related to metabolism and chemical toxicity. *Chem. Res. Toxicol.* **2001**, *14*, 611–650. [CrossRef] [PubMed]
10. Choudhury, S.R.; Ordaz, J.; Lo, C.L.; Damayanti, N.P.; Zhou, F.; Irudayaraj, J. ZnO nanoparticles induced reactive oxygen species promotes multimodal cyto- and epigenetic toxicity. *Toxicol. Sci.* **2017**, *156*, 261–274. [CrossRef] [PubMed]
11. Jimenez-Villarreal, J.; Rivas-Armendariz, D.I.; Perez-Vertti, R.D.A.; Calderon, E.O.; Garcia-Garza, R.; Betancourt-Martinez, N.D.; Serrano-Gallardo, L.B.; Moran-Martinez, J. Relationship between lymphocyte DNA fragmentation and dose of iron oxide (Fe_2O_3) and silicon oxide (SiO_2) nanoparticles. *Genet. Mol. Res.* **2017**. [CrossRef] [PubMed]
12. Milanowska, K.; Krwawicz, J.; Papaj, G.; Kosinski, J.; Poleszak, K.; Lesiak, J.; Osinska, E.; Rother, K.; Bujnicki, J.M. REPAIRtoire—A database of DNA repair pathways. *Nucleic Acids Res.* **2011**, *39*, D788–D792. [CrossRef] [PubMed]
13. Prasad, R.Y.; Chastain, P.D.; Nikolaishvili-Feinberg, N.; Smeester, L.; Kaufmann, W.K.; Fry, R.C. Titanium dioxide nanoparticles activate the ATM-Chk2 DNA damage response in human dermal fibroblasts. *Nanotoxicology* **2013**, *7*, 1111–1119. [CrossRef] [PubMed]
14. Asharani, P.; Sethu, S.; Lim, H.K.; Balaji, G.; Valiyaveettil, S.; Hande, M.P. Differential regulation of intracellular factors mediating cell cycle, DNA Repair and inflammation following exposure to silver nanoparticles in human cells. *Genome Integr.* **2012**, *3*, 2. [CrossRef] [PubMed]

15. Kovvuru, P.; Mancilla, P.E.; Shirode, A.B.; Murray, T.M.; Begley, T.J.; Reliene, R. Oral ingestion of silver nanoparticles induces genomic instability and DNA damage in multiple tissues. *Nanotoxicology* **2015**, *9*, 162–171. [CrossRef] [PubMed]

16. Asare, N.; Duale, N.; Slagsvold, H.H.; Lindeman, B.; Olsen, A.K.; Gromadzka-Ostrowska, J.; Meczynska-Wielgosz, S.; Kruszewski, M.; Brunborg, G.; Instanes, C. Genotoxicity and gene expression modulation of silver and titanium dioxide nanoparticles in mice. *Nanotoxicology* **2016**, *10*, 312–321. [CrossRef] [PubMed]

17. Satapathy, S.R.; Siddharth, S.; Das, D.; Nayak, A.; Kundu, C.N. Enhancement of cytotoxicity and inhibition of angiogenesis in oral cancer stem cells by a hybrid nanoparticle of bioactive quinacrine and silver: Implication of base excision repair cascade. *Mol. Pharm.* **2015**, *12*, 4011–4025. [CrossRef] [PubMed]

18. Van Berlo, D.; Hullmann, M.; Wessels, A.; Scherbart, A.M.; Cassee, F.R.; Gerlofs-Nijland, M.E.; Albrecht, C.; Schins, R.P. Investigation of the effects of short-term inhalation of carbon nanoparticles on brains and lungs of c57bl/6j and p47$^{Phox-/-}$ mice. *Neurotoxicology* **2014**, *43*, 65–72. [CrossRef] [PubMed]

19. Tang, S.; Cai, Q.; Chibli, H.; Allagadda, V.; Nadeau, J.L.; Mayer, G.D. Cadmium sulfate and CdTe-quantum dots alter DNA repair in zebrafish (*Danio rerio*) liver cells. *Toxicol. Appl. Pharmacol.* **2013**, *272*, 443–452. [CrossRef] [PubMed]

20. Tang, S.; Wu, Y.; Ryan, C.N.; Yu, S.; Qin, G.; Edwards, D.S.; Mayer, G.D. Distinct Expression Profiles of Stress Defense and DNA Repair Genes in Daphnia Pulex Exposed to Cadmium, Zinc, and Quantum Dots. *Chemosphere* **2015**, *120*, 92–99. [CrossRef] [PubMed]

21. Ahamed, M.; Siddiqui, M.A.; Akhtar, M.J.; Ahmad, I.; Pant, A.B.; Alhadlaq, H.A. Genotoxic potential of copper oxide nanoparticles in human lung epithelial cells. *Biochem. Biophys. Res. Commun.* **2010**, *396*, 578–583. [CrossRef] [PubMed]

22. Khatri, M.; Bello, D.; Pal, A.K.; Cohen, J.M.; Woskie, S.; Gassert, T.; Lan, J.; Gu, A.Z.; Demokritou, P.; Gaines, P. Evaluation of cytotoxic, genotoxic and inflammatory responses of nanoparticles from photocopiers in three human cell lines. *Part Fibre Toxicol.* **2013**, *10*, 42. [CrossRef] [PubMed]

23. El-Said, K.S.; Ali, E.M.; Kanehira, K.; Taniguchi, A. Molecular mechanism of DNA damage induced by titanium dioxide nanoparticles in toll-like receptor 3 or 4 expressing human hepatocarcinoma cell lines. *J. Nanobiotechnol.* **2014**, *12*, 48. [CrossRef] [PubMed]

24. Hanot-Roy, M.; Tubeuf, E.; Guilbert, A.; Bado-Nilles, A.; Vigneron, P.; Trouiller, B.; Braun, A.; Lacroix, G. Oxidative stress pathways involved in cytotoxicity and genotoxicity of titanium dioxide (TiO$_2$) nanoparticles on cells constitutive of alveolo-capillary barrier in vitro. *Toxicol. In Vitro* **2016**, *33*, 125–135. [CrossRef] [PubMed]

25. Pati, R.; Das, I.; Mehta, R.K.; Sahu, R.; Sonawane, A. Zinc-oxide nanoparticles exhibit genotoxic, clastogenic, cytotoxic and actin depolymerization effects by inducing oxidative stress responses in macrophages and adult mice. *Toxicol. Sci.* **2016**, *150*, 454–472. [CrossRef] [PubMed]

26. Chastain, P.D., II; Heffernan, T.P.; Nevis, K.R.; Lin, L.; Kaufmann, W.K.; Kaufman, D.G.; Cordeiro-Stone, M. Checkpoint regulation of replication dynamics in UV-irradiated human Cells. *Cell Cycle* **2006**, *5*, 2160–2167. [CrossRef] [PubMed]

27. Kansara, K.; Patel, P.; Shah, D.; Shukla, R.K.; Singh, S.; Kumar, A.; Dhawan, A. TiO2 nanoparticles induce DNA double strand breaks and cell cycle arrest in human alveolar cells. *Environ. Mol. Mutagen.* **2015**, *56*, 204–217. [CrossRef] [PubMed]

28. Hahm, S.H.; Park, J.H.; Ko, S.I.; Lee, Y.R.; Chung, I.S.; Chung, J.H.; Kang, L.W.; Han, Y.S. Knock-down of human MutY homolog (hMYH) decreases phosphorylation of checkpoint kinase 1 (Chk1) induced by hydroxyurea and UV treatment. *BMB Rep.* **2011**, *44*, 352–357. [CrossRef] [PubMed]

29. Hahm, S.H.; Chung, J.H.; Agustina, L.; Han, S.H.; Yoon, I.S.; Park, J.H.; Kang, L.W.; Park, J.W.; Na, J.J.; Han, Y.S. Human MutY homolog induces apoptosis in etoposide-treated HEK293 cells. *Oncol. Lett.* **2012**, *4*, 1203–1208. [PubMed]

30. Oka, S.; Ohno, M.; Tsuchimoto, D.; Sakumi, K.; Furuichi, M.; Nakabeppu, Y. Two distinct pathways of cell death triggered by oxidative damage to nuclear and mitochondrial DNAs. *EMBO J.* **2008**, *27*, 421–432. [CrossRef] [PubMed]

31. Millau, J.F.; Raffin, A.L.; Caillat, S.; Claudet, C.; Arras, G.; Ugolin, N.; Douki, T.; Ravanat, J.L.; Breton, J.; Oddos, T.; et al. A microarray to measure repair of damaged plasmids by cell lysates. *Lab Chip* **2008**, *8*, 1713–1722. [CrossRef] [PubMed]

32. Kumari, R.; Mondal, T.; Bhowmick, A.K.; Das, P. Impeded repair of abasic site damaged lesions in DNA adsorbed over functionalized multiwalled carbon nanotube and graphene oxide. *Mutat. Res. Genet. Toxicol. Environ. Mutagen.* **2016**, *803*, 39–46. [CrossRef] [PubMed]

33. Hartmann, A.; Speit, G. Effect of arsenic and cadmium on the persistence of mutagen-induced DNA lesions in human cells. *Environ. Mol. Mutagen.* **1996**, *27*, 98–104. [CrossRef]

34. Snyder, R.D.; Davis, G.F.; Lachmann, P.J. Inhibition by metals of X-ray and ultraviolet-induced DNA repair in human cells. *Biol. Trace Elem. Res.* **1989**, *21*, 389–398. [CrossRef] [PubMed]

35. Candeias, S.; Pons, B.; Viau, M.; Caillat, S.; Sauvaigo, S. Direct inhibition of excision/synthesis DNA repair activities by cadmium: Analysis on dedicated biochips. *Mutat. Res.* **2010**, *694*, 53–59. [CrossRef] [PubMed]

36. McNeill, D.R.; Narayana, A.; Wong, H.K.; Wilson, D.M., III. Inhibition of Ape1 nuclease activity by lead, iron, and cadmium. *Environ. Health Perspect.* **2004**, *112*, 799–804. [CrossRef] [PubMed]

37. Jugan, M.L.; Barillet, S.; Simon-Deckers, A.; Herlin-Boime, N.; Sauvaigo, S.; Douki, T.; Carriere, M. Titanium dioxide nanoparticles exhibit genotoxicity and impair DNA repair activity in A549 cells. *Nanotoxicology* **2012**, *6*, 501–513. [CrossRef] [PubMed]

38. Gebel, T.; Foth, H.; Damm, G.; Freyberger, A.; Kramer, P.J.; Lilienblum, W.; Rohl, C.; Schupp, T.; Weiss, C.; Wollin, K.M.; et al. Manufactured nanomaterials: Categorization and approaches to hazard assessment. *Arch. Toxicol.* **2014**, *88*, 2191–2211. [CrossRef] [PubMed]

39. Jones, R. It's not just about nanotoxicology. *Nat. Nanotechnol.* **2009**, *4*, 615. [CrossRef] [PubMed]

40. Carriere, M.; Sauvaigo, S.; Douki, T.; Ravanat, J.L. Impact of nanoparticles on DNA repair processes: Current knowledge and working hypotheses. *Mutagenesis* **2017**, *32*, 203–213. [CrossRef] [PubMed]

41. Lee, J.H.; Moon, M.C.; Lee, J.Y.; Yu, I.J. Challenges and perspectives of nanoparticle exposure assessment. *Toxicol. Res.* **2010**, *26*, 95–100. [CrossRef] [PubMed]

42. EPA. Nanomaterial Case Study: Nanoscale Silver in Disinfectant Spray. Available online: https://cfpub.epa.gov/ncea/risk/recordisplay.cfm?deid=241665 (Accessed on 9 March 2017).

43. Rim, K.T.; Song, S.W.; Kim, H.Y. Oxidative DNA damage from nanoparticle exposure and its application to workers' health: A literature review. *Saf. Health Work.* **2013**, *4*, 177–186. [CrossRef] [PubMed]

44. Petersen, E.J.; Henry, T.B.; Zhao, J.; MacCuspie, R.I.; Kirschling, T.L.; Dobrovolskaia, M.A.; Hackley, V.; Xing, B.S.; White, J.C. Identification and avoidance of potential artifacts and misinterpretations in nanomaterial ecotoxicity measurements. *Environ. Sci. Technol.* **2014**, *48*, 4226–4246. [CrossRef] [PubMed]

45. Alkilany, A.M.; Mahmoud, N.N.; Hashemi, F.; Hajipour, M.J.; Farvadi, F.; Mahmoudi, M. Misinterpretation in nanotoxicology: A personal perspective. *Chem. Res. Toxicol.* **2016**, *29*, 943–948. [CrossRef] [PubMed]

International Journal of
Molecular Sciences

MDPI

Article

Magnetic Hyperthermia and Oxidative Damage to DNA of Human Hepatocarcinoma Cells

Filippo Cellai [1], Armelle Munnia [1], Jessica Viti [1], Saer Doumett [2], Costanza Ravagli [2], Elisabetta Ceni [3], Tommaso Mello [3], Simone Polvani [3], Roger W. Giese [4], Giovanni Baldi [2], Andrea Galli [3] and Marco E. M. Peluso [1,*]

[1] Cancer Risk Factor Branch, Regional Cancer Prevention Laboratory, ISPO-Cancer Research and Prevention Institute, Florence 50139, Italy; f.cellai@ispo.toscana.it (F.C.); a.munnia@ispo.toscana.it (A.M.); jessicaviti@alice.it (J.V.)
[2] Nanobiotechnology Department, Colorobbia Consulting-Cericol, Sovigliana, Vinci 50053, Italy; doumetts@colorobbia.it (S.D.); ravaglic@colorobbia.it (C.R.); baldig@colorobbia.it (G.B.)
[3] Department of Experimental and Clinical Biomedical Sciences, University of Florence, Florence 50139, Italy; e.ceni@dfc.unifi.it (E.C.); tommaso.mello@unifi.it (T.M.); simone.polvani@unifi.it (S.P.); a.galli@dfc.unifi.it (A.G.)
[4] Department of Pharmaceutical Sciences in the Bouve College of Health Sciences, Barnett Institute, Northeastern University, Boston, MA 02115, USA; r.giese@neu.edu
* Correspondence: m.peluso@ispo.toscana.it; Fax: +39-055-3269-7879

Academic Editors: Ashis Basu and Takehiko Nohmi
Received: 1 March 2017; Accepted: 23 April 2017; Published: 29 April 2017

Abstract: Nanotechnology is addressing major urgent needs for cancer treatment. We conducted a study to compare the frequency of 3-(2-deoxy-β-D-erythro-pentafuranosyl)pyrimido[1,2-α] purin-10(3H)-one deoxyguanosine (M_1dG) and 8-oxo-7,8-dihydro-2′-deoxyguanosine (8-oxodG) adducts, biomarkers of oxidative stress and/or lipid peroxidation, on human hepatocarcinoma HepG2 cells exposed to increasing levels of Fe_3O_4-nanoparticles (NPs) versus untreated cells at different lengths of incubations, and in the presence of increasing exposures to an alternating magnetic field (AMF) of 186 kHz using ^{32}P-postlabeling. The levels of oxidative damage tended to increase significantly after ≥24 h of incubations compared to controls. The oxidative DNA damage tended to reach a steady-state after treatment with 60 μg/mL of Fe_3O_4-NPs. Significant dose–response relationships were observed. A greater adduct production was observed after magnetic hyperthermia, with the highest amounts of oxidative lesions after 40 min exposure to AMF. The effects of magnetic hyperthermia were significantly increased with exposure and incubation times. Most important, the levels of oxidative lesions in AMF exposed NP treated cells were up to 20-fold greater relative to those observed in nonexposed NP treated cells. Generation of oxidative lesions may be a mechanism by which magnetic hyperthermia induces cancer cell death.

Keywords: magnetic therapy; nanotoxicity; M_1dG; 8-oxodG; human hepatocarcinoma cells

---

1. Introduction

The use of nanotechnology in medicine is of interest for cancer diagnosis and treatment. In regards to treatment, a number of substances are currently under investigation for drug delivery, varying from liposomes to various polymers and magnetic metals [1–4]. In some cases, the characteristic properties of nanoparticles (NPs) result in their multifunctional potential for simultaneous early detection and treatment of malignant cell growth, such as those of Fe_3O_4-NPs [5–7]. For example, magnetic hyperthermia is a technique that utilizes the administration of Fe_3O_4-NPs into tumor sites and the exposure to alternating magnetic field (AMF) to kill cancer cells by heating. NP properties have been associated with a number of toxic effects, such as chromosomal aberrations, DNA strand breakage, oxidative damage to DNA and mutations [8,9]. Use of NPs along with AMF has been reported to enhance genotoxicity via oxidative stress [10]. In the current study, we focus on the type of genotoxicity in more detail.

A number of assays have been used for the detection of NP-related oxidative damaged to DNA, especially of 8-oxo-7,8-dihydro-2′-deoxyguanosine (8-oxodG), in various experimental cell lines [11]. The interest in 8-oxodG is due to the fact that it is a widely abundant oxidative lesion. Nevertheless, other kinds of oxidative lesions have also been identified in DNA, including the exocyclic 3-(2-deoxy-β-D-erythropentafuranosyl)pyrimido[1,2-α]purin-10(3H)-one–deoxyguanosine (M_1dG) adducts [12], that may be applied to nanotoxicology testing. In particular, the exposure to Fe_3O_4-NPs has been associated with enhanced oxidative stress [13]. This overproduction of reactive oxygen species (ROS) may induce lipid peroxidation (LPO) as well as M_1dG and 8-oxodG [12]. LPO is a process of oxidation of polyunsaturated fatty acids due to the presence of several double bonds in their structure and it involves production of peroxides, ROS, and other reactive species, such as malondialdehyde (MDA). MDA is a reactive LPO by-product, which interacts with DNA forming M_1dG adducts [14]. Both M_1dG and 8-oxodG, if not repaired, may induce mutations, such as base pair and frameshift mutations [15] or G:C to T:A transversions [16]. Furthermore, one of the more critical sites on DNA that can be damaged by oxidative damage is the *tumor suppressor P53* [17], a gene involved in cell cycle arrest, DNA repair, senescence, and apoptosis [18]. Early investigations have also indicated that great amounts of both M_1dG and/or 8-oxodG adducts are associated to cancer development and tumor progression [17,19–22], therefore these biomarkers are considered worthy indicators for carcinogenesis or mutagenesis events.

Herein, we report a toxicology study aimed at comparing the production of both exocyclic M_1dG and 8-oxodG adducts, biomarkers of oxidative stress and/or LPO [23], in human hepatocarcinoma HepG2 cells exposed to Fe_3O_4-NPs of 50 nm-size in presence or absence of increasing exposures to AMF and prolonged incubations compared to control cells. The levels of exocyclic M_1dG and 8-oxodG adducts were examined by ^{32}P-postlabeling [24–26]. Our aim was to investigate the mechanisms underlying the genotoxic effects caused by magnetic hyperthermia. Considering the technical development and the current and future use of this therapy in cancer treatment [27], an improved understanding of the mechanisms of action of magnetic hyperthermia is important.

2. Results

2.1. Fe₃O₄-Nanoparticles and Exocyclic M₁dG Adducts

One of the purpose of the present study was to evaluate the generation of exocyclic M_1dG adducts induced by the treatment with increasing amounts of Fe_3O_4-NPs, ranging from 30 to 60–90 μg/mL, at different incubation times, from 1 to 48 h. As shown in Table 1, no effects were observed at each concentration for dosages at 2 h of incubations. However, the production of exocyclic M_1dG adducts per 10^6 normal nucleotides was increased, up to about two-fold, in HepG2 cells which were treated with NP dosages for incubation times of $\geq$24 h compared to untreated samples. In our model, the difference between M_1dG adducts of cells exposed to increasing amounts of Fe_3O_4-NPs and basal adduct levels of control cells, i.e., 0.4 ± 0.1 (SE), was generally significant, all $p < 0.05$, after $\geq$24 h of incubation times, with exception of sample exposed to 30 μg/mL of NPs for 24 h ($p = 0.057$). Table 1 shows that the highest levels of M_1dG, i.e., 0.9 ± 0.1 (SE), were found in HepG2 cells treated with 60 μg/mL of Fe_3O_4-NPs. At higher dosages (90 μg/mL), the treatment was not associated with further adduct production, possibly due to the onset of excessive citoxicity. A significant dose–response relationship was found, with a p-value for the trend <0.001. The levels of NP-related M_1dG adducts were then compared with those caused by the xanthine/xanthine oxidase system, a ROS-generating system [28]. The frequency of exocyclic M_1dG adducts in HepG2 cells treated with Fe_3O_4-NPs was about three-fold lower compared to that induced by the ROS-generating system (Table 1).

Table 1. Mean levels of 3-(2-deoxy-β-D-erythro-pentafuranosyl)pyrimido[1,2-α]purin-10(3*H*)-one deoxyguanosine (M_1dG) adducts and 8-hydroxy-2′-deoxyguanosine (8-oxodG) per 10^6 normal nucleotides in HepG2 cells after treatment with 30, 60 or 90 µg/mL of Fe_3O_4-nanoparticles and prolonged incubation times compared to control cells. Adduct levels caused by a free radical-generating system in our model are reported, as positive internal control.

Exocyclic M_1dG and 8-oxodG Lesions					
	N	$M_1dG \pm SE$	*p*-Value	8-oxodG $\pm$ SE	*p*-Value
Adduct background					
Control cells					
Incubation times					
24 h [a]	10	0.4 ± 0.1		3.2 ± 0.2	
48 h [a]	10	0.4 ± 0.1		3.1 ± 0.2	
Adduct levels of magnetic nanoparticle treated cells					
Treated cells					
30 µg/mL					
Incubation times					
24 h	10	0.5 ± 0.1	0.057 [b]	11.2 ± 0.8	<0.05 [b]
48 h	10	0.6 ± 0.1	<0.05 [b]	14.7 ± 0.4	<0.05 [b]
60 µg/mL					
Incubation times					
24 h	10	0.8 ± 0.1	<0.05 [b]	18.4 ± 2.8	<0.05 [b]
48 h	10	0.9 ± 0.1	<0.05 [b]	19.4 ± 1.4	<0.05 [b]
90 µg/mL					
Incubation times					
24 h	10	0.7 ± 0.1	<0.05 [b]	17.1 ± 1.7	<0.05 [b]
48 h	10	0.8 ± 0.1	<0.05 [b]	18.0 ± 1.8	<0.05 [b]
Adduct levels of positive internal control					
Free radical-generating system					
0.2 mM xanthine/1.0 mU xanthine oxidase					
Incubation times					
24 h	10	1.9 ± 0.1	<0.05 [b]	4.6 ± 0.2	<0.05 [b]
48 h	10	2.2 ± 0.1	<0.05 [b]	5.4 ± 0.2	<0.05 [b]
0.2 mM xanthine/5.0 mU xanthine oxidase					
Incubation times					
24 h	10	2.9 ± 0.6	<0.05 [b]	8.3 ± 0.2	<0.05 [b]
48 h	10	2.7 ± 0.4	<0.05 [b]	9.3 ± 0.4	<0.05 [b]

[a] Reference levels; [b] *p*-values vs. appropriated controls.

2.2. Fe₃O₄-Nanoparticles and 8-oxodG Adducts

We next investigated the generation of 8-oxodG in our experimental model. Table 1 reports that there were no effects at each NP concentrations with ≤120 min incubations, whereas adduct production was significantly increased, up to six-fold, after ≥24 of incubations compared to untreated cells. The highest levels of 8-oxodG per 10^6 normal nucleotides, were found after treatment with 60–90 µg/mL of Fe_3O_4-NPs for period of incubation of ≥24, 18.4 ± 2.8 (SE), and 19.4 ± 1.4 (SE), respectively. The difference of 8-oxodG among HepG2 cells treated with various concentrations of Fe_3O_4-NPs relative to basal adduct levels, i.e., 3.2 ± 0.2 (SE), was always significant, all $p < 0.05$, after ≥24 h of incubations. The mean amounts of 8-oxodG detected in control cells (3.1–3.2 8-oxodG per 10^6 normal nucleotides) was in the range of 1.0 and 5.0 per 10^6 normal dG, which is typically observed in cultured cells [29]. As shown in Table 1, the association between 8-oxodG and NP levels was linear at low dosages, but tended to reach a steady-state at higher exposures. A significant dose–response relationship was observed, *p*-value for the trend <0.001. The amounts of NP-related 8-oxodG were then compared with those detected after treatment with the xanthine/xanthine oxidase system. Table 1 reports that the generation of 8-oxodG was about two-fold higher in HepG2 cells treated with the Fe_3O_4-NPs compared to those exposed to the ROS-generating system.

2.3. Magnetic Nanoparticles, Alternating Magnetic Field and Exocyclic M$_1$dG Adducts

We also evaluated the genotoxicity of Fe$_3$O$_4$-NPs (90 µg/mL) after increasing exposures, i.e., 20–40 min, to an AMF applied at 186 kHz frequency. Table 2 shows that a significant production of exocyclic M$_1$dG adducts, up to a 30-fold increment, was detected in treated cells after AMF exposures compared to controls. The genotoxic effect increased with both exposure and incubation times, with p-value for the trend <0.001. The highest value of the exocyclic M$_1$dG adducts of 15.9 ± 6.7 (SE) was detected in HepG2 cells containing 90 µg/mL of Fe$_3$O$_4$-NPs after 20 min of AMF exposure and 48 h of incubation. The difference of M$_1$dG was significant, all p < 0.05. Table 2 indicates that the levels of exocyclic M$_1$dG adducts were also higher, up to 20-fold, in NP treated cells after AMF exposures relative to nonexposed NP treated cells, all p < 0.05.

Table 2. Mean levels of 3-(2-deoxy-β-D-erythro-pentafuranosyl)pyrimido[1,2-α]purin-10(3*H*)-one deoxyguanosine (M$_1$dG) adducts and 8-hydroxy-2'-deoxyguanosine (8-oxodG) adducts per 10^6 normal nucleotides in 90 µg/mL Fe$_3$O$_4$-nanoparticle (NP) treated HepG2 cells in presence or absence of alternating magnetic field (AMF) exposures and prolonged incubation times compared to control cells.

		Exocyclic M$_1$dG and 8-oxodG Adducts			
	N	M$_1$dG ± SE	*p*-Value	8-oxodG ± SE	*p*-Value
Adduct background in presence or absence of AMF exposures					
Control cells					
Nonexposed to AMF					
Incubation times					
24 h [a]	10	0.4 ± 0.1		3.2 ± 0.2	
48 [a]	10	0.4 ± 0.1		3.1 ± 0.2	
Exposed to AMF (20 min)					
Incubation times					
24 h [a]	10	0.5 ± 0.1		3.2 ± 0.2	
48 [a]	10	0.5 ± 0.1		3.3 ± 0.3	
Exposed to AMF (40 min)					
Incubation times					
24 h [a]	10	0.5 ± 0.1		3.4 ± 0.2	
48 [a]	10	0.5 ± 0.1		3.9 ± 0.3	
Adduct levels of treated cells in presence or absence of AMF exposures					
NP treated cells					
Nonexposed to AMF					
Incubation times					
24 h [a]	10	0.7 ± 0.1	<0.05 [b]	16.7 ± 1.7	<0.05 [b]
48 [a]	10	0.8 ± 0.1	<0.05 [b]	21.2 ± 1.9	<0.05 [b]
Exposed to AMF (20 min)					
Incubation times					
24 h [a]	10	4.6 ± 0.5	<0.05 [b]	40.3 ± 3.3	<0.05 [b]
48 [a]	10	6.9 ± 0.6	<0.05 [b]	69.3 ± 1.4	<0.05 [b]
Exposed to AMF (40 min)					
Incubation times					
24 h [a]	10	6.4 ± 1.4	<0.05 [b]	36.5 ± 4.0	<0.05 [b]
48 [a]	10	15.9 ± 6.7	<0.05 [b]	79.3 ± 1.1	<0.05 [b]

[a] Reference levels; [b] *p*-values vs. appropriated referent cells.

2.4. Magnetic Nanoparticles, Alternating Magnetic Field and 8-oxodG Adducts

We investigated the association between the production of 8-oxodG in 90 µg/mL Fe_3O_4-NP treated cells after exposure (20 or 40 min) to an AMF applied at 186 kHz. As reported in Table 2, the generation of 8-oxodG was significantly increased, up to 25-fold, in treated cells after AMF exposure compared to controls. The effect was associated to both exposure and incubation, with *p*-value for the trend <0.001. The highest value of 8-oxodG of 79.3 ± 1.1 (SE) was detected in treated cells that were exposed to AMF for 40 min and incubated for 48 h. Table 2 shows that the levels of 8-oxodG were also higher, up to a five-fold increase, in treated cells after AMF exposure relative to nonexposed NP treated cells, all *p* < 0.05.

Since heating curves cannot be directly recorded during the experimental sessions because of sample sterility, kinetic pathways of temperature increase were recorded in solutions of Fe_3O_4-NPs containing the same amount of particles as the samples described in the manuscript (Figure 1). Temperature rise of the dispersion during AMF exposure was measured by an optical fiber probe (CEAM Vr18CR-PC) and it was recorded every second during the experimental measurements. The samples were inserted in a homemade polyethylene sample holder, surrounded by glass wool for thermal isolation, and then in the copper coil. The sample mass (m = 0–3 g) was chosen to minimize the effects of magnetic field inhomogeneity inside the sample.

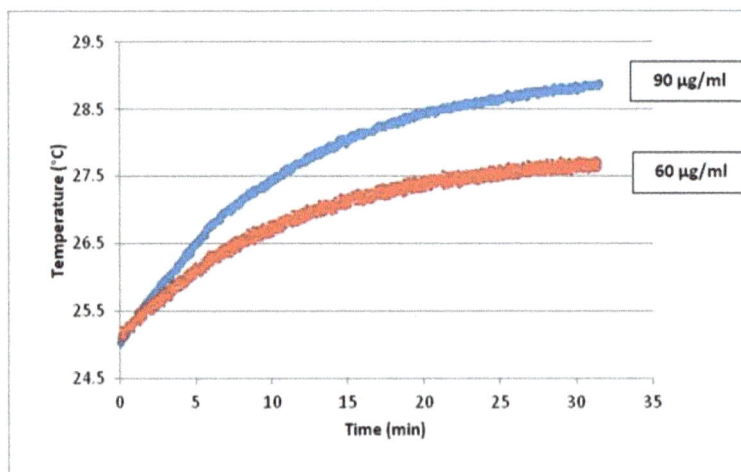

Figure 1. Kinetics of temperature increase versus exposure time for block polymer coated Fe_3O_4-NPs 90 µg/mL (blue curve) and 60 µg/mL (red curve).

2.5. Fe_3O_4-Nanoparticle Internalization

The uptake of Fe_3O_4-NPs conjugated with fluorescent dye Dylight650 was monitored by fluorescent microscopy. NP fluorescence inside cells was measured with background subtraction. As shown in Figure 2A–C, the NPs were internalized at the concentration of 90 µg/mL by HepG2 cells in sufficient quantity to be detected after 240 min (Figure 2C). A significant increase in fluorescence intensity was indeed observed only at that incubation time, *p* < 0.0001 (Figure 2D). The lack of genotoxic effects in Fe_3O_4-NPs treated cells at lower incubation time is apparently due to a lack of significant cellular penetration of NPs during this initial period.

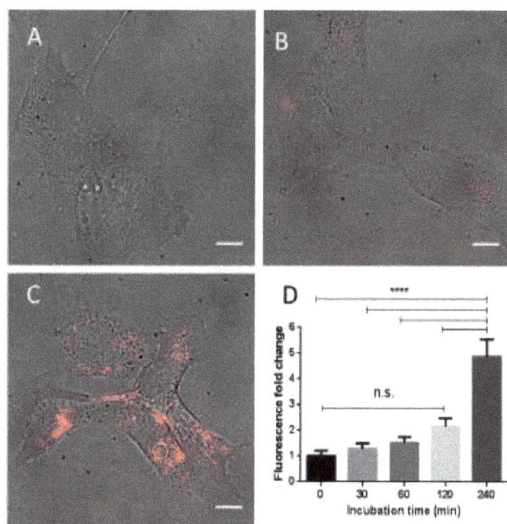

Figure 2. Human hepatocarcinoma (HepG2) cells incubated with 90 μg/mL Fe_3O_4-nanoparticles conjugated with fluorescent dye Dylight650. The images of treated cells at selected incubation times are shown: 0 min (**A**); 120 min (**B**); and 240 min (**C**). The increments in fluorescence intensity at each incubation times, mean ± standard error (SE), are reported in the insert (**D**). Scale bar: 10 μm; **** $p < 0.0001$; n.s. = not significant.

3. Discussion

The potential therapeutic benefits of Fe_3O_4-NPs [30] must consider their potential for off-target toxicity [8,9,31–33]. Fe_3O_4-NPs, if undirected, are mainly sequestered in liver tissue for elimination. Hence, we studied human hepatocarcinoma cells for NP-related toxicity. In particular, we investigated the pro-oxidant properties of Fe_3O_4-NPs in towards DNA. We asked whether experimental cells treated with increasing NP concentrations experienced genotoxic effects at various times of incubation. A further purpose of the present study was to investigate the generation of oxidative DNA lesions in Fe_3O_4-NP treated cells after AMF exposures. Two different biomarkers, i.e., exocyclic M_1dG and 8-oxodG adducts, of oxidative stress and/or LPO [23], were measured by ^{32}P-postlabeling.

High dosages of Fe_3O_4-NPs are expected to significantly increase cellular ROS and the production of oxidative damage to DNA [11]. This was found in our model, where the administration of increased NP concentrations (30, 60 or 90 μg/mL) was significantly associated with higher levels of both exocyclic M_1dG and 8-oxodG adducts in HepG2 cells after ≥24 h of incubation in comparison to untreated cells. No genotoxic effects were detected by each concentration dosages at ≤120 min of incubations, possibly due to a lack of penetration of NPs into cellular membrane of HepG2 cells during this initial period, as shown by fluorescent microscopy (Figure 2). We found a significant dose–response relationship between oxidized DNA lesions and Fe_3O_4-NPs. Nevertheless, while the associations with adduct production tended to be linear at low exposure, it was sublinear at high dosages, perhaps due to excessive cytotoxicity. We assume in this study that Fe_3O_4-NPs act as a pro-oxidant by causing iron-dependent Fenton reactions that generate first ·OH and, then, ·C radicals, through LPO induction, which are important intermediates that lead to the formation of oxidative DNA lesions. Consequently, the addition of Fe_3O_4-NPs to HepG2 cells increases the oxidation of genomic DNA, measured as promutagenic lesions. The toxic action of magnetic therapy may be mediated by several mechanisms, as shown in Figure 3.

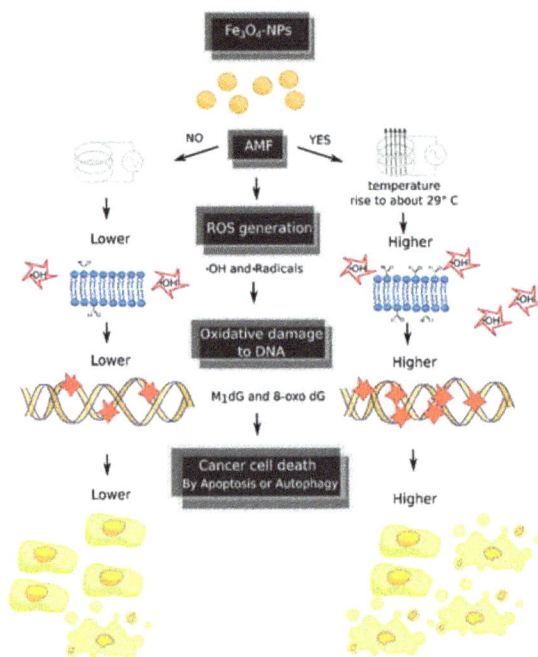

Figure 3. Potential mechanisms underlying the genotoxic effects caused by magnetic hyperthermia. AMF, alternating magnetic field; ROS, reactive oxygen species.

Our findings are in keeping with a series of earlier reports which have shown adverse effects associated with the exposure to Fe_3O_4-NPs in mammalian liver cells [11]. For example, Lin and coworkers [34] analyzed the toxicity of Fe_3O_4-NPs on human hepatocytes after 24 h of incubation using various assays including the comet assay and other biomarkers of oxidative stress, such as glutathione peroxidase, superoxide dismutase and MDA. In that study, the treatment with NPs was associated with both nuclear condensation and chromosomal DNA fragmentation. Significant reductions of cellular levels of glutathione peroxidase and superoxide dismutase levels as well as increased MDA were also observed. Sadeghi and coworkers [35] analyzed the effects of Fe_3O_4-NP exposures (75–100 µg/mL) in HepG2 cells after 12–24 h of treatment. Higher oxidative damage to DNA that was both concentration and time dependent was detected. In another study, Ma and coworkers [36] analyzed the effects of increasing dosages up to 40 mg/kg of Fe_3O_4-NPs on mice after seven days of treatment. Significant increments of both 8-oxodG and DNA-protein crosslinks were observed in liver tissue.

Magnetic hyperthermia is a technique based on the use of Fe_3O_4-NPs to remotely induce local heat in cancer cells when AMF is applied. Increased production of oxidative stress has been associated with this therapeutic approach [10,37]. We evaluated whether DNA-damaging effects induced by Fe_3O_4-NPs (90 µg/mL) are intensified by increasing exposures to AMF. Our most striking result shows that higher exposures to AMF induce a significant production of both exocyclic M_1dG and 8-oxodG adducts compared to controls cells. Thus, one important question is whether the increasing induction of DNA-modifications is correlated to a rising temperature. This is important, since as known chemical reactions can be promoted by higher temperatures. The effects, shown in the data reported in Figure 1, showed that oxidative lesions were induced with a temperature rise of 4 °C, but it is reasonable that the punctual heating of each single particle under AMF reaches even higher temperature that, to date, is not possible to be quantified. The genotoxic effects were also significantly associated with incubation times (24–48 h). The levels of oxidative DNA lesions in AMF exposed treated cells were also significantly higher in respect to nonexposed NP treated cells. Specifically, the levels of exocyclic M_1dG adducts in AMF exposed NP treated cells are up to 95% higher when compared to nonexposed NP treated cells. The amounts of 8-oxodG are up to 73% greater in respect to nonexposed controls. Nevertheless, the levels of M_1dG in AMF exposed NP treated HepG2 cells (30-fold increment vs. nonexposed control cells) was lower than that detected in A549 and BEAS-2B lung epithelial cells exposed to 10 μM of benzo(a)pyrene (54-fold increment vs. control cells) [38]. Our findings are in agreement with Shaw and coworkers [37], who observed enhancement of genotoxic effects by Fe_3O_4-NPs in cancer cells after magnetic flux using the Comet assay. In that study, magnetic field exposure increased the levels of DNA double and single strand breaks in cancer cells.

On the one hand, Fe_3O_4-NPs may absorb the energy from AMF and convert it into heat primarily through Brownian and Neel relaxations, when exposed to AMF [10]. The heat in turn could simply increase reaction rates for existing ROS. On the other hand, magnetic therapy could increase formation of ROS such as hydroxyl radicals at the surface of Fe_3O_4-NPs [39]. Increased levels of ferrous ions might form and react with H_2O_2 produced by mitochondria, inducing ROS generation through the Fenton reaction [40,41], which causes various kind of oxidative damage to DNA, such as exocyclic M_1dG and 8-oxodG adducts. Alternatively, ROS induction may be attributed to surface coating-cell interactions. Toxic effects of Fe_3O_4-NPs may partially derive from chemical reactions between coating and cell components during phagocytosis [42]. In support to this hypothesis, we showed an association between oxidative burst of activated cells and M_1dG generation [43,44].

The generation of oxidative DNA lesions, such as those caused from Fe_3O_4-NPs [45], has intensified interest in nanotoxicology because they are considered one way to induce of authophagy and apoptosis cell death of cancer cells [46]. When DNA is severely damaged or unrepaired, cells may remain quiescent and activate pro-survival mechanisms or undergo cell death. Once induced by DNA injury, autophagy may delay apoptosis in response to genetic damage by sustaining the mechanisms necessary for DNA repair, or may contribute to cell death when DNA is unrepaired and apoptosis is defective.

4. Material and Methods

4.1. Fe$_3$O$_4$-Nanoparticles, Cell Culture and Cell Treatments in Presence or Absence of Alternating Magnetic Field

The 50-nm hydrodynamic diameter size (by DLS measurements) block polymer coated Fe_3O_4-NPs of 3 mg/mL concentration were obtained from Colorobbia Consulting-Cericol, Italy. HepG2 cells were routinely cultivated under standard conditions. Cells were incubated in a humidified controlled atmosphere with a 95% to 5% ratio of air/CO_2, at 37 °C, with medium that was changed every 3 days. Trypsinized cells at 30–40% confluence were exposed to increasing levels of Fe_3O_4-NPs that were dispersed in cell culture medium to obtain the final concentrations of 30, 60 or 90 μg/mL. Increasing lengths of incubations were used, i.e., from 60–120 min to 24–48 h. The entire and general image of strategy of the present research is shown in Figures 4 and 5.

Specifically, the NP concentrations were comparable to those utilized in the studies of Kham and coworkers [47], and Alarifi and coworkers [48], who found increased toxic effects caused from the treatments with Fe_3O_4-NPs of comparable diameter size. Subsequently, the hepatocarcinoma cells, which were treated with 90 μg/mL of NPs, were exposed to 25 kA/m AMF at a frequency of 186 kHz with a copper coil of 3 turns and $\varnothing$ = 100 mm (Celes MP 6/400) for 20 or 40 min exposure times, and then returned to the incubator for 24 or 48 h of incubation. Cells not exposed to NPs served as negative control in each experiment. As positive internal control, experimental cells were treated with 0.2 mM xanthine plus 1.0 or 5.0 mU xanthine oxidase, a system capable of generating singlet oxygen and hydrogen peroxide [28], and incubated for 24–48 h.

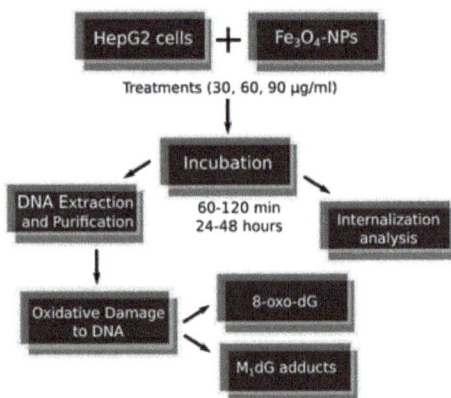

Figure 4. Strategy protocol of analytic research aimed to evaluate genotoxic activity of Fe_3O_4-nanoparticles with increasing dosages and incubation times in HepG2 cells.

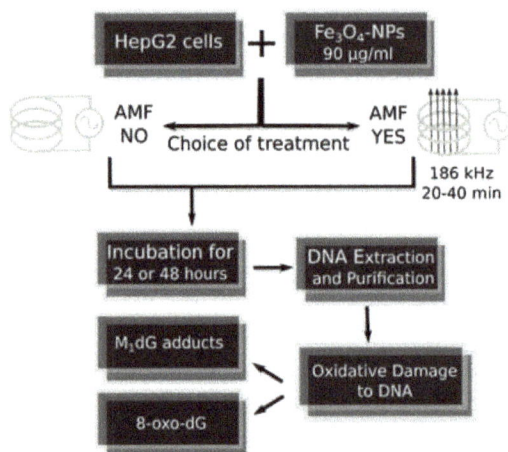

Figure 5. Strategy protocol of analytic research aimed to examine the genotoxic effects of Fe_3O_4-nanoparticles at select dosage and incubation times in presence or absence of magnetic therapy in HepG2 cells.

4.2. Preparation of M_1dG Reference Adduct Standard

A reference adduct standard was prepared: calf-thymus-DNA was treated with 10 mM MDA (ICN Biomedicals, Irvine, CA, USA), as previously reported [49]. MDA treated calf-thymus-DNA was diluted with untreated DNA to obtain decreasing levels of the reference adduct standard to generate a calibration curve.

4.3. Mass Spectrometry Analysis

The levels of exocyclic DNA adducts in MDA treated calf-thymus-DNA sample were analyzed by mass spectrometry (Voyager DE STR from Applied Biosystems, Framingham, MA, USA), as reported elsewhere [50].

4.4. Reference M_1dG Standard by ^{32}P-Postlabeling and Mass Spectrometry

The levels of the exocyclic M_1dG adducts were 5.0 ± 0.6 per 10^6 normal nucleotides in MDA treated calf-thymus-DNA using ^{32}P-postlabeling [24]. The presence of exocyclic M_1dG adducts in this sample was confirmed by matrix-assisted laser desorption/ionization time-of-flight-mass-spectrometry (MALDI-TOF-MS). A calibration curve was set up by diluting this sample with untreated calf-thymus-DNA and measuring the decreasing amount of M_1dG, $R^2 = 0.99$.

4.5. DNA Isolation and Hydrolysis

DNA from treated and untreated cells was extracted and purified from lysed cells using a method that requires digestion with ribonuclease A, ribonuclease T_1 and proteinase K treatment and extraction with saturated phenol, phenol/chloroform/isoamyl alcohol (25:24:1), chloroform/isoamyl alcohol (24:1) and ethanol precipitation [51]. DNA concentration and purity were determined using a spectrophotometer. Coded DNA samples were subsequently stored at $-80\,°C$ until laboratory analyses. Before adduct determination, DNA (2–5 µg) was hydrolyzed by incubation with micrococcal nuclease (21.45 mU/µL) and spleen phosphodiesterase (6.0 mU/µL) (Sigma Aldrich, St. Louis, MO, USA and Worthington, NJ, USA) in 5.0 mM Na succinate, 2.5 mM calcium chloride, pH 6.0 at $37\,°C$ for 4.5 h [52,53].

4.6. Exocyclic M_1dG Adduct Analysis

The generation of M_1dG adducts was analyzed by ^{32}P-postlabeling [24]. Hydrolyzed samples were incubated with nuclease P1 (0.1 U/µL) in 46.6 mM sodium acetate, pH 5.0, and 0.24 mM $ZnCl_2$ at $37\,°C$ for 30 min. After nuclease P1 treatment, 1.8 µL of 0.16 mM Tris base was added to the sample. The nuclease P1-resistant nucleotides were incubated with 7–25 µCi of carrier-free $[\gamma-^{32}P]ATP$ (3000 Ci/mM) and polynucleotide kinase T_4 (0.75 U/µL) to generate ^{32}P-labeled DNA adducts in bicine buffer, 20 mM bicine, 10 mM $MgCl_2$, 10 mM dithiotreithol, 0.5 mM spermidine, pH 9.0, at $37\,°C$ for 30 min. The generation of M_1dG adducts in Fe_3O_4-NPs treated hepatocarcinoma cells and control cells were measured using a modified version of ^{32}P-postlabeling [38]. Labeled samples were applied on polyethyleneimine cellulose thin-layer chromatography plates (Macherey-Nagel, Postfach, Germany). The chromatographic analysis of M_1dG adducts was performed using a low-urea solvent system known to be effective for the detection of low molecular weight and highly polar DNA adducts: 0.35 M $MgCl_2$ for the preparatory chromatography; and 2.1 M lithium formate, 3.75 M urea pH 3.75 and 0.24 M sodium phosphate, 2.4 M urea pH 6.4 for the two-dimensional chromatography.

Detection and quantification of modified and normal nucleotides, i.e., diluted samples that were not treated with nuclease P1, was performed by storage phosphor imaging techniques employing intensifying screens from Molecular Dynamics (Sunnyvale, CA, USA). The intensifying screens were scanned using a Typhoon 9210 (Amersham, UK). Software used to process the data was ImageQuant from Molecular Dynamics. After background subtraction, M_1dG adducts were expressed as relative adduct labeling (RAL) = pixel in adducted nucleotides/pixel in normal nucleotides. M_1dG levels were corrected across experiments based on the recovery of reference M_1dG adduct standard.

4.7. 8-oxodG Adduct Analysis

Digest DNA was diluted with ultrapure water to 20 ng/µL. Diluted DNA digest was incubated with 10 µCi of carrier-free $[\gamma-^{32}P]ATP$ (3000 Ci/mM) (Amersham, UK) and 2 U of polynucleotide kinase T4 (10 U/µL) (Roche Diagnostics, Indianapolis, IN, USA) to generate ^{32}P-labeled adducts in the reaction buffer (10×) at $37\,°C$ for 45 min [21,52]. ^{32}P-labeled samples were treated with 1.2 U of nuclease P1 (1.9 U/µL) (Sigma Aldrich, St. Louis, MO, USA) in 62.5 mM sodium acetate, pH 5.0, and 0.27 mM $ZnCl_2$ at $37\,°C$ for 60 min (final volume 10 µL) [25,26,53]. ^{32}P-labeled samples were applied on polyethyleneimine cellulose thin-layer chromatography plates (Macherey-Nagel, Postfach, Germany). The chromatographic analysis of 8-oxodG, a biomarker of oxidative stress and cancer risk [54], has been performed using a chromatographic system known to be effective for the detection and quantitative analysis of such oxidative lesions [25,26,55–58]. As previously reported [25], this procedure specifically detects 8-oxodG, which is the only adduct retained in chromatograms developed in acidic medium. Specifically, an aliquot (2.5 μL) of the labeled solution was applied to the origin of a chromatogram and developed overnight onto a 3 cm-long Whatman 1 paper wick with 1.5 M formic acid for the one-dimensional chromatography. For the two-dimensional chromatography, chromatographic plates were developed at the right angle to the previous development with 0.6 M ammonium formate, pH 6.0. Detection and quantification of 8-oxodG was obtained as described above. After background subtraction, the levels of 8-oxodG were expressed such as RAL = screen pixel in 8-oxodG spot/screen pixel in total normal nucleotides [25,26]. The levels of the normal nucleotides were measured as described above. The values measured for the 8-oxodG spot were corrected across experiments based on the recovery of the internal standard, e.g., 8-oxodG (Sigma Aldrich, St. Louis, MO, USA).

4.8. Fe_3O_4-Nanoparticle Internalization

To evaluate the dynamics of cellular internalization, Fe_3O_4-NPs were conjugated with the fluorescent dye DyLight650 and imaged by live-cell microscopy. Briefly, HepG2 cells exposed to 90 μg/mL of Fe_3O-NPs were cultured in FluoroBrite DMEM media supplemented with 10% FCS, and kept at 37 °C with 5% CO_2 during image acquisition. Images were acquired through a 63x HCX PLANAPO 1.2NA water-immersion objective, using a Leica AM6000 inverted microscope equipped with a DFC350FX camera. Images were collected before addition of NPs ($t = 0$) to analyze background and autofluorescence of experimental cells, and incubated at different incubation times, i.e., 30, 60, 120, or 240 min. Both transmitted and fluorescence images were collected. Quantification of intracellular fluorescence was performed using Fiji software [59] by manually drawing regions-of-interest (ROI) around cells in the brightfield channel and measuring the mean intensity fluorescence in each region. Mean background fluorescence was measured in each microscopic field (by drawing ROI not-containing cells) and subtracted from cellular fluorescence values. Fluorescence intensity values were normalized by the average fluorescence intensity at $t = 0$.

4.9. Statistical Analysis

The levels of oxidative DNA lesions were expressed as adducted nucleotides per 10^6 normal nucleotides. Adduct data were log transformed to stabilize the variance and normalize the distribution. Baseline characteristics between two groups were compared using independent t-test or Mann–Whitney U test for continues variables and chi square test for categorical variables. All statistical tests were two-sided and $p < 0.05$ was considered to be statistically significant. The data were analyzed using SPSS 13.0 (IBM SPSS Statistics, New York, NY, USA).

5. Conclusions

The greater production of Fe_3O_4-NPs-related oxidative DNA lesions after exposures to AMF may be one of the mechanisms by which magnetic therapy kills cancer cells. Concerning cancer cell death specifically, the hypothesis that increased levels of oxidative damage to DNA during magnetic therapy might contribute to kill cancer cells upon greater induction of apoptosis and autophagy seems to be fair enough. The generation of oxidative DNA lesions might be a further parameter that has to be maximized to optimize the action of this therapeutic approach.

Acknowledgments: This work was supported by the "Ente Cassa di Risparmio di Firenze", Florence, Italy.

Author Contributions: Marco E. M. Peluso, Andrea Galli, Filippo Cellai, Armelle Munnia, Saer Doumett, Costanza Ravagli, Elisabetta Ceni, Tommaso Mello and Giovanni Baldi conceived and designed the experiments; Filippo Cellai, Armelle Munnia, Jessica Viti, Saer Doumett, Costanza Ravagli, Elisabetta Ceni, Tommaso Mello, and Roger W. Giese performed the experiments; Marco E. M. Peluso, Filippo Cellai, Armelle Munnia,

Tommaso Mello, Simone Polvani and Saer Doumett analyzed the data; and Marco E. M. Peluso, Simone Polvani, Armelle Munnia and Filippo Cellai wrote the paper.

Conflicts of Interest: The authors declare no conflict of interest.

References

1. Guo, M.; Sun, Y.; Zhang, X.-D. Enhanced radiation therapy of gold nanoparticles in liver cancer. *Appl. Sci.* **2017**, *7*, 232. [CrossRef]
2. Ariga, K.; Minami, K.; Ebara, M.; Nakanishi, J. What are the emerging concepts and challenges in NANO? Nanoarchitectonics, hand-operating nanotechnology and mechanobiology. *Polym. J.* **2016**, *48*, 371–389. [CrossRef]
3. Chen, G.; Roy, I.; Yang, C.; Prasad, P.N. Nanochemistry and nanomedicine for nanoparticle-based diagnostics and therapy. *Chem. Rev.* **2016**, *116*, 2826–2885. [CrossRef] [PubMed]
4. Yamamoto, E.; Kuroda, K. Colloidal Mesoporous Silica Nanoparticles. *Bull. Chem. Soc. Jpn.* **2016**, *89*, 501–539. [CrossRef]
5. Nakanishi, W.; Minami, K.; Shrestha, L.K.; Ji, Q.; Hill, J.P.; Ariga, K. Bioactive nanocarbon assemblies: Nanoarchitectonics and applications. *Nano Today* **2014**, *9*, 378–394. [CrossRef]
6. Wicki, A.; Witzigmann, D.; Balasubramanian, V.; Huwyler, J. Nanomedicine in cancer therapy: Challenges, opportunities, and clinical applications. *J. Control. Release* **2015**, *200*, 138–157. [CrossRef] [PubMed]
7. Karimi, M.; Ghasemi, A.; Sahandi, Z.P.; Rahighi, R.; Moosavi, B.S.M.; Mirshekari, H.; Amiri, M.; Shafaei, P.Z.; Aslani, A.B.M.; Ghosh, D.; et al. Smart micro/nanoparticles in stimulus-responsive drug/gene delivery systems. *Chem. Soc. Rev.* **2016**, *45*, 1457–1501. [CrossRef] [PubMed]
8. Koedrith, P.; Boonprasert, R.; Kwon, J.Y.; Kim, I.; Seo, Y.R. Recent toxicological investigations of metal or metal oxide nanoparticles in mammalian models in vitro and in vivo: DNA damaging potential, and relevant physicochemical characteristics. *Mol. Toxicol.* **2014**, *10*, 107–126. [CrossRef]
9. Watanabe, M.; Yoneda, M.; Morohashi, A.; Hori, Y.; Okamoto, D.; Sato, A.; Kurioka, D.; Nittami, T.; Hirokawa, Y.; Shiraishi, T.; et al. Effects of Fe_3O_4 magnetic nanoparticles on A549 cells. *Int. J. Mol. Sci.* **2013**, *14*, 15546–15560. [CrossRef] [PubMed]
10. Wydra, R.J.; Oliver, C.E.; Anderson, K.W.; Dziubla, T.D.; Hilt, J.Z. Accelerated generation of free radicals by iron oxide nanoparticles in the presence of an alternating magnetic field. *RSC Adv.* **2015**, *5*, 18888–18893. [CrossRef] [PubMed]
11. Dissanayake, N.M.; Current, K.M.; Obare, S.O. Mutagenic effects of iron oxide nanoparticles on biological cells. *Int. J. Mol. Sci.* **2015**, *16*, 23482–23516. [CrossRef] [PubMed]
12. Marnett, L.J. Oxyradicals and DNA damage. *Carcinogenesis* **2000**, *21*, 361–370. [CrossRef] [PubMed]
13. Singh, N.; Jenkins, G.J.; Asadi, R.; Doak, S.H. Potential toxicity of superparamagnetic iron oxide nanoparticles (SPION). *Nano Rev.* **2010**, *1*, 1–15. [CrossRef] [PubMed]
14. Jeong, Y.C.; Swenberg, J.A. Formation of M_1G-dR from endogenous and exogenous ROS-inducing chemicals. *Free Radic. Biol. Med.* **2005**, *39*, 1021–1029. [CrossRef] [PubMed]
15. Zhou, X.; Taghizadeh, K.; Dedon, P.C. Chemical and biological evidence for base propenals as the major source of the endogenous M_1dG adduct in cellular DNA. *J. Biol. Chem.* **2005**, *280*, 25377–25382. [CrossRef] [PubMed]
16. Cooke, M.S.; Evans, M.D.; Dizdaroglu, M.; Lunec, J. Oxidative DNA damage: Mechanisms, mutation, and disease. *FASEB J.* **2003**, *17*, 1195–1214. [CrossRef] [PubMed]
17. Brancato, B.; Munnia, A.; Cellai, F.; Ceni, E.; Mello, T.; Bianchi, S.; Catarzi, S.; Risso, G.G.; Galli, A.; Peluso, M.E. 8-Oxo-7,8-dihydro-2'-deoxyguanosine and other lesions along the coding strand of the exon 5 of the tumour suppressor gene *P53* in a breast cancer case-control study. *DNA Res.* **2016**, *23*, 395–402. [CrossRef] [PubMed]
18. Bargonetti, J.; Manfredi, J.J. Multiple roles of the tumor suppressor p53. *Curr. Opin. Oncol.* **2002**, *14*, 86–91. [CrossRef]
19. Loft, S.; Moller, P. Oxidative DNA damage and human cancer: Need for cohort studies. *Antioxid. Redox Signal.* **2006**, *8*, 1021–1031. [CrossRef] [PubMed]

20. Peluso, M.; Munnia, A.; Risso, G.G.; Catarzi, S.; Piro, S.; Ceppi, M.; Giese, R.W.; Brancato, B. Breast fine-needle aspiration malondialdehyde deoxyguanosine adduct in breast cancer. *Free Radic. Res.* **2011**, *45*, 477–482. [CrossRef] [PubMed]

21. Munnia, A.; Amasio, M.E.; Peluso, M. Exocyclic malondialdehyde and aromatic DNA adducts in larynx tissues. *Free Radic. Biol. Med.* **2004**, *37*, 850–858. [CrossRef] [PubMed]

22. Wang, M.; Dhingra, K.; Hittelman, W.N.; Liehr, J.G.; de Andrade, M.; Li, D. Lipid peroxidation-induced putative malondialdehyde—DNA adducts in human breast tissues. *Cancer Epidemiol. Biomark. Prev.* **1996**, *5*, 705–710.

23. Sorensen, M.; Autrup, H.; Moller, P.; Hertel, O.; Jensen, S.S.; Vinzents, P.; Knudsen, L.E.; Loft, S. Linking exposure to environmental pollutants with biological effects. *Mutat. Res.* **2003**, *544*, 255–271. [CrossRef] [PubMed]

24. Vanhees, K.; van Schooten, F.J.; van Doorn-Khosrovani, S.B.; van Helden, S.; Munnia, A.; Peluso, M.; Briede, J.J.; Haenen, G.R.; Godschalk, R.W. Intrauterine exposure to flavonoids modifies antioxidant status at adulthood and decreases oxidative stress-induced DNA damage. *Free Radic. Biol. Med.* **2013**, *57*, 154–161. [CrossRef] [PubMed]

25. Izzotti, A.; Balansky, R.M.; Dagostini, F.; Bennicelli, C.; Myers, S.R.; Grubbs, C.J.; Lubet, R.A.; Kelloff, G.J.; de Flora, S. Modulation of biomarkers by chemopreventive agents in smoke-exposed rats. *Cancer Res.* **2001**, *61*, 2472–2479. [PubMed]

26. Izzotti, A.; Cartiglia, C.; De Flora, S.; Sacca, S. Methodology for evaluating oxidative DNA damage and metabolic genotypes in human trabecular meshwork. *Toxicol. Mech. Methods* **2003**, *13*, 161–168. [CrossRef] [PubMed]

27. Beik, J.; Abed, Z.; Ghoreishi, F.S.; Hosseini-Nami, S.; Mehrzadi, S.; Shakeri-Zadeh, A.; Kamrava, S.K. Nanotechnology in hyperthermia cancer therapy: From fundamental principles to advanced applications. *J. Control. Release* **2016**, *235*, 205–221. [CrossRef] [PubMed]

28. Link, E.M.; Riley, P.A. Role of hydrogen peroxide in the cytotoxicity of the xanthine/xanthine oxidase system. *Biochem. J.* **1988**, *249*, 391–399. [CrossRef] [PubMed]

29. European Standards Committee on Oxidative DNA Damage (ESCODD). Comparative analysis of baseline 8-oxo-7,8-dihydroguanine in mammalian cell DNA, by different methods in different laboratories: An approach to consensus. *Carcinogenesis* **2003**, *23*, 2129–2133.

30. Reimer, P.; Balzer, T. Ferucarbotran (Resovist): A new clinically approved RES-specific contrast agent for contrast-enhanced MRI of the liver: Properties, clinical development, and applications. *Eur. Radiol.* **2003**, *13*, 1266–1276. [PubMed]

31. Kim, J.S.; Yoon, T.J.; Yu, K.N.; Kim, B.G.; Park, S.J.; Kim, H.W.; Lee, K.H.; Park, S.B.; Lee, J.K.; Cho, M.H. Toxicity and tissue distribution of magnetic nanoparticles in mice. *Toxicol. Sci.* **2006**, *89*, 338–347. [CrossRef] [PubMed]

32. Malvindi, M.A.; De Matteis, V.; Galeone, A.; Brunetti, V.; Anyfantis, G.C.; Athanassiou, A.; Cingolani, R.; Pompa, P.P. Toxicity assessment of silica coated iron oxide nanoparticles and biocompatibility improvement by surface engineering. *PLoS ONE* **2014**, *9*, e85835. [CrossRef] [PubMed]

33. Valdiglesias, V.; Fernández-Bertólez, N.; Kiliç, G.; Costa, C.; Costa, S.; Fraga, S.; Bessa, M.J.; Pásaro, E.; Teixeira, J.P.; Laffon, B. Are iron oxide nanoparticles safe? Current knowledge and future perspectives. *J. Trace Elem. Med. Biol.* **2016**, *38*, 53–63. [CrossRef] [PubMed]

34. Lin, X.L.; Zhao, S.H.; Zhang, L.; Hu, G.Q.; Sun, Z.W.; Yang, W.S. Dose-dependent cytotoxicity and oxidative stress induced by "naked" Fe_3O_4 Nanoparticles in human hepatocyte. *Chem. Res. Chin. Univ.* **2012**, *28*, 114–118.

35. Sadeghi, L.; Tanwir, F.; Yousefi Babadi, V. In vitro toxicity of iron oxide nanoparticle: Oxidative damages on Hep G2 cells. *Exp. Toxicol. Pathol.* **2015**, *67*, 197–203. [CrossRef] [PubMed]

36. Ma, P.; Luo, Q.; Chen, J.; Gan, Y.; Du, J.; Ding, S.; Xi, Z.; Yang, X. Intraperitoneal injection of magnetic Fe_3O_4-nanoparticle induces hepatic and renal tissue injury via oxidative stress in mice. *Int. J. Nanomed.* **2012**, *7*, 4809–4818.

37. Shaw, J.; Raja, S.O.; Dasgupta, A.K. Modulation of cytotoxic and genotoxic effects of nanoparticles in cancer cells by external magnetic field. *Cancer Nanotechnol.* **2014**, *5*, 2. [CrossRef] [PubMed]

38. Van Helden, Y.G.; Keijer, J.; Heil, S.G.; Pico, C.; Palou, A.; Oliver, P.; Munnia, A.; Briede, J.J.; Peluso, M.; Franssen-van Hal, N.L.; et al. β-carotene affects oxidative stress-related DNA damage in lung epithelial cells and in ferret lung. *Carcinogenesis* **2009**, *30*, 2070–2076. [CrossRef] [PubMed]

39. Kim, D.H.; Kim, J.; Choi, W. Effect of magnetic field on the zero valent iron induced oxidation reaction. *J. Hazard. Mater.* **2011**, *192*, 928–931. [CrossRef] [PubMed]

40. Kadiiska, M.B.; Burkitt, M.J.; Xiang, Q.H.; Mason, R.P. Iron supplementation generates hydroxyl radical in vivo. An ESR spin-trapping investigation. *J. Clin. Investig.* **1995**, *96*, 1653–1657. [CrossRef] [PubMed]

41. Mello, T.; Zanieri, F.; Ceni, E.; Galli, A. Oxidative Stress in the Healthy and Wounded Hepatocyte: A Cellular Organelles Perspective. *Oxid. Med. Cell. Longev.* **2016**, *2016*, 832741. [CrossRef] [PubMed]

42. Orlando, A.; Colombo, M.; Prosperi, D.; Gregori, M.; Panariti, A.; Rivolta, I.; Masserini, M.; Cazzaniga, E. Iron oxide nanoparticles surface coating and cell uptake affect biocompatibility and inflammatory responses of endothelial cells and macrophages. *J. Nanopart. Res.* **2015**, *17*, 351. [CrossRef]

43. Gungor, N.; Knaapen, A.M.; Munnia, A.; Peluso, M.; Haenen, G.R.; Chiu, R.K.; Godschalk, R.W.; van Schooten, F.J. Genotoxic effects of neutrophils and hypochlorous acid. *Mutagenesis* **2010**, *25*, 149–154. [CrossRef] [PubMed]

44. Gungor, N.; Pennings, J.L.; Knaapen, A.M.; Chiu, R.K.; Peluso, M.; Godschalk, R.W.; van Schooten, F.J. Transcriptional profiling of the acute pulmonary inflammatory response induced by LPS: Role of neutrophils. *Respir. Res.* **2010**, *11*, 24. [CrossRef] [PubMed]

45. Parka, E.J.; Choia, D.H.; Kimb, H.; Leec, E.W.; Songc, J.; Chod, M.H.; Kima, J.H.; Kim, S.W. Magnetic iron oxide nanoparticles induce autophagy preceding apoptosis through mitochondrial damage and ER stress in RAW264.7 cells. *Toxicol. In Vitro* **2014**, 1402–1412. [CrossRef] [PubMed]

46. Filomeni, G.; De Zio, D.; Cecconi, F. Oxidative stress and autophagy: The clash between damage and metabolic needs. *Cell Death Differ.* **2015**, *22*, 377–388. [CrossRef] [PubMed]

47. Khan, M.; Mohammad, A.; Patil, G.; Naqvi, S.A.; Chauhan, L.K.; Ahmad, I. Induction of ROS, mitochondrial damage and autophagy in lung epithelial cancer cells by iron oxide nanoparticles. *Biomaterials* **2012**, *33*, 1477–1488. [CrossRef] [PubMed]

48. Alarifi, S.; Ali, D.; Alkahtani, S.; Alhader, M.S. Iron oxide nanoparticles induce oxidative stress, DNA damage, and caspase activation in the human breast cancer cell line. *Biol. Trace Elem. Res.* **2014**, *159*, 416–424. [CrossRef] [PubMed]

49. Bono, R.; Romanazzi, V.; Munnia, A.; Piro, S.; Allione, A.; Ricceri, F.; Guarrera, S.; Pignata, C.; Matullo, G.; Wang, P.; et al. Malondialdehyde-deoxyguanosine adduct formation in workers of pathology wards: The role of air formaldehyde exposure. *Chem. Res. Toxicol.* **2010**, *23*, 1342–1348. [CrossRef] [PubMed]

50. Wang, P.; Gao, J.; Li, G.; Shimelis, O.; Giese, R.W. Nontargeted analysis of DNA adducts by mass-tag MS: Reaction of p-benzoquinone with DNA. *Chem. Res. Toxicol.* **2012**, *25*, 2737–2743. [CrossRef] [PubMed]

51. Peluso, M.; Castegnaro, M.; Malaveille, C.; Talaska, G.; Vineis, P.; Kadlubar, F.; Bartsch, H. ^{32}P-postlabelling analysis of DNA adducted with urinary mutagens from smokers of black tobacco. *Carcinogenesis* **1990**, *11*, 1307–1311. [CrossRef] [PubMed]

52. Peluso, M.; Srivatanakul, P.; Munnia, A.; Jedpiyawongse, A.; Ceppi, M.; Sangrajrang, S.; Piro, S.; Boffetta, P. Malondialdehyde-deoxyguanosine adducts among workers of a Thai industrial estate and nearby residents. *Environ. Health Perspect.* **2010**, *118*, 55–59. [CrossRef] [PubMed]

53. Munnia, A.; Saletta, F.; Allione, A.; Piro, S.; Confortini, M.; Matullo, G.; Peluso, M. ^{32}P-Post-labelling method improvements for aromatic compound-related molecular epidemiology studies. *Mutagenesis* **2007**, *22*, 381–385. [CrossRef] [PubMed]

54. Valavanidis, A.; Vlachogianni, T.; Fiotakis, C. 8-hydroxy-2′-deoxyguanosine (8-OHdG): A critical biomarker of oxidative stress and carcinogenesis. *J. Environ. Sci. Health. Part C Environ. Carcinog. Ecotoxicol. Rev.* **2009**, *27*, 120–139. [CrossRef] [PubMed]

55. Izzotti, A.; De Flora, S.; Cartiglia, C.; Are, B.M.; Longobardi, M.; Camoirano, A.; Mura, I.; Dore, M.P.; Scanu, A.M.; Rocca, P.C.; et al. Interplay between Helicobacter pylori and host gene polymorphisms in inducing oxidative DNA damage in the gastric mucosa. *Carcinogenesis* **2007**, *28*, 892–898. [CrossRef] [PubMed]

56. Balansky, R.; Izzotti, A.; D'Agostini, F.; Longobardi, M.; Micale, R.T.; La Maestra, S.; Camoirano, A.; Ganchev, G.; Iltcheva, M.; Steele, V.E.; et al. Assay of lapatinib in murine models of cigarette smoke carcinogenesis. *Carcinogenesis* **2014**, *35*, 2300–2307. [CrossRef] [PubMed]

57. Micale, R.T.; La Maestra, S.; Di Pietro, A.; Visalli, G.; Baluce, B.; Balansky, R.; Steele, V.E.; de Flora, S. Oxidative stress in the lung of mice exposed to cigarette smoke either early in life or in adulthood. *Arch. Toxicol.* **2013**, *87*, 915–918. [CrossRef] [PubMed]

58. Devanaboyina, U.; Gupta, R.C. Sensitive detection of 8-hydroxy-2′deoxyguanosine in DNA by ^{32}P-postlabeling assay and the basal levels in rat tissues. *Carcinogenesis* **1996**, *17*, 917–924. [CrossRef] [PubMed]

59. Schindelin, J.; Arganda-Carreras, I.; Frise, E.; Kaynig, V.; Longair, M.; Pietzsch, T.; Preibisch, S.; Rueden, C.; Saalfeld, S.; Schmid, B.; et al. Fiji: An open-source platform for biological-image analysis. *Nat. Methods* **2012**, *9*, 676–682. [CrossRef] [PubMed]

International Journal of
Molecular Sciences

MDPI

Review
The Role of Resveratrol in Cancer Therapy

Jeong-Hyeon Ko [1], Gautam Sethi [2,3,4,*], Jae-Young Um [1], Muthu K Shanmugam [4], Frank Arfuso [5], Alan Prem Kumar [4], Anupam Bishayee [6] and Kwang Seok Ahn [1,*]

[1] College of Korean Medicine, Kyung Hee University, 24 Kyungheedae-ro, Dongdaemun-gu,
    Seoul 02447, Korea; gokjh1647@gmail.com (J.-H.K.); jyum@khu.ac.kr (J.-Y.U.)
[2] Department for Management of Science and Technology Development, Ton Duc Thang University,
    Ho Chi Minh City 700000, Vietnam; gautam.sethi@tdt.edu.vn
[3] Faculty of Pharmacy, Ton Duc Thang University, Ho Chi Minh City 700000, Vietnam;
    gautam.sethi@tdt.edu.vn
[4] Department of Pharmacology, Yong Loo Lin School of Medicine, National University of Singapore,
    Singapore 117600, Singapore; phcgs@nus.edu.sg (G.S.); phcsmk@nus.edu.sg (M.K.S.);
    csiapk@nus.edu.sg (A.P.K.)
[5] Stem Cell and Cancer Biology Laboratory, School of Biomedical Sciences, Curtin Health Innovation Research
    Institute, Curtin University, Perth WA 6009, Australia; frank.arfuso@curtin.edu.au
[6] Department of Pharmaceutical Sciences, College of Pharmacy, Larkin University, Miami, FL 33169, USA;
    abishayee@ularkin.org
* Correspondence: gautam.sethi@tdt.edu.vn or phcgs@nus.edu.sg (G.S.); ksahn@khu.ac.kr (K.S.A.);
    Tel.: +82-2-961-2316 (K.S.A)

Received: 15 November 2017; Accepted: 29 November 2017; Published: 1 December 2017

Abstract: Natural product compounds have recently attracted significant attention from the scientific community for their potent effects against inflammation-driven diseases, including cancer. A significant amount of research, including preclinical, clinical, and epidemiological studies, has indicated that dietary consumption of polyphenols, found at high levels in cereals, pulses, vegetables, and fruits, may prevent the evolution of an array of diseases, including cancer. Cancer development is a carefully orchestrated progression where normal cells acquires mutations in their genetic makeup, which cause the cells to continuously grow, colonize, and metastasize to other organs such as the liver, lungs, colon, and brain. Compounds that modulate these oncogenic processes can be considered as potential anti-cancer agents that may ultimately make it to clinical application. Resveratrol, a natural stilbene and a non-flavonoid polyphenol, is a phytoestrogen that possesses anti-oxidant, anti-inflammatory, cardioprotective, and anti-cancer properties. It has been reported that resveratrol can reverse multidrug resistance in cancer cells, and, when used in combination with clinically used drugs, it can sensitize cancer cells to standard chemotherapeutic agents. Several novel analogs of resveratrol have been developed with improved anti-cancer activity, bioavailability, and pharmacokinetic profile. The current focus of this review is resveratrol's in vivo and in vitro effects in a variety of cancers, and intracellular molecular targets modulated by this polyphenol. This is also accompanied by a comprehensive update of the various clinical trials that have demonstrated it to be a promising therapeutic and chemopreventive agent.

Keywords: Resveratrol; cancer; molecular targets; apoptosis; chemoprevention; therapy

1. Introduction

Cancer is one of the most commonly diagnosed diseases, and its related morbidity and mortality constitute a very significant health problem worldwide. Although great efforts have been made to discover a cure, cancer remains a very prominent cause of mortality in humans, and effective treatment remains a formidable challenge. An estimated 1.6 million new cancer diagnoses and approximately

600,000 cancer-related deaths are expected in the United States in 2017 alone [1]. Despite several novel improvements in diagnosis and surveillance, the overall cancer survival rate has not improved. Several personalized care medicines, such as targeted therapies, have emerged, providing improved clinical outcomes for cancer patients [2]. However, some of the recent advanced improvements in treating cancer have resulted in development of acquired resistance to chemotherapeutic agents [3]. Carcinogenesis is a multistep and multifactorial process involving the occurrence of clear and discrete molecular and cellular alterations; there are distinct but closely connected phases of initiation, promotion, and progression [4–6]. Current cancer therapies, e.g., chemotherapy, targeted agents, radiation, surgery, and immunosuppression, have limitations resulting from the development of resistance to the therapy [7]. The identification of protective molecules without side effects remains a primary objective in the fight against cancer. The other options aim at the early detection of cancer in the benign stage, which can help with its proper management [8].

Since ancient times, natural products have been used to prevent several chronic diseases, including cancer [9–18]. Revived interest in phytochemicals obtained from dietary or medicinal plant sources has provided an alternative source of bioactive compounds that can be used as preventive or therapeutic agents against a variety of diseases [19–23]. Phytochemicals such as phytoestrogens have been reported to modulate multiple cellular-signaling pathways, with no or minimal toxicity to normal cells [24,25]. The application of substances to prevent or delay the development of carcinogenesis has been termed chemoprevention [4], and there is burgeoning interest in the use of natural compounds as possible chemopreventive and therapeutic agents for human populations. Resveratrol is increasing in prominence because it has cancer-preventive and anti-cancer properties [25–28]. A non-flavonoid polyphenol, resveratrol (3,4′,5-trihydroxy-*trans*-stilbene) is a phytoalexin that naturally occurs in many species of plants, including peanuts, grapes, pines, and berries, and assists in the response against pathogen infections [29]. Interestingly, Chinese and Japanese traditional medicine also contain it, in the form of extracts such as those obtained from *Polygonum cuspidatum*, which can be used to treat inflammation, headaches, cancers, and amenorrhea.

The structure of resveratrol is stilbene-based and comprises two phenolic rings connected by a styrene double bond to produce 3,4′,5-trihydroxystilbene, which occurs in both the *trans*- and *cis*-isoforms (Figure 1). The *trans*-isoform is the major isoform, and represents the most extensively studied chemical form. Exposure to heat and ultraviolet radiation can cause the *trans*-isoform to convert into the *cis*-isoform, whose structure closely resembles that of the synthetic estrogen diethylstilbestrol. Because of this, resveratrol has also been classified as a phytoestrogen. Its biosynthetic pathway begins with a reaction between the malonyl CoA and coumaryl derivative, which is catalyzed by the enzyme stilbene synthase [30]. Resveratrol is easily available in a regular diet and has numerous health-augmenting properties, as well as some naturally occurring analogs, such as viniferins, pterostilbene, and piceid [31]. Additionally, some semi-synthetic resveratrol analogs have also been found to have certain pharmacological benefits, including chemoprevention actions, anti-oxidant effects, and anti-aging properties [32–34]. It had also been shown that resveratrol can reverse drug resistance in a variety of tumor cells by sensitizing them to chemotherapeutic agents [35,36]. In particular, it has been reported that *trans*-resveratrol and its glucoside have wide-ranging effects, including cardioprotective, anti-oxidative, anti-inflammatory, estrogenic/anti-estrogenic, and anti-tumor properties [37,38]. Moreover, the antimicrobial effects [39] of *trans*-resveratrol have been found to be useful in the management of cognitive disorders such as dementia [40,41]. This review, however, will concentrate primarily on resveratrol and discuss its diverse anti-cancer effects in various preclinical and clinical studies.

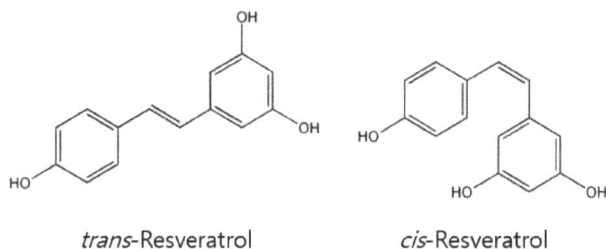

Figure 1. The chemical structure of two geometric isomers of resveratrol.

2. In Vitro Pharmacological Properties and Anti-Cancer Effects of Resveratrol

It has been shown that resveratrol possesses multifaceted salubrious properties, e.g., anti-inflammatory, anti-oxidative, and anti-aging qualities [42–44]. Resveratrol is a constituent of red wine, and therefore it is often postulated that resveratrol is a significant element in the French Paradox, the reduced risk of cardiovascular disease in French populations despite the high intake of saturated fats; that has been associated with high red wine consumption [45]. After Jang et al. [46] found that resveratrol inhibited carcinogenesis in a mouse-skin cancer model in 1997, a wealth of publications followed. It has been shown that resveratrol has in vitro cytotoxic effects against a large range of human tumor cells, including myeloid and lymphoid cancer cells, and breast, skin, cervix, ovary, stomach, prostate, colon, liver, pancreas, and thyroid carcinoma cells [25,47–49]. Resveratrol affects a variety of cancer stages from initiation and promotion to progression by affecting the diverse signal-transduction pathways that control cell growth and division, inflammation, apoptosis, metastasis, and angiogenesis.

3. Anti-Tumor-Initiation Activity

Neoplasia initiation concerns the alteration or mutation of genes resulting spontaneously from or caused by exposure to a carcinogenic agent, which finally results in mutagenesis [50]. Oxidative stress plays a dominant part in the causation of carcinogenesis [51]. Reactive oxygen species (ROS) can react with DNA in addition to chromatin proteins, resulting in several types of DNA damage [52,53]. In fact, chemical carcinogens cannot damage DNA until they are metabolized by phase-I biotransformation enzymes, especially cytochrome P450, in cells and converted to reactive electrophiles. In addition, carcinogen-DNA adduct formation gives rise to chemical carcinogenesis [54]. This initiation stage is irreversible but can be prevented by inhibiting the activity and expression of certain cytochrome P450 enzymes and augmenting the activity of phase-II detoxification enzymes, which transform carcinogens into less toxic and soluble products [55,56].

It has been found that resveratrol can inhibit events linked to the initiation of tumors. For instance, resveratrol treatment suppressed free radical formation induced by 12-*O*-tetradecanoylphorbol-13-acetate (TPA) in human leukemia HL-60 cells [57]. The diverse anti-oxidant properties of resveratrol have already been described previously [58,59]. Resveratrol is an excellent scavenger of hydroxyls and superoxides, as well as radicals induced by metals/enzymes and generated by cells [59]. It also protects against lipid peroxidation within cell membranes and damage to DNA resulting from ROS [59]. Furthermore, resveratrol functions as an anti-mutagen, as shown by its inhibition of the mutagenicity of N-methyl-N'-nitro-N-nitrosoguanidine in the *Salmonella typhimurium* strain TA100 [60]. It has been proposed that resveratrol can be a possible chemopreventive agent, and its anti-mutagenic and anti-carcinogenic properties have been demonstrated in several models [9,61,62].

In addition, resveratrol can inhibit 2,3,7,8-tetrachlorodibenzo-p-dioxin (TCDD)–induced expression of cytochrome P450 1A1 (CYP1A1) and 1B1 (CYP1B1), as well as their catalytic actions, in human breast epithelial Michigan cancer foundation (MCF)-10A cells [63]. Resveratrol can also abrogate the CYP1A activity induced by environmental aryl hydrocarbon benzo[a]pyrene (B[a]P) and catalyzed by directly suppressing the CYP1A1/1A2 enzyme activity and the signal-transduction pathway that up-regulates the expression of carcinogen-activating enzymes in human breast cancer MCF-7 and liver cancer HepG2 cells [64]. It has been reported that resveratrol also has inhibitory effects on aryl hydrocarbon receptor (AhR)–mediated activation of phase-I enzymes. The canonical AhR-dependent signaling pathway is thought to contribute to carcinogenic initiation by phase-I enzyme–activated polycyclic aromatic hydrocarbons (PAH). Briefly, PAH can bind to the AhR and facilitate its translocation into the nucleus, where the AhR develops into a heterodimer with AhR nuclear translocator (ARNT). The AhR/ARNT heterodimer then attaches to and transactivates xenobiotic response element–driven phase-I/II enzyme promoters, and initiates carcinogenesis. It has been postulated that resveratrol's inhibition of AhR signaling can suppresses this initiation process. For example, resveratrol caused inhibition of TCDD-induced recruitment of AhR and ARNT to the CYP1A1/1A2 and CYP1A1/1B1 promoter in HepG2 and MCF-7 cells, respectively, culminating in decreased expression [65]. Resveratrol also reduced TCDD-induced, AhR-mediated CYP1A1 expression in gastric cancer AGS cells [66]. Resveratrol could therefore modulate the activity and expression of some cytochrome P450 enzymes, and thereby help prevent cancer by limiting the activation of pro-carcinogens.

It has also been found that resveratrol increases both the activity and expression of NAD(P)H: quinone oxidoreductase-1 (NQO1), a carcinogen-detoxifying phase-II enzyme, in human leukemia K562 cells [67]. In addition, resveratrol was also found to induce the activity of the phase-II detoxifying metabolic enzyme quinone reductase (QR) within mouse liver-cancer Hepa 1c1c7 cells [68]. Within breast cancer cells, resveratrol induced QR expression via the estrogen receptor β (ER-β), thereby protecting against oxidative damage to DNA [69]. Resveratrol also augments the activity and expression of anti-oxidant and phase-II detoxifying enzymes through the activation of nuclear factor E2–related factor 2 (Nrf2). Nrf2 generally remains sequestered in the cytoplasm by binding Kelch-like ECH-associated protein 1 (Keap1). When Nrf2 is induced by dietary phytochemicals like resveratrol, it dissociates itself from Keap1 and translocates into the nucleus. Nrf2 thereafter attaches to the anti-oxidant response element (ARE) found in the promoters of several genes that encode phase-II enzymes, and thus regulates their transcriptional activation [70,71]. Resveratrol has been also shown to up-regulate the expression of heme oxygenase-1 (HO-1) via Nrf2 activation in PC12 cells. Resveratrol induction of the expression of NQO1 in TCDD-treated normal human breast epithelial MCF10F cells involved Nrf2, resulting in the formation of DNA adducts being suppressed [72].

Resveratrol also caused an increase in NQO1 after estradiol-3,4-quinone (E_2-3,4-Q) or 4-hydroxyestradiol (4-OHE$_2$) treatment in MCF10F cells [73]. In addition, resveratrol-induced Nrf2 signaling can lead to an increased expression of HO-1, NQO1, and the glutamate cysteine ligase (GCL) catalytic subunit in human bronchial epithelial HBE1 cells treated with cigarette-smoke extracts [74]. Resveratrol also restored glutathione levels in human lung cancer A549 cells treated with cigarette-smoke extracts, by Nrf2-induced GCL expression [75]. In leukemia K562 cells resveratrol increased NQO1 expression and induced Nrf2/Keap1/ARE binding to NQO1 promoter [67].

4. Anti-Tumor-Promotion Activity

Tumor promotion involves clonally enlarging initiated cells to create a continuously proliferating, premalignant lesion. Tumor promoters generally alter gene expression, resulting in increased cell proliferation and decreased death of cells [76]. Studies conducted in vitro have discovered that resveratrol exerts an anti-proliferative activity by inducing apoptosis. Of these, resveratrol modifies the balance of cyclins as well as cyclin-dependent kinases (CDKs), resulting in cell cycle inhibition at G0/G1 phase. For example, a link has been found between the inhibition of cyclin D1/CDK4 by resveratrol and cell cycle arrest in the G0/G1 phase within different cancer cells [77–80]. Resveratrol was also shown to increase the levels of cyclin A and E, with cell cycle arrest in the G2/M and S phases [81,82]. Similar findings have indicated that resveratrol causes the arrest of cell cycles and activation of the p53-dependent pathway [83–85].

p53, a tumor-suppressor protein, is an element critically linked to transcription, and is closely connected to the regulation of apoptosis and cell proliferation; and also acts as a key mediator in the prevention of carcinogenesis [86]. p53 that has been activated binds DNA and stimulates the expression of certain genes, e.g., *WAF1/CIP*1 encoding for p21, which belongs to the group of CDK inhibitors that are vital to the inhibition of cell growth [87]. Resveratrol reduced the development of human skin cancer A431 cells by downregulating the expression of cyclin D1, cyclin D2, and cyclin E, inhibiting the activities and/or expression of CDK2, CDK4, and CDK6, and upregulating the expression of p21. Resveratrol also suppressed the proliferation of breast cancer MCF-7 and human prostate cancer DU-145 cells [88] via modulating CDK4 and cyclin D1 expression, which have been linked to the induction of p21 and p53. When used to treat A549 cells, resveratrol caused S phase arrest, reduced retinoblastoma protein (Rb) phosphorylation, and induced p21 and p53 protein expression [89]. It has also been demonstrated that resveratrol limits the expression of Rb, another tumor-suppressor protein involved in the G1/S transition in normal conditions [79,82,85].

It has also been shown that resveratrol's anti-proliferative activity involves the stimulation of apoptosis within cancer cells [90–92]; it has been proposed that apoptosis activation could be a probable mechanism for chemotherapeutic agents to destroy cancerous cells [93,94]. In many human tumors, apoptosis has been found to be impaired, which suggests that the disruption of apoptotic functions significantly contributes to a normal cell being transformed into a tumor cell. Apoptosis is cell death that has been programmed, and a genetically regulated physiological mechanism to eliminate damaged or abnormal cells. It is also significant as a physiological-growth-control regulator and a tissue-homeostasis moderator in embryonic, fetal, and adult tissues. Apoptotic cells can be identified by regular biochemical and morphological properties, including membrane blebbing, cell shrinkage, nuclear DNA fragmentation, chromatin condensation, and formation of apoptotic bodies [95].

Apoptosis can be activated via two major pathways: the mitochondria-apoptosome-mediated intrinsic pathway and the death receptor–induced extrinsic pathway. [96,97]. The triggering of death receptors in the tumor necrosis factor (TNF) receptor superfamily, e.g., Fas (CD95/APO-1), or of TNF-related apoptosis-inducing ligand (TRAIL) receptors causes the initiator caspase-8 to be activated, which can mediate the apoptosis signal via direct cleavage of downstream effector caspases such as caspase-3 [98]. Caspases are an ubiquitous family of cysteine proteases, and have critical functions in apoptosis as upstream initiators and downstream effectors [99]. The intrinsic pathway is triggered by the dispensation of apoptogenic factors such as Omi/HtrA2, Smac/DIABLO, cytochrome c, apoptosis-inducing factors (AIFs), endonuclease G, caspase-2, or caspase-9 from the mitochondrial intermembrane space [100]. The dissemination of cytochrome c into the cytosol activates caspase-3 via the creation of the cytochrome c/apoptotic protease-activating factor-1 (Apaf-1)/caspase-9-containing apoptosome complex; Omi/HtrA2 and Smac/DIABLO encourage caspase activation by neutralizing the effects of inhibitors of apoptotic proteins (IAPs) [100,101].

Crosstalk also occurs between the two apoptotic pathways. For instance, Fas is connected to the intrinsic pathway that is regulated via the activation of caspase-8 to cause cleavage of the BID protein, causing cytochrome *c* to be released from the mitochondria [102,103]. Various apoptotic cell-death mechanisms have been propounded [104,105]. One logical approach to reducing the incidence of cancer appears to be the targeting of critical parts of apoptosis regulatory pathways, including the IAPs (in particular XIAP, cIAP1, and cIAP2), the anti-apoptotic Bcl-2 family of proteins, nuclear factor-kappa B (NF-κB), survivin, tyrosine kinases, caspases, and critical signaling pathways (phosphoinositide 3-kinase (PI3K)/AKT, STAT3/5, and MAPK pathways) [7,13,20,106–112]. Resveratrol prompts the death of tumor cells by modulating diverse signal transduction pathways via regulation of the levels of Fas and Fas-ligand (FasL) [113,114]. Resveratrol also enhances FasL expression in HL-60 cells, and the resveratrol-induced apoptosis is Fas signaling-dependent [113].

Similar outcomes have also been observed in breast [113] and colon cancer cells [114]. Mechanisms of cell death that are independent of Fas and caused by cytotoxic agents have also been propounded [115,116], and apoptosis induced by doxorubicin occurs through a Fas-independent pathway [116]. Likewise, it has been shown that resveratrol exhibits Fas-independent apoptosis in another leukemic THP-1 cell line [117]. It has also been observed that resveratrol induced the death of leukemia CEM-C7H2 cells in a Fas-independent manner, as demonstrated by the absence of apoptotic change in the presence of antibodies antagonistic to Fas or FasL [118]. Furthermore, resveratrol effectively triggered apoptosis in Fas-resistant Jurkat human leukemia cells [118].

It has been shown that resveratrol induces cell death in some cancer cells by changing the proteins of the Bcl-2 family [119]. The inhibition of anti-apoptotic proteins of the Bcl-2 family, and activation of pro-apoptotic proteins such as Bad, Bak or Bax, by resveratrol has also been shown to be a mechanism for caspase activation and cytochrome *c* release [120,121]. Interestingly, these effects may be correlated with p53 activation [122–125]. For instance, resveratrol increased the cytoplasmic concentration of calcium in human breast cancer MDA-MB-231 cells, which activated p53 and caused different pro-apoptotic genes to be transcribed [126].

It has also been shown that resveratrol induces apoptosis via inhibiting the PI3K/Akt/mTOR pathway [79,120,127–131], modulating the mitogen-activated protein kinase pathway (MAPK) [129,130,132], and inhibiting NF-κB activation [133,134]. Resveratrol triggered apoptosis within human T-cell acute lymphoblastic leukemia MOLT-4 cells by abrogating Akt phosphorylation, and subsequently preventing GSK3β from being activated [135]. Similarly, resveratrol induced apoptosis in ovarian, [136] breast, [137] uterine, [138] prostate, [120] and multiple myeloma cells [121], via inhibiting Akt phosphorylation. Chen et al. [139] determined that resveratrol inhibited the phosphorylation of PI3K/Akt (i.e., PI3K/Akt inactivation) in prostate cancer cells, resulting in decreased Forkhead box protein (FOXO) activation. Resveratrol's inhibition of the serine/threonine protein kinase Akt has been identified in anti-cancer activity modulated by the activation of FOXO3a in human breast cancer cells, because FOXO3a was not found to be activated by Akt [140].

It has been suggested that resveratrol interferes with the MAPK pathway. In cervical carcinoma cells, resveratrol inhibited the activation of p38, JNK1, and ERK2 [141]. Resveratrol activates ERK1/2 at low concentrations (1 pM–10 μM), but at higher concentrations (50–100 μM) can inhibit MAPK in human neuroblastoma SH-SY5Y cells [142]. In contrast, resveratrol activates ERK1/2 in prostate [143], breast [144,145], glial [146], head and neck [147], and ovarian cancer cells [148]. MAPKs in a constitutively active state are necessary to maintain the malignant state; however, short-term activation of MAPK may drive the cells to apoptosis [149]. It has also been reported that resveratrol causes activation of other kinases, like JNK and p38 [150]. Notably, it has been shown that the resveratrol's anti-tumor effects require p53 activation that is MAPK-induced, as well as the subsequent induction of apoptosis [151–153].

Resveratrol induces apoptosis in, and obstructs proliferation of, human multiple myeloma cells via inhibiting the constitutive activation of NF-κB through abrogating the IκB-α kinase activation, and thus down-regulating certain anti-apoptotic and pro-proliferation gene products, such as survivin, cIAP-2, cyclin D1, XIAP, Bcl-xL, Bfl-1/A1, Bcl-2, and TNF-α receptor-associated factor 2 (TRAF2) [121,154]. The constitutive activation of NF-κB, defined as the persistence of NF-κB within the nucleus, is apparent in a wide range of cancer cells [155–158]. Active NF-κB drives the expression of a plethora of genes that guard against apoptotic cell death and maintain cell proliferation [158]. Deregulation of the NF-κB signaling pathway can cause increased apoptosis as NF-κB modulates anti-apoptotic genes, e.g., TRAF1 and TRAF2, and thus changes the activities of caspases critical to the majority of apoptotic processes [159]. It has been determined that resveratrol can suppress NF-κB-regulated gene products connected with inflammation matrix metalloproteinase (MMP)-3, MMP-9, cyclooxygenase-2 (COX-2), and vascular endothelial growth factor (VEGF), inhibit anti-apoptotic proteins (Bcl-xL, Bcl-2, and TRAF1), and activate cleaved-caspase-3 [160].

Resveratrol also causes inhibition of signal transducers and activators of transcription 3 (STAT3), which adds to its pro-apoptotic and anti-proliferative potential [121]. STAT3 is a critical element in inflammation-related tumorigenesis as it promotes the proliferation, survival, invasion, angiogenesis, and metastasis of tumor cells [112,161]. The activation of NF-κB also promotes inflammation, proliferation, and tumorigenesis [162]. STAT3 and NF-κB are two central transcriptional factors linking tumorigenesis and inflammation; both of them can be activated as a response to certain stimuli, such as cytokines, growth factors, and stress signals. Abnormal signaling of STAT3 or NF-κB in malignant cells is therefore a promising target of therapy. STAT3 and NF-κB are activated via distinct pathways, and move to the nucleus to effect transcriptional activity. STAT3 and NF-κB that are constitutively activated by acetylation and/or phosphorylation in tumor cells, have been closely linked to both cancer development and progression [163,164]. Kim et al. reported that resveratrol caused inhibition of the nuclear translocation of STAT3 in renal cell carcinoma [165].

Interestingly, Wen et al. showed that inhibiting NF-κB nuclear translocation caused apoptosis in resveratrol-treated medulloblastoma cells [166]. It has been suggested that cross-talk occurs between the STAT3 and NF-κB pathways, because of the release of IL-6 and other cytokines, and because of the activation of cytokine receptors. STAT3 and NF-κB actually co-regulate many inflammatory and oncogenic genes, like *IL-1β*, *Bcl-xL*, *Myc*, *COX-2*, and *cyclin D1* [161]. By their possible functional interaction, STAT3 and NF-κB collaboratively promote the development of tumors via inducing the expression of pro-tumorigenic genes [167]. The dysregulation of these genes because of the constant activation of both STAT3 and NF-κB in tumors and the tumor microenvironment is critical to tumor progression. Inflammation can regulate angiogenesis and cellular proliferation, and inhibits apoptosis [168]. It has also been reported that resveratrol inhibits the processes of several inflammatory enzymes in vitro, e.g., COXs and lipoxygenases (LOXs) [169,170]. It was shown in a recent study that resveratrol could radiosensitize and block the STAT3 signaling pathway by inducing SOCS-1, thereby reducing STAT3 phosphorylation and proliferation in head and neck tumor cells [171].

5. Anti-Tumor-Progression Activity

Tumor progression involves several processes such as that lead to tumor metastasis. Several genes are mutated or deleted that sustain the development of aggressive tumors. The invasion and metastasis of cancer cells involve the destruction of the extracellular matrix (ECM) and basement membrane, by proteolytic enzymes, such as matrix metalloproteinases (MMPs). Of these enzymes, MMP-2 and MMP-9 are overexpressed within a variety of malignant tumors modulating cell invasion and metastasis [172]. Tissue inhibitor metalloproteinase proteins (TIMPs), on the other hand, are a protein group comprising TIMP-1, -2, -3, and -4 acting as natural MMP inhibitors [173]. To sustain their swift growth, invasive tumors also need to grow new blood vessels via a process called angiogenesis. During angiogenesis, endothelial cells can be stimulated by various growth factors, including fibroblast growth factor (FGF) and VEGF, and travel to where the new blood vessels are required. Blocking the development of new blood vessels causes the supply of nutrients and oxygen to be reduced and, as a result, the size of the tumor and metastasis may also be reduced.

It has been suggested that resveratrol plays a role in inhibiting the expression of MMP (mainly MMP-9) [174–177] and angiogenesis markers such as VEGF, EGFR or FGF-2 [79,178]. Resveratrol reduced the phorbo-12-myristate 13-acetate (PMA)-induced migration and invasion ability of liver cancer HepG2 and Hep3B cells. In HepG2 cells, resveratrol up-regulated TIMP-1 protein expression and down-regulated MMP-9 activity, while the activities of MMP-2 and MMP-9 were decreased, along with a rise in the protein-expression level of TIMP-2 in Hep3B cells [175]. HepG2 cells treated with TNF-α expressed a high level of MMP-9, which resveratrol suppressed considerably via down-regulating the expression of NF-κB, resulting in the expression of MMP-9 protein being suppressed and the invasive capability of HepG2 cells being diminished [174]. Resveratrol treatment of breast cancer MDA-MB231 cells caused inhibition of the epidermal growth factor (EGF)-induced elevation of cell migration, and of the expression of MMP-9. Resveratrol also reduced a subunit of the mammalian mediator complex for transcription (called MED28, and whose over-expression can increase migration), via the EGFR/PI3K signaling pathways [176]. Both VEGF and hypoxia-inducible factor-1α (HIF-1α) are over-expressed in several human tumors and their metastases, and are closely linked to a more aggressive tumor phenotype. It has been reported that resveratrol suppresses the expression of VEGF and HIF-1α in human ovarian cancer cells via abrogating the activation of the PI3K/Akt and MAPK signaling pathways [179]. Resveratrol caused inhibition of the expression of these molecules, which suggests that it could be part of an efficacious anti-cancer therapy for preventing cancer and its metastasis [180–182].

Malignant transformation may be linked to signaling pathways during tumorigenesis, thereby promoting epithelial-to-mesenchymal transition (EMT), which may in turn increase the invasiveness and motility of cancer cells, and trigger cancer metastasis [183,184]. Many studies have shown that resveratrol suppresses the development of tumor invasion and metastasis through inhibiting signaling pathways associated with EMT [185]. Transforming growth factor-beta (TGF-β) is a widely known cytokine that encourages invasion, proliferation, EMT, and angiogenesis of cancer cells, and the TGF-β/Smad signaling pathway can activate EMT during cancer metastasis [186,187]. Resveratrol (20 μM) inhibited TGF-β-induced EMT in A549 lung cancer cells by augmenting the expression of E-cadherin and attenuating the expression of vimentin and fibronectin, as well as the EMT-inducing transcription factors Slug and Snail [188]. Qing Ji et al. showed that resveratrol inhibited EMT induced by TGF-β, as well as the invasion and metastasis of colorectal cancer, via reducing Smad2/3 expression [189]. NF-κB can also promote EMT, in addition to cancer migration and invasion [190–192].

Several studies have shown that NF-κB is a significant EMT regulator for different types of cells [190–194]. The roles for NF-κB have been found to be linked to the expression of various genes related to EMT, such as *ZEB1*, *Snail*, *E-cadherin*, *MMP-7*, *MMP-9*, and *MMP-13* [192,193,195,196]. NF-κB can also be activated through PI3K/Akt signaling pathway to drive EMT and cancer-cell metastasis. Resveratrol suppressed the metastatic potential of pancreatic cancer PANC-1 cells in vitro by regulating factors related to EMT (vimentin, E-cadherin, N-cadherin, MMP-2, and MMP-9) and modulating the activation of PI3K/Akt/NF-κB pathways [197].

6. Pre-Clinical Studies

Resveratrol has also been reported to possess a significant anti-cancer property in various preclinical animal models (Table 1).

Table 1. In vivo anti-cancer effects of resveratrol.

Cancer Model	Animal Model	Dose	Outcome	References
Skin	DMBA/TPA model in female CD-1 mice	1, 5, 10, 25 μmol topically twice/week for 18 weeks	Incidence↓ Number of tumors per mouse↓	[46]
	Mouse xenograft models of A431 cells	10, 20, 40 μg i.p. for 14 days	Xenograft volume↓ Free radical scavenging Incidence↓ Number of tumors per mouse↓	[198]
	DMBA-initiated and TPA-promoted papillomas in female ICR mice	85 nmol/L for 21 days; topical application	Prevent onset of skin tumor	[199]
	DMBA/TPA model in CD-1 mice	1, 5, 10, 25 μmol Twice/week, for 18 Wk; topical application	Skin tumor incidence↓ Apoptosis↑; p53↑; Bax↑; cytochrome C↑; APAF↑; Bcl2↓	[200]
	DMBA-TPA–model in male Swiss albino mice	50 μmol/mouse for 3–24 week; topical application	Inhibits photocarcinogenesis; Cox2↓; lipid peroxidation↓; ODC↓	[201]
	UVB-mediated photocarcinogenesis in female SKH-1 mice	25 μmol/mouse; topical application	Decrease hyperplasia; p53↑; Cox2↓; ODC↓; survivin↓ mRNA and protein	[202]
	UVB-induced skin hyperplasia in female SKH-1 mice	10 μmol/mouse; 7 times, on alternate days; topical application	Skin tumor incidence↓ ↑Survivin mRNA and protein; ↑ phospho-survivin; ↓Smac/DIABLO	[203,204].
	UVB-induced skin tumorigenesis in female SKH-1 mice	25, 50 μmol/mouse; twice/week for 28 weeks; topical application	Suppresses melanoma tumor growth	[205]
	C57Bl/6N mice transplanted with B16-BL6 melanoma cells	50 mg/kg b.w.; i.p. for 19 days		[206]

Table 1. *Cont.*

Cancer Model	Animal Model	Dose	Outcome	References
Breast	Spontaneous mammary tumor in female FVB/N HER-2/neu mice	4 μg/mouse/day in drinking water for 2 months	Onset of tumorigenesis↓ Tumor volume↓ Multiplicity↓ Apoptosis↑	[207]
	Female athymic mice xenograft models of MDA-MB-231 cells	25 mg/kg/day i.p. daily for 3 weeks	Tumor volume↓ TUNEL staining↓ Microvessel density↓	[208]
	Female Balb/c mice xenograft with cigarette smoke condensate-transformed, MCF-10A-Tr cells in mammary fat pad	40 mg/kg/day orally for 30 days	Tumor volume↓	[209]
	DMBA-induced mammary carcinogenesis in female Sprague-Dawley rats	10 ppm mixed in diet; for 127 days	Suppressed tumor growth NF-κB↓;Cox2↓; MMP9↓	[210]
	DMBA-induced mammary carcinogenesis in female Sprague-Dawley rats	100 mg/kg b.w. mixed in diet; for 25 weeks	Suppressed tumor growth Cell proliferation↓ Apoptosis↑	[211]
	MNU-induced mammary tumorigenesis in female Sprague-Dawley rats	100 mg/kg b.w. by oral gavage for 127 days	Estrogen modulation Reduces tumor growth	[212]
	MDA-MB-231 breast tumor xenograft model	25 mg/kg b.w, by i.p., for 3 weeks	Inhibits tumor growth Apoptosis↑ Angiogenesis↓	[208]
	Female HER-2/neu transgenic mice model	0.2 mg/kg b.w in drinking water for 2 months	Delays the development and reduces the metastatic growth of spontaneous mammary tumors Apoptosis↑ ↓HER-2/neu mRNA and protein	[207]
	MDA-MB-231 breast tumor xenograft model in female athymic nu/nu mice	5 and 25 mg/kg b.w., thrice a week by oral gavage for 117 days,	In combination with quercetin and catechin retards the growth of tumor	[213]
Prostate	Athymic nude mice xenograft models of PC-3 cells	30 mg/kg/day Thrice/week, total 6 weeks	Tumor volume↓ Cell proliferation↓ Apoptosis↑ Number of blood vessels↓	[214]
	Male nude mice xenograft models with Du145-EV-Luc or Du145-MTA1 shRNA-Luc in anterior prostate	50 mg/kg/day i.p. daily 14 days after implantation, total 6 weeks	Tumor growth↓ Progression, local invasion↓ Spontaneous metastasis↓ Angiogenesis↓ Apoptosis↑	[215]
	Transgenic adenocarcinoma of mouse prostate (TRAMP) model	625 mg/kg mixed in diet for 7–23 weeks	ER-β ↑; IGF-I ↑; ↓phospho-ERK-1;↓ERK-2	[216]
	Transgenic rat adenocarcinoma of prostate (TRAP) model	50, 100 or 200 μg/ml in drinking water for 7 weeks	Apoptosis ↑; ↓AR; ↓GK11 mRNA	[217]
Lung	Female C57BL/6 mice xenograft models of LLC tumors	0.6, 2.5 or 10 mg/kg/day i.p. daily for 21 days	Tumor volume/weight↓ Metastasis to lung↓	[218]
	Nude mice xenograft models of A549	15, 30 or 60 mg/kg i.v. daily for 15 days	Tumor volume↓	[219]
	C57BL/6 mice implanted with Lewis lung carcinoma lung tumor model	5 and 25 mg/kg, i.p. for 15 days	Metastasis↓ Angiogenesis↓	[220]
	C57BL/6 mice implanted with Lewis lung carcinoma lung tumor model	20 mg/kg, i.p. for 17 days	Angiogenesis↓ Apoptosis ↑	[221]

Table 1. *Cont.*

Cancer Model	Animal Model	Dose	Outcome	References
Colon	DMH models in male Wistar rats	8 mg/kg/day orally daily for 30 weeks	Incidence↓, Tumor volume↓, Tumor burden/rat↓ Histopathological lesions DMH↓	[222]
	BP models in male Apc^Min mice	45 µg/kg/day orally, for 60 days	Number of colon adenomas↓ Dysplasia occurrence↓	[223]
	AOM induced colon cancer in male F344 rats	200 µg/kg b.w. in drinking water	Bax↑; p21↑	[224]
	ApcMin/+ mice model	0.01% in drinking water for 7 weeks	Reduce formation of tumor in small intestine cyclin D1 and D2↓	[225]
	ApcMin/+ mice model	240 mg/kg b.w. mixed in diet for 10–14 weeks	Suppress intestinal adenoma formation Cox1 and 2↓; PGE2↓	[226]
Liver	Male Donryu rats xenograft models of AH109A cells	10, 50 ppm in diet for 20 days	Tumor weight↓ Metastasis↓	[227]
	Male Wistar rats implanted with AH-130 hepatoma cells	1 mg/kg; 7 days; i.p.	Tumor weight↓ Apoptosis↑ ↑cells at G2/M	[228]
	BALB/c mice implanted with H22 hepatoma cells	500, 1000, 1500 mg/kg; 10 days; abdominal injection	Immunomodulatory activity↑	[229]
	BALB/c mice implanted with H22 hepatoma cells	5, 10, 15 mg/kg; 10 days; abdominal injection	Tumor volume↓ Apoptosis↑ cyclin B1↓; p34cdc2↓	[230]
	BALB/c mice implanted with H22 hepatoma cells	5, 10, 15 mg/kg; 10 days; abdominal injection	Synergistic anti-tumor effect in combination with 5-FU; S-phase arrest	[231]
	Female BALB/c mice implanted with HepG2 cells	15 mg/kg; every alternate day for 21 days; i.p.	Tumor growth↓ Apoptosis↑ Caspase 3↑	[232]
	DENA-initiated GST-P-positive hepatic pre-neoplastic foci in male Sprague–Dawley rats	15% (*w/w*) grape extract in diet; 11 weeks	Tumor growth↓ Lipid peroxidation↓ Fas ↓	[233]
	DENA-initiated and PB-promoted hepatocyte nodule formation in female Sprague–Dawley rats	50, 100, 300 mg/kg; 20 weeks; diet	Tumor growth↓ Apoptosis↑ Cell proliferation↓ Bcl2↓; Bax↑	[234]

↓: downregulated; ↑: upregulated; UVB: ultraviolet B; DMBA: 7,12-Dimethylbenz[a]anthracene; MNU: methyl-N-nitrosourea; AOM: azoxymethane; DENA: diethylnitrosamine; GST-P: glutathione S-transferase; PB: phenobarbital ; p53: tumor protein p53; Bax: Bcl-2-associated-X-protein; APAF: Apoptotic protease activating factor 1; Bcl2: B-cell lymphoma 2; Cox: cyclooxygenase; ODC: ornithine decarboxylase; Smac/DIABLO: Second mitochondriaderived activator of caspases /Diablo homolog; TUNEL: Terminal deoxynucleotidyl transferase dUTP nick end labeling; NF-κB: nuclear factor kappa-light-chain-enhancer of activated B cells; MMP9: matrix metalloproteinase nine; HER-2: human epidermal growth factor receptor 2; ER-β: estrogen receptor beta; IGF-I: insulin-like growth factor 1ERK: extracellular regulated kinase; AR: androgen receptor; GK11: glandular kallikrein 11; DMH: 1,2-dimethylhydrazine ; PGE2: prostaglandin E2; 5-FU: 5-fluorouracil.

7. Skin Cancer

The first preclinical study of the anti-cancer or chemopreventive effect of resveratrol was reported in a two-stage, 7,12-Dimethylbenz[a]anthracene (DMBA)-initiated and 12-O-tetradecanoyl-13-acetate (TPA)-promoted mouse-skin carcinogenesis model [46]. Thereafter, several in vivo skin cancer studies have been performed with DMBA/TPA [46,199–201,235,236], DMBA alone, [237–239], TPA alone [240–242], ultraviolet B radiation (UVB) exposure [202–204,243], benzo[a]pyrene (BP) [237], and xenograft models [198]. In the DMBA/TPA models, resveratrol treatment reduced the incidence [46,199–201,235], multiplicity [46,199,201,235], and tumor volume [201,235,236], and delayed the onset of tumorigenesis [201].

Resveratrol prevented DMBA/TPA-induced skin cancer from developing in mice, and was effective at all stages of carcinogenesis.

Soleas et al. discovered that resveratrol was somewhat efficacious in reducing the rate of tumor formation and the number of animals that developed skin tumors induced by DMBA [200]. Resveratrol inhibited tumor promotion in the DMBA–TPA mouse-skin carcinogenesis model, possibly because (at least in part) of its anti-oxidant properties [199]. Resveratrol administration restored glutathione (GSH) levels, superoxide dismutase (SOD), GSH peroxidase, and catalase activities to control values (mice without UVB irradiation) [244]. Furthermore, resveratrol exerted an anti-oxidant effect with a reduction in H_2O_2 and lipid peroxidation in the skin [202]. It has been shown that the anti-proliferative effects of this stilbene can be regulated by cell-cycle regulatory proteins such as the expression of CDK2, 4, and 6, cyclin D1 and D2, and proliferating cell nuclear antigen (PCNA), while the expression of p21 was increased [203].

Resveratrol effectively hindered the development of DMBA/TPA-induced mouse-skin tumors by inducing apoptosis, which was indicated by the induction of cytochrome *c* release, the expression of Bax, p53, and Apaf-1, and the inhibition of Bcl-2 [201]. Afaq et al. determined that resveratrol had the ability to reduce edema and inflammation resulting from short-term UVB exposure in the skin of SKH-1 hairless mice, possibly because of the inhibition of ornithine decarboxylase (ODC) [202]. Treatment with resveratrol both before and after exposure to UVB suppressed development of skin tumor [204]. Resveratrol's anti-tumor properties have also been linked to lower expression levels of TGF-β1 and augmented expression levels of E-cadherin [243]. Oral gavage of resveratrol hindered the development of a mouse melanoma (B16BL6 cell line) xenograft carried in mice, with decreased expression of Akt [245]. In a murine model of the human cutaneous skin squamous carcinoma A431 cell-line xenograft, resveratrol treatment reduced the volume of the tumor, raised the expression levels of ERK and p53, and lowered the expression level of survivin [198]. Nevertheless, resveratrol did not reduce the tumor growth of other melanoma cell lines, including A375, B16M, and DM738 xenografts in mice [246,247].

8. Breast Cancer

Resveratrol has exhibited anti-cancer and chemopreventive properties in various animal breast cancer models. Models of chemically induced mammary-gland carcinogenesis using N-methyl-nitrosourea (MNU) [212], estradiol [248], or DMBA [46], in addition to models of spontaneous mammary tumors with HER-2/neu-overexpressed [207] or Brca1-mutated (K14cre; Brca1F/F; p53F/F) mice [249], have been employed to determine resveratrol's preventive or curative effects. Oral administration of resveratrol was also found to reduce tumorigenesis induced by N-nitroso-N-methylurea (NMU) in rats [212,250].

Resveratrol, in a xenograft animal model, inhibited the development of ER-β–positive MDA-MB-231 and estrogen receptor (ER)-α–negative tumor explants, raised apoptosis, and lowered angiogenesis in nude mice [208]. However, resveratrol did not affect the in vivo development and metastasis of transplanted ER-α–negative 4T1 murine mammary cancer cells in nude mice [251]. Bove et al. studied resveratrol's in vivo effect with doses of 1–5 mg/kg per day administered intraperitoneally, and proposed that this ineffectiveness may have been the result of an insufficient dose of resveratrol. In another study, oral resveratrol at 100 or 200 mg/kg inhibited the development of 4T1 cells and metastasis in mouse lungs [252]. These findings were linked to both the reduced activity and expression of MMP-9. These data suggest that resveratrol's effects on breast cancer hinge on the dose and route of administration.

With breast cancer cell–implanted fat-pad models employing cigarette smoke condensate–transformed MCF-10ATr cells [209] or SUM159 cells [253], resveratrol caused down-regulation of the expression of various proteins linked to survival and cell proliferation (cyclin D1, PI3K, PCNA, and β-catenin), proteins related to DNA repair (Fen-1, DNA-ligase-I, Pol-δ, and Pol-ε), and an anti-apoptotic protein (Bcl-xL). It also caused an up-regulation of the pro-apoptotic protein Bax and tumor-suppressor gene p21 in mouse mammary tissue [209,253]. When used to supplement drinking water, resveratrol delayed the growth of spontaneous mammary tumors in HER-2/neu transgenic mice, and lowered the mean size and number of mammary tumors by causing down-regulation of the HER-2/neu gene expression and raising apoptosis in the mammary glands of these mice [207].

9. Prostate Cancer

Dietary resveratrol considerably lowered the incidence of prostatic adenocarcinoma in the transgenic adenocarcinoma mouse prostate (TRAMP) model [216]. Resveratrol suppressed prostate cancer growth via down-regulating the androgen receptor (AR) expression in the TRAMP model of prostate cancer. Additionally, besides down-regulating the AR expression, resveratrol also suppressed the mRNA level of androgen-responsive glandular kallikrein 11, which has been determined to be an ortholog of the human prostate specific antigen (PSA) [217]. In a xenograft model, resveratrol delayed the development of AR-positive LNCaP tumors and inhibited the expression of steroid hormone response markers [254].

With the use of AR-negative PC-3 human prostate cancer–cell xenografts in the flank regions of mice, post-treatment with oral resveratrol (30 mg/kg/day) decreased the volume of tumors, with lowered tumor-cell proliferation and neovascularization, and induced apoptosis [214]. Intraperitoneal post-treatment with resveratrol (25 mg/kg/day) also decreased the tumor volume of PC-3 cell xenografts in mouse prostates [255]. Additionally, intraperitoneal post-treatment of resveratrol (50 mg/kg/day) in the orthotopic DU-145 prostate cancer model decreased the growth, progression, local invasion, and spontaneous metastasis of tumors [215].

10. Colorectal Cancer

Colorectal cancers arise due to several factors such as diet rich in red meat and processed meat and other lifestyle factors such as smoking and drinking alcohol [256],. Resveratrol's in vivo effectiveness has been tested with colorectal cancer models employing genetically modified animals such as Apc$^{Pirc/+}$ rats and Apc$^{Min/+}$ mice. Colon cancer can be induced by chemical carcinogens, which include azoxymethane (AOM), AOM plus dextran sulfate sodium (DSS), 2-amino-1-methyl-6-phenylimidazo[4,5-b]pyridine, 2-amino-3-methylimidazo[4,5-f]quinoline, and 1,2-dimethylhydrazine (DMH) [257,258]. The pathophysiological and histopathological features/ manifestations of colon cancer include aberrant crypt foci (ACF), hyperplasia, adenocarcinoma, and adenoma [258]. In models induced with AOM or AOM plus DSS, the oral administration (in the gavage or diet) of resveratrol decreased the incidence [259,260], individual size [224], and multiplicity [224,259,261] of ACF in rodent models, and triggered biomarker alterations.

Resveratrol augmented the expression of Bax [224], p53, and p-p53 at Ser15 [259], HO-1 [261], glutathione reductase (GR) [261], and Nrf2 [261], and reduced the expression of COX-2 [259,261], inducible nitric oxide synthase (iNOS) [259,261], TNF-α [259], aldose reductase [261], NF-κB [261], and p-protein kinase C-β2 (PKC-β2) [261]. It has been propounded that resveratrol down-regulates the aldose reductase–dependent activation of NF-κB and PKC-β2, with an ensuing lowering of the expression levels of COX-2 and iNOS [261]. In models induced with DMH, resveratrol decreased the incidence, [222] size [222,262], and multiplicity of ACF [222,262,263], as well as histopathological lesions [222] and DNA damage in leukocytes [264]. When used against colon carcinogenesis, the anti-tumor effects of resveratrol were found to be accompanied by alterations in the activities of enzymes. In rat models, the processes of anti-oxidant enzymes, including catalase (CAT) and SOD in the intestine/colon [262], liver [265], and erythrocytes [264], were augmented, and the processes of biotransforming enzymes, including β-glucosidase, β-glucuronidase, β-galactosidase, nitroreductase, and mucinase, in fresh fecal and colonic mucosal samples were reduced [222]. Resveratrol lowered the expression levels of ODC, COX-2, Mucin 1, cell surface associated (MUC1), heat-shock protein (Hsp)27, and Hsp70 in colonic mucosa [266], and increased the expression levels of caspase-3 in the colonic mucosa [266], and increased glutathione in the reduced state (GSH) in the liver, intestine/colon, plasma, and erythrocytes [262,264,265].

In models with genetically modified mice (e.g., Apc$^{Min/+}$ mice [223,225,226]), and in mice with the APC locus knockout and activated *KRAS* [267], resveratrol supplementation inhibited the development of colon tumors [223,225,226,267,268] and occurrence of dysplasia [223].

11. Liver Cancer

The anti-cancer potential of resveratrol in liver carcinogenesis was exemplified by a decreased incidence and smaller numbers of nodules in models of animals employing chemical carcinogens [e.g., diethylnitrosamine (DENA) [269], DENA plus phenobarbital [234,270], and DENA plus 2-acetylaminofluorene (2-AAF) [271] or transgenic mice (e.g., hepatitis B virus X protein (HBx)-expressing transgenic mice) [272]. Additionally, resveratrol's anti-tumor effects have been reported in xenograft models using hepatoma cell lines (e.g., H22, AH-130, HepG2, and AH109A) [227–229,232]. Dietary resveratrol completely prevented DENA-induced lipid peroxidation and enhanced protein carbonyl formation, which indicates that it may also attenuate oxidative stress in the liver. Resveratrol also elevated the expression of hepatic Nrf2 and reduced the expression of iNOS. That study reported that the attenuation of oxidative and nitrosative stress and the alleviation of the inflammatory response could be mediated through the transcriptional and translational regulation of Nrf2 signaling [273]. Recent studies with Nrf2-deficient mice have shown that Nrf2 plays a role in protecting the liver from xenobiotic-initiated hepatocarcinogenesis [274].

Rajasekaran et al. have studied resveratrol's ability to prevent or treat hepatocellular carcinoma by administering resveratrol, starting at the time of DENA injection or for 15 days after the development of hepatocellular carcinoma [269]. Resveratrol treatment at both time points also reduced cell crowding and alteration in the cellular architecture, and decreased the liver size compared with control rats treated with DENA [269]. In the DENA-induced hepatocellular carcinoma model, administration of resveratrol inhibited the formation of hepatocyte nodules via down-regulating Hsp70 and COX-2 expression, through lowering the translocation of NF-κB from the cytoplasm to the nucleus [275]. Another study using the same administered dose of resveratrol also determined that the levels and expressions of hepatic TNF-α, IL-1β, and IL-6 induced by DENA can be reversed [276]. Resveratrol also exhibited a remarkable anti-angiogenic effect during the development of DENA-induced hepatocellular carcinogenesis, perhaps by blocking VEGF expression via the down-regulation of HIF-1α [277].

Resveratrol considerably lowered the cell count of a swiftly growing tumor (Yoshida AH-130 ascites hepatoma) injected into rats, thereby triggering apoptosis and cell accumulation in the G2/M phase [228]. It was further demonstrated that the inhibition of cell cycle progression involved reductions in the expression of p34cdc2 and cyclin B1 in murine transplantable liver tumors after resveratrol administration [230]. It has also been reported that resveratrol had anti-tumor-growth and anti-metastasis effects in Donryu rats that had an ascites AH109A hepatoma cell line subcutaneously implanted [227].

In another study, resveratrol inhibited tumor growth and angiogenesis in a hepatoma xenograft mouse model [278]. Salado et al. used B16 melanoma (B16M) cells to study the effects of resveratrol treatment on hepatic metastasis caused mainly by the production of pro-inflammatory cytokines [279]. Lin et al. investigated the effects of treatment with resveratrol on the precancerous stage of liver carcinogenesis in spontaneously induced hepatocellular carcinoma in HBx transgenic mice [272]. Resveratrol supplementation significantly reduced the incidence of hepatocellular carcinoma and increased the latency of tumor formation. Resveratrol inhibited hepatic lipogenesis and intracellular ROS, and the results from liver cancer models have been consistently positive, indicating the potential benefit of resveratrol in hepatocellular carcinoma prevention and/or therapy.

12. Pancreatic Cancer

Several lines of evidence suggests that age, being overweight, pancreatitis and family history of pancreatic cancer are the major risk factor for the development of pancreatic cancer. Within a xenograft mouse model, resveratrol delayed or suppressed the promotion of pancreatic cancer via inhibiting the activity of leukotriene A4 hydrolase (LTA$_4$H), which stimulates the generation of pro-inflammatory cytokines and mediators [280], and also stimulates cancer cell proliferation [281,282]. Resveratrol blocked the tumor development of PANC-1 cells orthotopically implanted in nude mice, with augmented expression of apoptosis/cell cycle arrest proteins including Bim, p27, and cleaved caspase-3, and reduced cell survival/proliferation markers including PCNA expression and the phosphorylation of PI3K, ERK, Akt, FOXO3a (Ser253), and p-FOXO1 (Ser256) in tumor tissues [283]. Resveratrol treatment inhibited the formation and development of pancreatic cancer in Kras^{G12D} transgenic mice that spontaneously develop pancreatic tumors [284]. However, dietary resveratrol had no anti-carcinogenic effect on BOP (N-nitrosobis(2-oxopropyl)amine)-induced pancreatic carcinogenesis in hamsters [285]. Further studies are necessary for additional preclinical evaluation of the efficacy of resveratrol in treating pancreatic cancer.

13. Lung Cancer

In preclinical models, lung carcinogenesis is known to be induced by a variety of agents, including diethylnitrosamine (DEN), nitrosamine 4-(methyl-nitrosamino)-1-(3-pyridyl)-1-butanone (NNK), uracil mustard, vinyl carbamate, urethane, MNU, and BP [11]. In the BP-induced mouse lung carcinogenesis model, resveratrol treatment lowered the level of BP diolepoxide (BPDE)-DNA adducts [286], improved the ultrahistoarchitecture [287], and reduced the size of tumor nodules by increasing pulmonary caspase-3 and -9 activity. It also abrogated glucose uptake/turnover, reduced the serum lactate dehydrogenase (LDH) activity (which is heightened in cancer cells), and lowered the p-p53 levels at Ser15 (the hyperphosphorylation of which can result in the inactivation of p53) [288]. In Lewis lung carcinoma cell xenograft models, treatment with resveratrol reduced the growth of tumors [218,221]. It has been also discovered that treatment with resveratrol reduced the development of A549 and MSTO-211H xenografts in mice [219,289,290].

Resveratrol's anti-tumor effects in A549 xenografts were reduced in Forkhead box protein C2 (FOXC2)-overexpressing A549 xenografts, which suggests that resveratrol possibly induces anti-tumor activity through FOXC2 [289]. Another study discovered that resveratrol did not affect the development of Lewis lung carcinoma implanted in mice, but demonstrated an evident anti-metastatic effect, decreasing both the weight and number of lung metastases [220]. However, resveratrol used to supplement the diet did not affect lung tumor multiplicity in BP plus NNK-induced lung carcinogenesis in A/J mice [291]. Similarly, in BP-induced lung carcinogenesis, resveratrol did not cause a change in the expression levels of BP-metabolizing genes (such as CYP1A1 and CYP1B1) and the number of B[a]P-protein adducts in lung tissues [292]. Another study found that both the natural Egr-1 promoter and the synthetic promoter triggered the expression of GADD45α when used with resveratrol, and then suppressed the proliferation of A549 lung cancer cells and induced apoptosis [293].

14. Other Cancers

Resveratrol provides considerable protection against the induction of cancer within the oral cavity [294] and the esophagus [295], among other tissues. Its cancer chemopreventive activity aside, resveratrol can also inhibit the development and/or induce the regression of established tumors in xenograft models for cancers of the ovaries [296], urinary bladder [79], stomach [297], and head and neck [298,299]. Resveratrol treatment effectively suppressed the growth rate of and augmented apoptosis in neuroblastoma; this was accompanied by the up-regulation of cyclin E and the down-regulation of p21 [300]. It has recently been demonstrated that resveratrol considerably reduced tumor growth via inducing apoptosis, which involved direct activation of the mitochondrial intrinsic apoptotic pathway in the NGP and SK-N-AS xenograft models of human neuroblastoma [301]. Resveratrol caused significant inhibition of cerebral tumors through inducing apoptosis and inhibiting angiogenesis induced by glioma [302]. Rats that had undergone resveratrol treatment had lower growth rates of glioma, which correlated with the blood flow of tumors (signified by the color Doppler vascularity index) and density of microvessels.

Resveratrol's anti-angiogenic effect has caused researchers to investigate if it could inhibit the development of a murine fibrosarcoma; water supplemented with resveratrol indeed significantly inhibited the development of T241fibrosarcoma in mice via suppressing angiogenesis [303]. Resveratrol's in vivo anti-cancer effects were studied in N-nitrosomethyl-benzylamine (NMBA)-induced esophageal tumorigenesis in rats. Resveratrol suppressed both the size and number of NMBA-induced esophageal tumors per rat through targeting prostaglandin E2 and COXs [304]. In a gastric cancer xenograft nude mouse model, resveratrol inhibited the growth of tumors, with reductions in the expression of cyclin D1, Ki67, CDK4, and CDK6, and increases in the expression of p16, p21, and β-Gal [305]. Resveratrol considerably inhibited carcinoma development when it was injected in close proximity to the carcinoma in a tumor model created by transplanting human primary gastric cancer cells into the subcutaneous tissue of nude mice [297]. Resveratrol induced apoptosis in implanted tumor cells via down-regulation of the apoptosis-regulated gene Bcl-2 and up-regulation of the apoptosis-regulated gene Bax. For the anti-tumor effects in head and neck cancer, resveratrol suppressed tumor stemness via lowering the expression of mesenchymal-like protein (Vimentin) and stemness markers (Oct4 and Nestin), inducing epithelial protein expression (E-cadherin) [299], and increasing γ-histone 2AX (a DNA damage marker) and cleaved caspase-3 expression [298]. In an ovarian cancer model, resveratrol abrogated the development of NuTu-19 ovarian cancer cells in vitro. However, in vivo, when NuTu-19 cells were injected into the ovarian bursa of rats and the rats were fed with resveratrol (100 mg/kg) mixed in their diet for 28 days, the growth of the ovarian tumors was not significantly inhibited [306].

15. Clinical Trials with Resveratrol

Although it is clear that resveratrol has shown excellent anti-cancer properties, most of the studies were performed in cell-culture and pre-clinical models. These physiological effects of resveratrol were also investigated in humans because it cannot be assumed that the results of tests in animal models will hold true for humans, because of differences in genetics and metabolism profile. The pharmacokinetics, metabolism, and toxicity of resveratrol have been assessed in healthy volunteers and cancer patients [307–309]. Resveratrol is metabolized swiftly, mainly into glucuronide and sulfate conjugates that are excreted via the urine. Because of the poor bioavailability of resveratrol due to its extensive metabolism, large doses (up to a maximum of 5 g/day) have been utilized by researchers. These studies have shown that resveratrol seems to be well tolerated and safe. However, adverse effects including diarrhea, nausea, and abdominal pain were observed in subjects taking more than 1 g of resveratrol daily [307]. Subsequent clinical trials are currently investigating this dose limit [307,310]. Resveratrol's poor bioavailability is a significant issue with regard to extrapolating its effects to humans, and various approaches have been created to enhance its bioavailability [311], including consuming it with various foods [312], using it in combination with an additional phytochemical piperine [313], and using a prodrug approach [314], micronized powders [315,316], or nanotechnological formulations [317–319].

The effect of resveratrol in cancer patients has been investigated in a few clinical trials (Table 2). The first clinical trial dealing with resveratrol and cancer was performed by Nguyen et al. [320]. They examined the effects of freeze-dried grape powder (GP) (containing resveratrol and resveratrol derived from plants) on the Wnt signaling pathway, which is known to be involved in colon carcinogenesis [321], in regular colon cancer and colonic mucosa. GP administration (80 g/day containing 0.07 mg of resveratrol) for two weeks resulted in decreased Wnt target gene expression within regular mucosa, but had no effect on cancerous mucosa. This indicates that GP or resveratrol may play a beneficial part in the prevention of colon cancer, rather than in the treatment of established colon cancer. Patel et al.

studied the effects of the administration of resveratrol at 0.5 or 1 g/day for eight days on proliferation marker Ki-67 expression in colorectal tissue, and reported a 5% decrease in the proliferation of tumor cells [322]. In colorectal cancer patients with hepatic metastasis, SRT501 (a micronized resveratrol formulation manufactured by Sirtris Pharmaceuticals, a GSK Company, Cambridge, MA, USA) supplementation at 5 g/day for two weeks increased the amount of cleaved caspase-3 within hepatic tissue, which suggests that there was increased apoptosis of cancerous tissue compared with subjects treated with a placebo [315].

In a muscadine grape skin extract phase 1 study with biochemically recurrent prostate cancer patients who were assigned to a high dose (4000 mg/patient) of pulverized muscadine grape (*Vitis rotundifolia*) skin that contains ellagic acid, quercetin, and resveratrol was found to be safe and warrants further investigation in dose-evaluating phase II trial [323]. In another randomized placebo controlled clinical study using two doses of resveratrol (150 mg or 1000 mg resveratrol daily) for 4 months was found to significantly lowered serum levels of androstenedione, dehydroepiandrosterone and dehydroepiandrosterone-sulphate, whereas prostate size was unaffected in benign prostate hyperplasia patients [324].

Table 2. Selected clinical trials evaluating the effect of resveratrol in cancer patients.

Participants	Resveratrol Formulation and Dosages	Outcome	References
Colorectal cancer patients (*n* = 8)	Grape powder (80 or 120 g/day) or Resveratrol (20 or 80 mg/day) for 2 weeks	Inhibition of Wnt target gene expression in normal colonic mucosa.	[320]
Colorectal cancer patients (*n* = 20)	Resveratrol (0.5 or 1g) for 8 days	Reduction of Ki-67 levels by 5 and 7% in cancerous and normal tissue, respectively.	[322]
Colorectal cancer patients with hepatic metastasis (*n* = 6)	Micronized resveratrol (SRT5001, 5 g) for 14 days	Detection of resveratrol in hepatic tissue and increased (39%) content of cleaved caspase-3 in malignant hepatic tissue.	[315]
Multiple myeloma patients (*n* = 24)	Micronized resveratrol (SRT5001, 5 g) for 20 days in a 21 day cycle up to 12 cycles	Unacceptable safety profile and minimal efficacy in patients with relapsed/refractory multiple myeloma highlighting the risks of novel drug development in such populations.	[316]
Biochemically recurrent prostate cancer patients (*n* = 14)	Pulverized muscadine grape skin extract (MPX) 4000 mg/patient	MPX was found to be safe and warrants further investigation in dose-evaluating phase II trial	[323]
Benign prostate hyperplasis patients (*n* = 66)	Two doses of resveratrol (150 mg or 1000 mg resveratrol daily) for 4 months	Significantly lowered the serum levels of androgens with no changes in prostate tumor growth.	[324]

Primary protein carbonylation has been found to be increased several folds in presence of high levels of reactive oxygen species (ROS) such as superoxide anion free radical ($O_2{}^-$) and nitric oxide free radical (NO) and other reactive free radicals, such as hydrogen peroxide (H_2O_2), hydroxyl radical (HO), and peroxynitrite anion ($ONOO^-$). There are several sources of ROS in the digestive tract and several microbes present in the colon produce a large amount of ROS inside the cells are by products of mitochondrial respiration in aerobic metabolism, and in chronic inflammation, a large amount of ROS is produced by neutrophil phagocytosis of bacteria, granular materials, or soluble irritants [325,326]. The oxidative decomposition of polyunsaturated fatty acids can initiate chain reactions that lead to the formation of a variety of carbonyl species (three to nine carbons in length), the most reactive and cytotoxic being α,β-unsaturated aldehydes also referred to as electrophilic carbonyls. These include acrolein, glyoxal, methylglyoxal, crotonaldehyde, malondialdehyde, and 4-hydroxynonenal. Reactive ketones or aldehydes that can be reacted by 2,4-dinitrophenylhydrazine (DNPH) to form 2,4-dinitrophenylhydrazone (DNP). Ulcerative colitis (UC) is a type of chronic inflammatory bowel disease (IBD) in which oxidative stress plays a critical role in its pathogenesis and malignant progression to colorectal cancer (CRC) [327,328]. Oxidative activation of transcription factors NF-κB stimulates expression of a variety of pro-inflammatory cytokines in the intestinal epithelial cells, such as TNF-α, IL-1, IL-8, and COX-2, and promotes inflammation and carcinogenesis. Oxidative stress also activates mitogen-activated protein (MAP) kinase (MAPK) signaling pathways. The human gastrointestinal tract is exposed to carbonyl threats such as consumption red meat, alcoholic beverages and smoking increases protein carbonylation, inflammation and initiation of tumor development. However, dietary intake of green leafy vegetables, fruits, fish and wine has shown to decrease protein carbonylation [329]. It has also been reported that resveratrol supplementation at 5 mg/day for six days increased the degree of protein carbonyl concentrations and cytoprotective enzyme NQO1 in colorectal mucosa tissues from patients with colorectal cancer, compared with their control subjects [330]. However, contrary to these positive findings, some evidence that resveratrol supplementation may have adverse effects in certain cancer patients also exist. In a phase II clinical trial involving multiple myeloma patients, SRT501 supplementation at 5 g/day caused several unexpected adverse effects, including nephrotoxicity, which may have led to the death of one patient [316]. However, this high dose of SRT501 was determined to be safe in other clinical trials involving several healthy and diseased populations [315,316]. There are very low amounts of human data regarding the efficacy of resveratrol in cancer treatment. Since most of these clinical trials have had a small patient sample size and used different doses and different routes of resveratrol, the data from human clinical studies have shown inconsistent outcomes of resveratrol administration.

In addition to the effects in subjects with cancer, the effect of resveratrol in subjects with a higher cancer risk has also been demonstrated. For instance, resveratrol supplementation at 50 mg two times per day for 12 weeks reduced the DNA methylation of the tumor-suppressor gene *Ras* association domain-containing protein 1 (RASSF1A) in the breasts of women with higher risk of breast cancer [331]. It has also been shown that resveratrol supplementation at 1 g/day for 12 weeks increases the concentrations of sex steroid hormone binding globulin (SHBG), which has been linked to a reduction in the risk of breast cancer [332], and has favorable effects on estrogen metabolism; thus, it can lower risk factors for breast cancer in obese and overweight postmenopausal women [333]. Another clinical study concentrated on resveratrol's effects on potential biomarkers for cancer risk reduction. Circulating concentrations of insulin-like growth factor (IGF-1) and IGF-binding protein 3 (IGFBP-3) are linked to a higher risk of common cancers [334]. Brown et al. showed that resveratrol administration at 2.5 g/day for 29 days resulted in a reduction of the circulating levels of IGF-1 and IGFBP-3 in healthy volunteers [335]. Their research suggests that resveratrol's ability to decrease circulating IGF-1 and IGFBP-3 in humans may constitute an anti-carcinogenic mechanism. In another study, Chow et al. found that resveratrol administration at 1 g/day for four weeks modulated phase I isoenzymes (cytochrome P450) and phase II detoxification enzymes involved in carcinogen activation and detoxification [310]. However, these beneficial effects are mostly minimal and sometimes controversial. Nevertheless, it seems that resveratrol has had some beneficial effects with regard to the prevention and treatment of cancer. Therefore, the efficacy and safety of resveratrol in human trials must be further investigated to better understand and develop its therapeutic potential for cancer patients.

16. Conclusions and Future Perspectives

Using a variety of in vivo and in vitro models, it has been proven that resveratrol is capable of attenuating the various stages of carcinogenesis, some of which are briefly described in Figure 2. A vast body of experimental in vivo and in vitro studies and a few clinical trials has presented evidence of resveratrol's great potential as an anti-cancer agent, both for the prevention and therapy of a large range of cancers. Resveratrol has a very low toxicity, and, although it has multiple molecular targets, it acts on different protective and common pathways that are usually altered in a great number of tumors. This suggests that resveratrol may be more suitable for use as an anti-carcinogen and it can also effectively exert it antineoplastic effects in conjunction with diverse chemotherapeutics and targeted therapies. The ability to prevent carcinogenesis includes the inhibition of oxidative stress, inflammation, and cancer-cell proliferation, and the activation of tightly regulated cell-death mechanisms. Due to the complexity and number of cellular processes involved, however, more studies must be performed to completely understand how resveratrol could be used to prevent the development of cancer. Moreover, resveratrol's poor bioavailability in humans has been a critical concern with regard to the translation of basic research findings to the development of therapeutic agents. Although human clinical trials have produced positive findings, many conflicting results remain, which may be partly because of the dosing protocols employed. To augment resveratrol's bioavailability and as a potential adjuvant, active research should be focused on resveratrol delivery systems, formulations, and modulation of resveratrol metabolism, and resveratrol's possible interactions with other compounds, as well as the development of more bioavailable analogs of the compound.

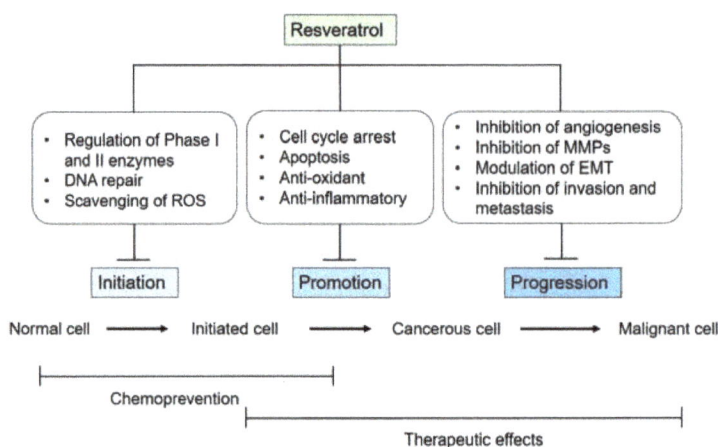

Figure 2. A schematic diagram summarizing the potential mechanism(s) underlying the anticancer effects of resveratrol.

Acknowledgments: This work was supported by a grant from the National Research Foundation of Korea (NRF) funded by the Korean government (MSIP) (NRF-2015R1A4A1042399 and NRF-2016R1A6A3A11930941).

Author Contributions: Jeong-Hyeon Ko, Jae-Young Um and Muthu K Shanmugam designed and wrote the manuscript; Frank Arfuso, Alan Prem Kumar, Anupam Bishayee, Gautam Sethi, and Kwang Seok Ahn edited and finalized the manuscript for submission.

Conflicts of Interest: The authors declare no competing financial interests.

References

1.   Siegel, R.L.; Miller, K.D.; Jemal, A. Cancer Statistics, 2017. *CA Cancer J. Clin.* **2017**, *67*, 7–30. [CrossRef] [PubMed]
2.   Okimoto, R.A.; Bivona, T.G. Recent advances in personalized lung cancer medicine. *Pers. Med.* **2014**, *11*, 309–321. [CrossRef]
3.   Krepler, C.; Xiao, M.; Sproesser, K.; Brafford, P.A.; Shannan, B.; Beqiri, M.; Liu, Q.; Xu, W.; Garman, B.; Nathanson, K.L.; et al. Personalized Preclinical Trials in BRAF Inhibitor-Resistant Patient-Derived Xenograft Models Identify Second-Line Combination Therapies. *Clin. Cancer Res.* **2016**, *22*, 1592–1602. [CrossRef] [PubMed]
4.   Hong, W.K.; Sporn, M.B. Recent advances in chemoprevention of cancer. *Science* **1997**, *278*, 1073–1077. [CrossRef] [PubMed]

5. Sethi, G.; Shanmugam, M.K.; Ramachandran, L.; Kumar, A.P.; Tergaonkar, V. Multifaceted link between cancer and inflammation. *Biosci. Rep.* **2012**, *32*, 1–15. [CrossRef] [PubMed]
6. Chai, E.Z.; Siveen, K.S.; Shanmugam, M.K.; Arfuso, F.; Sethi, G. Analysis of the intricate relationship between chronic inflammation and cancer. *Biochem. J.* **2015**, *468*, 1–15. [CrossRef] [PubMed]
7. Sethi, G.; Tergaonkar, V. Potential pharmacological control of the NF-κB pathway. *Trends Pharmacol. Sci.* **2009**, *30*, 313–321. [CrossRef] [PubMed]
8. Janakiram, N.B.; Mohammed, A.; Madka, V.; Kumar, G.; Rao, C.V. Prevention and treatment of cancers by immune modulating nutrients. *Mol. Nutr. Food Res.* **2016**, *60*, 1275–1294. [CrossRef] [PubMed]
9. Shanmugam, M.K.; Kannaiyan, R.; Sethi, G. Targeting cell signaling and apoptotic pathways by dietary agents: Role in the prevention and treatment of cancer. *Nutr. Cancer* **2011**, *63*, 161–173. [CrossRef] [PubMed]
10. Aggarwal, B.B.; Van Kuiken, M.E.; Iyer, L.H.; Harikumar, K.B.; Sung, B. Molecular targets of nutraceuticals derived from dietary spices: Potential role in suppression of inflammation and tumorigenesis. *Exp. Biol. Med.* **2009**, *234*, 825–849. [CrossRef] [PubMed]
11. Aggarwal, B.B.; Vijayalekshmi, R.V.; Sung, B. Targeting inflammatory pathways for prevention and therapy of cancer: Short-term friend, long-term foe. *Clin. Cancer Res.* **2009**, *15*, 425–430. [CrossRef] [PubMed]
12. Yang, S.F.; Weng, C.J.; Sethi, G.; Hu, D.N. Natural bioactives and phytochemicals serve in cancer treatment and prevention. *Evid.-Based Complement. Altern. Med.* **2013**, *2013*, 698190. [CrossRef] [PubMed]
13. Tang, C.H.; Sethi, G.; Kuo, P.L. Novel medicines and strategies in cancer treatment and prevention. *BioMed Res. Int.* **2014**, *2014*, 474078. [CrossRef] [PubMed]
14. Shanmugam, M.K.; Rane, G.; Kanchi, M.M.; Arfuso, F.; Chinnathambi, A.; Zayed, M.E.; Alharbi, S.A.; Tan, B.K.; Kumar, A.P.; Sethi, G. The multifaceted role of curcumin in cancer prevention and treatment. *Molecules* **2015**, *20*, 2728–2769. [CrossRef] [PubMed]
15. Kannaiyan, R.; Shanmugam, M.K.; Sethi, G. Molecular targets of celastrol derived from Thunder of God Vine: Potential role in the treatment of inflammatory disorders and cancer. *Cancer Lett.* **2011**, *303*, 9–20. [CrossRef] [PubMed]
16. Hsieh, Y.S.; Yang, S.F.; Sethi, G.; Hu, D.N. Natural bioactives in cancer treatment and prevention. *BioMed Res. Int.* **2015**, *2015*, 182835. [CrossRef] [PubMed]
17. Bishayee, A.; Sethi, G. Bioactive natural products in cancer prevention and therapy: Progress and promise. *Semin. Cancer Biol.* **2016**, *40–41*, 1–3. [CrossRef] [PubMed]
18. Shrimali, D.; Shanmugam, M.K.; Kumar, A.P.; Zhang, J.; Tan, B.K.; Ahn, K.S.; Sethi, G. Targeted abrogation of diverse signal transduction cascades by emodin for the treatment of inflammatory disorders and cancer. *Cancer Lett.* **2013**, *341*, 139–149. [CrossRef] [PubMed]
19. Shanmugam, M.K.; Nguyen, A.H.; Kumar, A.P.; Tan, B.K.; Sethi, G. Targeted inhibition of tumor proliferation, survival, and metastasis by pentacyclic triterpenoids: Potential role in prevention and therapy of cancer. *Cancer Lett.* **2012**, *320*, 158–170. [CrossRef] [PubMed]
20. Shanmugam, M.K.; Lee, J.H.; Chai, E.Z.; Kanchi, M.M.; Kar, S.; Arfuso, F.; Dharmarajan, A.; Kumar, A.P.; Ramar, P.S.; Looi, C.Y.; et al. Cancer prevention and therapy through the modulation of transcription factors by bioactive natural compounds. *Semin. Cancer Biol.* **2016**, *40–41*, 35–47. [CrossRef] [PubMed]
21. Shanmugam, M.K.; Arfuso, F.; Kumar, A.P.; Wang, L.; Goh, B.C.; Ahn, K.S.; Bishayee, A.; Sethi, G. Modulation of diverse oncogenic transcription factors by thymoquinone, an essential oil compound isolated from the seeds of *Nigella sativa* Linn. *Pharmacol. Res.* **2017**. [CrossRef] [PubMed]
22. Prasannan, R.; Kalesh, K.A.; Shanmugam, M.K.; Nachiyappan, A.; Ramachandran, L.; Nguyen, A.H.; Kumar, A.P.; Lakshmanan, M.; Ahn, K.S.; Sethi, G. Key cell signaling pathways modulated by zerumbone: Role in the prevention and treatment of cancer. *Biochem. Pharmacol.* **2012**, *84*, 1268–1276. [CrossRef] [PubMed]
23. Shanmugam, M.K.; Warrier, S.; Kumar, A.P.; Sethi, G.; Arfuso, F. Potential role of natural compounds as anti-angiogenic agents in cancer. *Curr. Vasc. Pharmacol.* **2017**, *15*, 503–519. [CrossRef] [PubMed]
24. Newman, D.J.; Cragg, G.M. Natural Products as Sources of New Drugs from 1981 to 2014. *J. Nat. Prod.* **2016**, *79*, 629–661. [CrossRef] [PubMed]
25. Aggarwal, B.B.; Bhardwaj, A.; Aggarwal, R.S.; Seeram, N.P.; Shishodia, S.; Takada, Y. Role of resveratrol in prevention and therapy of cancer: Preclinical and clinical studies. *Anticancer Res.* **2004**, *24*, 2783–2840. [PubMed]
26. Bishayee, A. Cancer prevention and treatment with resveratrol: From rodent studies to clinical trials. *Cancer Prev. Res.* **2009**, *2*, 409–418. [CrossRef] [PubMed]

27. Bishayee, A.; Politis, T.; Darvesh, A.S. Resveratrol in the chemoprevention and treatment of hepatocellular carcinoma. *Cancer Treat. Rev.* **2010**, *36*, 43–53. [CrossRef] [PubMed]

28. Sinha, D.; Sarkar, N.; Biswas, J.; Bishayee, A. Resveratrol for breast cancer prevention and therapy: Preclinical evidence and molecular mechanisms. *Semin. Cancer Boil.* **2016**, *40–41*, 209–232. [CrossRef] [PubMed]

29. Cucciolla, V.; Borriello, A.; Oliva, A.; Galletti, P.; Zappia, V.; Della Ragione, F. Resveratrol: From basic science to the clinic. *Cell Cycle* **2007**, *6*, 2495–2510. [CrossRef] [PubMed]

30. Soleas, G.J.; Diamandis, E.P.; Goldberg, D.M. Resveratrol: A molecule whose time has come? And gone? *Clin. Biochem.* **1997**, *30*, 91–113. [CrossRef]

31. Jeandet, P.; Douillet-Breuil, A.C.; Bessis, R.; Debord, S.; Sbaghi, M.; Adrian, M. Phytoalexins from the Vitaceae: Biosynthesis, phytoalexin gene expression in transgenic plants, antifungal activity, and metabolism. *J. Agric. Food Chem.* **2002**, *50*, 2731–2741. [CrossRef] [PubMed]

32. Cai, Y.J.; Wei, Q.Y.; Fang, J.G.; Yang, L.; Liu, Z.L.; Wyche, J.H.; Han, Z. The 3,4-dihydroxyl groups are important for trans-resveratrol analogs to exhibit enhanced antioxidant and apoptotic activities. *Anticancer Res.* **2004**, *24*, 999–1002. [PubMed]

33. Colin, D.; Lancon, A.; Delmas, D.; Lizard, G.; Abrossinow, J.; Kahn, E.; Jannin, B.; Latruffe, N. Antiproliferative activities of resveratrol and related compounds in human hepatocyte derived HepG2 cells are associated with biochemical cell disturbance revealed by fluorescence analyses. *Biochimie* **2008**, *90*, 1674–1684. [CrossRef] [PubMed]

34. Moran, B.W.; Anderson, F.P.; Devery, A.; Cloonan, S.; Butler, W.E.; Varughese, S.; Draper, S.M.; Kenny, P.T. Synthesis, structural characterisation and biological evaluation of fluorinated analogues of resveratrol. *Bioorg. Med. Chem.* **2009**, *17*, 4510–4522. [CrossRef] [PubMed]

35. Mondal, A.; Bennett, L.L. Resveratrol enhances the efficacy of sorafenib mediated apoptosis in human breast cancer MCF7 cells through ROS, cell cycle inhibition, caspase 3 and PARP cleavage. *Biomed. Pharmacother.* **2016**, *84*, 1906–1914. [CrossRef] [PubMed]

36. Lee, Y.J.; Lee, G.J.; Yi, S.S.; Heo, S.H.; Park, C.R.; Nam, H.S.; Cho, M.K.; Lee, S.H. Cisplatin and resveratrol induce apoptosis and autophagy following oxidative stress in malignant mesothelioma cells. *Food Chem. Toxicol.* **2016**, *97*, 96–107. [CrossRef] [PubMed]

37. Stagos, D.; Amoutzias, G.D.; Matakos, A.; Spyrou, A.; Tsatsakis, A.M.; Kouretas, D. Chemoprevention of liver cancer by plant polyphenols. *Food Chem. Toxicol.* **2012**, *50*, 2155–2170. [CrossRef] [PubMed]

38. Carter, L.G.; D'Orazio, J.A.; Pearson, K.J. Resveratrol and cancer: Focus on in vivo evidence. *Endocr. Relat. Cancer* **2014**, *21*, R209–R225. [CrossRef] [PubMed]

39. Stagos, D.; Portesis, N.; Spanou, C.; Mossialos, D.; Aligiannis, N.; Chaita, E.; Panagoulis, C.; Reri, E.; Skaltsounis, L.; Tsatsakis, A.M.; et al. Correlation of total polyphenolic content with antioxidant and antibacterial activity of 24 extracts from Greek domestic *Lamiaceae* species. *Food Chem. Toxicol.* **2012**, *50*, 4115–4124. [CrossRef] [PubMed]

40. Mazzanti, G.; Di Giacomo, S. Curcumin and Resveratrol in the Management of Cognitive Disorders: What is the Clinical Evidence? *Molecules* **2016**, *21*, 1243. [CrossRef] [PubMed]

41. Molino, S.; Dossena, M.; Buonocore, D.; Ferrari, F.; Venturini, L.; Ricevuti, G.; Verri, M. Polyphenols in dementia: From molecular basis to clinical trials. *Life Sci.* **2016**, *161*, 69–77. [CrossRef] [PubMed]

42. Wadsworth, T.L.; Koop, D.R. Effects of the wine polyphenolics quercetin and resveratrol on pro-inflammatory cytokine expression in RAW 264.7 macrophages. *Biochem. Pharmacol.* **1999**, *57*, 941–949. [CrossRef]

43. Ray, P.S.; Maulik, G.; Cordis, G.A.; Bertelli, A.A.; Bertelli, A.; Das, D.K. The red wine antioxidant resveratrol protects isolated rat hearts from ischemia reperfusion injury. *Free Radic. Boil. Med.* **1999**, *27*, 160–169. [CrossRef]

44. Baur, J.A.; Pearson, K.J.; Price, N.L.; Jamieson, H.A.; Lerin, C.; Kalra, A.; Prabhu, V.V.; Allard, J.S.; Lopez-Lluch, G.; Lewis, K.; et al. Resveratrol improves health and survival of mice on a high-calorie diet. *Nature* **2006**, *444*, 337–342. [CrossRef] [PubMed]

45. Renaud, S.; de Lorgeril, M. Wine, alcohol, platelets, and the French paradox for coronary heart disease. *Lancet* **1992**, *339*, 1523–1526. [CrossRef]

46. Jang, M.; Cai, L.; Udeani, G.O.; Slowing, K.V.; Thomas, C.F.; Beecher, C.W.; Fong, H.H.; Farnsworth, N.R.; Kinghorn, A.D.; Mehta, R.G.; et al. Cancer chemopreventive activity of resveratrol, a natural product derived from grapes. *Science* **1997**, *275*, 218–220. [CrossRef] [PubMed]

47. Tome-Carneiro, J.; Larrosa, M.; Gonzalez-Sarrias, A.; Tomas-Barberan, F.A.; Garcia-Conesa, M.T.; Espin, J.C. Resveratrol and clinical trials: The crossroad from in vitro studies to human evidence. *Curr. Pharm. Des.* **2013**, *19*, 6064–6093. [CrossRef] [PubMed]

48. Kundu, J.K.; Surh, Y.J. Cancer chemopreventive and therapeutic potential of resveratrol: Mechanistic perspectives. *Cancer Lett.* **2008**, *269*, 243–261. [CrossRef] [PubMed]

49. Harikumar, K.B.; Kunnumakkara, A.B.; Sethi, G.; Diagaradjane, P.; Anand, P.; Pandey, M.K.; Gelovani, J.; Krishnan, S.; Guha, S.; Aggarwal, B.B. Resveratrol, a multitargeted agent, can enhance antitumor activity of gemcitabine in vitro and in orthotopic mouse model of human pancreatic cancer. *Int. J. Cancer* **2010**, *127*, 257–268. [PubMed]

50. Minamoto, T.; Mai, M.; Ronai, Z. Environmental factors as regulators and effectors of multistep carcinogenesis. *Carcinogenesis* **1999**, *20*, 519–527. [CrossRef] [PubMed]

51. Khansari, N.; Shakiba, Y.; Mahmoudi, M. Chronic inflammation and oxidative stress as a major cause of age-related diseases and cancer. *Recent Pat. Inflamm. Allergy Drug Discov.* **2009**, *3*, 73–80. [CrossRef] [PubMed]

52. Barzilai, A.; Yamamoto, K. DNA damage responses to oxidative stress. *DNA Repair* **2004**, *3*, 1109–1115. [CrossRef] [PubMed]

53. Fruehauf, J.P.; Meyskens, F.L., Jr. Reactive oxygen species: A breath of life or death? *Clin. Cancer Res.* **2007**, *13*, 789–794. [CrossRef] [PubMed]

54. Windmill, K.F.; McKinnon, R.A.; Zhu, X.; Gaedigk, A.; Grant, D.M.; McManus, M.E. The role of xenobiotic metabolizing enzymes in arylamine toxicity and carcinogenesis: Functional and localization studies. *Mutat. Res.* **1997**, *376*, 153–160. [CrossRef]

55. Galati, G.; Teng, S.; Moridani, M.Y.; Chan, T.S.; O'Brien, P.J. Cancer chemoprevention and apoptosis mechanisms induced by dietary polyphenolics. *Drug Metab. Drug Interact.* **2000**, *17*, 311–349. [CrossRef]

56. Guengerich, F.P. Metabolism of chemical carcinogens. *Carcinogenesis* **2000**, *21*, 345–351. [CrossRef] [PubMed]

57. Sharma, S.; Stutzman, J.D.; Kelloff, G.J.; Steele, V.E. Screening of potential chemopreventive agents using biochemical markers of carcinogenesis. *Cancer Res.* **1994**, *54*, 5848–5855. [PubMed]

58. Martinez, J.; Moreno, J.J. Effect of resveratrol, a natural polyphenolic compound, on reactive oxygen species and prostaglandin production. *Biochem. Pharmacol.* **2000**, *59*, 865–870. [CrossRef]

59. Leonard, S.S.; Xia, C.; Jiang, B.H.; Stinefelt, B.; Klandorf, H.; Harris, G.K.; Shi, X. Resveratrol scavenges reactive oxygen species and effects radical-induced cellular responses. *Biochem. Biophys. Res. Commun.* **2003**, *309*, 1017–1026. [CrossRef] [PubMed]

60. Kim, H.J.; Chang, E.J.; Bae, S.J.; Shim, S.M.; Park, H.D.; Rhee, C.H.; Park, J.H.; Choi, S.W. Cytotoxic and antimutagenic stilbenes from seeds of *Paeonia lactiflora*. *Arch. Pharm. Res.* **2002**, *25*, 293–299. [CrossRef] [PubMed]

61. Sgambato, A.; Ardito, R.; Faraglia, B.; Boninsegna, A.; Wolf, F.I.; Cittadini, A. Resveratrol, a natural phenolic compound, inhibits cell proliferation and prevents oxidative DNA damage. *Mutat. Res.* **2001**, *496*, 171–180. [CrossRef]

62. Attia, S.M. Influence of resveratrol on oxidative damage in genomic DNA and apoptosis induced by cisplatin. *Mutat. Res.* **2012**, *741*, 22–31. [CrossRef] [PubMed]

63. Chen, Z.H.; Hurh, Y.J.; Na, H.K.; Kim, J.H.; Chun, Y.J.; Kim, D.H.; Kang, K.S.; Cho, M.H.; Surh, Y.J. Resveratrol inhibits TCDD-induced expression of CYP1A1 and CYP1B1 and catechol estrogen-mediated oxidative DNA damage in cultured human mammary epithelial cells. *Carcinogenesis* **2004**, *25*, 2005–2013. [CrossRef] [PubMed]

64. Ciolino, H.P.; Yeh, G.C. Inhibition of aryl hydrocarbon-induced cytochrome P-450 1A1 enzyme activity and CYP1A1 expression by resveratrol. *Mol. Pharmacol.* **1999**, *56*, 760–767. [PubMed]

65. Beedanagari, S.R.; Bebenek, I.; Bui, P.; Hankinson, O. Resveratrol inhibits dioxin-induced expression of human CYP1A1 and CYP1B1 by inhibiting recruitment of the aryl hydrocarbon receptor complex and RNA polymerase II to the regulatory regions of the corresponding genes. *Toxicol. Sci.* **2009**, *110*, 61–67. [CrossRef] [PubMed]

66. Peng, T.L.; Chen, J.; Mao, W.; Song, X.; Chen, M.H. Aryl hydrocarbon receptor pathway activation enhances gastric cancer cell invasiveness likely through a c-Jun-dependent induction of matrix metalloproteinase-9. *BMC Cell Boil.* **2009**, *10*, 27. [CrossRef] [PubMed]

67. Hsieh, T.C.; Lu, X.; Wang, Z.; Wu, J.M. Induction of quinone reductase NQO1 by resveratrol in human K562 cells involves the antioxidant response element ARE and is accompanied by nuclear translocation of transcription factor Nrf2. *Med. Chem.* **2006**, *2*, 275–285. [PubMed]

68. Heo, Y.H.; Kim, S.; Park, J.E.; Jeong, L.S.; Lee, S.K. Induction of quinone reductase activity by stilbene analogs in mouse Hepa 1c1c7 cells. *Arch. Pharm. Res.* **2001**, *24*, 597–600. [CrossRef] [PubMed]

69. Bianco, N.R.; Chaplin, L.J.; Montano, M.M. Differential induction of quinone reductase by phytoestrogens and protection against oestrogen-induced DNA damage. *Biochem. J.* **2005**, *385*, 279–287. [CrossRef] [PubMed]

70. Lee, J.S.; Surh, Y.J. Nrf2 as a novel molecular target for chemoprevention. *Cancer Lett.* **2005**, *224*, 171–184. [CrossRef] [PubMed]

71. Kensler, T.W.; Wakabayashi, N. Nrf2: Friend or foe for chemoprevention? *Carcinogenesis* **2010**, *31*, 90–99. [CrossRef] [PubMed]

72. Lu, F.; Zahid, M.; Wang, C.; Saeed, M.; Cavalieri, E.L.; Rogan, E.G. Resveratrol prevents estrogen-DNA adduct formation and neoplastic transformation in MCF-10F cells. *Cancer Prev. Res.* **2008**, *1*, 135–145. [CrossRef] [PubMed]

73. Zahid, M.; Gaikwad, N.W.; Ali, M.F.; Lu, F.; Saeed, M.; Yang, L.; Rogan, E.G.; Cavalieri, E.L. Prevention of estrogen-DNA adduct formation in MCF-10F cells by resveratrol. *Free Radic. Boil. Med.* **2008**, *45*, 136–145. [CrossRef] [PubMed]

74. Zhang, H.; Shih, A.; Rinna, A.; Forman, H.J. Exacerbation of tobacco smoke mediated apoptosis by resveratrol: An unexpected consequence of its antioxidant action. *Int. J. Biochem. Cell Boil.* **2011**, *43*, 1059–1064. [CrossRef] [PubMed]

75. Kode, A.; Rajendrasozhan, S.; Caito, S.; Yang, S.R.; Megson, I.L.; Rahman, I. Resveratrol induces glutathione synthesis by activation of Nrf2 and protects against cigarette smoke-mediated oxidative stress in human lung epithelial cells. *Am. J. Physiol. Lung Cell. Mol. Physiol.* **2008**, *294*, L478–L488. [CrossRef] [PubMed]

76. Klaunig, J.E.; Kamendulis, L.M. The role of oxidative stress in carcinogenesis. *Annu. Rev. Pharmacol. Toxicol.* **2004**, *44*, 239–267. [CrossRef] [PubMed]

77. Wolter, F.; Akoglu, B.; Clausnitzer, A.; Stein, J. Downregulation of the cyclin D1/Cdk4 complex occurs during resveratrol-induced cell cycle arrest in colon cancer cell lines. *J. Nutr.* **2001**, *131*, 2197–2203. [PubMed]

78. Benitez, D.A.; Pozo-Guisado, E.; Alvarez-Barrientos, A.; Fernandez-Salguero, P.M.; Castellon, E.A. Mechanisms involved in resveratrol-induced apoptosis and cell cycle arrest in prostate cancer-derived cell lines. *J. Androl.* **2007**, *28*, 282–293. [CrossRef] [PubMed]

79. Bai, Y.; Mao, Q.Q.; Qin, J.; Zheng, X.Y.; Wang, Y.B.; Yang, K.; Shen, H.F.; Xie, L.P. Resveratrol induces apoptosis and cell cycle arrest of human T24 bladder cancer cells in vitro and inhibits tumor growth in vivo. *Cancer Sci.* **2010**, *101*, 488–493. [CrossRef] [PubMed]

80. Gatouillat, G.; Balasse, E.; Joseph-Pietras, D.; Morjani, H.; Madoulet, C. Resveratrol induces cell-cycle disruption and apoptosis in chemoresistant B16 melanoma. *J. Cell. Biochem.* **2010**, *110*, 893–902. [CrossRef] [PubMed]

81. Ferry-Dumazet, H.; Garnier, O.; Mamani-Matsuda, M.; Vercauteren, J.; Belloc, F.; Billiard, C.; Dupouy, M.; Thiolat, D.; Kolb, J.P.; Marit, G.; et al. Resveratrol inhibits the growth and induces the apoptosis of both normal and leukemic hematopoietic cells. *Carcinogenesis* **2002**, *23*, 1327–1333. [CrossRef] [PubMed]

82. Filippi-Chiela, E.C.; Villodre, E.S.; Zamin, L.L.; Lenz, G. Autophagy interplay with apoptosis and cell cycle regulation in the growth inhibiting effect of resveratrol in glioma cells. *PLoS ONE* **2011**, *6*, e20849. [CrossRef] [PubMed]

83. Liao, P.C.; Ng, L.T.; Lin, L.T.; Richardson, C.D.; Wang, G.H.; Lin, C.C. Resveratrol arrests cell cycle and induces apoptosis in human hepatocellular carcinoma Huh-7 cells. *J. Med. Food* **2010**, *13*, 1415–1423. [CrossRef] [PubMed]

84. Rashid, A.; Liu, C.; Sanli, T.; Tsiani, E.; Singh, G.; Bristow, R.G.; Dayes, I.; Lukka, H.; Wright, J.; Tsakiridis, T. Resveratrol enhances prostate cancer cell response to ionizing radiation. Modulation of the AMPK, Akt and mTOR pathways. *Radiat. Oncol.* **2011**, *6*, 144. [CrossRef] [PubMed]

85. Hsieh, T.C.; Wong, C.; John Bennett, D.; Wu, J.M. Regulation of p53 and cell proliferation by resveratrol and its derivatives in breast cancer cells: An in silico and biochemical approach targeting integrin αvβ3. *Int. J. Cancer* **2011**, *129*, 2732–2743. [CrossRef] [PubMed]

86. Farnebo, M.; Bykov, V.J.; Wiman, K.G. The p53 tumor suppressor: A master regulator of diverse cellular processes and therapeutic target in cancer. *Biochem. Biophys. Res. Commun.* **2010**, *396*, 85–89. [CrossRef] [PubMed]

87. Gartel, A.L.; Tyner, A.L. The role of the cyclin-dependent kinase inhibitor p21 in apoptosis. *Mol. Cancer Ther.* **2002**, *1*, 639–649. [PubMed]

88. Kim, Y.A.; Rhee, S.H.; Park, K.Y.; Choi, Y.H. Antiproliferative effect of resveratrol in human prostate carcinoma cells. *J. Med. Food* **2003**, *6*, 273–280. [CrossRef] [PubMed]

89. Kim, Y.A.; Lee, W.H.; Choi, T.H.; Rhee, S.H.; Park, K.Y.; Choi, Y.H. Involvement of p21WAF1/CIP1, pRB, Bax and NF-κB in induction of growth arrest and apoptosis by resveratrol in human lung carcinoma A549 cells. *Int. J. Oncol.* **2003**, *23*, 1143–1149. [CrossRef] [PubMed]

90. Ahmad, N.; Adhami, V.M.; Afaq, F.; Feyes, D.K.; Mukhtar, H. Resveratrol causes WAF-1/p21-mediated G(1)-phase arrest of cell cycle and induction of apoptosis in human epidermoid carcinoma A431 cells. *Clin. Cancer Res.* **2001**, *7*, 1466–1473. [PubMed]

91. Dorrie, J.; Gerauer, H.; Wachter, Y.; Zunino, S.J. Resveratrol induces extensive apoptosis by depolarizing mitochondrial membranes and activating caspase-9 in acute lymphoblastic leukemia cells. *Cancer Res.* **2001**, *61*, 4731–4739. [PubMed]

92. Tinhofer, I.; Bernhard, D.; Senfter, M.; Anether, G.; Loeffler, M.; Kroemer, G.; Kofler, R.; Csordas, A.; Greil, R. Resveratrol, a tumor-suppressive compound from grapes, induces apoptosis via a novel mitochondrial pathway controlled by Bcl-2. *FASEB J.* **2001**, *15*, 1613–1615. [CrossRef] [PubMed]

93. Naik, P.; Karrim, J.; Hanahan, D. The rise and fall of apoptosis during multistage tumorigenesis: Down-modulation contributes to tumor progression from angiogenic progenitors. *Genes Dev.* **1996**, *10*, 2105–2116. [CrossRef] [PubMed]

94. Deigner, H.P.; Kinscherf, R. Modulating apoptosis: Current applications and prospects for future drug development. *Curr. Med. Chem.* **1999**, *6*, 399–414. [PubMed]

95. Hengartner, M.O. The biochemistry of apoptosis. *Nature* **2000**, *407*, 770–776. [CrossRef] [PubMed]

96. Okada, H.; Mak, T.W. Pathways of apoptotic and non-apoptotic death in tumour cells. *Nat. Rev. Cancer* **2004**, *4*, 592–603. [CrossRef] [PubMed]

97. Hu, W.; Kavanagh, J.J. Anticancer therapy targeting the apoptotic pathway. *Lancet. Oncol.* **2003**, *4*, 721–729. [CrossRef]

98. Ashkenazi, A.; Dixit, V.M. Apoptosis control by death and decoy receptors. *Curr. Opin. Cell Boil.* **1999**, *11*, 255–260. [CrossRef]

99. Nicholson, D.W.; Thornberry, N.A. Caspases: Killer proteases. *Trends Biochem. Sci.* **1997**, *22*, 299–306. [CrossRef]

100. Van Loo, G.; Saelens, X.; van Gurp, M.; MacFarlane, M.; Martin, S.J.; Vandenabeele, P. The role of mitochondrial factors in apoptosis: A Russian roulette with more than one bullet. *Cell Death Differ.* **2002**, *9*, 1031–1042. [CrossRef] [PubMed]

101. Du, C.; Fang, M.; Li, Y.; Li, L.; Wang, X. Smac, a mitochondrial protein that promotes cytochrome c-dependent caspase activation by eliminating IAP inhibition. *Cell* **2000**, *102*, 33–42. [CrossRef]

102. Sun, X.M.; MacFarlane, M.; Zhuang, J.; Wolf, B.B.; Green, D.R.; Cohen, G.M. Distinct caspase cascades are initiated in receptor-mediated and chemical-induced apoptosis. *J. Boil. Chem.* **1999**, *274*, 5053–5060. [CrossRef]

103. Korsmeyer, S.J.; Wei, M.C.; Saito, M.; Weiler, S.; Oh, K.J.; Schlesinger, P.H. Pro-apoptotic cascade activates BID, which oligomerizes BAK or BAX into pores that result in the release of cytochrome c. *Cell Death Differ.* **2000**, *7*, 1166–1173. [CrossRef] [PubMed]

104. Fulda, S.; Debatin, K.M. Exploiting death receptor signaling pathways for tumor therapy. *Biochim. Biophys. Acta* **2004**, *1705*, 27–41. [CrossRef] [PubMed]

105. Cummings, J.; Ward, T.H.; Ranson, M.; Dive, C. Apoptosis pathway-targeted drugs—From the bench to the clinic. *Biochim. Biophys. Acta* **2004**, *1705*, 53–66. [CrossRef] [PubMed]

106. Neergheen, V.S.; Bahorun, T.; Taylor, E.W.; Jen, L.S.; Aruoma, O.I. Targeting specific cell signaling transduction pathways by dietary and medicinal phytochemicals in cancer chemoprevention. *Toxicology* **2010**, *278*, 229–241. [CrossRef] [PubMed]

107. Ahn, K.S.; Sethi, G.; Aggarwal, B.B. Nuclear factor-κB: From clone to clinic. *Curr. Mol. Med.* **2007**, *7*, 619–637. [CrossRef] [PubMed]

108. Li, F.; Zhang, J.; Arfuso, F.; Chinnathambi, A.; Zayed, M.E.; Alharbi, S.A.; Kumar, A.P.; Ahn, K.S.; Sethi, G. NF-κB in cancer therapy. *Arch. Toxicol.* **2015**, *89*, 711–731. [CrossRef] [PubMed]

109. Chai, E.Z.; Shanmugam, M.K.; Arfuso, F.; Dharmarajan, A.; Wang, C.; Kumar, A.P.; Samy, R.P.; Lim, L.H.; Wang, L.; Goh, B.C.; et al. Targeting transcription factor STAT3 for cancer prevention and therapy. *Pharmacol. Ther.* **2016**, *162*, 86–97. [CrossRef] [PubMed]

110. Sethi, G.; Sung, B.; Aggarwal, B.B. Nuclear factor-κB activation: From bench to bedside. *Exp. Biol. Med.* **2008**, *233*, 21–31. [CrossRef] [PubMed]

111. Singh, S.S.; Yap, W.N.; Arfuso, F.; Kar, S.; Wang, C.; Cai, W.; Dharmarajan, A.M.; Sethi, G.; Kumar, A.P. Targeting the PI3K/Akt signaling pathway in gastric carcinoma: A reality for personalized medicine? *World J. Gastroenterol.* **2015**, *21*, 12261–12273. [CrossRef] [PubMed]

112. Siveen, K.S.; Sikka, S.; Surana, R.; Dai, X.; Zhang, J.; Kumar, A.P.; Tan, B.K.; Sethi, G.; Bishayee, A. Targeting the STAT3 signaling pathway in cancer: Role of synthetic and natural inhibitors. *Biochim. Biophys. Acta* **2014**, *1845*, 136–154. [CrossRef] [PubMed]

113. Clement, M.V.; Hirpara, J.L.; Chawdhury, S.H.; Pervaiz, S. Chemopreventive agent resveratrol, a natural product derived from grapes, triggers CD95 signaling-dependent apoptosis in human tumor cells. *Blood* **1998**, *92*, 996–1002. [PubMed]

114. Delmas, D.; Rebe, C.; Lacour, S.; Filomenko, R.; Athias, A.; Gambert, P.; Cherkaoui-Malki, M.; Jannin, B.; Dubrez-Daloz, L.; Latruffe, N.; et al. Resveratrol-induced apoptosis is associated with Fas redistribution in the rafts and the formation of a death-inducing signaling complex in colon cancer cells. *J. Boil. Chem.* **2003**, *278*, 41482–41490. [CrossRef] [PubMed]

115. Micheau, O.; Solary, E.; Hammann, A.; Dimanche-Boitrel, M.T. Fas ligand-independent, FADD-mediated activation of the Fas death pathway by anticancer drugs. *J. Boil. Chem.* **1999**, *274*, 7987–7992. [CrossRef]

116. Petak, I.; Tillman, D.M.; Harwood, F.G.; Mihalik, R.; Houghton, J.A. Fas-dependent and -independent mechanisms of cell death following DNA damage in human colon carcinoma cells. *Cancer Res.* **2000**, *60*, 2643–2650. [PubMed]

117. Tsan, M.F.; White, J.E.; Maheshwari, J.G.; Bremner, T.A.; Sacco, J. Resveratrol induces Fas signalling-independent apoptosis in THP-1 human monocytic leukaemia cells. *Br. J. Haematol.* **2000**, *109*, 405–412. [CrossRef] [PubMed]

118. Bernhard, D.; Tinhofer, I.; Tonko, M.; Hubl, H.; Ausserlechner, M.J.; Greil, R.; Kofler, R.; Csordas, A. Resveratrol causes arrest in the S-phase prior to Fas-independent apoptosis in CEM-C7H2 acute leukemia cells. *Cell Death Differ.* **2000**, *7*, 834–842. [CrossRef] [PubMed]

119. Delmas, D.; Solary, E.; Latruffe, N. Resveratrol, a phytochemical inducer of multiple cell death pathways: Apoptosis, autophagy and mitotic catastrophe. *Curr. Med. Chem.* **2011**, *18*, 1100–1121. [CrossRef] [PubMed]

120. Aziz, M.H.; Nihal, M.; Fu, V.X.; Jarrard, D.F.; Ahmad, N. Resveratrol-caused apoptosis of human prostate carcinoma LNCaP cells is mediated via modulation of phosphatidylinositol 3′-kinase/Akt pathway and Bcl-2 family proteins. *Mol. Cancer Ther.* **2006**, *5*, 1335–1341. [CrossRef] [PubMed]

121. Bhardwaj, A.; Sethi, G.; Vadhan-Raj, S.; Bueso-Ramos, C.; Takada, Y.; Gaur, U.; Nair, A.S.; Shishodia, S.; Aggarwal, B.B. Resveratrol inhibits proliferation, induces apoptosis, and overcomes chemoresistance through down-regulation of STAT3 and nuclear factor-κB-regulated antiapoptotic and cell survival gene products in human multiple myeloma cells. *Blood* **2007**, *109*, 2293–2302. [CrossRef] [PubMed]

122. Casanova, F.; Quarti, J.; da Costa, D.C.; Ramos, C.A.; da Silva, J.L.; Fialho, E. Resveratrol chemosensitizes breast cancer cells to melphalan by cell cycle arrest. *J. Cell. Biochem.* **2012**, *113*, 2586–2596. [CrossRef] [PubMed]

123. Gokbulut, A.A.; Apohan, E.; Baran, Y. Resveratrol and quercetin-induced apoptosis of human 232B4 chronic lymphocytic leukemia cells by activation of caspase-3 and cell cycle arrest. *Hematology* **2013**, *18*, 144–150. [CrossRef] [PubMed]

124. Frazzi, R.; Valli, R.; Tamagnini, I.; Casali, B.; Latruffe, N.; Merli, F. Resveratrol-mediated apoptosis of hodgkin lymphoma cells involves SIRT1 inhibition and FOXO3a hyperacetylation. *Int. J. Cancer* **2013**, *132*, 1013–1021. [CrossRef] [PubMed]

125. Shankar, S.; Chen, Q.; Siddiqui, I.; Sarva, K.; Srivastava, R.K. Sensitization of TRAIL-resistant LNCaP cells by resveratrol (3, 4′, 5 tri-hydroxystilbene): Molecular mechanisms and therapeutic potential. *J. Mol. Signal.* **2007**, *2*, 7. [CrossRef] [PubMed]

126. Van Ginkel, P.R.; Yan, M.B.; Bhattacharya, S.; Polans, A.S.; Kenealey, J.D. Natural products induce a G protein-mediated calcium pathway activating p53 in cancer cells. *Toxicol. Appl. Pharmacol.* **2015**, *288*, 453–462. [CrossRef] [PubMed]

127. Faber, A.C.; Dufort, F.J.; Blair, D.; Wagner, D.; Roberts, M.F.; Chiles, T.C. Inhibition of phosphatidylinositol 3-kinase-mediated glucose metabolism coincides with resveratrol-induced cell cycle arrest in human diffuse large B-cell lymphomas. *Biochem. Pharmacol.* **2006**, *72*, 1246–1256. [CrossRef] [PubMed]

128. Wang, Y.; Romigh, T.; He, X.; Orloff, M.S.; Silverman, R.H.; Heston, W.D.; Eng, C. Resveratrol regulates the PTEN/AKT pathway through androgen receptor-dependent and -independent mechanisms in prostate cancer cell lines. *Hum. Mol. Genet.* **2010**, *19*, 4319–4329. [CrossRef] [PubMed]

129. Banerjee Mustafi, S.; Chakraborty, P.K.; Raha, S. Modulation of Akt and ERK1/2 pathways by resveratrol in chronic myelogenous leukemia (CML) cells results in the downregulation of Hsp70. *PLoS ONE* **2010**, *5*, e8719. [CrossRef] [PubMed]

130. Parekh, P.; Motiwale, L.; Naik, N.; Rao, K.V. Downregulation of cyclin D1 is associated with decreased levels of p38 MAP kinases, Akt/PKB and Pak1 during chemopreventive effects of resveratrol in liver cancer cells. *Exp. Toxicol. Pathol.* **2011**, *63*, 167–173. [CrossRef] [PubMed]

131. He, X.; Wang, Y.; Zhu, J.; Orloff, M.; Eng, C. Resveratrol enhances the anti-tumor activity of the mTOR inhibitor rapamycin in multiple breast cancer cell lines mainly by suppressing rapamycin-induced AKT signaling. *Cancer Lett.* **2011**, *301*, 168–176. [CrossRef] [PubMed]

132. Colin, D.; Limagne, E.; Jeanningros, S.; Jacquel, A.; Lizard, G.; Athias, A.; Gambert, P.; Hichami, A.; Latruffe, N.; Solary, E.; et al. Endocytosis of resveratrol via lipid rafts and activation of downstream signaling pathways in cancer cells. *Cancer Prev. Res.* **2011**, *4*, 1095–1106. [CrossRef] [PubMed]

133. Pozo-Guisado, E.; Merino, J.M.; Mulero-Navarro, S.; Lorenzo-Benayas, M.J.; Centeno, F.; Alvarez-Barrientos, A.; Fernandez-Salguero, P.M. Resveratrol-induced apoptosis in MCF-7 human breast cancer cells involves a caspase-independent mechanism with downregulation of Bcl-2 and NF-κB. *Int. J. Cancer* **2005**, *115*, 74–84. [CrossRef] [PubMed]

134. Benitez, D.A.; Hermoso, M.A.; Pozo-Guisado, E.; Fernandez-Salguero, P.M.; Castellon, E.A. Regulation of cell survival by resveratrol involves inhibition of NF κB-regulated gene expression in prostate cancer cells. *Prostate* **2009**, *69*, 1045–1054. [CrossRef] [PubMed]

135. Cecchinato, V.; Chiaramonte, R.; Nizzardo, M.; Cristofaro, B.; Basile, A.; Sherbet, G.V.; Comi, P. Resveratrol-induced apoptosis in human T-cell acute lymphoblastic leukaemia MOLT-4 cells. *Biochem. Pharmacol.* **2007**, *74*, 1568–1574. [CrossRef] [PubMed]

136. Kueck, A.; Opipari, A.W., Jr.; Griffith, K.A.; Tan, L.; Choi, M.; Huang, J.; Wahl, H.; Liu, J.R. Resveratrol inhibits glucose metabolism in human ovarian cancer cells. *Gynecol. Oncol.* **2007**, *107*, 450–457. [CrossRef] [PubMed]

137. Li, Y.; Liu, J.; Liu, X.; Xing, K.; Wang, Y.; Li, F.; Yao, L. Resveratrol-induced cell inhibition of growth and apoptosis in MCF7 human breast cancer cells are associated with modulation of phosphorylated Akt and caspase-9. *Appl. Biochem. Biotechnol.* **2006**, *135*, 181–192. [CrossRef]

138. Sexton, E.; Van Themsche, C.; LeBlanc, K.; Parent, S.; Lemoine, P.; Asselin, E. Resveratrol interferes with AKT activity and triggers apoptosis in human uterine cancer cells. *Mol. Cancer* **2006**, *5*, 45. [CrossRef] [PubMed]

139. Chen, Q.; Ganapathy, S.; Singh, K.P.; Shankar, S.; Srivastava, R.K. Resveratrol induces growth arrest and apoptosis through activation of FOXO transcription factors in prostate cancer cells. *PLoS ONE* **2010**, *5*, e15288. [CrossRef] [PubMed]

140. Su, J.L.; Yang, C.Y.; Zhao, M.; Kuo, M.L.; Yen, M.L. Forkhead proteins are critical for bone morphogenetic protein-2 regulation and anti-tumor activity of resveratrol. *J. Boil. Chem.* **2007**, *282*, 19385–19398. [CrossRef] [PubMed]

141. Yu, R.; Hebbar, V.; Kim, D.W.; Mandlekar, S.; Pezzuto, J.M.; Kong, A.N. Resveratrol inhibits phorbol ester and UV-induced activator protein 1 activation by interfering with mitogen-activated protein kinase pathways. *Mol. Pharmacol.* **2001**, *60*, 217–224. [PubMed]

142. Miloso, M.; Bertelli, A.A.; Nicolini, G.; Tredici, G. Resveratrol-induced activation of the mitogen-activated protein kinases, ERK1 and ERK2, in human neuroblastoma SH-SY5Y cells. *Neurosci. Lett.* **1999**, *264*, 141–144. [CrossRef]

143. Lin, H.Y.; Shih, A.; Davis, F.B.; Tang, H.Y.; Martino, L.J.; Bennett, J.A.; Davis, P.J. Resveratrol induced serine phosphorylation of p53 causes apoptosis in a mutant p53 prostate cancer cell line. *J. Urol.* **2002**, *168*, 748–755. [CrossRef]

144. Zhang, S.; Cao, H.J.; Davis, F.B.; Tang, H.Y.; Davis, P.J.; Lin, H.Y. Oestrogen inhibits resveratrol-induced post-translational modification of p53 and apoptosis in breast cancer cells. *Br. J. Cancer* **2004**, *91*, 178–185. [CrossRef] [PubMed]

145. Bergh, J.J.; Lin, H.Y.; Lansing, L.; Mohamed, S.N.; Davis, F.B.; Mousa, S.; Davis, P.J. Integrin αVβ3 contains a cell surface receptor site for thyroid hormone that is linked to activation of mitogen-activated protein kinase and induction of angiogenesis. *Endocrinology* **2005**, *146*, 2864–2871. [CrossRef] [PubMed]

146. Lin, H.Y.; Tang, H.Y.; Keating, T.; Wu, Y.H.; Shih, A.; Hammond, D.; Sun, M.; Hercbergs, A.; Davis, F.B.; Davis, P.J. Resveratrol is pro-apoptotic and thyroid hormone is anti-apoptotic in glioma cells: Both actions are integrin and ERK mediated. *Carcinogenesis* **2008**, *29*, 62–69. [CrossRef] [PubMed]

147. Lin, H.Y.; Sun, M.; Tang, H.Y.; Simone, T.M.; Wu, Y.H.; Grandis, J.R.; Cao, H.J.; Davis, P.J.; Davis, F.B. Resveratrol causes COX-2- and p53-dependent apoptosis in head and neck squamous cell cancer cells. *J. Cell. Biochem.* **2008**, *104*, 2131–2142. [CrossRef] [PubMed]

148. Lin, C.; Crawford, D.R.; Lin, S.; Hwang, J.; Sebuyira, A.; Meng, R.; Westfall, J.E.; Tang, H.Y.; Lin, S.; Yu, P.Y.; et al. Inducible COX-2-dependent apoptosis in human ovarian cancer cells. *Carcinogenesis* **2011**, *32*, 19–26. [CrossRef] [PubMed]

149. Lassus, P.; Roux, P.; Zugasti, O.; Philips, A.; Fort, P.; Hibner, U. Extinction of Rac1 and Cdc42Hs signalling defines a novel p53-dependent apoptotic pathway. *Oncogene* **2000**, *19*, 2377–2385. [CrossRef] [PubMed]

150. Dong, Z. Molecular mechanism of the chemopreventive effect of resveratrol. *Mutat. Res.* **2003**, *523–524*, 145–150. [CrossRef]

151. She, Q.B.; Huang, C.; Zhang, Y.; Dong, Z. Involvement of c-jun NH(2)-terminal kinases in resveratrol-induced activation of p53 and apoptosis. *Mol. Carcinog.* **2002**, *33*, 244–250. [CrossRef] [PubMed]

152. Shimizu, T.; Nakazato, T.; Xian, M.J.; Sagawa, M.; Ikeda, Y.; Kizaki, M. Resveratrol induces apoptosis of human malignant B cells by activation of caspase-3 and p38 MAP kinase pathways. *Biochem. Pharmacol.* **2006**, *71*, 742–750. [CrossRef] [PubMed]

153. She, Q.B.; Bode, A.M.; Ma, W.Y.; Chen, N.Y.; Dong, Z. Resveratrol-induced activation of p53 and apoptosis is mediated by extracellular-signal-regulated protein kinases and p38 kinase. *Cancer Res.* **2001**, *61*, 1604–1610. [PubMed]

154. Jazirehi, A.R.; Bonavida, B. Resveratrol modifies the expression of apoptotic regulatory proteins and sensitizes non-Hodgkin's lymphoma and multiple myeloma cell lines to paclitaxel-induced apoptosis. *Mol. Cancer Ther.* **2004**, *3*, 71–84. [PubMed]

155. Mann, A.P.; Verma, A.; Sethi, G.; Manavathi, B.; Wang, H.; Fok, J.Y.; Kunnumakkara, A.B.; Kumar, R.; Aggarwal, B.B.; Mehta, K. Overexpression of tissue transglutaminase leads to constitutive activation of nuclear factor-κB in cancer cells: Delineation of a novel pathway. *Cancer Res.* **2006**, *66*, 8788–8795. [CrossRef] [PubMed]

156. Qiao, L.; Zhang, H.; Yu, J.; Francisco, R.; Dent, P.; Ebert, M.P.; Rocken, C.; Farrell, G. Constitutive activation of NF-κB in human hepatocellular carcinoma: Evidence of a cytoprotective role. *Hum. Gene Ther.* **2006**, *17*, 280–290. [CrossRef] [PubMed]

157. Baby, J.; Pickering, B.F.; Vashisht Gopal, Y.N.; Van Dyke, M.W. Constitutive and inducible nuclear factor-κB in immortalized normal human bronchial epithelial and non-small cell lung cancer cell lines. *Cancer Lett.* **2007**, *255*, 85–94. [CrossRef] [PubMed]

158. Lenz, G.; Davis, R.E.; Ngo, V.N.; Lam, L.; George, T.C.; Wright, G.W.; Dave, S.S.; Zhao, H.; Xu, W.; Rosenwald, A.; et al. Oncogenic CARD11 mutations in human diffuse large B cell lymphoma. *Science* **2008**, *319*, 1676–1679. [CrossRef] [PubMed]

159. Sughra, K.; Birbach, A.; de Martin, R.; Schmid, J.A. Interaction of the TNFR-receptor associated factor TRAF1 with I-κB kinase-2 and TRAF2 indicates a regulatory function for NF-κB signaling. *PLoS ONE* **2010**, *5*, e12683. [CrossRef] [PubMed]

160. Csaki, C.; Mobasheri, A.; Shakibaei, M. Synergistic chondroprotective effects of curcumin and resveratrol in human articular chondrocytes: Inhibition of IL-1β-induced NF-κB-mediated inflammation and apoptosis. *Arthritis Res. Ther.* **2009**, *11*, R165. [CrossRef] [PubMed]

161. Yu, H.; Pardoll, D.; Jove, R. STATs in cancer inflammation and immunity: A leading role for STAT3. *Nat. Rev. Cancer* **2009**, *9*, 798–809. [CrossRef] [PubMed]

162. Hoesel, B.; Schmid, J.A. The complexity of NF-κB signaling in inflammation and cancer. *Mol. Cancer* **2013**, *12*, 86. [CrossRef] [PubMed]

163. Johnston, P.A.; Grandis, J.R. STAT3 signaling: Anticancer strategies and challenges. *Mol. Interv.* **2011**, *11*, 18–26. [CrossRef] [PubMed]

164. Yenari, M.A.; Han, H.S. Influence of hypothermia on post-ischemic inflammation: Role of nuclear factor kappa B (NFκB). *Neurochem. Int.* **2006**, *49*, 164–169. [CrossRef] [PubMed]

165. Kim, C.; Baek, S.H.; Um, J.Y.; Shim, B.S.; Ahn, K.S. Resveratrol attenuates constitutive STAT3 and STAT5 activation through induction of PTPepsilon and SHP-2 tyrosine phosphatases and potentiates sorafenib-induced apoptosis in renal cell carcinoma. *BMC Nephrol.* **2016**, *17*, 19. [CrossRef] [PubMed]

166. Wen, S.; Li, H.; Wu, M.L.; Fan, S.H.; Wang, Q.; Shu, X.H.; Kong, Q.Y.; Chen, X.Y.; Liu, J. Inhibition of NF-κB signaling commits resveratrol-treated medulloblastoma cells to apoptosis without neuronal differentiation. *J. Neuro-Oncol.* **2011**, *104*, 169–177. [CrossRef] [PubMed]

167. Fan, Y.; Mao, R.; Yang, J. NF-κB and STAT3 signaling pathways collaboratively link inflammation to cancer. *Protein Cell* **2013**, *4*, 176–185. [CrossRef] [PubMed]

168. Steele, V.E.; Hawk, E.T.; Viner, J.L.; Lubet, R.A. Mechanisms and applications of non-steroidal anti-inflammatory drugs in the chemoprevention of cancer. *Mutat. Res.* **2003**, *523–524*, 137–144. [CrossRef]

169. Donnelly, L.E.; Newton, R.; Kennedy, G.E.; Fenwick, P.S.; Leung, R.H.; Ito, K.; Russell, R.E.; Barnes, P.J. Anti-inflammatory effects of resveratrol in lung epithelial cells: Molecular mechanisms. *Am. J. Physiol. Lung Cell. Mol. Physiol.* **2004**, *287*, L774–L783. [CrossRef] [PubMed]

170. Pinto, M.C.; Garcia-Barrado, J.A.; Macias, P. Resveratrol is a potent inhibitor of the dioxygenase activity of lipoxygenase. *J. Agric. Food Chem.* **1999**, *47*, 4842–4846. [CrossRef] [PubMed]

171. Baek, S.H.; Ko, J.H.; Lee, H.; Jung, J.; Kong, M.; Lee, J.W.; Lee, J.; Chinnathambi, A.; Zayed, M.E.; Alharbi, S.A.; et al. Resveratrol inhibits STAT3 signaling pathway through the induction of SOCS-1: Role in apoptosis induction and radiosensitization in head and neck tumor cells. *Phytomedicine* **2016**, *23*, 566–577. [CrossRef] [PubMed]

172. Nelson, A.R.; Fingleton, B.; Rothenberg, M.L.; Matrisian, L.M. Matrix metalloproteinases: Biologic activity and clinical implications. *J. Clin. Oncol.* **2000**, *18*, 1135–1149. [CrossRef] [PubMed]

173. Jinga, D.C.; BLiDARu, A.; Condrea, I.; Ardeleanu, C.; Dragomir, C.; Szegli, G.; Stefanescu, M.; Matache, C. MMP-9 and MMP-2 gelatinases and TIMP-1 and TIMP-2 inhibitors in breast cancer: Correlations with prognostic factors. *J. Cell. Mol. Med.* **2006**, *10*, 499–510. [CrossRef] [PubMed]

174. Yu, H.; Pan, C.; Zhao, S.; Wang, Z.; Zhang, H.; Wu, W. Resveratrol inhibits tumor necrosis factor-alpha-mediated matrix metalloproteinase-9 expression and invasion of human hepatocellular carcinoma cells. *Biomed. Pharmacother.* **2008**, *62*, 366–372. [CrossRef] [PubMed]

175. Weng, C.J.; Wu, C.F.; Huang, H.W.; Wu, C.H.; Ho, C.T.; Yen, G.C. Evaluation of anti-invasion effect of resveratrol and related methoxy analogues on human hepatocarcinoma cells. *J. Agric. Food Chem.* **2010**, *58*, 2886–2894. [CrossRef] [PubMed]

176. Lee, M.F.; Pan, M.H.; Chiou, Y.S.; Cheng, A.C.; Huang, H. Resveratrol modulates MED28 (Magicin/EG-1) expression and inhibits epidermal growth factor (EGF)-induced migration in MDA-MB-231 human breast cancer cells. *J. Agric. Food Chem.* **2011**, *59*, 11853–11861. [CrossRef] [PubMed]

177. Castino, R.; Pucer, A.; Veneroni, R.; Morani, F.; Peracchio, C.; Lah, T.T.; Isidoro, C. Resveratrol reduces the invasive growth and promotes the acquisition of a long-lasting differentiated phenotype in human glioblastoma cells. *J. Agric. Food Chem.* **2011**, *59*, 4264–4272. [CrossRef] [PubMed]

178. Vergara, D.; Valente, C.M.; Tinelli, A.; Siciliano, C.; Lorusso, V.; Acierno, R.; Giovinazzo, G.; Santino, A.; Storelli, C.; Maffia, M. Resveratrol inhibits the epidermal growth factor-induced epithelial mesenchymal transition in MCF-7 cells. *Cancer Lett.* **2011**, *310*, 1–8. [CrossRef] [PubMed]

179. Cao, Z.; Fang, J.; Xia, C.; Shi, X.; Jiang, B.H. trans-3,4,5′-Trihydroxystibene inhibits hypoxia-inducible factor 1α and vascular endothelial growth factor expression in human ovarian cancer cells. *Clin. Cancer Res.* **2004**, *10*, 5253–5263. [CrossRef] [PubMed]

180. Trapp, V.; Parmakhtiar, B.; Papazian, V.; Willmott, L.; Fruehauf, J.P. Anti-angiogenic effects of resveratrol mediated by decreased VEGF and increased TSP1 expression in melanoma-endothelial cell co-culture. *Angiogenesis* **2010**, *13*, 305–315. [CrossRef] [PubMed]

181. Zhang, M.; Li, W.; Yu, L.; Wu, S. The suppressive effect of resveratrol on HIF-1α and VEGF expression after warm ischemia and reperfusion in rat liver. *PLoS ONE* **2014**, *9*, e109589. [CrossRef] [PubMed]

182. Seong, H.; Ryu, J.; Jeong, J.Y.; Chung, I.Y.; Han, Y.S.; Hwang, S.H.; Park, J.M.; Kang, S.S.; Seo, S.W. Resveratrol suppresses vascular endothelial growth factor secretion via inhibition of CXC-chemokine receptor 4 expression in ARPE-19 cells. *Mol. Med. Rep.* **2015**, *12*, 1479–1484. [CrossRef] [PubMed]

183. Thiery, J.P.; Acloque, H.; Huang, R.Y.; Nieto, M.A. Epithelial-mesenchymal transitions in development and disease. *Cell* **2009**, *139*, 871–890. [CrossRef] [PubMed]

184. Chaffer, C.L.; Weinberg, R.A. A perspective on cancer cell metastasis. *Science* **2011**, *331*, 1559–1564. [CrossRef] [PubMed]

185. Xu, Q.; Zong, L.; Chen, X.; Jiang, Z.; Nan, L.; Li, J.; Duan, W.; Lei, J.; Zhang, L.; Ma, J.; et al. Resveratrol in the treatment of pancreatic cancer. *Ann. N. Y. Acad. Sci.* **2015**, *1348*, 10–19. [CrossRef] [PubMed]

186. Blobe, G.C.; Schiemann, W.P.; Lodish, H.F. Role of transforming growth factor β in human disease. *N. Engl. J. Med.* **2000**, *342*, 1350–1358. [CrossRef] [PubMed]

187. Heldin, C.H.; Vanlandewijck, M.; Moustakas, A. Regulation of EMT by TGFβ in cancer. *FEBS Lett.* **2012**, *586*, 1959–1970. [CrossRef] [PubMed]

188. Wang, H.; Zhang, H.; Tang, L.; Chen, H.; Wu, C.; Zhao, M.; Yang, Y.; Chen, X.; Liu, G. Resveratrol inhibits TGF-β1-induced epithelial-to-mesenchymal transition and suppresses lung cancer invasion and metastasis. *Toxicology* **2013**, *303*, 139–146. [CrossRef] [PubMed]

189. Ji, Q.; Liu, X.; Han, Z.; Zhou, L.; Sui, H.; Yan, L.; Jiang, H.; Ren, J.; Cai, J.; Li, Q. Resveratrol suppresses epithelial-to-mesenchymal transition in colorectal cancer through TGF-β1/Smads signaling pathway mediated Snail/E-cadherin expression. *BMC Cancer* **2015**, *15*, 97. [CrossRef] [PubMed]

190. Huber, M.A.; Azoitei, N.; Baumann, B.; Grunert, S.; Sommer, A.; Pehamberger, H.; Kraut, N.; Beug, H.; Wirth, T. NF-κB is essential for epithelial-mesenchymal transition and metastasis in a model of breast cancer progression. *J. Clin. Investig.* **2004**, *114*, 569–581. [CrossRef] [PubMed]

191. Huber, M.A.; Beug, H.; Wirth, T. Epithelial-mesenchymal transition: NF-κB takes center stage. *Cell Cycle* **2004**, *3*, 1477–1480. [CrossRef] [PubMed]

192. Maier, H.J.; Schmidt-Strassburger, U.; Huber, M.A.; Wiedemann, E.M.; Beug, H.; Wirth, T. NF-κB promotes epithelial-mesenchymal transition, migration and invasion of pancreatic carcinoma cells. *Cancer Lett.* **2010**, *295*, 214–228. [CrossRef] [PubMed]

193. Chua, H.L.; Bhat-Nakshatri, P.; Clare, S.E.; Morimiya, A.; Badve, S.; Nakshatri, H. NF-κB represses E-cadherin expression and enhances epithelial to mesenchymal transition of mammary epithelial cells: Potential involvement of ZEB-1 and ZEB-2. *Oncogene* **2007**, *26*, 711–724. [CrossRef] [PubMed]

194. Min, C.; Eddy, S.F.; Sherr, D.H.; Sonenshein, G.E. NF-κB and epithelial to mesenchymal transition of cancer. *J. Cell. Biochem.* **2008**, *104*, 733–744. [CrossRef] [PubMed]

195. Barbera, M.J.; Puig, I.; Dominguez, D.; Julien-Grille, S.; Guaita-Esteruelas, S.; Peiro, S.; Baulida, J.; Franci, C.; Dedhar, S.; Larue, L.; et al. Regulation of Snail transcription during epithelial to mesenchymal transition of tumor cells. *Oncogene* **2004**, *23*, 7345–7354. [CrossRef] [PubMed]

196. Bloomston, M.; Zervos, E.E.; Rosemurgy, A.S., II. Matrix metalloproteinases and their role in pancreatic cancer: A review of preclinical studies and clinical trials. *Ann. Surg. Oncol.* **2002**, *9*, 668–674. [CrossRef] [PubMed]

197. Li, W.; Ma, J.; Ma, Q.; Li, B.; Han, L.; Liu, J.; Xu, Q.; Duan, W.; Yu, S.; Wang, F.; et al. Resveratrol inhibits the epithelial-mesenchymal transition of pancreatic cancer cells via suppression of the PI-3K/Akt/NF-κB pathway. *Curr. Med. Chem.* **2013**, *20*, 4185–4194. [CrossRef] [PubMed]

198. Hao, Y.; Huang, W.; Liao, M.; Zhu, Y.; Liu, H.; Hao, C.; Liu, G.; Zhang, G.; Feng, H.; Ning, X.; et al. The inhibition of resveratrol to human skin squamous cell carcinoma A431 xenografts in nude mice. *Fitoterapia* **2013**, *86*, 84–91. [CrossRef] [PubMed]

199. Kapadia, G.J.; Azuine, M.A.; Tokuda, H.; Takasaki, M.; Mukainaka, T.; Konoshima, T.; Nishino, H. Chemopreventive effect of resveratrol, sesamol, sesame oil and sunflower oil in the Epstein-Barr virus early antigen activation assay and the mouse skin two-stage carcinogenesis. *Pharmacol. Res.* **2002**, *45*, 499–505. [CrossRef] [PubMed]

200. Soleas, G.J.; Grass, L.; Josephy, P.D.; Goldberg, D.M.; Diamandis, E.P. A comparison of the anticarcinogenic properties of four red wine polyphenols. *Clin. Biochem.* **2002**, *35*, 119–124. [CrossRef]

201. Kalra, N.; Roy, P.; Prasad, S.; Shukla, Y. Resveratrol induces apoptosis involving mitochondrial pathways in mouse skin tumorigenesis. *Life Sci.* **2008**, *82*, 348–358. [CrossRef] [PubMed]

202. Afaq, F.; Adhami, V.M.; Ahmad, N. Prevention of short-term ultraviolet B radiation-mediated damages by resveratrol in SKH-1 hairless mice. *Toxicol. Appl. Pharmacol.* **2003**, *186*, 28–37. [CrossRef]

203. Reagan-Shaw, S.; Afaq, F.; Aziz, M.H.; Ahmad, N. Modulations of critical cell cycle regulatory events during chemoprevention of ultraviolet B-mediated responses by resveratrol in SKH-1 hairless mouse skin. *Oncogene* **2004**, *23*, 5151–5160. [CrossRef] [PubMed]

204. Aziz, M.H.; Reagan-Shaw, S.; Wu, J.; Longley, B.J.; Ahmad, N. Chemoprevention of skin cancer by grape constituent resveratrol: Relevance to human disease? *FASEB J.* **2005**, *19*, 1193–1195. [CrossRef] [PubMed]

205. Aziz, M.H.; Afaq, F.; Ahmad, N. Prevention of ultraviolet-B radiation damage by resveratrol in mouse skin is mediated via modulation in survivin. *Photochem. Photobiol.* **2005**, *81*, 25–31. [CrossRef] [PubMed]

206. Caltagirone, S.; Rossi, C.; Poggi, A.; Ranelletti, F.O.; Natali, P.G.; Brunetti, M.; Aiello, F.B.; Piantelli, M. Flavonoids apigenin and quercetin inhibit melanoma growth and metastatic potential. *Int. J. Cancer* **2000**, *87*, 595–600. [CrossRef]

207. Provinciali, M.; Re, F.; Donnini, A.; Orlando, F.; Bartozzi, B.; Di Stasio, G.; Smorlesi, A. Effect of resveratrol on the development of spontaneous mammary tumors in HER-2/neu transgenic mice. *Int. J. Cancer* **2005**, *115*, 36–45. [CrossRef] [PubMed]

208. Garvin, S.; Ollinger, K.; Dabrosin, C. Resveratrol induces apoptosis and inhibits angiogenesis in human breast cancer xenografts in vivo. *Cancer Lett.* **2006**, *231*, 113–122. [CrossRef] [PubMed]

209. Mohapatra, P.; Satapathy, S.R.; Das, D.; Siddharth, S.; Choudhuri, T.; Kundu, C.N. Resveratrol mediated cell death in cigarette smoke transformed breast epithelial cells is through induction of p21Waf1/Cip1 and inhibition of long patch base excision repair pathway. *Toxicol. Appl. Pharmacol.* **2014**, *275*, 221–231. [CrossRef] [PubMed]

210. Banerjee, S.; Bueso-Ramos, C.; Aggarwal, B.B. Suppression of 7,12-dimethylbenz(a)anthracene-induced mammary carcinogenesis in rats by resveratrol: Role of nuclear factor-κB, cyclooxygenase 2, and matrix metalloprotease 9. *Cancer Res.* **2002**, *62*, 4945–4954. [PubMed]

211. Whitsett, T.; Carpenter, M.; Lamartiniere, C.A. Resveratrol, but not EGCG, in the diet suppresses DMBA-induced mammary cancer in rats. *J. Carcinog.* **2006**, *5*, 15. [CrossRef] [PubMed]

212. Bhat, K.P.; Lantvit, D.; Christov, K.; Mehta, R.G.; Moon, R.C.; Pezzuto, J.M. Estrogenic and antiestrogenic properties of resveratrol in mammary tumor models. *Cancer Res.* **2001**, *61*, 7456–7463. [PubMed]

213. Schlachterman, A.; Valle, F.; Wall, K.M.; Azios, N.G.; Castillo, L.; Morell, L.; Washington, A.V.; Cubano, L.A.; Dharmawardhane, S.F. Combined resveratrol, quercetin, and catechin treatment reduces breast tumor growth in a nude mouse model. *Transl. Oncol.* **2008**, *1*, 19–27. [CrossRef] [PubMed]

214. Ganapathy, S.; Chen, Q.; Singh, K.P.; Shankar, S.; Srivastava, R.K. Resveratrol enhances antitumor activity of TRAIL in prostate cancer xenografts through activation of FOXO transcription factor. *PLoS ONE* **2010**, *5*, e15627. [CrossRef] [PubMed]

215. Li, K.; Dias, S.J.; Rimando, A.M.; Dhar, S.; Mizuno, C.S.; Penman, A.D.; Lewin, J.R.; Levenson, A.S. Pterostilbene acts through metastasis-associated protein 1 to inhibit tumor growth, progression and metastasis in prostate cancer. *PLoS ONE* **2013**, *8*, e57542. [CrossRef] [PubMed]

216. Harper, C.E.; Patel, B.B.; Wang, J.; Arabshahi, A.; Eltoum, I.A.; Lamartiniere, C.A. Resveratrol suppresses prostate cancer progression in transgenic mice. *Carcinogenesis* **2007**, *28*, 1946–1953. [CrossRef] [PubMed]

217. Seeni, A.; Takahashi, S.; Takeshita, K.; Tang, M.; Sugiura, S.; Sato, S.Y.; Shirai, T. Suppression of prostate cancer growth by resveratrol in the transgenic rat for adenocarcinoma of prostate (TRAP) model. *APJCP* **2008**, *9*, 7–14. [PubMed]

218. Kimura, Y.; Okuda, H. Resveratrol isolated from *Polygonum cuspidatum* root prevents tumor growth and metastasis to lung and tumor-induced neovascularization in Lewis lung carcinoma-bearing mice. *J. Nutr.* **2001**, *131*, 1844–1849. [PubMed]

219. Yin, H.T.; Tian, Q.Z.; Guan, L.; Zhou, Y.; Huang, X.E.; Zhang, H. In vitro and in vivo evaluation of the antitumor efficiency of resveratrol against lung cancer. *APJCP* **2013**, *14*, 1703–1706. [CrossRef] [PubMed]

220. Busquets, S.; Ametller, E.; Fuster, G.; Olivan, M.; Raab, V.; Argiles, J.M.; Lopez-Soriano, F.J. Resveratrol, a natural diphenol, reduces metastatic growth in an experimental cancer model. *Cancer Lett.* **2007**, *245*, 144–148. [CrossRef] [PubMed]

221. Lee, E.O.; Lee, H.J.; Hwang, H.S.; Ahn, K.S.; Chae, C.; Kang, K.S.; Lu, J.; Kim, S.H. Potent inhibition of Lewis lung cancer growth by heyneanol A from the roots of Vitis amurensis through apoptotic and anti-angiogenic activities. *Carcinogenesis* **2006**, *27*, 2059–2069. [CrossRef] [PubMed]

222. Sengottuvelan, M.; Nalini, N. Dietary supplementation of resveratrol suppresses colonic tumour incidence in 1,2-dimethylhydrazine-treated rats by modulating biotransforming enzymes and aberrant crypt foci development. *Br. J. Nutr.* **2006**, *96*, 145–153. [CrossRef] [PubMed]

223. Huderson, A.C.; Myers, J.N.; Niaz, M.S.; Washington, M.K.; Ramesh, A. Chemoprevention of benzo(a)pyrene-induced colon polyps in ApcMin mice by resveratrol. *J. Nutr. Biochem.* **2013**, *24*, 713–724. [CrossRef] [PubMed]

224. Tessitore, L.; Davit, A.; Sarotto, I.; Caderni, G. Resveratrol depresses the growth of colorectal aberrant crypt foci by affecting bax and p21(CIP) expression. *Carcinogenesis* **2000**, *21*, 1619–1622. [CrossRef] [PubMed]

225. Schneider, Y.; Duranton, B.; Gosse, F.; Schleiffer, R.; Seiler, N.; Raul, F. Resveratrol inhibits intestinal tumorigenesis and modulates host-defense-related gene expression in an animal model of human familial adenomatous polyposis. *Nutr. Cancer* **2001**, *39*, 102–107. [CrossRef] [PubMed]

226. Sale, S.; Tunstall, R.G.; Ruparelia, K.C.; Potter, G.A.; Steward, W.P.; Gescher, A.J. Comparison of the effects of the chemopreventive agent resveratrol and its synthetic analog trans 3,4,5,4'-tetramethoxystilbene (DMU-212) on adenoma development in the Apc(Min+) mouse and cyclooxygenase-2 in human-derived colon cancer cells. *Int. J. Cancer* **2005**, *115*, 194–201. [CrossRef] [PubMed]

227. Miura, D.; Miura, Y.; Yagasaki, K. Hypolipidemic action of dietary resveratrol, a phytoalexin in grapes and red wine, in hepatoma-bearing rats. *Life Sci.* **2003**, *73*, 1393–1400. [CrossRef]

228. Carbo, N.; Costelli, P.; Baccino, F.M.; Lopez-Soriano, F.J.; Argiles, J.M. Resveratrol, a natural product present in wine, decreases tumour growth in a rat tumour model. *Biochem. Biophys. Res. Commun.* **1999**, *254*, 739–743. [CrossRef] [PubMed]

229. Liu, H.S.; Pan, C.E.; Yang, W.; Liu, X.M. Antitumor and immunomodulatory activity of resveratrol on experimentally implanted tumor of H22 in Balb/c mice. *World J. Gastroenterol.* **2003**, *9*, 1474–1476. [CrossRef] [PubMed]

230. Yu, L.; Sun, Z.J.; Wu, S.L.; Pan, C.E. Effect of resveratrol on cell cycle proteins in murine transplantable liver cancer. *World J. Gastroenterol.* **2003**, *9*, 2341–2343. [CrossRef] [PubMed]

231. Wu, S.L.; Sun, Z.J.; Yu, L.; Meng, K.W.; Qin, X.L.; Pan, C.E. Effect of resveratrol and in combination with 5-FU on murine liver cancer. *World J. Gastroenterol.* **2004**, *10*, 3048–3052. [CrossRef] [PubMed]

232. Yang, H.L.; Chen, W.Q.; Cao, X.; Worschech, A.; Du, L.F.; Fang, W.Y.; Xu, Y.Y.; Stroncek, D.F.; Li, X.; Wang, E.; et al. Caveolin-1 enhances resveratrol-mediated cytotoxicity and transport in a hepatocellular carcinoma model. *J. Transl. Med.* **2009**, *7*, 22. [CrossRef] [PubMed]

233. Kweon, S.; Kim, Y.; Choi, H. Grape extracts suppress the formation of preneoplastic foci and activity of fatty acid synthase in rat liver. *Exp. Mol. Med.* **2003**, *35*, 371–378. [CrossRef] [PubMed]

234. Bishayee, A.; Dhir, N. Resveratrol-mediated chemoprevention of diethylnitrosamine-initiated hepatocarcinogenesis: Inhibition of cell proliferation and induction of apoptosis. *Chem.-Biol. Interact.* **2009**, *179*, 131–144. [CrossRef] [PubMed]

235. Boily, G.; He, X.H.; Pearce, B.; Jardine, K.; McBurney, M.W. SirT1-null mice develop tumors at normal rates but are poorly protected by resveratrol. *Oncogene* **2009**, *28*, 2882–2893. [CrossRef] [PubMed]

236. Kowalczyk, M.C.; Junco, J.J.; Kowalczyk, P.; Tolstykh, O.; Hanausek, M.; Slaga, T.J.; Walaszek, Z. Effects of combined phytochemicals on skin tumorigenesis in SENCAR mice. *Int. J. Oncol.* **2013**, *43*, 911–918. [CrossRef] [PubMed]

237. Szaefer, H.; Krajka-Kuzniak, V.; Baer-Dubowska, W. The effect of initiating doses of benzo[a]pyrene and 7,12-dimethylbenz[a]anthracene on the expression of PAH activating enzymes and its modulation by plant phenols. *Toxicology* **2008**, *251*, 28–34. [CrossRef] [PubMed]

238. Roy, P.; Kalra, N.; Prasad, S.; George, J.; Shukla, Y. Chemopreventive potential of resveratrol in mouse skin tumors through regulation of mitochondrial and PI3K/AKT signaling pathways. *Pharm. Res.* **2009**, *26*, 211–217. [CrossRef] [PubMed]

239. Yusuf, N.; Nasti, T.H.; Meleth, S.; Elmets, C.A. Resveratrol enhances cell-mediated immune response to DMBA through TLR4 and prevents DMBA induced cutaneous carcinogenesis. *Mol. Carcinog.* **2009**, *48*, 713–723. [CrossRef] [PubMed]

240. Jang, M.; Pezzuto, J.M. Effects of resveratrol on 12-O-tetradecanoylphorbol-13-acetate-induced oxidative events and gene expression in mouse skin. *Cancer Lett.* **1998**, *134*, 81–89. [CrossRef]

241. Kundu, J.K.; Chun, K.S.; Kim, S.O.; Surh, Y.J. Resveratrol inhibits phorbol ester-induced cyclooxygenase-2 expression in mouse skin: MAPKs and AP-1 as potential molecular targets. *BioFactors* **2004**, *21*, 33–39. [CrossRef] [PubMed]

242. Cichocki, M.; Paluszczak, J.; Szaefer, H.; Piechowiak, A.; Rimando, A.M.; Baer-Dubowska, W. Pterostilbene is equally potent as resveratrol in inhibiting 12-O-tetradecanoylphorbol-13-acetate activated NFκB, AP-1, COX-2, and iNOS in mouse epidermis. *Mol. Nutr. Food Res.* **2008**, *52*, S62–S70. [CrossRef] [PubMed]

243. Kim, K.H.; Back, J.H.; Zhu, Y.; Arbesman, J.; Athar, M.; Kopelovich, L.; Kim, A.L.; Bickers, D.R. Resveratrol targets transforming growth factor-β2 signaling to block UV-induced tumor progression. *J. Investig. Dermatol.* **2011**, *131*, 195–202. [CrossRef] [PubMed]

244. Sirerol, J.A.; Feddi, F.; Mena, S.; Rodriguez, M.L.; Sirera, P.; Aupi, M.; Perez, S.; Asensi, M.; Ortega, A.; Estrela, J.M. Topical treatment with pterostilbene, a natural phytoalexin, effectively protects hairless mice against UVB radiation-induced skin damage and carcinogenesis. *Free Radic. Boil. Med.* **2015**, *85*, 1–11. [CrossRef] [PubMed]

245. Bhattacharya, S.; Darjatmoko, S.R.; Polans, A.S. Resveratrol modulates the malignant properties of cutaneous melanoma through changes in the activation and attenuation of the antiapoptotic protooncogenic protein Akt/PKB. *Melanoma Res.* **2011**, *21*, 180–187. [CrossRef] [PubMed]

246. Asensi, M.; Medina, I.; Ortega, A.; Carretero, J.; Bano, M.C.; Obrador, E.; Estrela, J.M. Inhibition of cancer growth by resveratrol is related to its low bioavailability. *Free Radic. Boil. Med.* **2002**, *33*, 387–398. [CrossRef]

247. Niles, R.M.; Cook, C.P.; Meadows, G.G.; Fu, Y.M.; McLaughlin, J.L.; Rankin, G.O. Resveratrol is rapidly metabolized in athymic (nu/nu) mice and does not inhibit human melanoma xenograft tumor growth. *J. Nutr.* **2006**, *136*, 2542–2546. [PubMed]

248. Qin, W.; Zhang, K.; Clarke, K.; Weiland, T.; Sauter, E.R. Methylation and miRNA effects of resveratrol on mammary tumors vs. normal tissue. *Nutr. Cancer* **2014**, *66*, 270–277. [CrossRef] [PubMed]

249. Zander, S.A.; Kersbergen, A.; Sol, W.; Gonggrijp, M.; van de Wetering, K.; Jonkers, J.; Borst, P.; Rottenberg, S. Lack of ABCG2 shortens latency of BRCA1-deficient mammary tumors and this is not affected by genistein or resveratrol. *Cancer Prev. Res.* **2012**, *5*, 1053–1060. [CrossRef] [PubMed]

250. Sato, M.; Pei, R.J.; Yuri, T.; Danbara, N.; Nakane, Y.; Tsubura, A. Prepubertal resveratrol exposure accelerates N-methyl-N-nitrosourea-induced mammary carcinoma in female Sprague-Dawley rats. *Cancer Lett.* **2003**, *202*, 137–145. [CrossRef] [PubMed]

251. Bove, K.; Lincoln, D.W.; Tsan, M.F. Effect of resveratrol on growth of 4T1 breast cancer cells in vitro and in vivo. *Biochem. Biophys. Res. Commun.* **2002**, *291*, 1001–1005. [CrossRef] [PubMed]

252. Lee, H.S.; Ha, A.W.; Kim, W.K. Effect of resveratrol on the metastasis of 4T1 mouse breast cancer cells in vitro and in vivo. *Nutr. Res. Pract.* **2012**, *6*, 294–300. [CrossRef] [PubMed]

253. Fu, Y.; Chang, H.; Peng, X.; Bai, Q.; Yi, L.; Zhou, Y.; Zhu, J.; Mi, M. Resveratrol inhibits breast cancer stem-like cells and induces autophagy via suppressing Wnt/β-catenin signaling pathway. *PLoS ONE* **2014**, *9*, e102535. [CrossRef] [PubMed]

254. Wang, T.T.; Hudson, T.S.; Wang, T.C.; Remsberg, C.M.; Davies, N.M.; Takahashi, Y.; Kim, Y.S.; Seifried, H.; Vinyard, B.T.; Perkins, S.N.; et al. Differential effects of resveratrol on androgen-responsive LNCaP human prostate cancer cells in vitro and in vivo. *Carcinogenesis* **2008**, *29*, 2001–2010. [CrossRef] [PubMed]

255. Brizuela, L.; Dayon, A.; Doumerc, N.; Ader, I.; Golzio, M.; Izard, J.C.; Hara, Y.; Malavaud, B.; Cuvillier, O. The sphingosine kinase-1 survival pathway is a molecular target for the tumor-suppressive tea and wine polyphenols in prostate cancer. *FASEB J.* **2010**, *24*, 3882–3894. [CrossRef] [PubMed]

256. Willett, W.C. Diet, nutrition, and avoidable cancer. *Environ. Health Perspect.* **1995**, *103*, 165–170. [CrossRef] [PubMed]

257. Ruggeri, B.A.; Camp, F.; Miknyoczki, S. Animal models of disease: Pre-clinical animal models of cancer and their applications and utility in drug discovery. *Biochem. Pharmacol.* **2014**, *87*, 150–161. [CrossRef] [PubMed]

258. Washington, M.K.; Powell, A.E.; Sullivan, R.; Sundberg, J.P.; Wright, N.; Coffey, R.J.; Dove, W.F. Pathology of rodent models of intestinal cancer: Progress report and recommendations. *Gastroenterology* **2013**, *144*, 705–717. [CrossRef] [PubMed]

259. Cui, X.; Jin, Y.; Hofseth, A.B.; Pena, E.; Habiger, J.; Chumanevich, A.; Poudyal, D.; Nagarkatti, M.; Nagarkatti, P.S.; Singh, U.P.; et al. Resveratrol suppresses colitis and colon cancer associated with colitis. *Cancer Prev. Res.* **2010**, *3*, 549–559. [CrossRef] [PubMed]

260. Liao, W.; Wei, H.; Wang, X.; Qiu, Y.; Gou, X.; Zhang, X.; Zhou, M.; Wu, J.; Wu, T.; Kou, F.; et al. Metabonomic variations associated with AOM-induced precancerous colorectal lesions and resveratrol treatment. *J. Proteome Res.* **2012**, *11*, 3436–3448. [CrossRef] [PubMed]

261. Chiou, Y.S.; Tsai, M.L.; Nagabhushanam, K.; Wang, Y.J.; Wu, C.H.; Ho, C.T.; Pan, M.H. Pterostilbene is more potent than resveratrol in preventing azoxymethane (AOM)-induced colon tumorigenesis via activation of the NF-E2-related factor 2 (Nrf2)-mediated antioxidant signaling pathway. *J. Agric. Food Chem.* **2011**, *59*, 2725–2733. [CrossRef] [PubMed]

262. Sengottuvelan, M.; Senthilkumar, R.; Nalini, N. Modulatory influence of dietary resveratrol during different phases of 1,2-dimethylhydrazine induced mucosal lipid-peroxidation, antioxidant status and aberrant crypt foci development in rat colon carcinogenesis. *Biochim. Biophys. Acta* **2006**, *1760*, 1175–1183. [CrossRef] [PubMed]

263. Alfaras, I.; Juan, M.E.; Planas, J.M. trans-Resveratrol reduces precancerous colonic lesions in dimethylhydrazine-treated rats. *J. Agric. Food Chem.* **2010**, *58*, 8104–8110. [CrossRef] [PubMed]

264. Sengottuvelan, M.; Deeptha, K.; Nalini, N. Resveratrol ameliorates DNA damage, prooxidant and antioxidant imbalance in 1,2-dimethylhydrazine induced rat colon carcinogenesis. *Chem.-Biol. Interact.* **2009**, *181*, 193–201. [CrossRef] [PubMed]

265. Sengottuvelan, M.; Viswanathan, P.; Nalini, N. Chemopreventive effect of trans-resveratrol—A phytoalexin against colonic aberrant crypt foci and cell proliferation in 1,2-dimethylhydrazine induced colon carcinogenesis. *Carcinogenesis* **2006**, *27*, 1038–1046. [CrossRef] [PubMed]

266. Sengottuvelan, M.; Deeptha, K.; Nalini, N. Influence of dietary resveratrol on early and late molecular markers of 1,2-dimethylhydrazine-induced colon carcinogenesis. *Nutrition* **2009**, *25*, 1169–1176. [CrossRef] [PubMed]

267. Saud, S.M.; Li, W.; Morris, N.L.; Matter, M.S.; Colburn, N.H.; Kim, Y.S.; Young, M.R. Resveratrol prevents tumorigenesis in mouse model of Kras activated sporadic colorectal cancer by suppressing oncogenic Kras expression. *Carcinogenesis* **2014**, *35*, 2778–2786. [CrossRef] [PubMed]

268. Schneider, Y.; Vincent, F.; Duranton, B.; Badolo, L.; Gosse, F.; Bergmann, C.; Seiler, N.; Raul, F. Anti-proliferative effect of resveratrol, a natural component of grapes and wine, on human colonic cancer cells. *Cancer Lett.* **2000**, *158*, 85–91. [CrossRef]

269. Rajasekaran, D.; Elavarasan, J.; Sivalingam, M.; Ganapathy, E.; Kumar, A.; Kalpana, K.; Sakthisekaran, D. Resveratrol interferes with N-nitrosodiethylamine-induced hepatocellular carcinoma at early and advanced stages in male Wistar rats. *Mol. Med. Rep.* **2011**, *4*, 1211–1217. [PubMed]

270. Luther, D.J.; Ohanyan, V.; Shamhart, P.E.; Hodnichak, C.M.; Sisakian, H.; Booth, T.D.; Meszaros, J.G.; Bishayee, A. Chemopreventive doses of resveratrol do not produce cardiotoxicity in a rodent model of hepatocellular carcinoma. *Investig. New Drugs* **2011**, *29*, 380–391. [CrossRef] [PubMed]

271. Wu, X.; Li, C.; Xing, G.; Qi, X.; Ren, J. Resveratrol Downregulates Cyp2e1 and Attenuates Chemically Induced Hepatocarcinogenesis in SD Rats. *J. Toxicol. Pathol.* **2013**, *26*, 385–392. [CrossRef] [PubMed]

272. Lin, H.C.; Chen, Y.F.; Hsu, W.H.; Yang, C.W.; Kao, C.H.; Tsai, T.F. Resveratrol helps recovery from fatty liver and protects against hepatocellular carcinoma induced by hepatitis B virus X protein in a mouse model. *Cancer Prev. Res.* **2012**, *5*, 952–962. [CrossRef] [PubMed]

273. Bishayee, A.; Barnes, K.F.; Bhatia, D.; Darvesh, A.S.; Carroll, R.T. Resveratrol suppresses oxidative stress and inflammatory response in diethylnitrosamine-initiated rat hepatocarcinogenesis. *Cancer Prev. Res.* **2010**, *3*, 753–763. [CrossRef] [PubMed]

274. Kitamura, Y.; Umemura, T.; Kanki, K.; Kodama, Y.; Kitamoto, S.; Saito, K.; Itoh, K.; Yamamoto, M.; Masegi, T.; Nishikawa, A.; et al. Increased susceptibility to hepatocarcinogenicity of Nrf2-deficient mice exposed to 2-amino-3-methylimidazo[4,5-f]quinoline. *Cancer Sci.* **2007**, *98*, 19–24. [CrossRef] [PubMed]

275. Bishayee, A.; Waghray, A.; Barnes, K.F.; Mbimba, T.; Bhatia, D.; Chatterjee, M.; Darvesh, A.S. Suppression of the inflammatory cascade is implicated in resveratrol chemoprevention of experimental hepatocarcinogenesis. *Pharm. Res.* **2010**, *27*, 1080–1091. [CrossRef] [PubMed]

276. Mbimba, T.; Awale, P.; Bhatia, D.; Geldenhuys, W.J.; Darvesh, A.S.; Carroll, R.T.; Bishayee, A. Alteration of hepatic proinflammatory cytokines is involved in the resveratrol-mediated chemoprevention of chemically-induced hepatocarcinogenesis. *Curr. Pharm. Biotechnol.* **2012**, *13*, 229–234. [CrossRef] [PubMed]

277. Bishayee, A.; Petit, D.M.; Samtani, K. Angioprevention is Implicated in Resveratrol Chemoprevention of Experimental Hepatocarcinogenesis. *J. Carcinog. Mutagen.* **2010**, *1*, 102. [CrossRef]

278. Yu, H.B.; Zhang, H.F.; Zhang, X.; Li, D.Y.; Xue, H.Z.; Pan, C.E.; Zhao, S.H. Resveratrol inhibits VEGF expression of human hepatocellular carcinoma cells through a NF-κB-mediated mechanism. *Hepato-Gastroenterology* **2010**, *57*, 1241–1246. [PubMed]

279. Salado, C.; Olaso, E.; Gallot, N.; Valcarcel, M.; Egilegor, E.; Mendoza, L.; Vidal-Vanaclocha, F. Resveratrol prevents inflammation-dependent hepatic melanoma metastasis by inhibiting the secretion and effects of interleukin-18. *J. Transl. Med.* **2011**, *9*, 59. [CrossRef] [PubMed]

280. Byrum, R.S.; Goulet, J.L.; Snouwaert, J.N.; Griffiths, R.J.; Koller, B.H. Determination of the contribution of cysteinyl leukotrienes and leukotriene B4 in acute inflammatory responses using 5-lipoxygenase- and leukotriene A4 hydrolase-deficient mice. *J. Immunol.* **1999**, *163*, 6810–6819. [PubMed]

281. Bortuzzo, C.; Hanif, R.; Kashfi, K.; Staiano-Coico, L.; Shiff, S.J.; Rigas, B. The effect of leukotrienes B and selected HETEs on the proliferation of colon cancer cells. *Biochim. Biophys. Acta* **1996**, *1300*, 240–246. [CrossRef]

282. Tong, W.G.; Ding, X.Z.; Hennig, R.; Witt, R.C.; Standop, J.; Pour, P.M.; Adrian, T.E. Leukotriene B4 receptor antagonist LY293111 inhibits proliferation and induces apoptosis in human pancreatic cancer cells. *Clin. Cancer Res.* **2002**, *8*, 3232–3242. [PubMed]

283. Roy, S.K.; Chen, Q.; Fu, J.; Shankar, S.; Srivastava, R.K. Resveratrol inhibits growth of orthotopic pancreatic tumors through activation of FOXO transcription factors. *PLoS ONE* **2011**, *6*, e25166. [CrossRef] [PubMed]

284. Shankar, S.; Nall, D.; Tang, S.N.; Meeker, D.; Passarini, J.; Sharma, J.; Srivastava, R.K. Resveratrol inhibits pancreatic cancer stem cell characteristics in human and KrasG12D transgenic mice by inhibiting pluripotency maintaining factors and epithelial-mesenchymal transition. *PLoS ONE* **2011**, *6*, e16530. [CrossRef] [PubMed]

285. Kuroiwa, Y.; Nishikawa, A.; Kitamura, Y.; Kanki, K.; Ishii, Y.; Umemura, T.; Hirose, M. Protective effects of benzyl isothiocyanate and sulforaphane but not resveratrol against initiation of pancreatic carcinogenesis in hamsters. *Cancer Lett.* **2006**, *241*, 275–280. [CrossRef] [PubMed]

286. Revel, A.; Raanani, H.; Younglai, E.; Xu, J.; Rogers, I.; Han, R.; Savouret, J.F.; Casper, R.F. Resveratrol, a natural aryl hydrocarbon receptor antagonist, protects lung from DNA damage and apoptosis caused by benzo[a]pyrene. *JAT* **2003**, *23*, 255–261. [CrossRef] [PubMed]

287. Malhotra, A.; Nair, P.; Dhawan, D.K. Premature mitochondrial senescence and related ultrastructural changes during lung carcinogenesis modulation by curcumin and resveratrol. *Ultrastruct. Pathol.* **2012**, *36*, 179–184. [CrossRef] [PubMed]

288. Malhotra, A.; Nair, P.; Dhawan, D.K. Study to evaluate molecular mechanics behind synergistic chemo-preventive effects of curcumin and resveratrol during lung carcinogenesis. *PLoS ONE* **2014**, *9*, e93820. [CrossRef] [PubMed]

289. Yu, Y.H.; Chen, H.A.; Chen, P.S.; Cheng, Y.J.; Hsu, W.H.; Chang, Y.W.; Chen, Y.H.; Jan, Y.; Hsiao, M.; Chang, T.Y.; et al. MiR-520h-mediated FOXC2 regulation is critical for inhibition of lung cancer progression by resveratrol. *Oncogene* **2013**, *32*, 431–443. [CrossRef] [PubMed]

290. Lee, K.A.; Lee, Y.J.; Ban, J.O.; Lee, Y.J.; Lee, S.H.; Cho, M.K.; Nam, H.S.; Hong, J.T.; Shim, J.H. The flavonoid resveratrol suppresses growth of human malignant pleural mesothelioma cells through direct inhibition of specificity protein 1. *Int. J. Mol. Med.* **2012**, *30*, 21–27. [PubMed]

291. Hecht, S.S.; Kenney, P.M.; Wang, M.; Trushin, N.; Agarwal, S.; Rao, A.V.; Upadhyaya, P. Evaluation of butylated hydroxyanisole, myo-inositol, curcumin, esculetin, resveratrol and lycopene as inhibitors of benzo[a]pyrene plus 4-(methylnitrosamino)-1-(3-pyridyl)-1-butanone-induced lung tumorigenesis in A/J mice. *Cancer Lett.* **1999**, *137*, 123–130. [CrossRef]

292. Berge, G.; Ovrebo, S.; Eilertsen, E.; Haugen, A.; Mollerup, S. Analysis of resveratrol as a lung cancer chemopreventive agent in A/J mice exposed to benzo[a]pyrene. *Br. J. Cancer* **2004**, *91*, 1380–1383. [CrossRef] [PubMed]

293. Shi, Q.; Geldenhuys, W.; Sutariya, V.; Bishayee, A.; Patel, I.; Bhatia, D. CArG-driven GADD45α activated by resveratrol inhibits lung cancer cells. *Genes Cancer* **2015**, *6*, 220–230. [PubMed]

294. Berta, G.N.; Salamone, P.; Sprio, A.E.; Di Scipio, F.; Marinos, L.M.; Sapino, S.; Carlotti, M.E.; Cavalli, R.; Di Carlo, F. Chemoprevention of 7,12-dimethylbenz[a]anthracene (DMBA)-induced oral carcinogenesis in hamster cheek pouch by topical application of resveratrol complexed with 2-hydroxypropyl-beta-cyclodextrin. *Oral Oncol.* **2010**, *46*, 42–48. [CrossRef] [PubMed]

295. Woodall, C.E.; Li, Y.; Liu, Q.H.; Wo, J.; Martin, R.C. Chemoprevention of metaplasia initiation and carcinogenic progression to esophageal adenocarcinoma by resveratrol supplementation. *Anti-Cancer Drugs* **2009**, *20*, 437–443. [CrossRef] [PubMed]

296. Lee, M.H.; Choi, B.Y.; Kundu, J.K.; Shin, Y.K.; Na, H.K.; Surh, Y.J. Resveratrol suppresses growth of human ovarian cancer cells in culture and in a murine xenograft model: Eukaryotic elongation factor 1A2 as a potential target. *Cancer Res.* **2009**, *69*, 7449–7458. [CrossRef] [PubMed]

297. Zhou, H.B.; Chen, J.J.; Wang, W.X.; Cai, J.T.; Du, Q. Anticancer activity of resveratrol on implanted human primary gastric carcinoma cells in nude mice. *World J. Gastroenterol.* **2005**, *11*, 280–284. [CrossRef] [PubMed]

298. Tyagi, A.; Gu, M.; Takahata, T.; Frederick, B.; Agarwal, C.; Siriwardana, S.; Agarwal, R.; Sclafani, R.A. Resveratrol selectively induces DNA Damage, independent of Smad4 expression, in its efficacy against human head and neck squamous cell carcinoma. *Clin. Cancer Res.* **2011**, *17*, 5402–5411. [CrossRef] [PubMed]

299. Hu, F.W.; Tsai, L.L.; Yu, C.H.; Chen, P.N.; Chou, M.Y.; Yu, C.C. Impairment of tumor-initiating stem-like property and reversal of epithelial-mesenchymal transdifferentiation in head and neck cancer by resveratrol treatment. *Mol. Nutr. Food Res.* **2012**, *56*, 1247–1258. [CrossRef] [PubMed]

300. Chen, Y.; Tseng, S.H.; Lai, H.S.; Chen, W.J. Resveratrol-induced cellular apoptosis and cell cycle arrest in neuroblastoma cells and antitumor effects on neuroblastoma in mice. *Surgery* **2004**, *136*, 57–66. [CrossRef] [PubMed]

301. Van Ginkel, P.R.; Sareen, D.; Subramanian, L.; Walker, Q.; Darjatmoko, S.R.; Lindstrom, M.J.; Kulkarni, A.; Albert, D.M.; Polans, A.S. Resveratrol inhibits tumor growth of human neuroblastoma and mediates apoptosis by directly targeting mitochondria. *Clin. Cancer Res.* **2007**, *13*, 5162–5169. [CrossRef] [PubMed]

302. Tseng, S.H.; Lin, S.M.; Chen, J.C.; Su, Y.H.; Huang, H.Y.; Chen, C.K.; Lin, P.Y.; Chen, Y. Resveratrol suppresses the angiogenesis and tumor growth of gliomas in rats. *Clin. Cancer Res.* **2004**, *10*, 2190–2202. [CrossRef] [PubMed]

303. Brakenhielm, E.; Cao, R.; Cao, Y. Suppression of angiogenesis, tumor growth, and wound healing by resveratrol, a natural compound in red wine and grapes. *FASEB J.* **2001**, *15*, 1798–1800. [CrossRef] [PubMed]

304. Li, Z.G.; Hong, T.; Shimada, Y.; Komoto, I.; Kawabe, A.; Ding, Y.; Kaganoi, J.; Hashimoto, Y.; Imamura, M. Suppression of N-nitrosomethylbenzylamine (NMBA)-induced esophageal tumorigenesis in F344 rats by resveratrol. *Carcinogenesis* **2002**, *23*, 1531–1536. [CrossRef] [PubMed]

305. Yang, Q.; Wang, B.; Zang, W.; Wang, X.; Liu, Z.; Li, W.; Jia, J. Resveratrol inhibits the growth of gastric cancer by inducing G1 phase arrest and senescence in a Sirt1-dependent manner. *PLoS ONE* **2013**, *8*, e70627. [CrossRef] [PubMed]

306. Stakleff, K.S.; Sloan, T.; Blanco, D.; Marcanthony, S.; Booth, T.D.; Bishayee, A. Resveratrol exerts differential effects in vitro and in vivo against ovarian cancer cells. *APJCP* **2012**, *13*, 1333–1340. [CrossRef] [PubMed]

307. Patel, K.R.; Scott, E.; Brown, V.A.; Gescher, A.J.; Steward, W.P.; Brown, K. Clinical trials of resveratrol. *Ann. N. Y. Acad. Sci.* **2011**, *1215*, 161–169. [CrossRef] [PubMed]

308. Cottart, C.H.; Nivet-Antoine, V.; Laguillier-Morizot, C.; Beaudeux, J.L. Resveratrol bioavailability and toxicity in humans. *Mol. Nutr. Food Res.* **2010**, *54*, 7–16. [CrossRef] [PubMed]

309. Gescher, A.; Steward, W.P.; Brown, K. Resveratrol in the management of human cancer: How strong is the clinical evidence? *Ann. N. Y. Acad. Sci.* **2013**, *1290*, 12–20. [CrossRef] [PubMed]

310. Chow, H.H.; Garland, L.L.; Hsu, C.H.; Vining, D.R.; Chew, W.M.; Miller, J.A.; Perloff, M.; Crowell, J.A.; Alberts, D.S. Resveratrol modulates drug- and carcinogen-metabolizing enzymes in a healthy volunteer study. *Cancer Prev. Res.* **2010**, *3*, 1168–1175. [CrossRef] [PubMed]

311. Smoliga, J.M.; Blanchard, O. Enhancing the delivery of resveratrol in humans: If low bioavailability is the problem, what is the solution? *Molecules* **2014**, *19*, 17154–17172. [CrossRef] [PubMed]

312. La Porte, C.; Voduc, N.; Zhang, G.; Seguin, I.; Tardiff, D.; Singhal, N.; Cameron, D.W. Steady-State pharmacokinetics and tolerability of trans-resveratrol 2000 mg twice daily with food, quercetin and alcohol (ethanol) in healthy human subjects. *Clin. Pharmacokinet.* **2010**, *49*, 449–454. [CrossRef] [PubMed]

313. Johnson, J.J.; Nihal, M.; Siddiqui, I.A.; Scarlett, C.O.; Bailey, H.H.; Mukhtar, H.; Ahmad, N. Enhancing the bioavailability of resveratrol by combining it with piperine. *Mol. Nutr. Food Res.* **2011**, *55*, 1169–1176. [CrossRef] [PubMed]

314. Liang, L.; Liu, X.; Wang, Q.; Cheng, S.; Zhang, S.; Zhang, M. Pharmacokinetics, tissue distribution and excretion study of resveratrol and its prodrug 3,5,4′-tri-O-acetylresveratrol in rats. *Phytomedicine* **2013**, *20*, 558–563. [CrossRef] [PubMed]

315. Howells, L.M.; Berry, D.P.; Elliott, P.J.; Jacobson, E.W.; Hoffmann, E.; Hegarty, B.; Brown, K.; Steward, W.P.; Gescher, A.J. Phase I randomized, double-blind pilot study of micronized resveratrol (SRT501) in patients with hepatic metastases—Safety, pharmacokinetics, and pharmacodynamics. *Cancer Prev. Res.* **2011**, *4*, 1419–1425. [CrossRef] [PubMed]

316. Popat, R.; Plesner, T.; Davies, F.; Cook, G.; Cook, M.; Elliott, P.; Jacobson, E.; Gumbleton, T.; Oakervee, H.; Cavenagh, J. A phase 2 study of SRT501 (resveratrol) with bortezomib for patients with relapsed and or refractory multiple myeloma. *Br. J. Haematol.* **2013**, *160*, 714–717. [CrossRef] [PubMed]

317. Ansari, K.A.; Vavia, P.R.; Trotta, F.; Cavalli, R. Cyclodextrin-based nanosponges for delivery of resveratrol: In vitro characterisation, stability, cytotoxicity and permeation study. *AAPS PharmSciTech* **2011**, *12*, 279–286. [CrossRef] [PubMed]

318. Pangeni, R.; Sahni, J.K.; Ali, J.; Sharma, S.; Baboota, S. Resveratrol: Review on therapeutic potential and recent advances in drug delivery. *Exp. Opin. Drug Deliv.* **2014**, *11*, 1285–1298. [CrossRef] [PubMed]

319. Wang, S.; Su, R.; Nie, S.; Sun, M.; Zhang, J.; Wu, D.; Moustaid-Moussa, N. Application of nanotechnology in improving bioavailability and bioactivity of diet-derived phytochemicals. *J. Nutr. Biochem.* **2014**, *25*, 363–376. [CrossRef] [PubMed]

320. Nguyen, A.V.; Martinez, M.; Stamos, M.J.; Moyer, M.P.; Planutis, K.; Hope, C.; Holcombe, R.F. Results of a phase I pilot clinical trial examining the effect of plant-derived resveratrol and grape powder on Wnt pathway target gene expression in colonic mucosa and colon cancer. *Cancer Managen. Res.* **2009**, *1*, 25–37.

321. Robbins, D.H.; Itzkowitz, S.H. The molecular and genetic basis of colon cancer. *Med. Clin. N. Am.* **2002**, *86*, 1467–1495. [CrossRef]

322. Patel, K.R.; Brown, V.A.; Jones, D.J.; Britton, R.G.; Hemingway, D.; Miller, A.S.; West, K.P.; Booth, T.D.; Perloff, M.; Crowell, J.A.; et al. Clinical pharmacology of resveratrol and its metabolites in colorectal cancer patients. *Cancer Res.* **2010**, *70*, 7392–7399. [CrossRef] [PubMed]

323. Paller, C.J.; Rudek, M.A.; Zhou, X.C.; Wagner, W.D.; Hudson, T.S.; Anders, N.; Hammers, H.J.; Dowling, D.; King, S.; Antonarakis, E.S.; et al. A phase I study of muscadine grape skin extract in men with biochemically recurrent prostate cancer: Safety, tolerability, and dose determination. *Prostate* **2015**, *75*, 1518–1525. [CrossRef] [PubMed]

324. Kjaer, T.N.; Ornstrup, M.J.; Poulsen, M.M.; Jorgensen, J.O.; Hougaard, D.M.; Cohen, A.S.; Neghabat, S.; Richelsen, B.; Pedersen, S.B. Resveratrol reduces the levels of circulating androgen precursors but has no effect on, testosterone, dihydrotestosterone, PSA levels or prostate volume. A 4-month randomised trial in middle-aged men. *Prostate* **2015**, *75*, 1255–1263. [CrossRef] [PubMed]

325. Geiszt, M.; Lekstrom, K.; Brenner, S.; Hewitt, S.M.; Dana, R.; Malech, H.L.; Leto, T.L. NAD(P)H oxidase 1, a product of differentiated colon epithelial cells, can partially replace glycoprotein 91phox in the regulated production of superoxide by phagocytes. *J. Immunol.* **2003**, *171*, 299–306. [CrossRef] [PubMed]

326. Dutta, S.; Rittinger, K. Regulation of NOXO1 activity through reversible interactions with p22 and NOXA1. *PLoS ONE* **2010**, *5*, e10478. [CrossRef] [PubMed]

327. Okur, H.; Kucukaydin, M.; Kose, K.; Kontas, O.; Dogam, P.; Kazez, A. Hypoxia-induced necrotizing enterocolitis in the immature rat: The role of lipid peroxidation and management by vitamin E. *J. Pediatr. Surg.* **1995**, *30*, 1416–1419. [CrossRef]

328. Chan, K.L.; Hui, C.W.; Chan, K.W.; Fung, P.C.; Wo, J.Y.; Tipoe, G.; Tam, P.K. Revisiting ischemia and reperfusion injury as a possible cause of necrotizing enterocolitis: Role of nitric oxide and superoxide dismutase. *J. Pediatr. Surg.* **2002**, *37*, 828–834. [CrossRef] [PubMed]

329. Wang, Z.; Li, S.; Cao, Y.; Tian, X.; Zeng, R.; Liao, D.F.; Cao, D. Oxidative Stress and Carbonyl Lesions in Ulcerative Colitis and Associated Colorectal Cancer. *Oxid. Med. Cell. Longev.* **2016**, *2016*, 9875298. [CrossRef] [PubMed]

330. Cai, H.; Scott, E.; Kholghi, A.; Andreadi, C.; Rufini, A.; Karmokar, A.; Britton, R.G.; Horner-Glister, E.; Greaves, P.; Jawad, D.; et al. Cancer chemoprevention: Evidence of a nonlinear dose response for the protective effects of resveratrol in humans and mice. *Sci. Transl. Med.* **2015**, *7*, 298ra117. [CrossRef] [PubMed]

331. Zhu, W.; Qin, W.; Zhang, K.; Rottinghaus, G.E.; Chen, Y.C.; Kliethermes, B.; Sauter, E.R. Trans-resveratrol alters mammary promoter hypermethylation in women at increased risk for breast cancer. *Nutr. Cancer* **2012**, *64*, 393–400. [CrossRef] [PubMed]

332. Key, T.; Appleby, P.; Barnes, I.; Reeves, G.; Endogenous, H.; Breast Cancer Collaborative, G. Endogenous sex hormones and breast cancer in postmenopausal women: Reanalysis of nine prospective studies. *J. Natl. Cancer Inst.* **2002**, *94*, 606–616. [PubMed]

333. Chow, H.H.; Garland, L.L.; Heckman-Stoddard, B.M.; Hsu, C.H.; Butler, V.D.; Cordova, C.A.; Chew, W.M.; Cornelison, T.L. A pilot clinical study of resveratrol in postmenopausal women with high body mass index: Effects on systemic sex steroid hormones. *J. Transl. Med.* **2014**, *12*, 223. [CrossRef] [PubMed]

334. Renehan, A.G.; Zwahlen, M.; Minder, C.; O'Dwyer, S.T.; Shalet, S.M.; Egger, M. Insulin-like growth factor (IGF)-I, IGF binding protein-3, and cancer risk: Systematic review and meta-regression analysis. *Lancet* **2004**, *363*, 1346–1353. [CrossRef]

335. Brown, V.A.; Patel, K.R.; Viskaduraki, M.; Crowell, J.A.; Perloff, M.; Booth, T.D.; Vasilinin, G.; Sen, A.; Schinas, A.M.; Piccirilli, G.; et al. Repeat dose study of the cancer chemopreventive agent resveratrol in healthy volunteers: Safety, pharmacokinetics, and effect on the insulin-like growth factor axis. *Cancer Res.* **2010**, *70*, 9003–9011. [CrossRef] [PubMed]

International Journal of
Molecular Sciences

MDPI

Article

Supercritical-Carbon Dioxide Fluid Extract from *Chrysanthemum indicum* Enhances Anti-Tumor Effect and Reduces Toxicity of Bleomycin in Tumor-Bearing Mice

Hong-Mei Yang [1], Chao-Yue Sun [1], Jia-Li Liang [1], Lie-Qiang Xu [1], Zhen-Biao Zhang [1], Dan-Dan Luo [1], Han-Bin Chen [1], Yong-Zhong Huang [1], Qi Wang [2], David Yue-Wei Lee [3], Jie Yuan [1,4,*] and Yu-Cui Li [1,*]

[1] Guangdong Provincial Key Laboratory of New Drug Development and Research of Chinese Medicine, Guangzhou University of Chinese Medicine, Guangzhou 510006, China; yanghongmei326@gmail.com (H.-M.Y.); yuhongl0822@gmail.com (C.-Y.S.); xing623853793@gmail.com (J.-L.L.); xulieqiang123@gmail.com (L.-Q.X.); Lauyh0822@gmail.com (Z.-B.Z.); zhuojiany38193@gmail.com (D.-D.L.); chenhanbin999@gmail.com (H.-B.C.); jakin3305@gmail.com (Y.-Z.H.)
[2] Guangdong New South Artepharm, Co., Ltd., Guangzhou 510006, China; l1393318376@gmail.com
[3] Department of McLean Hospital, Harvard Medical School, Belmont, CA 02478-9106, USA; ywlee228@gmail.com
[4] Dongguan Mathematical Engineering Academy of Chinese Medicine, Guangzhou University of Chinese Medicine, Dongguan 523000, China
[*] Correspondence: yuanjie@gzucm.edu.cn (J.Y.); liyucui@gzucm.edu.cn (Y.-C.L.); Tel.: +86-20-3935-8517 (J.Y. & Y.-C.L.); Fax: +86-20-3935-8390 (J.Y. & Y.-C.L.)

Academic Editors: Ashis Basu and Takehiko Nohmi
Received: 13 December 2016; Accepted: 13 February 2017; Published: 24 February 2017

Abstract: Bleomycin (BLM), a family of anti-tumor drugs, was reported to exhibit severe side effects limiting its usage in clinical treatment. Therefore, finding adjuvants that enhance the anti-tumor effect and reduce the detrimental effect of BLM is a prerequisite. *Chrysanthemum indicum*, an edible flower, possesses abundant bioactivities; the supercritical-carbon dioxide fluid extract from flowers and buds of *C. indicum* (CI_{SCFE}) have strong anti-inflammatory, anti-oxidant, and lung protective effects. However, the role of CI_{SCFE} combined with BLM treatment on tumor-bearing mice remains unclear. The present study aimed to investigate the potential synergistic effect and the underlying mechanism of CI_{SCFE} combined with BLM in the treatment of hepatoma 22 (H22) tumor-bearing mice. The results suggested that the oral administration of CI_{SCFE} combined with BLM could markedly prolong the life span, attenuate the BLM-induced pulmonary fibrosis, suppress the production of pro-inflammatory cytokines (interleukin-6), tumor necrosis factor-α, activities of myeloperoxidase, and malondiadehyde. Moreover, CI_{SCFE} combined with BLM promoted the ascites cell apoptosis, the activities of caspases 3 and 8, and up-regulated the protein expression of p53 and down-regulated the transforming growth factor-$\beta 1$ by activating the gene expression of miR-29b. Taken together, these results indicated that CI_{SCFE} could enhance the anti-cancer activity of BLM and reduce the BLM-induced pulmonary injury in H22 tumor-bearing mice, rendering it as a potential adjuvant drug with chemotherapy after further investigation in the future.

Keywords: supercritical-carbon dioxide fluid of *C. indicum* (CI_{SCFE}); BLM; anti-tumor effect; pulmonary fibrosis; synergism effect

Int. J. Mol. Sci. **2017**, *18*, 465

1. Introduction

Bleomycin (BLM), a glycopetide originally isolated from *Streptomyces verticillus* [1], is a clinical anti-cancer drug primarily used for the treatment of hepatocellular carcinoma (HCC) and nasopharyngeal carcinoma (NPC). The anti-tumor mechanism mainly consists of inducing DNA damage and has been demonstrated to be mediated through the induction of oxidative stress [1]. Several studies revealed that BLM is vital in the clinical treatment of HCC. However, BLM exhibits the main side effect of dose-dependent pulmonary toxicity, which affects 20% of treated individuals. Pulmonary fibrosis is a severe form of lung toxicity, which was induced by BLM [2]. However, the etiology and mechanism of pulmonary fibrosis have not yet been elucidated. A number of studies have reported that the combination therapy can not only enhance the anticancer effect, but also attenuate the toxicity side-effects to the organs [3,4]. Therefore, the development of a drug that confers lung protection during BLM treatment and improves the chemotherapeutic efficacy of BLM in cancer is essential.

The integration of different signaling pathways plays a critical role in the normal development and tissue homeostasis of metazoans. When one or more signals fail to integrate, the entire signaling network might collapse resulting in diseases, especially cancer [5]. The loss of cross-talk among the two most critical pathways—tumor suppressor Trp53 (p53) and tumor growth factor beta (TGF-β) signaling leads to many kinds of tumors and organ fibrosis [6,7]. p53 can suppress the TGF-β signal, thereby inhibiting the microRNAs (miRNAs) such as miR-17-92/miR-106b-25 clusters to retain the integrity of the antitumor signals. Loss of p53 can lead to the loss of TGF-β receptor 2 (TGFBR2) and miR-34a expression, resulting in attenuated antiproliferative signals [8]. Sun et al. reported that p53 was essential for doxorubicin-induced apoptosis via the TGF-β signaling pathway in osteosarcoma-derived cells [9]. On the other hand, p53 was required for the expression of plasminogen activator inhibitor-1 (PAI-1), a major TGF-β1 target gene and a key causative element in fibrotic disorders [10]. Moreover, Wang et al. found that astaxanthin ameliorated lung fibrosis in rat by regulating the cross-talk between p53 and TGF-β signaling [11]. Thus, substances with an effect on the regulation of p53 and TGF-β signaling pathways may be beneficial for improving the chemotherapeutic efficacy of BLM or alleviating the pulmonary toxicity induced by BLM.

Chrysanthemum indicum (*C. indicum*) Linné, a traditional medicinal and edible flower, is widely used as herbal tea, alcoholic beverage, and food additive or directly used to treat several infectious diseases and ailments, such as headache, eye diseases, and various immune-related disorders with high efficacy and low toxicity [12–14]. Moreover, the essential oil from the flowers possesses anti-bacterial and anti-cancer properties [15]. Importantly, the supercritical-carbon dioxide fluid extract from flowers and buds of *C. indicum* (CI$_{SCFE}$) have been extensively applied not only in many classical prescription, but also used in daily life as functional foods, cosmetics, and beverages [16]. Pongjit et al. reported that CI$_{SCFE}$ has a strong ability to protect against the chemotherapy-induced renal cell damage [17]. In addition, our previous study demonstrated that CI$_{SCFE}$ has a protective effect against lipopolysaccharide-induced lung injury and UV-induced skin injury [18,19]. However, the effective antitumor activity of CI$_{SCFE}$ combined with BLM in vivo remains unclear.

In the present study, we used the classical H22 ascites tumor-bearing mice model [20,21] to explore the potential synergistic effect of CI$_{SCFE}$ combined with BLM and investigate the underlying mechanism in the treatment of cancer.

2. Results

2.1. Anti-Tumor Activities of CI$_{SCFE}$, BLM, and Their Combination on H22 Tumor-Bearing Mice

To better understand the anti-tumor activities of CI$_{SCFE}$, BLM, and their combination, we evaluated the life span of the H22 tumor-bearing mice model. As shown in Figure 1, compared with the model group, CI$_{SCFE}$ (L: 240 mg/kg, M: 360 mg/kg, H: 480 mg/kg) alone groups exhibited no significant influence on the life-span of the tumor-bearing mice ($p > 0.05$), the BLM alone group could prolong the survival time ($p < 0.05$), while the mice treated with BLM + CI$_{SCFE}$ (M: 360 mg/kg, H: 480 mg/kg) for seven days could significantly prolong the life span as compared to BLM alone ($p < 0.05$). These data suggested that CI$_{SCFE}$-M, H doses could improve the BLM anti-tumor effect. Thus, CI$_{SCFE}$ at a middle dose of 360 mg/kg was used in the subsequent studies.

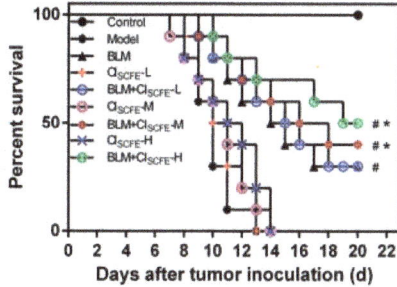

Figure 1. Survivals curve of CI$_{SCFE}$, bleomycin (BLM), and their combination on tumor-bearing mice. The survival rate was followed-up until 22 days after inoculation. Each group comprised of eight mice. [#] $p < 0.05$ vs. model group; [*] $p < 0.05$ vs. BLM group.

2.2. Synergistic Effect of CI$_{SCFE}$ Combined with BLM on Tumor Growth

Figure 2 summarizes the effect of CI$_{SCFE}$ combined with BLM on tumor growth. The weight, abnormal diameter, as well as ascites of mice in the model group increased rapidly compared to the control group. On the contrary, the weights, abnormal diameters, and ascites of mice significantly decreased in the BLM alone and BLM + CI$_{SCFE}$-M groups (M: 360 mg/kg), compared to the model group during seven days ($p < 0.05$). The mice in the treated groups showed greater vitality and were in good order, while the combination was more effective than individual treatment. Additionally, no obvious differences were observed in body weight and abnormal diameter between CI$_{SCFE}$-M alone group and model group ($p > 0.05$). These results demonstrated that CI$_{SCFE}$ had little or no effect on tumor-bearing mice; however, it significantly enhanced the anti-tumor activity of BLM.

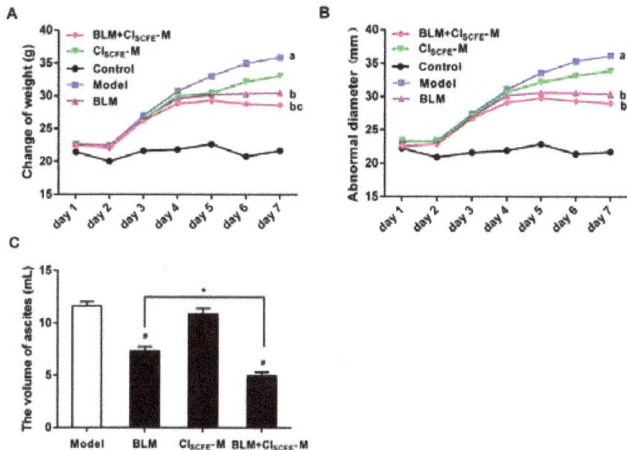

Figure 2. Effect of CI$_{SCFE}$-M, BLM, and their combination on the change of weight (**A**); abnormal diameter (**B**); and ascites (**C**) of tumor-bearing mice. Data represent mean $\pm$ SEM ($n = 10$). [#] $p < 0.05$ vs. model group; [*] $p < 0.05$ vs. BLM group. a < 0.05 vs. control group; b < 0.05 vs. model group, c < 0.05 vs. BLM group.

2.3. Synergistic Effect of BLM with CI$_{SCFE}$ in Inducing H22 Ascites Cell Apoptosis

To evaluate whether the combination of CI$_{SCFE}$ and BLM can potentially enhance the efficacy of BLM on H22 ascites cell apoptosis, flow cytometry was utilized to assess the rate of apoptosis. As shown in Figure 3, in the BLM treatment alone and BLM + CI$_{SCFE}$-M combined group, the rate of apoptotic cells (Annexin V+/PI− + Annexin V+/PI+) was notably increased as compared to the model group (all $p < 0.05$); the rate of apoptotic cells was obviously increased in the BLM + CI$_{SCFE}$-M combined group as compared to BLM alone ($p < 0.05$). All results suggested that CI$_{SCFE}$ combined with BLM could remarkably increase the H22 ascites cell apoptosis.

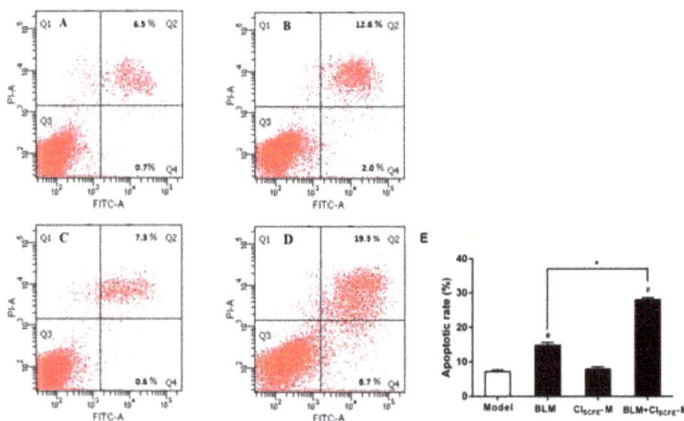

Figure 3. The synergistic effect of BLM with CI$_{SCFE}$ on H22 ascites cell apoptosis induction was analyzed by flow cytometry. (**A**) model group; (**B**) BLM group; (**C**) CI$_{SCFE}$-M group; (**D**) BLM + CI$_{SCFE}$-M group; (**E**) apoptotic rate. The cell populations of Annexin V+/PI− and Annexin V+/PI+ were estimated to represent the total number of apoptotic cells. Data are expressed as mean ± SEM (n = 3). [#] $p < 0.05$ vs. model group; [*] $p < 0.05$ vs. BLM group.

2.4. CI$_{SCFE}$ Enhanced the Anti-Tumor Effect of BLM by Modulating the Activities of Caspase 3 and Caspase 8

Previous studies have demonstrated that caspase families play a vital role in tumor cell apoptosis, including caspase 3 and caspase 8. Figure 4 shows that the activities of caspase 3 (A) and caspase 8 (B) were significantly up-regulated in the BLM alone group and the BLM + CI$_{SCFE}$-M combined group as compared to the model group (all $p < 0.05$). Furthermore, the combination of BLM + CI$_{SCFE}$-M had a statistically stronger effect than BLM alone (all $p < 0.05$). These results suggested that CI$_{SCFE}$ combined with BLM enhanced the effect of BLM on caspase 3 and caspase 8 activities.

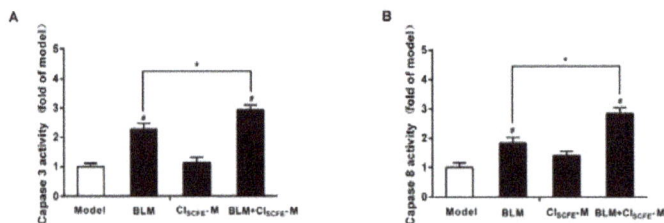

Figure 4. CI$_{SCFE}$ enhanced the anti-tumor effect of BLM by modulating the activities of caspase 3 and caspase 8. The activities of apoptotic performer caspase 3 (**A**); and caspase 8 (**B**) were measured. Data are presented as mean ± SEM of the changes compared to the model group (n = 8). [#] $p < 0.05$ vs. model group; [*] $p < 0.05$ vs. BLM group.

2.5. Effect of CI$_{SCFE}$ Attenuated BLM-Induced Lung Fibrosis

As shown in Figure 5, the lung tissues presented normal structure with no inflammatory, pathological, or collagen deposition in the control group (Figure 5A,F). In comparison with the control group, the model group (Figure 5B,G) and the CI$_{SCFE}$ alone group (Figure 5C,H) showed no obvious pathological changes, whereas in the BLM alone group (Figure 5D,I), haematoxylin-eosin (H&E) staining presented obvious pulmonary injury, including alveolar wall, alveolar, vascular congestion, and inflammatory cell infiltration. Moreover, Masson's trichrome staining suggested that BLM alone group had massive collagen deposition in the lung interstitium and around the bronchioles as compared to the model group. On the other hand, the combination of BLM + CI$_{SCFE}$-M had remarkably attenuated the pulmonary inflammatory damage and fibrosis as compared to the BLM alone group (Figure 5E,J). In addition, the severity of lung injury was analyzed by H&E staining. As shown in Figure 5K, the model group displayed no obvious pulmonary injury as compared to the control group ($p > 0.05$).

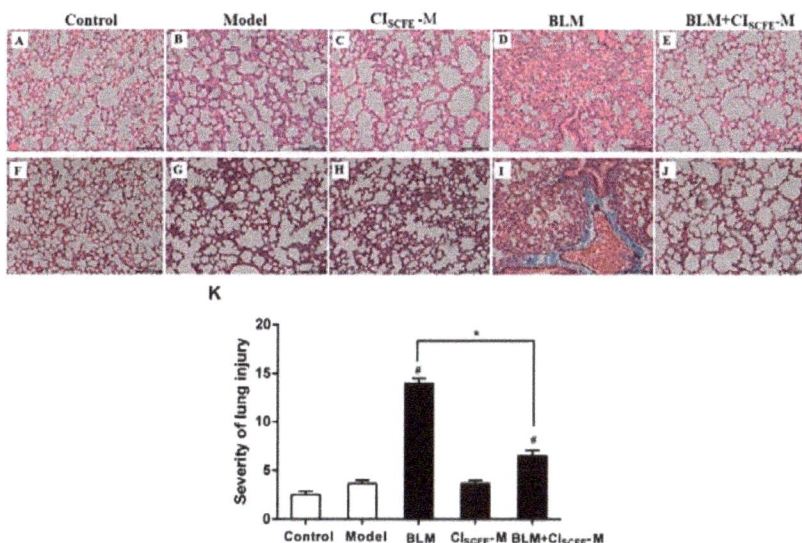

Figure 5. Effect of CI$_{SCFE}$ attenuated BLM-induced lung fibrosis. Lung tissue sections were stained with haematoxylin-eosin (H&E) (**A–E**) for pathological examination (200×); Masson's (**F–J**) for collagen deposition (200×); (**K**) severity scores of lung injury. The slides were histopathologically evaluated using a semiquantitative scoring method. The total lung injury score was estimated by adding up the individual scores of each category. Scale bar indicates 100 μm. Data are expressed as mean ± SEM ($n = 8$). $^#$ $p < 0.05$ vs. model group; * $p < 0.05$ vs. BLM group.

2.6. Effect of CI$_{SCFE}$ on Cytokine Production Induced by BLM in the Lung Tissues

To evaluate the extent of inflammation in lung tissues, the productions of tumor necrosis factor-alpha (TNF-α) and interleukin (IL-6) were measured. Figure 6 demonstrated a remarkable increase in the levels of TNF-α (A) and IL-6 (B) in the BLM-induced lung injuries as compared to the model group ($p < 0.05$, respectively). Conversely, the combination group dramatically decreased the production of these cytokines as compared to the BLM alone group ($p < 0.05$, respectively). No significant differences were observed between the model and CI$_{SCFE}$ alone groups.

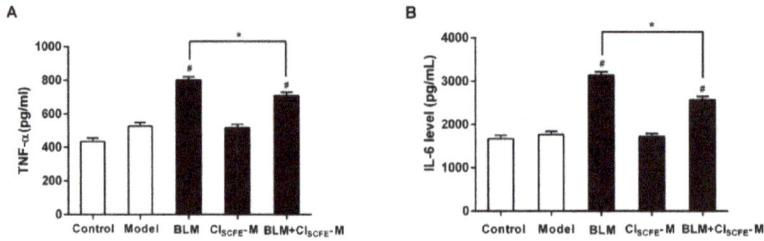

Figure 6. Effect of CI$_{SCFE}$ on BLM-induced cytokines productions in lung tissues. The activities of tumor necrosis factor-alpha (TNF-α) (**A**); and interleukin (IL-6) (**B**) in lungs are presented as the mean ± SEM (*n* = 8). [#] *p* < 0.05 vs. model group; * *p* < 0.05 vs. BLM group.

2.7. Effect of CI$_{SCFE}$ on BLM-Induced Oxidative Stress

To explore whether the protection effect of CI$_{SCFE}$ against BLM-induced lung injury was related to anti-oxidative effect, the activities of myeloperoxidase (MPO) and malondialdehyde (MDA) were assayed. As shown in Figure 7, treatment with BLM alone notably increased the levels of MPO (A) and MDA (B) as compared to the model group (*p* < 0.05, respectively). However, coupling BLM with CI$_{SCFE}$ remarkably decreased the MPO and MDA activities when compared with BLM alone (*p* < 0.05, respectively), indicating that CI$_{SCFE}$ could decrease the BLM-induced oxidative stress.

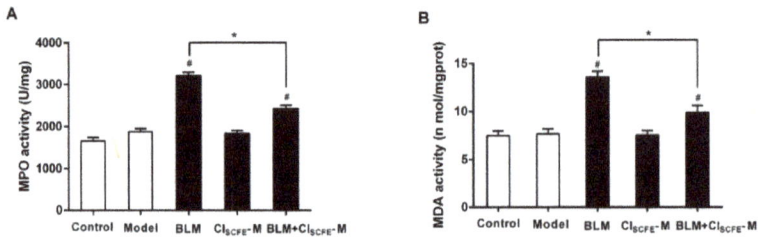

Figure 7. Effect of CI$_{SCFE}$ on BLM-induced oxidative stress. (**A**) Myeloperoxidase (MPO) activity; (**B**) malondialdehyde (MDA) activity. Data are represented as mean ± SEM (*n* = 8). [#] *p* < 0.05 vs. model group; * *p* < 0.05 vs. BLM group.

2.8. Effect of CI$_{SCFE}$ and BLM Treatments on p53 and TGF-β1 Expressions

As shown in Figure 8, when compared with the model group, the protein expression of p53 (A) in the ascites cells of BLM-treated mice was apparently up-regulated (*p* < 0.05), whereas the combination of BLM with CI$_{SCFE}$ significantly increased the p53 expression as compared to the BLM alone group (*p* < 0.05). On the other hand, the combination of BLM with CI$_{SCFE}$ could significantly inhibit the TGF-β1 (B) expression in lungs when compared with the BLM alone group (*p* < 0.05).

2.9. Expression of miR-29b in Ascites Cells and Lung Tissues

miR-29b is a well-established vital tumor suppressor and fibrosis modulator [22,23], playing a key role in cancer with visceral fibrosis. Thus, we attempted to evaluate whether CI$_{SCFE}$ could modulate the miR-29b expression of BLM-treated tumor-bearing mice. As shown in Figure 9, the treatment with CI$_{SCFE}$ or BLM alone did not exhibit any distinct effect on miR-29b expression in the ascites cells and lung tissues as compared to the model group (*p* < 0.05, respectively). However, the combination of BLM with CI$_{SCFE}$ significantly enhanced the expression of miR-29b in ascites cells and lung tissues as compared to BLM alone (*p* < 0.05).

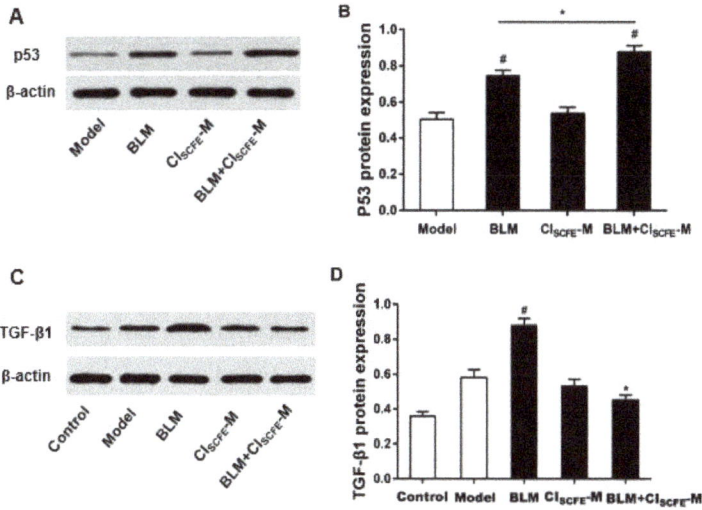

Figure 8. Effect of CI$_{SCFE}$ and BLM treatment on protein-related expression in H22 tumor-bearing mice. The protein expressions of p53 (**A**) and tumor growth factor-beta (TGF-β)1 (**C**) were determined by Western blot. Data (**B,D**) are presented as mean $\pm$ SEM ($n = 4$). [#] $p < 0.05$ vs. model group; [*] $p < 0.05$ vs. BLM group.

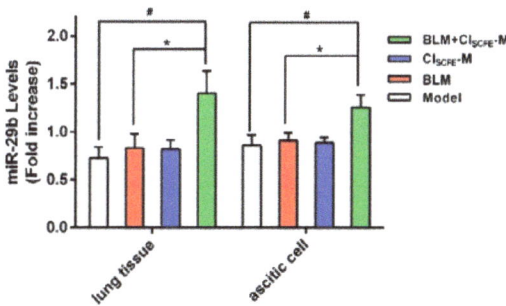

Figure 9. miR-29b expression in lung tissues and ascites cells of H22 tumor-bearing mice. Data are mean $\pm$ SEM ($n = 4$). [#] $p < 0.05$ vs. model group; [*] $p < 0.05$ vs. BLM group.

3. Discussion

The clinical usage of chemotherapeutics is well-known to exert severe side-effects [24]. Therefore, discovering and developing adjunctive agents with physiological activities has become an international topic of intensive medical research [25]. BLM is an established anti-cancer drug mainly used in the treatment of HCC and NPC. It has also been demonstrated to be associated with other cytotoxic reagents for the treatment of testis cancer and Hodgkin disease, and these two diseases have a high cure rate obtained by chemotherapy [26]. Interestingly, it is crucial that the main advantage of BLM is neither immunosuppression nor myelosuppression. Thus, the development of an adjuvant is imperative to not only improve the antitumor effect but also attenuate the side-effects of BLM. *C. indicum*, a traditional medicinal and edible flower, possesses heat clearing and toxin-removal activities. Modern pharmacological research displayed that CI$_{SCFE}$ has anti-bacterial, anti-virus, anti-inflammation, anti-sympathetic, anti-oxidant, and anti-neoplastic functions [27,28]. These data suggested that CI$_{SCFE}$ alone exerted no apparent anti-tumor effect in the mice. However, CI$_{SCFE}$

combined with BLM groups markedly prolonged the life-span and significantly decreased the change of weight, abnormal diameter, as well as ascites fluid compared to the BLM alone group. These results suggested that CI$_{SCFE}$ could be a potential adjuvant for BLM to enhance the anti-cancer effect in tumor-bearing mice.

In order to further understand the mechanism of CI$_{SCFE}$ enhanced anti-cancer activity of BLM, we sought to investigate the tumor cell apoptosis. Apoptosis is involved in many physiological processes, which can exclude the abnormal or damaged cells, promoting cancer cell apoptosis as an efficient method to control the tumor growth [29]. Previous studies have revealed that CI$_{SCFE}$ can induce apoptosis and inhibit cell proliferation through signal transducers and activators of transcription factors-3 (STAT-3) and NF-κB signaling pathways in different cancer cell lines [30,31]. Our results showed that CI$_{SCFE}$ combined with BLM could remarkably increase the H22 ascites cell apoptosis. Caspase 8 as a critical pro-apoptotic molecule, which, once activated, can trigger the downstream caspase cascade, including the major cell apoptosis executor caspase 3 [32], Thus, the caspase 3 with caspase 8 activities were measured in the present study to investigate the underlying mechanism of the effect of CI$_{SCFE}$. Furthermore, p53 is a crucial apoptotic protein that directly mediates the downstream caspase 3 and caspase 8 to exert its effect on cell apoptosis. Furthermore, p53 is also a vital tumor suppressor, inactivated in most human cancers with high mutations [33], and can induce the down-regulation of specific proteins. Here, we found that in the CI$_{SCFE}$ combined with BLM groups, the expression of p53 was notably enhanced as compared to BLM alone, with the markedly increased promotion of apoptosis in cancer cells, accompanied by enhanced activities of caspases 3 and 8. These findings imply that the mechanism of CI$_{SCFE}$ combined with BLM in inducing apoptosis of H22 ascites tumor cells might also involve the activation of the p53 apoptotic pathway.

Based on above findings, CI$_{SCFE}$ alone group had no obvious effect on cells apoptosis; however, when coupled with BLM, CI$_{SCFE}$ remarkably promoted the survival rate of H22 tumor-bearing mice. Could it also be caused by reducing the detrimental effect of BLM? It is well-known that one of the most serious side-effects of BLM is pulmonary fibrosis, which in turn increases the mortality rate of BLM-treated patients [34]. A pulmonary pathological slide of lung tissue revealed an obvious pulmonary fibrosis for BLM treatment alone after seven days, whereas the symptoms alleviated in the slides from the combined group of BLM with CI$_{SCFE}$. Pulmonary fibrosis is a chronic fibrosis interstitial lung disease with poor prognosis and unknown etiology. The pro-inflammatory cytokines are known to play a significant role in the processing of pulmonary fibrosis, including TNF-α and IL-6. Furthermore, the excessive free radicals will result in oxidative stress or chronic inflammation [1,2,35], the related enzymes have a significant function in the processing of lung fibrosis, and thus, the activities of the oxidant enzymes (MPO and MDA) were also measured in this study. The results showed that CI$_{SCFE}$ combined with BLM could distinctly decrease the levels of inflammatory cytokines (TNF-α, IL-6) and oxidant enzymes (MPO, MDA). Moreover, previous data reported that TGF-β1 suppressed the release of pro-inflammatory cytokines including TNF-α, IL-6, and itself. In turn, TNF-α and IL-6 stimulated the activity of the TGF-β1 [36]. Consecutively, TGF-β1 inhibited and adjusted the balance of MDA levels with MPO activity in lung tissues [37]. In the present study, the expression of TGF-β1 was evidently decreased in the CI$_{SCFE}$ combined BLM group as compared to BLM alone, thereby proving that CI$_{SCFE}$ could relieve the side-effects of BLM.

P53 and TGF-β1 signaling pathways are equally important in the initiation of cancer, due to abundant cross-talk [38]. In the tumor progression, TGF-β1 and p53 cooperate to regulate anti-proliferation and apoptotic effects under normal conditions. TGF-β1 signaling pathway exhibited a major role in the early anti-tumor function and late-promoting effects [39]. If p53 harbors mutations, the expression of TGF-β1 will be abnormally enhanced and result in cancer. Previous studies have demonstrated that the balance of p53 and TGF-β1 signaling pathways was modulated by the integration of different factors, including other signaling pathways, proteins, chemokines, and miRNAs, of which the participation of miRNAs' has become increasingly important in recent studies.

miR-29b, a vital tumor suppressor, can suppress the tumor cell growth by regulating the expression of p53 [40], thereby playing a critical role in the process of fibrosis diseases in various tissues, including liver [41], lung [42], kidney [43], and heart [44]. In the tumor procession, miR-29b could regulate the balance of p53 and TGF-β1 signaling pathways simultaneously. The TGF-β1 signaling pathway could, in turn, modulate the activity of miR-29b [8]. Intriguingly, miR-29b could upregulate the level of p53 [40], and consequently activate downstream the p53 pathway including caspases 3 and 8, which eventually induce the apoptosis of tumor cells [45]. In addition, the development of tumors often accompanies chronic inflammation and oxidant stress [46], and thus, the inflammatory cytokines and oxidant enzymes also play a significant role in tumor development. The normal activities of oxidant enzymes (MPO, MDA) could promote cell proliferation; however, the abnormal expression would upregulate the expression of p53 and miR-29b, inducing cell apoptosis [47]. The inflammatory cytokines also play a pivotal role in the procession and metastasis in the tumor, whereby the excess inflammatory factors would upregulate the miR-29b, p53, and caspases 3 and 8 levels to stimulate the tumor cell apoptosis [47–49]. Cui et al. [46] demonstrated that the expression of p53 was upregulated in inflamed tissues, and then, p53 negatively regulated the pro-inflammatory factors. Moreover, the pro-inflammatory factors stimulated the expression of miR-29b and enhanced the anti-tumor effect mediated by enhancing the cell apoptosis [8]. On the other hand, miR-29b suppressed the expression of TGF-β1 to mediate the procession of pulmonary fibrosis [22]. The present data substantiated that the treatment with CI_{SCFE} or BLM alone had no obvious effect on miR-29b expression in the ascites cells and lung tissues. Interestingly, the expression of miR-29b levels was dramatically enhanced when CI_{SCFE} was coupled with BLM. These results indicated that CI_{SCFE} combined with BLM affected the miR-29b expression, and regulated the balance between p53 and TGF-β1 signaling pathways in H22 tumor-bearing mice.

4. Experimental Section

4.1. Materials

Bleomycin (BLM) hydrochloride was purchased from Haizheng Pharmaceuticals (Zhejiang, China). Mouse tumor necrosis factor-α (TNF-α) and interleukin-6 (IL-6) the enzyme-linked immunosorbent assay (ELISA) reagents were purchased from eBioscience (San Diego, CA, USA); Myeloperoxidase (MPO) and malondiadehyde (MDA) Colorimetric Activity Assay Kits were obtained from Jiancheng Institution of Biotechnology (Nanjing, China). Medium RPMI 1640 and fetal bovine serum (FBS) were purchased from Gibco (Grand Island, NY, USA); Penicillin-Streptomycin were obtained from Hyclone (Logan, UT, USA); The Annexin V-fluorescein isothiocyanate (FITC) apoptosis kit was offered by Keygen Biotech (Nanjing, China); TRIzol reagent was offered by Invitrogen Life Technologies (Shanghai, China). All other chemicals and reagents used in the study were of analytical grade.

4.2. Preparation of CI_{SCFE}

The supercritical fluid CO_2 extract of *Chrysanthemum indicum* (CI_{SCFE}) was prepared and offered by the Institute of New Drug Research & Development Guangzhou University of Chinese Medicine (Lot. 20121104) [19]. According to our previous report, the composition analysis was done by combining Gas Chromatography-Mass Spectrometer (GC-MS) and high-performance liquid chromatography with Photodiode Array Detector (HPLC-PAD). Thirty compounds were detected by GC-MS, four compounds were identified by HPLC-PAD (the brief analysis methods and chemical profile of CI_{SCFE} are presented in the Supplementary Materials). In the present study, the CI_{SCFE} was suspended in 3% Tween 80 and confected into different concentration solution. BLM was dissolved in normal saline.

4.3. Cell Culture

The mouse H22 hepatocellular carcinoma cells used in this study were purchased from American Type Culture Collection (Rockville, MD, USA) and were revived at 37 °C, then maintained at Dulbecco's Modified Eagle's medium (obtained RPMI-1640 medium supplemented with 10% FBS and 1% Penicillin-Streptomycin) in a humidified atmosphere with 5% CO_2. The cells were incubated in RPMI1640 medium until they reached approximately 2×10^6 cells/mL and 80% viability.

4.4. Animals

Male Kunming (KM) mice (18–22 g) were purchased from the Experimental Animal Center, Institute of Guangzhou University of Chinese Medicine (Certificate number SCXK2008-0020; Ethical permission date was September 21, 2015, Guangzhou, China). The animals were housed in a 12-h light/dark cycle under a constant temperature of 24 °C and relative humidity of 65% ± 15%, and fed with standard diet and tap water. The animal experiments were conducted according to the guidelines established by the National Institutes of Health (NIH) Guide for the Care and Use of Laboratory Animals. The procedures were approved by the Animal Care and Welfare Committee of Guangzhou University of Chinese Medicine.

4.5. Animal Experiments

H22 cells (2×10^6 cells/mL) were inoculated through abdomen into male KM mice and the ascites cells were passaged three times in the mice, after seven days, the ascites fluid was extracted and diluted with normal saline; the cell concentration was adjusted to 2×10^6 cells/mL and injected into each animal. After five days, 90 mice were randomly divided into nine groups with 10 mice in each group: the control group (normal saline, intraperitoneal (ip) injection), model group (normal saline, ip), BLM alone group (7.5 mg/kg, ip), CI_{SCFE}-L, M, H doses alone group (240, 360, 480 mg/kg, respectively, intragastrical (ig) administration), and BLM (7.5 mg/kg, ip) combined with CI_{SCFE}-L, M, H doses group (240, 360, 480 mg/kg, respectively, ig). After 24 h, the control and model were intraperitoneally injected normal saline, and BLM alone group with BLM. The CI_{SCFE} alone group were respectively gavaged CI_{SCFE}-L, M, H solvent; the BLM combined with CI_{SCFE} groups were respectively intraperitoneally injected BLM and gavaged CI_{SCFE}-L, M, H solvent once per day for a total of seven consecutive days. All the mice were allowed free access to water and food until death, and the survival rate was calculated.

Another 50 mice were randomly divided into five groups with 10 mice in each group: the control and model group (normal saline, ip), BLM alone group (7.5 mg/kg, ip), CI_{SCFE}-M doses alone group (360 mg/kg, ig), and BLM (7.5 mg/kg, ip) combined with CI_{SCFE}-M dose group (360 mg/kg, ig). After 24 h, the control and model groups were intraperitoneally injected with normal saline and the BLM alone group was intraperitoneally injected with BLM; the CI_{SCFE} alone group was gavaged CI_{SCFE}-M solvent; the BLM combined with CI_{SCFE} groups was intraperitoneally injected BLM and gavaged CI_{SCFE}-M solvent once per day for a total of seven consecutive days. At day 8, 10 mice of each group were executed and ascites were collected with lung tissues for the subsequent tests. A portion of the ascites was solubilized in the TRIzol reagent for the extraction of total RNA, and the other portion was used for Western blotting analysis. The lung tissue was rapidly removed and washed in ice-cold normal saline, snap-frozen in liquid nitrogen, and stored at −80 °C until further analysis.

4.6. Histopathological Examination

Lung tissues were fixed in 10% neutral buffered formalin and embedded in paraffin wax, cut into 5 μm thick slices, and subjected to haematoxylin-eosin (H&E) staining and Masson's trichrome staining to detect inflammation or collagen deposition, respectively. The lung injuries were ranked from 0 (normal) to 4 (severe) for four categories: congestion, edema, interstitial inflammation, and inflammatory cell infiltration; the overall lung injury score was calculated by adding up the individual scores of each category [50].

4.7. Determination of MPO, MDA Activities

Lung tissues (0.2 g) were homogenized (1000 rpm, 30 s) with steel shot in nine volumes of cold normal saline (4 °C), followed by centrifugation at $3000 \times g$ for 10 min at 4 °C. The total supernatant was extracted, and subsequently, the enzyme-linked immunosorbent assay (ELISA) kits (Jianchen Institution of Biotechnology, Nanjing, China) were utilized for the estimation of activities of MPO and MDA in the lung tissues.

4.8. Evaluation of the Expression of TNF-α and IL-6

Other lung tissues (0.3 g) were homogenized (1000 rpm, 30 s) with steel shot in 9 volumes of cold PBS (4 °C) and centrifuged at $3000 \times g$ for 10 min at 4 °C. The total supernatant was extracted to estimate the secreted TNF-α as well as IL-6 by the operation sequences utilizing ELISA kits.

4.9. Flow Cytometry Analysis (FACS)

The ascites fluid was centrifuged (1000 rpm, 3 min) and washed twice using the pre-chilled PBS. The cell concentration was adjusted to 1×10^6 cells/mL PBS. Then, 400 μL Annexin V-FITC and 5 μL Annexin V-FITC were integrated with the cells, respectively, and incubated for 15 min in the dark (4 °C). Subsequently, 10 μL propidium iodide (PI) was lightly mixed with the cells and incubated for 5 min (4 °C), followed by FACS (Becton Dickinson FACS Calibur). The apoptotic cells were quantified as the percentage of sub-G1 DNA content in each sample.

4.10. Western Blot Analysis

The ascites cells were washed with cold PBS three times and centrifuged at $1000 \times g$ for 5 min at 4 °C, the total supernatant was aspirated and 200 μL radio immunoprecipitation assay buffer (RIPA) lysis buffer added and agitated to completely to crack the cells on the ice, followed by centrifugation at $12,000 \times g$ for 5 min at 4 °C. The total protein was collected in the supernatant. Next, the nuclear and cell plasma protein extraction kits were utilized to isolate the respective protein fractions. The protein concentration was estimated by bicinchoninic acid assay method, an equivalent amount separated on 10% SDS-polyacrylamide gel electrophoresis (PAGE), and transferred to polyvinylidene fluoride (PVDF) membranes. The membranes were blocked for 1 h with PBS containing 5% dried milk powder and incubated overnight at 4 °C with rabbit anti-Mcl-1 (1:100, Santa Cruz Biotechnology, Inc., Shanghai, China), rabbit anti-BAG3 (1:1000), or mouse anti-MMP2 (1:200, DaiichiFine Chemical Co., Ltd., Shanghai, China). Then, the membranes were washed in TBST, and the appropriate horseradish peroxidase (HRP)-conjugated secondary antibodies diluted in TBST were added. The lung tissues and ascites cells were washed two to three times with cold PBS to remove the blood and then homogenized. Approximately, 10 volumes of RIPA lysis buffer was added and mixed vigorously complete cracking of the cells on ice, followed by centrifugation at $12,000 \times g$ for 5 min at 4 °C; the total protein was collected in the supernatant.

4.11. Caspase 3 and Caspase 8 Activities Assay

The activities of caspase 3 and caspase 8 in the H22 ascites cells were evaluated by the respective caspase activity assay kit purchased from Keygen Biotech (Nanjing, China). Both activities were correspondingly assayed by spectrophotometry ($\lambda = 405$ nm), according to the manufacture's protocols.

4.12. Real-Time Polymerase Chain Reaction Analysis

H22 ascites fluid, as well as lung tissues, were washed with PBS; the ascites fluid was extracted with TRIzol reagent (Invitrogen, Carlsbad, CA, USA) to isolate the total RNA. Total RNA (1.5 μg)

Int. J. Mol. Sci. **2017**, *18*, 465

was reverse transcribed using Kit (Applied Biosystems, Branch burg, NJ, USA) to yield cDNA. The reaction was run at 50 °C for 2 min, 95 °C for 10 min, followed by 40 cycles at 95 °C for 15 s and 60 °C for 1 min, the solubility curve 75–95 °C, heat up 1 °C each 20 s on Applied Biosystems Step-One Fast Real-Time PCR system. Glyceraldehyde-3-phosphate dehydrogenase (GAPDH) was used as an internal control. Fold change = $2^{-\Delta\Delta Ct}$, $\Delta\Delta C_t = (C_{tSample} - C_{tGAPDH}) - (C_{tControl} - C_{tGAPDH})$. The primers sequences of gene including miR-29b, GAPDH was synthesized by Invitrogen, the siRNA sequences were listed as follows: 5′-CTCAACTGGTGTCGTGGAGTCGGCAATTCAGTTGAGTCTAAACC-3′; 5′-ACACTCCAGCTGGGGCTGGTTTCATATGGTGG-3′ for miRNA-29b and 5′-CTCGCTTCGG CAGCACA-3′ and 5′-AACGCTTCACGAATTTGCGT-3′ for GAPDH control.

4.13. Statistical Analysis

All data were assessed by one-way ANOVA method, Least Significant Difference (LSD), and Dunnett's T3 (3) test for comparison between any two means; $p < 0.05$ was considered statistically significant. The comparison of survival curves was determined by the log-rank (Mantel-Cox) test. All data were analyzed using the statistical analysis software (SPSS 13.0, New York, NY, USA).

5. Conclusions

Our studies confirmed that CI_{SCFE} could enhance the anti-cancer activity of BLM in H22 tumor-bearing mice. The synergistic effect of BLM was related to CI_{SCFE} by inducing apoptosis of H22 ascites tumor cells and reducing the pulmonary injury induced by BLM. Thus, the possible underlying mechanism was associated with the regulation of the balance of p53 and TGF-β1 signaling pathways. These results indicated that CI_{SCFE} could serve as a putative adjuvant drug with chemotherapy in the future.

Supplementary Materials: Supplementary materials can be found at www.mdpi.com/1422-0067/18/3/465/s1.

Acknowledgments: This research was supported by grants from the National Natural Science Foundation of China (No. 81403169), Guangdong Natural Science Foundation (No. 2014A030310224), Yang Fan Innovative And Entrepreneurial Research Team Project (No. 2014YT02S008), Macao and Taiwan Science and Technology Cooperation Program of China (No. 2014DFH30010), and Science and Technology Planning Project of Guangdong Province of China (No. 2013B090600026 and 2016A020226049).

Author Contributions: Jie Yuan and Yu-Cui Li conceived and designed the experiments; Hong-Mei Yang, Chao-Yue Sun, Jia-Li Liang, and Lie-Qiang Xu performed the experiments; Zhen-Biao Zhang, Dan-Dan Luo, and Han-Bin Chen analyzed the data; Yong-Zhong Huang, Qi Wang, and David Yue-Wei Lee contributed materials/analysis tools; Hong-Mei Yang and Yu-Cui Li wrote the paper.

Conflicts of Interest: The authors declare that there is no conflict of interest.

References

1. Bugaut, H.; Bruchard, M.; Berger, H.; Derangere, V.; Odoul, L.; Euvrard, R.; Ladoire, S.; Chalmin, F.; Vegran, F.; Rebe, C.; et al. Bleomycin exerts ambivalent antitumor immune effect by triggering both immunogenic cell death and proliferation of regulatory T cells. *PLoS ONE* **2013**, *8*, e65181. [CrossRef] [PubMed]
2. Burgy, O.; Wettstein, G.; Bellaye, P.S.; Decologne, N.; Racoeur, C.; Goirand, F.; Beltramo, G.; Hernandez, J.F.; Kenani, A.; Camus, P.; et al. Deglycosylated bleomycin has the antitumor activity of bleomycin without pulmonary toxicity. *Sci. Transl. Med.* **2016**, *8*, 320–326. [CrossRef] [PubMed]
3. Yunos, N.M.; Beale, P.; Yu, J.Q.; Huq, F. Synergism from sequenced combinations of curcumin and epigallocatechin-3-gallate with cisplatin in the killing of human ovarian cancer cells. *Anticancer Res.* **2011**, *31*, 1131–1140. [PubMed]
4. Li, F.F.; Zhang, N. Ceramide: Therapeutic potential in combination therapy for cancer treatment. *Curr. Drug Metab.* **2016**, *17*, 37–51. [CrossRef]
5. Martin, G.S. Cell signaling and cancer. *Cancer Cell* **2003**, *4*, 167–174. [CrossRef]
6. Zhou, L.; Wang, L.; Lu, L.; Jiang, P.; Sun, H.; Wang, H. Inhibition of miR-29 by TGF-β-Smad3 signaling through dual mechanisms promotes transdifferentiation of mouse myoblasts into myofibroblasts. *PLoS ONE* **2012**, *7*, e33766. [CrossRef] [PubMed]

7. Brosh, R.; Shalgi, R.; Liran, A.; Landan, G.; Korotayev, K.; Nguyen, G.H.; Enerly, E.; Johnsen, H.; Buganim, Y.; Solomon, H.; et al. p53-repressed miRNAs are involved with E2F in a feed-forward loop promoting proliferation. *Mo. Syst. Biol.* **2008**, *4*, 229. [CrossRef] [PubMed]

8. Sivadas, V.P.; Kannan, S. The microrna networks of TGF-β signaling in cancer. *Tumour Biol.* **2014**, *35*, 2857–2869. [CrossRef] [PubMed]

9. Sun, Y.; Xia, P.; Zhang, H.; Liu, B.; Shi, Y. p53 is required for doxorubicin-induced apoptosis via the TGF-β signaling pathway in osteosarcoma-derived cells. *Am. J. Cancer Res.* **2016**, *6*, 114–125. [PubMed]

10. Overstreet, J.M.; Samarakoon, R.; Meldrum, K.K.; Higgins, P.J. Redox control of p53 in the transcriptional regulation of TGF-β1 target genes through smad cooperativity. *Cell. Signal.* **2014**, *26*, 1427–1436. [CrossRef] [PubMed]

11. Wang, M.; Zhang, J.; Song, X.; Liu, W.; Zhang, L.; Wang, X.; Lv, C. Astaxanthin ameliorates lung fibrosis in vivo and in vitro by preventing transdifferentiation, inhibiting proliferation, and promoting apoptosis of activated cells. *Food Chem. Toxicol.* **2013**, *56*, 450–458. [CrossRef] [PubMed]

12. Matsuda, H.; Morikawa, T.; Toguchida, I.; Harima, S.; Yoshikawa, M. Medicinal flowers. VI. Absolute stereostructures of two new flavanone glycosides and a phenylbutanoid glycoside from the flowers of *Chrysanthemum indicum* L.: Their inhibitory activities for rat lens aldose reductase. *Chem. Pharm. Bull.* **2002**, *50*, 972–975. [CrossRef] [PubMed]

13. Seo, D.W.; Cho, Y.R.; Kim, W.; Eom, S.H. Phytochemical linarin enriched in the flower of *Chrysanthemum indicum* inhibits proliferation of A549 human alveolar basal epithelial cells through suppression of the AKT-dependent signaling pathway. *J. Med. Food* **2013**, *16*, 1086–1094. [CrossRef] [PubMed]

14. Cheng, W.M.; Li, J.; You, T.P.; Hu, C.M. Anti-inflammatory and immunomodulatory activities of the extracts from the inflorescence of *Chrysanthemum indicum Linne*. *J. Ethnopharmacol.* **2005**, *101*, 334–337. [CrossRef] [PubMed]

15. Li, Z.F.; Wang, Z.D.; Ji, Y.Y.; Zhang, S.; Huang, C.; Li, J.; Xia, X.M. Induction of apoptosis and cell cycle arrest in human HCC MHCC97H cells with *Chrysanthemum indicum* extract. *World. J. Gastroenterol.* **2009**, *15*, 4538–4546. [CrossRef] [PubMed]

16. Wu, X.L.; Li, C.W.; Chen, H.M.; Su, Z.Q.; Zhao, X.N.; Chen, J.N.; Lai, X.P.; Zhang, X.J.; Su, Z.R. Anti-inflammatory effect of supercritical-carbon dioxide fluid extract from flowers and buds of *Chrysanthemum indicum Linnen*. *Evid. Based Complement. Altern. Med.* **2013**, *2013*, 413237. [CrossRef] [PubMed]

17. Pongjit, K.; Ninsontia, C.; Chaotham, C.; Chanvorachote, P. Protective effect of glycine max and *Chrysanthemum indicum* extracts against cisplatin-induced renal epithelial cell death. *Hum. Exp. Toxicol.* **2011**, *30*, 1931–1944. [CrossRef] [PubMed]

18. Zhang, X.; Xie, Y.L.; Yu, X.T.; Su, Z.Q.; Yuan, J.; Li, Y.C.; Su, Z.R.; Zhan, J.Y.; Lai, X.P. Protective effect of super-critical carbon dioxide fluid extract from flowers and buds of *Chrysanthemum indicum* linnen against ultraviolet-induced photo-aging in mice. *Rejuvenation Res.* **2015**, *18*, 437–448. [CrossRef] [PubMed]

19. Wu, X.L.; Feng, X.X.; Li, C.W.; Zhang, X.J.; Chen, Z.W.; Chen, J.N.; Lai, X.P.; Zhang, S.X.; Li, Y.C.; Su, Z.R. The protective effects of the supercritical-carbon dioxide fluid extract of *Chrysanthemum indicum* against lipopolysaccharide-induced acute lung injury in mice via modulating toll-like receptor 4 signaling pathway. *Mediat. Inflamm.* **2014**, *2014*, 246407. [CrossRef] [PubMed]

20. Yoshiji, H.; Kuriyama, S.; Hicklin, D.J.; Huber, J.; Yoshii, J.; Ikenaka, Y.; Noguchi, R.; Nakatani, T.; Tsujinoue, H.; Fukui, H. The vascular endothelial growth factor receptor KDR/Flk-1 is a major regulator of malignant ascites formation in the mouse hepatocellular carcinoma model. *Hepatology* **2001**, *33*, 841–847. [CrossRef] [PubMed]

21. Zhang, J.Z.; Peng, D.G.; Lu, H.J.; Liu, Q.L. Attenuating the toxicity of cisplatin by using selenosulfate with reduced risk of selenium toxicity as compared with selenite. *Toxicol. Appl. Pharmacol.* **2008**, *226*, 251–259. [CrossRef] [PubMed]

22. Cushing, L.; Kuang, P.P.; Qian, J.; Shao, F.Z.; Wu, J.J.; Little, F.; Thannickal, V.J.; Cardoso, W.V.; Lu, J.N. miR-29 is a major regulator of genes associated with pulmonary fibrosis. *Am. J. Respir. Cell Mol. Biol.* **2011**, *45*, 287–294. [CrossRef] [PubMed]

23. Yan, B.; Guo, Q.; Fu, F.J.; Wang, Z.; Yin, Z.; Wei, Y.B.; Yang, J.R. The role of miR-29b in cancer: Regulation, function, and signaling. *OncoTargets Ther.* **2015**, *8*, 539–548.

24. Mattheolabakis, G.; Ling, D.; Ahmad, G.; Amiji, M. Enhanced anti-tumor efficacy of lipid-modified platinum derivatives in combination with survivin silencing sirna in resistant non-small cell lung cancer. *Pharm. Res.* **2016**, *33*, 2943–2953. [CrossRef] [PubMed]

25. Lu, B.; Li, M.; Yin, R. Phytochemical content, health benefits, and toxicology of common edible flowers: A review (2000–2015). *Crit. Rev. Food Sci. Nutr.* **2016**, *56*, 130–148. [CrossRef] [PubMed]

26. Froudarakis, M.; Hatzimichael, E.; Kyriazopoulou, L.; Lagos, K.; Pappas, P.; Tzakos, A.G.; Karavasilis, V.; Daliani, D.; Papandreou, C.; Briasoulis, E. Revisiting bleomycin from pathophysiology to safe clinical use. *Crit. Rev. Oncol. Hematol.* **2013**, *87*, 90–100. [CrossRef] [PubMed]

27. Akihisa, T.; Tokuda, H.; Ichiishi, E.; Mukainaka, T.; Toriumi, M.; Ukiya, M.; Yasukawa, K.; Nishino, H. Anti-tumor promoting effects of multiflorane-type triterpenoids and cytotoxic activity of karounidiol against human cancer cell lines. *Cancer Lett.* **2001**, *173*, 9–14. [CrossRef]

28. Yanez, J.; Vicente, V.; Alcaraz, M.; Castillo, J.; Benavente-Garcia, O.; Canteras, M.; Teruel, J.A. Cytotoxicity and antiproliferative activities of several phenolic compounds against three melanocytes cell lines: Relationship between structure and activity. *Nutr. Cancer* **2004**, *49*, 191–199. [CrossRef] [PubMed]

29. Xu, B.; Huang, Y.Q.; Niu, X.B.; Tao, T.; Jiang, L.; Tong, N.; Chen, S.Q.; Liu, N.; Zhu, W.D.; Chen, M. Hsa-miR-146a-5p modulates androgen-independent prostate cancer cells apoptosis by targeting ROCK1. *Prostate* **2015**, *75*, 1896–1903. [CrossRef] [PubMed]

30. Kim, C.; Kim, M.C.; Kim, S.M.; Nam, D.; Choi, S.H.; Kim, S.H.; Ahn, K.S.; Lee, E.H.; Jung, S.H.; Ahn, K.S. *Chrysanthemum indicum* L. Extract induces apoptosis through suppression of constitutive STAT3 activation in human prostate cancer DU145 cells. *Phytother. Res.* **2013**, *27*, 30–38. [CrossRef] [PubMed]

31. Kim, J.E.; Jun, S.; Song, M.; Kim, J.H.; Song, Y.J. The extract of *Chrysanthemum indicum* linne inhibits EBV LMP1-induced NF-κB activation and the viability of EBV-transformed lymphoblastoid cell lines. *Food Chem. Toxicol.* **2012**, *50*, 1524–1528. [CrossRef] [PubMed]

32. Shu, G.; Yang, T.; Wang, C.; Su, H.; Xiang, M. Gastrodin stimulates anticancer immune response and represses transplanted H22 hepatic ascitic tumor cell growth: Involvement of NF-κB signaling activation in CD4$^+$ T cells. *Toxicol. Appl. Pharmacol.* **2013**, *269*, 270–279. [CrossRef] [PubMed]

33. Vousden, K.H.; Prives, C. p53 and prognosis: New insights and further complexity. *Cell.* **2005**, *120*, 7–10. [CrossRef] [PubMed]

34. Raghu, G. Idiopathic pulmonary fibrosis: Guidelines for diagnosis and clinical management have advanced from consensus-based in 2000 to evidence-based in 2011. *Eur. Respir. J.* **2011**, *37*, 743–746. [CrossRef] [PubMed]

35. Du, L.; Yang, Y.-H.; Wang, Y.-M.; Xue, C.-H.; Kurihara, H.; Takahashi, K. Antitumour activity of EPA-enriched phospholipids liposomes against S180 ascitic tumour-bearing mice. *J. Funct. Foods* **2015**, *19*, 970–982. [CrossRef]

36. Zhao, L.; Wang, X.; Chang, Q.; Xu, J.; Huang, Y.; Guo, Q.; Zhang, S.; Wang, W.; Chen, X.; Wang, J. Neferine, a bisbenzylisoquinline alkaloid attenuates bleomycin-induced pulmonary fibrosis. *Eur. J. Pharmacol.* **2010**, *627*, 304–312. [CrossRef] [PubMed]

37. Altintas, N.; Erboga, M.; Aktas, C.; Bilir, B.; Aydin, M.; Sengul, A.; Ates, Z.; Topcu, B.; Gurel, A. Protective effect of infliximab, a tumor necrosis factor-α inhibitor, on bleomycin-induced lung fibrosis in rats. *Inflammation* **2016**, *39*, 65–78. [CrossRef] [PubMed]

38. Hanahan, D.; Weinberg, R.A. Hallmarks of cancer: The next generation. *Cell* **2011**, *144*, 646–674. [CrossRef] [PubMed]

39. Derynck, R.; Akhurst, R.J.; Balmain, A. TGFβ signaling in tumor suppression and cancer progression. *Nat. Genet.* **2001**, *29*, 117–129. [CrossRef] [PubMed]

40. Park, S.Y.; Lee, J.H.; Ha, M.; Nam, J.W.; Kim, V.N. miR-29 miRNAs activate p53 by targeting p85α and CDC42. *Nat. Struct. Mol. Biol.* **2009**, *16*, 23–29. [CrossRef] [PubMed]

41. Wang, J.; Chu, E.S.; Chen, H.Y.; Man, K.; Go, M.Y.; Huang, X.R.; Lan, H.Y.; Sung, J.J.; Yu, J. MicroRNA-29b prevents liver fibrosis by attenuating hepatic stellate cell activation and inducing apoptosis through targeting PI3K/AKT pathway. *Oncotarget* **2015**, *6*, 7325–7338. [CrossRef] [PubMed]

42. He, Y.; Huang, C.; Lin, X.; Li, J. MicroRNA-29 family, a crucial therapeutic target for fibrosis diseases. *Biochimie* **2013**, *95*, 1355–1359. [CrossRef] [PubMed]

43. Qin, W.; Chung, A.C.; Huang, X.R.; Meng, X.M.; Hui, D.S.; Yu, C.M.; Sung, J.J.; Lan, H.Y. TGF-β/Smad3 signaling promotes renal fibrosis by inhibiting miR-29. *J. Am. Soc. Nephrol.* **2011**, *22*, 1462–1474. [CrossRef] [PubMed]

44. Van Rooij, E.; Sutherland, L.B.; Thatcher, J.E.; DiMaio, J.M.; Naseem, R.H.; Marshall, W.S.; Hill, J.A.; Olson, E.N. Dysregulation of microRNAs after myocardial infarction reveals a role of miR-29 in cardiac fibrosis. *Proc. Natl. Acad. Sci. USA* **2008**, *105*, 13027–13032. [CrossRef] [PubMed]

45. Rebbaa, A.; Chou, P.M.; Emran, M.; Mirkin, B.L. Doxorubicin-induced apoptosis in caspase-8-deficient neuroblastoma cells is mediated through direct action on mitochondria. *Cancer Chemother. Pharmacol.* **2001**, *48*, 423–428. [CrossRef] [PubMed]

46. Cui, Y.; Guo, G. Immunomodulatory function of the tumor suppressor p53 in host immune response and the tumor microenvironment. *Int. J. Mol. Sci.* **2016**, *17*, 16. [CrossRef] [PubMed]

47. Federico, A.; Morgillo, F.; Tuccillo, C.; Ciardiello, F.; Loguercio, C. Chronic inflammation and oxidative stress in human carcinogenesis. *Int. J. Cancer* **2007**, *121*, 2381–2386. [CrossRef] [PubMed]

48. Reuter, S.; Gupta, S.C.; Chaturvedi, M.M.; Aggarwal, B.B. Oxidative stress, inflammation, and cancer: How are they linked? *Free Radic. Biol. Med.* **2010**, *49*, 1603–1616. [CrossRef] [PubMed]

49. Mathe, E.; Nguyen, G.H.; Funamizu, N.; He, P.; Moake, M.; Croce, C.M.; Hussain, S.P. Inflammation regulates microRNA expression in cooperation with p53 and nitric oxide. *Int. J. Cancer* **2012**, *131*, 760–765. [CrossRef] [PubMed]

50. Liu, Y.; Wu, H.; Nie, Y.C.; Chen, J.L.; Su, W.W.; Li, P.B. Naringin attenuates acute lung injury in LPS-treated mice by inhibiting NF-κB pathway. *Int. Immunopharmacol.* **2011**, *11*, 1606–1612. [CrossRef] [PubMed]

International Journal of
Molecular Sciences

MDPI

Review

Significance of Wild-Type p53 Signaling in Suppressing Apoptosis in Response to Chemical Genotoxic Agents: Impact on Chemotherapy Outcome

Razmik Mirzayans *, Bonnie Andrais, Piyush Kumar and David Murray

Department of Oncology, University of Alberta, Cross Cancer Institute, Edmonton, AB T6G 1Z2, Canada;
bonnie.andrais@ahs.ca (B.A.); pkumar@ualberta.ca (P.K.); david.murray5@ahs.ca (D.M.)
* Correspondence: razmik.mirzayans@ahs.ca; Tel.: +1-780-432-8897

Academic Editors: Ashis Basu and Takehiko Nohmi
Received: 20 March 2017; Accepted: 25 April 2017; Published: 28 April 2017

Abstract: Our genomes are subject to potentially deleterious alterations resulting from endogenous sources (e.g., cellular metabolism, routine errors in DNA replication and recombination), exogenous sources (e.g., radiation, chemical agents), and medical diagnostic and treatment applications. Genome integrity and cellular homeostasis are maintained through an intricate network of pathways that serve to recognize the DNA damage, activate cell cycle checkpoints and facilitate DNA repair, or eliminate highly injured cells from the proliferating population. The wild-type p53 tumor suppressor and its downstream effector $p21^{WAF1}$ (p21) are key regulators of these responses. Although extensively studied for its ability to control cell cycle progression, p21 has emerged as a multifunctional protein capable of downregulating p53, suppressing apoptosis, and orchestrating prolonged growth arrest through stress-induced premature senescence. Studies with solid tumors and solid tumor-derived cell lines have revealed that such growth-arrested cancer cells remain viable, secrete growth-promoting factors, and can give rise to progeny with stem-cell-like properties. This article provides an overview of the mechanisms by which p53 signaling suppresses apoptosis following genotoxic stress, facilitating repair of genomic injury under physiological conditions but having the potential to promote tumor regrowth in response to cancer chemotherapy.

Keywords: chemical genotoxic agents; p53 signaling; $p21^{WAF1}$ (CDKN1A); DNAJB9; multinucleated giant cells; premature senescence; apoptosis; mutational processes

---

1. Introduction

Our cells are continuously exposed to potentially deleterious genotoxic events from both endogenous and exogenous sources that jeopardize genome integrity. The plethora of DNA lesions include DNA strand breaks and base alterations induced by ionizing radiation and chemical agents that generate reactive oxygen species, DNA alkylation and formation of abasic sites induced by alkylating agents, bulky DNA lesions induced by ultraviolet light (UV), DNA interstrand crosslinks induced by bifunctional alkylating agents and platinum drugs, and DNA-protein crosslinks arising from a wide range of chemicals, such as chemotherapeutic drugs and formaldehyde [1–4]. Constitutively available DNA repair processes deal with low levels of genomic injury and assist in ameliorating the detrimental effects of such agents. An increase in DNA damage above a threshold level activates the DNA damage surveillance network, which involves multiple signaling pathways that protect against genomic instability and restrict aberrant cell growth in response to genotoxic stress [5]. The wild-type p53 tumor suppressor functions at the hub of this network [6,7].

In the mid 1990s it was proposed that the principal role of p53 in determining cell fate following genotoxic stress is to either promote survival by activating cell cycle checkpoints and facilitating

DNA repair or induce apoptotic cell death. This two-armed model of the DNA damage surveillance network—namely, repair and survive, or die through apoptosis—provided the impetus for extensive research directed towards modulating p53 in an attempt to improve the outcome of conventional cancer therapies. However, it soon became clear that p53's function extends beyond canonical cell cycle and apoptotic signaling, and impacts additional diverse biological processes including senescence and metabolism [8,9]. Murine cancer models have been employed to investigate the impact of p53 activation in the response of oncogene-driven cancers. As pointed out by Stegh [9], *"confirming important roles of p53 in cancer suppression, these studies showed that reactivation of p53 in established tumors can temporarily stop tumor growth; the precise cellular mechanism is cancer type-specific, as lymphomas die by apoptosis, whereas p53 restoration in sarcomas and liver carcinomas leads to growth arrest and senescence. p53-driven apoptosis and senescence responses associated with temporary p53 reactivation led to prolonged survival. Although cancer remission was not permanent, and p53-resistant tumors emerged . . . "* The promises, challenges and perils of targeting p53 in cancer therapy have been extensively discussed [8–10].

A growing body of evidence suggests that the primary response triggered by moderate, clinically relevant doses of cancer therapeutic agents is a sustained proliferation block and not apoptosis in most human cell types (e.g., dermal fibroblasts, solid tumor-derived cells) [6,11], with activation of p53 signaling suppressing (rather than promoting) apoptosis [12–14]. Such growth-arrested cells remain viable for long times (months) post-treatment, secrete a myriad of biologically active factors, and can give rise to progeny exhibiting stem-cell-like properties.

Herein we briefly review the mechanisms by which wild-type p53 suppresses apoptosis following genotoxic stress, focusing on the roles played by DNAJ homolog subfamily B member 9 (DNAJB9) and p21^{WAF1} (p21; also called CDKN1A). In addition, we discuss the significance of p53-mediated protection against apoptosis under physiological conditions, and the dark side of this function of p53 in the context of cancer chemotherapy.

2. Biological Outputs Orchestrated by Wild-Type p53

2.1. p53 Functions

Wild-type p53 is a multifunctional tumor suppressor capable of activating transient cell cycle checkpoints, accelerating DNA repair processes including nucleotide excision repair and rejoining of DNA double strand-breaks (DSBs), and eliminating highly injured cells from the proliferating population by inducing stress-induced premature senescence (SIPS) or apoptotic cell death [6]. p53 exerts these effects both directly, through protein–protein interaction (e.g., interacting with key mediators of DNA repair and apoptosis [6,15]), and indirectly by transcriptionally activating p21 and other key players in the DNA damage surveillance network [6,16].

SIPS is a sustained growth arrested state resembling replicative senescence, a hallmark of mammalian cell aging [17]. Both events are characterized by the acquisition of flattened and enlarged cell morphology and expression of the marker senescence associated β-galactosidase (SA-β-gal) in cells that retain viability and exhibit metabolic activity. Unlike replicative senescence, which is triggered by erosion and dysfunction of telomeres, SIPS is induced by DNA damage and other types of genotoxic stress but is not dependent on telomere status and telomerase function [17]. Both events are largely (but not always) dependent on wild-type p53 signaling in general, and sustained nuclear accumulation of p21 in particular.

SIPS, triggered by DNA-damaging agents, is a prominent response of normal human fibroblasts and solid tumor-derived cell lines that express wild-type p53 [6]. In addition, Li–Fraumeni syndrome fibroblasts [18] and some lung carcinoma cell lines [19] that lack wild-type p53 function also exhibit a high degree of SIPS in response to genotoxic stress (ionizing radiation). SIPS in p53-deficient cells correlated with induction of p16^{INK4A} (p16) but not of p21, leading us to propose that p16 might function in a redundant pathway of senescence (both replicative senescence and SIPS), triggering this process only in the absence of wild-type p53 activity [18]. Interestingly, p16 has been reported to be repressed in a p53-dependent manner. Hernández-Vargas et al. [20], for example, reported that p53 transcriptionally activates the helix-loop-helix transcriptional regulator protein Id1, a well-known repressor of *p16^INK4A^* [21,22]. In addition, Leong et al. [23] demonstrated that p53 downregulates p16 through Id1-independent mechanisms.

2.2. p53 Regulation in the Absence of Genotoxic Stress

In normal, unstressed cells, the wild-type p53 protein undergoes rapid turnover and is thus maintained at low steady state levels that restrict its function [6,7]. Turnover of p53 is controlled by several ubiquitin ligases, some of which are regulated in a p53-dependent manner. MDM2 (murine double minute-2 homologue; also known as HDM2 in human) is the most intensively studied regulator of p53 stability and function. In the absence of DNA damage, MDM2 binds to the N-terminal region of p53 and inhibits its activity by blocking p53-mediated transactivation, exporting p53 from the nucleus to the cytoplasm, and promoting the proteasomal degradation of p53. MDM2-mediated mono-ubiquitination of p53 triggers its cytoplasmic sequestration, whereas poly-ubiquitination results in p53 degradation.

2.3. p53 Regulation Following Genotoxic Stress

Recent studies have revealed that a threshold level of genotoxic stress must be reached to trigger the DNA damage surveillance network [5]. This response is initiated by rapid stabilization of p53, its nuclear accumulation, and activation of its transcriptional and biological functions [24]. Stabilization and activation of p53 is largely a consequence of phosphorylation of the molecule on different residues, which can be mediated by various protein kinases, including ATM (ataxia telangiectasia mutated), ATR (ATM and RAD3-related), checkpoint kinase 1 (CHK1), checkpoint kinase 2 (CHK2), and p38 mitogen-activated protein kinase (MAPK) [25–28]. In response to DNA damage, phosphorylation of p53 on Ser20 and of MDM2 on Ser395, mediated by kinases such as ATM, interrupts the p53–MDM2 interaction, resulting in p53 accumulation, subcellular shuttling and activation [7].

Rapid activation of the DNA damage surveillance network in response to genotoxic stress must be followed by restoration of the cell to its pre-stress state to allow the maintenance of cell homeostasis and resumption of normal growth. This critical function is largely accomplished by WIP1 (wild-type p53-induced phosphatase 1), a p53-regulated type 2C serine/threonine phosphatase [29].

2.4. p53 Dynamics Following Genotoxic Stress

The mechanism by which a single tumor suppressor, p53, orchestrates complex responses to DNA damage has been the subject of extensive research. Much attention has been focused on the function of p53 and its downstream programs at relatively short times (within hours) after genotoxic insult. In 2004, Lahav and associates [30] reported studies with the MCF7 breast carcinoma cell line demonstrating that the temporal dynamics of p53 following DNA damage constitutes another potential level of regulation for different biological outcomes. Immunoblot and single-cell observation methods revealed that p53 levels rise and fall in a wavelike or "pulsed" manner in response to DNA double-strand breaks induced by ionizing radiation. Both MDM2 [30] and WIP1 [31] were shown to contribute to the negative regulation of p53 at various p53 waves. These observations led the authors to propose a model in which the initial p53 waves would allow the cells to activate cell cycle checkpoints to facilitate repair, and the subsequent waves to determine cell fate.

These ground-breaking discoveries provided an impetus for a number of studies involving mathematical simulations that were designed to uncover the basis for the "digital" p53 response and the biological consequences of different p53 waves. As discussed previously [6,32], most such studies assumed that the ultimate cell fate might reflect apoptosis, even in MCF7 cells which are relatively insensitive to undergoing apoptosis consequent to therapeutic exposures [33–35]. Purvis et al. [36], however, determined the predominant cell fate resulting from p53 dynamics post-irradiation and showed this to be SIPS in MCF7 cells. We have reported a similar outcome with the A172 malignant glioma cell line [32].

Advances and perspectives regarding the dynamics and mathematical models of p53 signaling in response to different types of DNA damage, together with insight into the biological functions of such dynamics, have been extensively reviewed [32,37] and will not be considered further.

2.5. A Threshold Mechanism Determines the Choice Between p53-Mediated Growth Arrest versus Apoptosis

The biological output of p53 signaling in response to genotoxic stress in terms of sustained growth arrest or apoptotic cell death depends on several factors, including the amount and type of genotoxic insult and the genetic background of the cells [38–40]. As extensively discussed recently [6,32], in most human cell types (e.g., non-cancerous skin fibroblast strains and solid tumor-derived cell lines), exposure to moderate doses of genotoxic agents (e.g., ionizing radiation, UV, chemotherapeutic drugs) promotes a high degree of SIPS but only marginal (if any) apoptosis. Moderate doses refer to those that are typically used in the in vitro colony formation assay and are relevant to in vivo therapeutic studies with animal models. Exposure to extremely high doses of such agents, resulting in <1% clonogenic survival, triggers apoptosis in a significant proportion (~50%) of the cells. (The importance of the apoptotic threshold for exposure to cancer chemotherapeutic agents will be considered in Section 5.1.)

Recently, Kracikova et al. [41] determined the influence of p53 expression levels on biological outcomes in the absence of genotoxic stress. These authors used two approaches to achieve conditions where the only variable is the level of p53: (i) an inducible system with human epithelial cells that allows tight regulation of p53 expression; and (ii) human cancer cells treated with the p53 activator nutlin-3. Both approaches demonstrated that low and high p53 expression triggered growth arrest and apoptosis, respectively. Consistent with these observations, real-time PCR, microarray and ChIP analyses showed that p53 binds to and transcriptionally activates both pro-arrest and pro-apoptotic target genes proportionally to its expression levels. However, low levels of p53 pro-arrest proteins initiated the growth-arrested response, whereas low levels of pro-apoptotic proteins failed to trigger apoptosis. The authors concluded that their observations *"suggest a mechanism whereby the biological outcome of p53 activation is determined by different cellular thresholds for arrest and apoptosis. Lowering the apoptotic threshold was sufficient to switch the p53 cell fate from arrest to apoptosis, which has important implications for the effectiveness of p53-based cancer therapy."* Growth arrest in these experiments was judged from accumulation of cells in the G0/G1 phase of the cell cycle [41]. In other studies, nutlin-3-triggered activation of p53 signaling was shown to result in marginal apoptosis but a high degree of growth arrest through SIPS in p53 wild-type cancer cells [42–45].

2.6. Anti-Apoptotic Property of p53 Signaling under Physiological Conditions

Under some conditions, p53 is known to activate apoptotic signaling (but not necessarily cell death) both directly, through its proline-rich region, and indirectly by inducing the expression of pro-apoptotic proteins such a BAX (BCL-2-associated X protein), PUMA (p53 upregulated modulator of apoptosis) and NOXA (the Latin word for damage) [6,46]. Simultaneously, under the same conditions, p53 also transcriptionally activates a host of anti-apoptotic proteins, including p21, 14-3-3δ, WIP1 and DNAJB9 [6,46,47]. Thus, in most cell types (e.g., cells derived from solid tumors), activation of p53 signaling not only fails to promote apoptotic cell death (i.e., cell demise), it actually protects against this response. As extensively discussed by Jänicke et al. [46], the anti-apoptotic and transient growth inhibitory properties of p53 *"are surely essential for normal development and maintenance of a healthy organism, but may easily turn into the dark side of the tumor suppressor p53 contributing to tumorigenesis."* In their article, which was published in 2008 [46], these authors considered approximately 40 p53-regulated proteins that exhibit anti-apoptotic properties. Below we will limit our discussion to DNAJB9 and p21, both of which participate in a negative regulatory loop with p53 (Figure 1).

2.7. p53–DNAJB9 Regulatory Loop: Impact on Apoptosis

DNAJB9 (DNAJ homolog subfamily B member 9) functions in many cellular processes by regulating the ATPase activities of the 70 kDa heat shock proteins (Hsp70s). Recently, Lee et al. [47] identified DNAJB9 as a transcriptional target of p53 in human cancer cell lines. Employing Western and Northern blot analyses, p53-dependent expression of DNAJB9 was demonstrated in both overexpression studies with EJ-p53, a human bladder carcinoma cell line that expresses p53 under the control of a tetracycline-regulated promoter, and with p53 wild-type cancer cell lines (e.g., SKNSH neuroblastoma) after treatment with doxorubicin and other chemotherapeutic agents. Immunofluorescence experiments demonstrated that DNAJB9 co-localizes with p53 in both the cytoplasm and nucleus after genotoxic stress. DNAJB9 depletion and overexpression studies demonstrated that this p53-regulated protein inhibits the pro-apoptotic function of p53 through a physical interaction. Thus, DNABJ9 suppresses apoptosis in response to chemotherapeutic agents by forming a negative regulatory loop with p53.

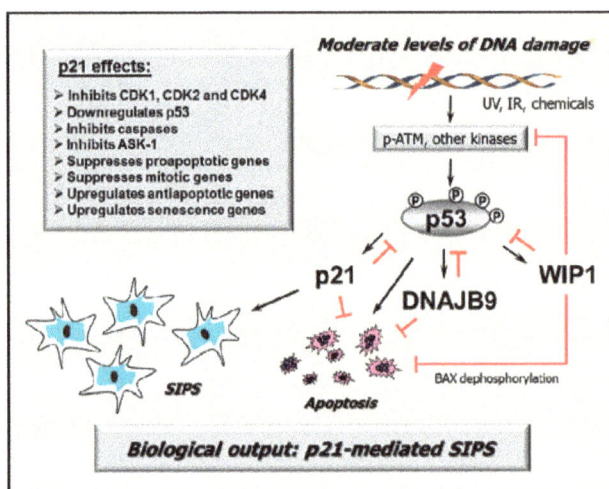

Figure 1. A partial schematic of the DNA damage surveillance network illustrating the importance of negative regulation of p53 by p21, DNAJ homolog subfamily B member 9 (DNAJB9), and wild-type p53-induced phosphatase 1 (WIP1) in suppressing apoptosis as discussed in this article. Arrows indicate stimulation and T-shaped lines indicate inhibition. Multiple functions of p21 in the DNA damage surveillance network are indicated.

2.8. Multiple Functions of p21: Downregulating p53 and More

The p21 protein was discovered by different groups in the early 1990s and was variously called WAF1 (for wild-type p53-activated fragment 1), CIP1 (for CDK-interacting protein 1), and SDI1 (for senescent cell-derived inhibitor 1) [48,49]. It has been extensively studied for its ability to influence cell cycle progression by inhibiting the activity of cyclin/cyclin dependent kinase (CDK) complexes (e.g., CDK1, 2 and 4). In 2002, Javelaud and Besançon [50] reported an additional function for p21 in the DNA damage surveillance network. Disruption of p21 expression in HCT116 colorectal carcinoma cells, either by gene targeting or gene silencing by using antisense oligonucleotides, resulted in an increase in p53 steady-state levels in the absence of genotoxic treatment. Elevated expression of p53 in p21-depleted HCT116 cells correlated with high expression of p14^{ARF}, the product of an alternative transcript of the *INK4A* locus, which is known to promote p53 stability through binding to its negative regulator, MDM2 [51,52]. In addition, elevated expression of p53 in p21-depleted cells resulted in marked sensitivity to chemotherapeutic drug-induced cytotoxicity through activation of the mitochondrial pathway of apoptosis. Thus, p21 may indirectly participate in the regulation of p53 protein stability through preventing p14^{ARF}-mediated MDM2 breakdown, resulting in marked resistance towards stress-induced apoptosis.

Since then, numerous reports have established the broad-acting functions of p21 beyond its influence on the cell cycle. For example, we recently demonstrated that one mechanism by which p21 exerts its inhibitory effects on p53 and apoptosis is through regulating WIP1, an oncogenic phosphatase that inactivates p53 and its upstream kinases [53]. Other groups have demonstrated that the anti-apoptotic property of p21 also relies on its ability to inhibit the activity of proteins directly involved in the induction of apoptosis, including the caspase cascade, stress-activated protein kinases (SAPKs) and apoptosis signal-regulating kinase 1 (ASK1) [54–56], and to control transcription, resulting in downregulation of pro-apoptotic genes [56] and upregulation of genes that encode secreted factors with anti-apoptotic activities [55,56].

It is noteworthy that in a review article published in 2012 [32], we suggested that p21 might function as a positive regulator of p53 in the DNA damage surveillance network. This notion was based on a report suggesting that loss of p21 in the HCT116 cell line led to cytoplasmic sequestration of p53 and inhibition of its transcriptional activity [57]. This observation, however, was not confirmed by us [53] and others [58]. On the contrary, we found that loss of p21 in this cell line results in robust accumulation of p53, and that p53 molecules are phosphorylated (e.g., on Ser15) and accumulated in the nucleus even in the absence of exogenous stress [53]. Accordingly, we and others have concluded that p21 downregulates p53, at least in the HCT116 colon carcinoma [50,53,58,59], MCF7 breast carcinoma [53], and HT1080 fibrosarcoma [59] cell lines.

In addition to its strong anti-apoptotic properties, p21 also plays a key role in orchestrating the complex SIPS program in cells expressing wild-type p53 [6,60]. Studies with cancer cell lines treated with chemotherapeutic agents demonstrated that p21 forms a positive regulatory loop with ATM and that this interaction is essential for the maintenance of the growth-arrested response, a hallmark of SIPS [32,61]; pharmacological targeting of either p21 or ATM triggers apoptosis of growth-arrested cancer cells [61].

Some authors use the term "arrest" without clearly distinguishing between transient G1/S checkpoint activation and SIPS. As discussed recently [6], these two responses are uncoupled, at least in human skin fibroblast strains and solid tumor-derived cell lines. In these cell types, G1/S checkpoint activation following exposure to DNA-damaging agents is an early event required to provide time for the repair of genomic injury before resumption of the cell cycle, whereas SIPS is manifested at late times (several days) post-treatment. Multiple factors contribute to the regulation of SIPS, including p21-mediated expression of a battery of genes involved in growth arrest, senescence, and aging, coupled with p21-mediated downregulation of numerous genes that control mitosis [17,55].

To summarize, the pivotal role of p21 in determining cell fate in response to genotoxic stress is not only through activating the G1/S cell cycle checkpoint, but also through controlling gene expression, suppressing apoptosis by acting at different levels of the death cascade, and promoting growth arrest through SIPS.

3. Activation of Apoptotic Signaling Does Not Always Lead to Cell Death: Impact on Chemosensitivity Assessment

It is now widely accepted that transcriptional activation of pro-apoptotic proteins (e.g., PUMA, NOXA, BAX) might not inevitably lead to cell death as a result of concomitant activation of a host of anti-apoptotic proteins that maintain p53 under the apoptotic threshold level (e.g., MDM2, p21, WIP1, DNAJB9) [6,47], sequester pro-apoptotic factors such as BAX (e.g., 14-3-3δ) [62–66], and inhibit ASK1 and the caspase cascade (e.g., p21) [6]. Similarly, while caspase 3 functions as a key apoptosis executioner under some conditions, such as in the development and maintenance of the hematopoietic system, under other conditions it reveals its dark side by promoting tumor growth [67–77]. For these and several other reasons, the Nomenclature Committee on Cell Death (NCCD) has cautioned the scientific community about the use/misuse of terminologies and concepts in the area of cell death research. Notably, in their 2009 article, the NCCD pointed out that bona fide "dead cells" would be different from "dying cells" that have not crossed the point of no return and have not concluded their demise [78]. It is worth noting that radiosensitivity and chemosensitivity, assessed by the widely-used multi-well plate colorimetric assays, which determine the inhibition of cell growth (resulting from the combined impact of checkpoint activation, growth inhibition and cytotoxicity), have often been misinterpreted to reflect loss of viability and hence cell death.

Recently we reviewed the current knowledge on responses induced by ionizing radiation that can lead to cancer cell death or survival depending on the context [11]. These include activation of caspases (e.g., caspase 3), growth arrest through SIPS, and creation of polyploid/multinucleated giant cells (hereafter called MNGCs) (also see Figure 2). Such potentially pro-survival responses are triggered not only by ionizing radiation, but also by chemotherapeutic drugs [74,79–84] and hypoxia [72,85–89].

Caspase 3 is extensively studied for its role in the execution phase of apoptosis [90]. Accordingly, the activated (cleaved) form of caspase 3 has often been used as a molecular marker of apoptosis. Paradoxically, in recent years caspase 3 has also been demonstrated to function as a survival factor, promoting the growth of tumor-repopulating cells [68–77]. This pro-survival effect of caspase 3 has been attributed to secretion of prostaglandin E_2 (PGE_2) [70,72,74]. The caspase 3-PGE_2 survival pathway is triggered by various stimuli, including ionizing radiation [70,74], chemotherapeutic drugs [72,74] and hypoxia [72]. Interestingly, the biological outcome associated with caspase 3 activation is in part dependent on p21 (reviewed in [32]). Thus, p21-mediated inhibition of caspase 3 activity results in suppression of apoptosis in response to genotoxic stress. On the other hand, caspase 3-mediated cleavage of p21 generates a 15 kDa fragment of p21 that appears to positively regulate apoptosis by forming a complex with active caspase 3. It is currently unknown whether p21 might play a role in the regulation of the caspase 3-PGE_2 survival pathway.

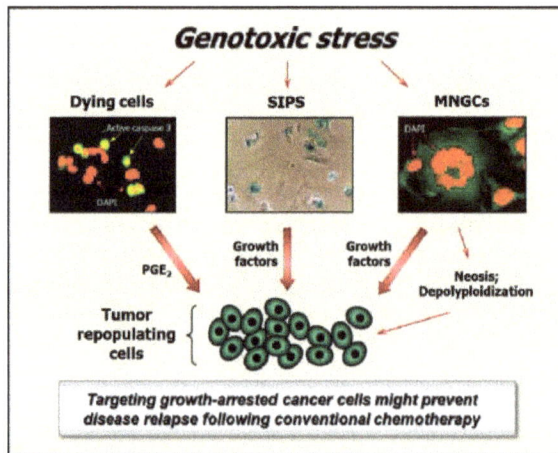

Figure 2. Examples of genotoxic stress-induced responses associated with cancer cell death or survival depending on context: Activation of caspase 3, induction of stress-induced premature senescence (SIPS), and creation of multinucleated giant cells (MNGCs). SIPS is a genetically-controlled process, mediated by p21 or p16, depending on the p53 status of the cells [6,11]. MNGCs can be created through different routes, including endoreduplication (replication of chromosomes without subsequent cell division) and homotypic cell fusions [11].

Whether caspase 3 plays a role in growth-arrested cancer cells also remains to be elucidated. However, it is well known that cancer cells undergoing SIPS remain viable and acquire the ability to secrete factors that can promote proliferation and invasiveness in cell culture models and tumor development in vivo [91,92]. This so-called "senescence-associated secretory phenotype" (SASP) includes several families of soluble and insoluble factors that can affect surrounding cells by activating various cell surface receptors and corresponding signal transduction pathways [91,92].

While some authors consider SASP to be the "dark" side of senescence [91–96], others have proposed that induction of senescence (SIPS) might be advantageous for cancer treatment [97–101]. As pointed out by Maier et al. [101], although *"accumulation of senescent cancer cells leads to an increased secretion of inflammatory cytokines, which might cause age-related pathologies, like secondary cancers, in the long term, the primary aim of cancer treatment leading to a longer overall survival should always take preference. Thus, the hypothetical possibility that senescent cells may be dormant with an intrinsic capability to reawaken years after the treatment is of secondary concern, similar to the risk of inducing second cancers."* However, aside from SASP, there is now compelling evidence that cancer cells undergoing SIPS can themselves give rise to stem-cell-like progeny, thereby contributing to cancer relapse following therapy [11,102,103].

Like cells undergoing SIPS, MNGCs also remain viable and secrete cell-growth promoting factors [11]. This property of MNGCs was first reported over 60 years ago for HeLa cervical carcinoma cells exposed to ionizing radiation [104–106]. HeLa cells harbor wild-type alleles of *TP53*, but are infected with human papillomavirus (HPV) 18, the E6 protein of which disables the p53–p21 axis [107]. This observation prompted Puck and Marcus to develop the feeder layer clonogenic assay, in which a "lawn" of heavily-irradiated feeder cells (which encompass MNGCs) is inoculated into a culture dish to promote the growth of test cells given graded doses of genotoxic agents [105]. Recently, we demonstrated that exposure of a panel of p53-deficient or p21-deficient solid tumor-derived cell lines to moderate doses of ionizing radiation (e.g., 8 Gy) results in the development of MNGCs that remain adherent to the culture dish, retain viability, metabolize 3-(4,5-dimethylthiazol-2-yl)-2,5-diphenyl-tetrazolium bromide (MTT), and exhibit DNA synthesis for long times (e.g., three weeks) post-irradiation [108].

Collectively, these observations underscore the importance of distinguishing between dead cells and growth arrested cells that might be mistakenly scored as "dead" in the colony formation and other cell-based radiosensitivity/chemosensitivity assays. As pointed out recently [108], the creation of viable growth arrested cells (e.g., MNGCs) complicates the interpretation of data obtained with multi-well plate colorimetric tests routinely used in anti-cancer drug-screening endeavors.

4. Extrapolating Results Obtained in Overexpression Studies to Clinically Relevant Conditions

The preceding discussion raises a fundamental question with respect to p53 regulation and function. As discussed by Uversky [7], p53 undergoes extensive post-translational modifications (e.g., phosphorylation, acetylation) that are critical for its stabilization and activation. Such modifications result in accumulation of p53 in the nucleus and the formation of p53 tetramers, which then bind to the promoters of target genes and trigger their expression. Genotoxic stress activates factors such as ATM and ATR that initiate the DNA damage surveillance network by mediating p53 posttranslational modifications (also see Figure 1). Despite the wealth of knowledge regarding the importance of genotoxic stress (e.g., DNA damage) in activating the p53-mediated transcriptional program, this same transcriptional program has also been reported to be activated by ectopic expression of wild-type p53 without exposure to exogenous stress [41]. Does this indicate that stress-triggered p53 posttranslational modifications are not needed for activation of its transcriptional program, which appears to be highly unlikely, or does p53 overexpression by itself create a non-physiological condition that it is sufficient to trigger the stress response?

It is important to note that many reports suggesting a positive role for wild-type p53 in triggering apoptosis, either with or without exposure to genotoxic agents, involved overexpression experiments with a variety of transformed/malignant cell types (e.g., T-cell leukemia cell lines). In addition, many authors did not follow the NCCD recommendations to distinguish between "dying cells" (i.e., exhibiting transient activation of a death-related biochemical pathway) and cells that are irreversibly committed to die. Taken together with the observations of Kracikova et al. [41] described above (Section 2.5.), demonstrating that different expression levels of exogenous p53 yield different outcomes (G1 arrest versus apoptotic signaling), caution should be exercised in extrapolating results obtained in overexpression studies to clinically relevant conditions (e.g., cancer chemotherapy) particularly when it pertains to p53-directed cancer cell death.

5. Fate of Growth-Arrested Cancer Cells

5.1. Clinically Relevant Doses of Chemotherapeutic Agents Predominantly Trigger Cancer Cell Dormancy Rather Than Cell Death

Studies with some cell types (e.g., solid tumor-derived cell lines) have shown that many (if not all) chemotherapeutic agents predominantly trigger growth arrest but not cell death when administered at clinically achievable concentrations [11]. Below, we mainly focus on cisplatin and solid tumors.

Berndtsson et al. [79] examined numerous articles, published between 2002 and 2005, which reported apoptosis after cisplatin treatment in a wide range of cell lines. The mean cisplatin concentration used to induce apoptosis was 52 µM; when examined, concentrations below 20 µM did not induce apoptosis but triggered growth arrest [79]. These and more recent studies have reported IC_{50} values (50% inhibiting concentrations) of >40 µM and <2 µM for induction of apoptosis and growth arrest by cisplatin, respectively [11]. We have observed a similar trend with a panel of cancer cell lines expressing wild-type p53 (HCT116, A549, MCF7), mutant p53 (MDA-MD-231, SUM159) or no p53 (HCT116 p53 knockout). In all cell lines, a 3-day incubation with 10 µM cisplatin resulted in growth inhibition of all cells (IC_{50} values ranging from 0.5 to 2 µM), but did not induce cytotoxicity when evaluated by the vital dye (trypan blue) exclusion and other assays (data not shown).

The finding that very high concentrations of cisplatin are required to induce apoptosis in solid tumor-derived cell lines is not surprising given that this effect has been reported to primarily reflect cisplatin-induced injury to mitochondria rather than to nuclear DNA [79]. This raises the important question as to whether high, apoptosis-triggering concentrations of cisplatin are relevant for in vivo studies and, by inference, for treating cancer patients. Puig et al. [82] have addressed this question using a rat colon carcinoma cell line grown both in vitro and in vivo. Treatment of animals with cisplatin concentrations corresponding to those which induced apoptosis in the cell-based (tissue culture) experiments caused major toxic side effects on the gastrointestinal tract, bone marrow and kidney. When administered at tolerated concentrations (corresponding to ≤10 µM in cell culture experiments), cisplatin induced tumor cell dormancy (through SIPS and multinucleation) but did not kill tumor cells.

5.2. Hypoxia and the Creation of MNGCs

Hypoxia is one of the most important pathological features of solid tumors, and represents a major obstacle in cancer therapy [109,110]. Hypoxia constitutes a physiological selective pressure promoting tumor aggressiveness, which is largely associated with the maintenance and formation of cancer stem cells, promoting their phenotype and tumorigenesis. Many of the cellular responses to hypoxia are controlled by the transcription factor hypoxia-inducible factor-1 (HIF-1), which is a heterodimer composed of α and β subunits. HIF-1α contains two oxygen dependent degradation domains. Under normoxic conditions these domains are continuously hydroxylated by prolyl hydroxylases, resulting in HIF-1α degradation. Hypoxic conditions result in stabilization of HIF-1α. Stabilized HIF-1α accumulates in the nucleus where it binds to HIF-1β subunit, forming a transcription factor capable of activating the expression of numerous target genes, including those involved in energy production, angiogenesis, and metabolic adaptation to hypoxia [109,111].

Cobalt chloride ($CoCl_2$) is used as a hypoxia mimicking agent when administered under normoxic conditions. $CoCl_2$ stabilizes HIF-1α by inhibiting prolyl hydroxylase enzymes [112–114]. Several reports have demonstrated that treatment of human cancer cells with $CoCl_2$ induces the formation of

MNGCs through endoreduplication and/or cell fusion. MNGCs exhibit resistance to genotoxic agents (e.g., doxorubicin [115]) and give rise to tumor repopulating progeny through splitting, budding, or burst-like mechanisms (see also Section 5.3). CoCl$_2$-triggered creation of MNGCs and emergence of their proliferating progeny has been reported for ovarian [88,116], breast [87] and colon [83] carcinoma cells. Studies with cancer cell lines as well as tumor tissues from cancer patients have identified several factors that appear to promote the survival of MNGCs and control their fate. These include the cell cycle regulatory proteins cyclin E, S-phase kinase-associated protein 2 (SKP2), and stathmin [88], as well as the epithelial–mesenchymal transition (EMT)-related proteins E-cadherin, N-cadherin, and vimentin [87].

5.3. Genome Reduction and Neosis of MNGCs

The observation that MNGCs created in response to genotoxic stress (ionizing radiation) remain viable and secrete cell growth-promoting factors was reported over half a century ago [104–106]. This seminal discovery was largely overlooked and with time the induction of massive genetic anomalies seen in MNGCs was often assumed to be associated with cell death through mitotic catastrophe. In 2000, Erenpreisa et al. [117] and Illidge et al. [118] reported that MNGCs that develop in heavily irradiated p53-deficient human cell cultures undergo a complex breakdown and sub-nuclear reorganization, ultimately giving rise to rapidly propagating daughter cells. The authors proposed that the development of MNGCs might represent a unique mechanism of "repair" enabling p53-deficient cancer cells to maintain proliferative capacity despite experiencing extensive genomic instability [117,118]. These initial experiments involved a Burkitt's lymphoma cell line. The creation of MNGCs capable of generating proliferating daughter cells has now been reported from different laboratories for various cell types in response to a wide range of genotoxic agents, including chemotherapeutic drugs. This so-called "endopolyploidy-stemness" route of cancer-cell survival consequent to therapeutic exposure has also been documented with short-term (2–3 weeks) cultures of primary human breast cancer specimens [119].

One mechanism of depolyploidization of MNGCs is through genome reduction division (reviewed in [120]). In this process, MNGCs first undergo a ploidy cycle, which is regulated by key mediators of mitosis (e.g., aurora B kinase), meiosis (e.g., MOS), and self-renewal (e.g., OCT4), ultimately giving rise to a para-diploid progeny, containing a near-diploid number of chromosomes, that exhibit mitotic propagation. In 2004, Sundaram et al. [121] reported an alternative mechanism of genome reduction in MNGCs. Computerized video time-lapse microscopy revealed that, although MNGCs may cease to divide, each giant cell might produce numerous (50 or more) small cells with low cytoplasmic content (karyoplasts) via the nuclear budding process of "neosis" that resembles the parasexual mode of somatic reduction division of simple organisms like fungi. The resultant budding daughter cells begin to divide by mitosis and transiently display stem cell properties, and subsequently experience a complex life cycle eventually leading to the development of highly metastatic and therapy resistant descendants. This parasexual mode of somatic reduction division of MNGCs (neosis) has been reported from different laboratories for human ovarian [88,116], breast [87] and colon carcinoma cell lines [83] as well as other biological systems [122–128].

5.4. Is SIPS Reversible?

The creation of MNGCs and their proliferating daughter cells appears to be a general feature of cells that lack wild-type p53 function. Under some conditions, p53-proficient cells that undergo SIPS can also follow the polyploidy stemness route [11]. As an example, doxorubicin treatment of HCT116 colon carcinoma cell cultures resulted in growth-arrested cells that were positive in the SA-β-gal assay, but with time these same cells became polyploid and generated growing progeny [129]. In the same study, MCF7 breast cancer cells also exhibited growth arrest coupled with SA-β-gal staining following doxorubicin treatment, but this response was not accompanied by polyploidy and generation of proliferating progeny [129]. Although the basis for reversibility of SIPS in some contexts is not known, it is interesting to note that HCT116 cells are caspase 3 proficient, whereas MCF7 cells do not express caspase 3.

6. Targeting Growth-Arrested Cancer Cells as a Potential Therapeutic Strategy

The importance of MNGCs in the failure of cancer therapy has been largely overlooked. In part, this may be because the creation of such cells has been considered to be rare and also multinucleation has often been assumed to reflect death through "mitotic catastrophe" or other mechanisms. As discussed in this article, when administered at clinically relevant doses, cancer chemotherapeutic agents trigger a high proportion of MNGCs in solid tumor-derived cell lines (especially those lacking wild-type p53 function) that remain viable and can give rise to tumor repopulating progeny. Shockingly, only a single multinucleated giant cancer cell has been shown to be sufficient to cause metastatic disease when grafted under the skin of an animal [115]. Cancer cells undergoing SIPS in response to chemotherapeutic agents can also escape from the growth-arrested state and give rise to tumor-repopulating progeny.

Accordingly, targeting growth-arrested cancer cells might represent an effective therapeutic strategy. To this end, Crescenzi et al. [61] reported that downregulating either ATM or p21 in cancer cells that have undergone SIPS in response to chemotherapeutic drugs results in their demise. For targeting MNGCs for destruction, different approaches have been reported to be effective. These include viral infection [104] and treatment with pharmacological inhibitors of different members of the BCL-XL/BCL-2 pathway [130]. In addition, we have recently demonstrated that the apoptosis activators sodium salicylate (an inhibitor of the p38 MAPK) or dichloroacetate (a modulator of glucose metabolism) also kill MNGCs under conditions that have little or no effect on parental (mono-nucleated) cells [108]. The results of these proof-of-principle in vitro experiments are encouraging and warrant further studies with animal models.

7. Mutational Signatures in Human Cancers

The aforementioned reports concluding that the creation of MNGCs following chemotherapy might represent a survival mechanism for cancer cells involved studies not only with cultured cells and animal models, but also with patient specimens. Other studies, however, also reporting extensive experimental and clinical data, have concluded that this response might reflect a favorable therapeutic outcome (e.g., [131]). Similarly, the conclusions that SIPS might represent a favorable [97–101] or unfavorable [97–103] therapeutic outcomes have also been based on extensive experimental/clinical data. Such apparently conflicting observations might not be entirely unexpected when considering the distinct mutational types in aging and cancer.

The advent of next-generation sequencing technologies has enabled large-scale sequencing of all protein-coding exons (whole-exome sequencing) or even whole cancer genomes (whole-genome sequencing) in a single experiment [132–138]. These sequencing efforts have enabled the identification of many thousands of mutations per cancer which provided sufficient power to detect different mutational patterns or "signatures." Each biological perturbation or "mutational process" (e.g., tobacco smoke, sunlight exposure, deamination of DNA bases) is shown to leave a characteristic "mark" or mutational signature on the cancer genome (reviewed in [135]) (Figure 3).

Each mutational signature is defined by: (i) the type of genomic injury that has occurred as a result of a diversity of exogenous and endogenous genotoxic stresses; (ii) the integrity of DNA repair and other aspects of the DNA damage surveillance network that were successively activated; and (iii) the strength and duration of exposure to each mutational process. Additionally, as pointed out by Helleday et al. [135], "*cancers are likely to comprise different cell populations (that is, subclonal populations), which can be variably exposed to each mutational process; this promotes the complexity of the final landscape of somatic mutations in a cancer genome. The final "mutational portrait," which is obtained after a cancer has been removed by surgery and then sequenced, is therefore a composite of multiple mutational signatures.*"

Thus, as previously anticipated, these large-scale sequencing technologies coupled with bioinformatic and computational tools for deciphering the "scars" (signatures) of mutational processes have demonstrated significant variability in the mutation landscape in cancers of the same histological type. Application of such approaches might similarly unfold the molecular basis for the fate of growth-arrested cancer cells in terms of death versus survival. This might in turn set the stage for designing novel therapeutic strategies for specifically targeting growth-arrested cancer cells before they will have the opportunity to generate tumor-repopulating progeny.

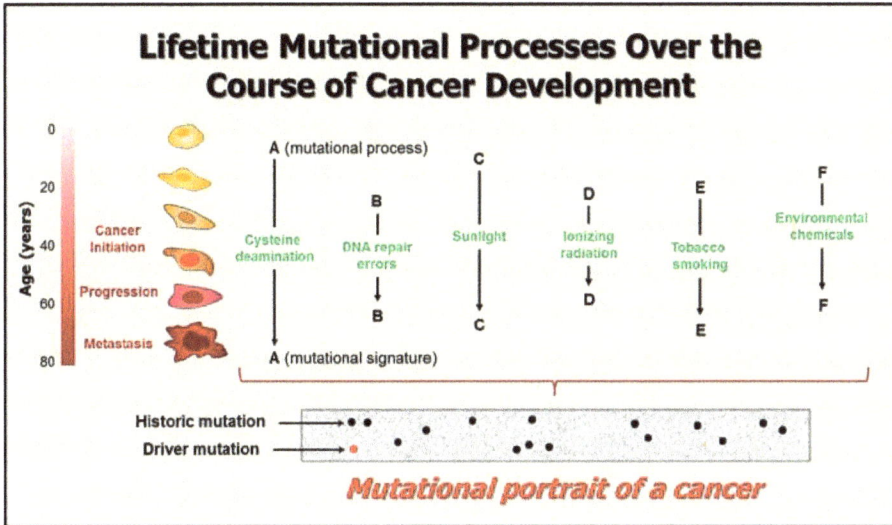

Figure 3. Cartoon showing mutational processes that can "scar" the genome during different periods of a person's life span. The various mutations found in a tumor are grouped into "driver" mutations, which are ongoing and confer selective cancer phenotypes, and "historic" (or passenger) mutations which are far more numerous and hitchhike with driver mutations, but do not appear to be causative of cancer development. For details concerning ionizing radiation and other stimuli, consult [138] and [132–135], respectively. Adapted from Helleday et al. [135].

8. Conclusions

Inhibition of cell growth is an important response to genotoxic stress, either under physiological conditions or in cancer therapy. This response is fundamental for the maintenance of genomic stability and cellular homeostasis under physiological conditions. On the other hand, stress-induced growth arrest in cancer cells—reflecting either SIPS (predominantly in p53 wild-type cells) or the creation of MNGCs (predominantly in p53-deficient cells)—can provide a "survival" mechanism, ultimately resulting in the emergence of cancer repopulating progeny. Selective targeting of growth-arrested cancer cells (e.g., MNGCs) could represent a promising strategy for improving the outcome of conventional chemotherapy.

Acknowledgments: This work was supported by the Canadian Breast Cancer Foundation—Prairies/North West Territories region, the Alberta Innovates-Health Solutions (grant 101201164) and the Alberta Cancer Foundation-Transformative Program (file 26603).

Conflicts of Interest: The authors declare no conflict of interest.

Abbreviations

UV	Ultraviolet light
DSBs	DNA double-strand breaks
SA-β-gal	Senescence-associated β-galactosidase
SIPS	Stress-induced premature senescence
DNAJB9	DNAJ homolog subfamily B member 9
MDM2	Murine double minute-2 homologue
ATM	Ataxia telangiectasia mutated
ATR	ATM and RAD3-related
CHK	Checkpoint kinase
WIP1	Wild-type p53-induced phosphatase 1
BAX	BCL-2-associated X protein
PUMA	p53 upregulated modulator of apoptosis
WAF1	Wild-type p53-activated fragment 1
CIP1	CDK-interacting protein 1
SDI1	Senescent cell-derived inhibitor 1
CDK	Cyclin dependent kinase
SAPK	Stress-activated protein kinase
ASK1	Apoptosis signal-regulating kinase 1
HPV	Human papillomavirus
PGE_2	Prostaglandin E_2
SASP	Senescence-associated secretory phenotype
SKP2	S-phase kinase-associated protein 2
EMT	Epithelial-mesenchymal transition
MTT	3-(4,5-Dimethylthiazol-2-yl)-2,5-diphenyl-tetrazolium bromide
IC_{50}	Inhibiting concentration, 50%
NCCD	Nomenclature Committee on Cell Death

References

1. De Bont, R.; van Larebeke, N. Endogenous DNA damage in humans: A review of quantitative data. *Mutagenesis* **2004**, *19*, 169–185. [CrossRef] [PubMed]
2. Hou, L.; Zhang, X.; Wang, D.; Baccarelli, A. Environmental chemical exposures and human epigenetics. *Int. J. Epidemiol.* **2012**, *41*, 79–105. [CrossRef] [PubMed]
3. Coelho, M.M.V.; Matos, T.R.; Apetato, M. The dark side of the light: Mechanisms of photocarcinogenesis. *Clin. Dermatol.* **2016**, *34*, 563–570. [CrossRef] [PubMed]
4. Lai, Y.; Yu, R.; Hartwell, H.J.; Moeller, B.C.; Bodnar, W.M.; Swenberg, J.A. Measurement of endogenous versus exogenous formaldehyde-induced DNA-protein crosslinks in animal tissues by stable isotope labeling and ultrasensitive mass spectrometry. *Cancer Res.* **2016**, *76*, 2652–2661. [CrossRef] [PubMed]
5. Saintigny, Y.; Chevalier, F.; Bravard, A.; Dardillac, E.; Laurent, D.; Hem, S.; Dépagne, J.; Radicella, J.P.; Lopez, B.S. A threshold of endogenous stress is required to engage cellular response to protect against mutagenesis. *Sci. Rep.* **2016**, *6*, 29412. [CrossRef] [PubMed]
6. Mirzayans, R.; Andrais, B.; Scott, A.; Wang, Y.W.; Murray, D. Ionizing radiation-induced responses in human cells with differing *TP53* status. *Int. J. Mol. Sci.* **2013**, *14*, 22409–22435. [CrossRef] [PubMed]
7. Uversky, V.N. p53 proteoforms and intrinsic disorder: An illustration of the protein structure-function continuum concept. *Int. J. Mol. Sci.* **2016**, *17*, 1874. [CrossRef] [PubMed]
8. Vousden, K.H.; Prives, C. Blinded by the light: The growing complexity of p53. *Cell* **2009**, *137*, 413–431. [CrossRef] [PubMed]
9. Stegh, A.H. Targeting the p53 signaling pathway in cancer therapy—The promises, challenges, and perils. *Expert Opin. Ther. Targets* **2012**, *16*, 67–83. [CrossRef] [PubMed]
10. Desilet, N.; Campbell, T.N.; Choy, F.Y.M. p53-based anti-cancer therapies: An empty promise? *Curr. Issues Mol. Biol.* **2010**, *12*, 143–146. [PubMed]

11. Mirzayans, R.; Andrais, B.; Kumar, P.; Murray, D. The growing complexity of cancer cell response to DNA-damaging agents: Caspase 3 mediates cell death or survival? *Int. J. Mol. Sci.* **2016**, *17*, 708. [CrossRef] [PubMed]

12. Zuco, V.; Zunino, F. Cyclic pifithrin-α sensitizes wild type p53 tumor cells to antimicrotubule agent-induced apoptosis. *Neoplasia* **2008**, *10*, 587–596. [CrossRef] [PubMed]

13. Amin, A.R.; Thakur, V.S.; Gupta, K.; Jackson, M.W.; Harada, H.; Agarwal, M.K.; Shin, D.M.; Wald, D.N.; Agarwal, M.L. Restoration of p53 functions protects cells from concanavalin a-induced apoptosis. *Mol. Cancer Ther.* **2010**, *9*, 471–479. [CrossRef] [PubMed]

14. Waye, S.; Naeem, A.; Choudhry, M.U.; Parasido, E.; Tricoli, L.; Sivakumar, A.; Mikhaiel, J.P.; Yenugonda, V.; Rodriguez, O.C.; Karam, S.D.; et al. The p53 tumor suppressor protein protects against chemotherapeutic stress and apoptosis in human medulloblastoma cells. *Aging* **2015**, *7*, 854–868. [CrossRef] [PubMed]

15. Sengupta, S.; Harris, C.C. p53: Traffic cop at the crossroads of DNA repair and recombination. *Nat. Rev. Mol. Cell Biol.* **2005**, *6*, 44–55. [CrossRef] [PubMed]

16. Levine, A.J.; Hu, W.; Feng, Z. The p53 pathway: What questions remain to be explored? *Cell Death Differ.* **2006**, *13*, 1027–1036. [CrossRef] [PubMed]

17. Roninson, I.B. Tumor cell senescence in cancer treatment. *Cancer Res.* **2003**, *63*, 2705–2715. [PubMed]

18. Mirzayans, R.; Andrais, B.; Scott, A.; Paterson, M.C.; Murray, D. Single-cell analysis of p16[INK4a] and p21[WAF1] expression suggests distinct mechanisms of senescence in normal human and Li-Fraumeni Syndrome fibroblasts. *J. Cell. Physiol.* **2010**, *223*, 57–67. [PubMed]

19. Wang, M.; Morsbach, F.; Sander, D.; Gheorghiu, L.; Nanda, A.; Benes, C.; Kriegs, M.; Krause, M.; Dikomey, E.; Baumann, M.; et al. EGF receptor inhibition radiosensitizes NSCLC cells by inducing senescence in cells sustaining DNA double-strand breaks. *Cancer Res.* **2011**, *71*, 6261–6269. [CrossRef] [PubMed]

20. Hernández-Vargas, H.; Ballestar, E.; Carmona-Saez, P.; von Kobbe, C.; Bañón-Rodriguez, I.; Esteller, M.; Moreno-Bueno, G.; Palacios, J. Transcriptional profiling of MCF7 breast cancer cells in response to 5-Fluorouracil: Relationship with cell cycle changes and apoptosis, and identification of novel targets of p53. *Int. J. Cancer* **2006**, *119*, 1164–1175. [CrossRef] [PubMed]

21. Alani, R.M.; Young, A.Z.; Shifflett, C.B. Id1 regulation of cellular senescence through transcriptional repression of p16/Ink4a. *Proc. Natl. Acad. Sci. USA* **2001**, *98*, 7812–7816. [CrossRef] [PubMed]

22. Polsky, D.; Young, A.Z.; Busam, K.J.; Alani, R.M. The transcriptional repressor of p16/Ink4a, Id1, is up-regulated in early melanomas. *Cancer Res.* **2001**, *61*, 6008–6011. [PubMed]

23. Leong, W.F.; Chau, J.F.L.; Li, B. p53 deficiency leads to compensatory up-regulation of p16[INK4a]. *Mol. Cancer Res.* **2009**, *7*, 354–360. [CrossRef] [PubMed]

24. Ljungman, M. Dial 9-1-1 for p53: Mechanisms of p53 activation by cellular stress. *Neoplasia* **2000**, *2*, 208–225. [CrossRef] [PubMed]

25. Efeyan, A.; Serrano, M. p53: Guardian of the genome and policeman of the oncogenes. *Cell Cycle* **2007**, *6*, 1006–1010. [CrossRef] [PubMed]

26. Bulavin, D.V.; Saito, S.; Hollander, M.C.; Sakaguchi, K.; Anderson, C.W.; Appella, E.; Fornace, A.J., Jr. Phosphorylation of human p53 by p38 kinase coordinates N-terminal phosphorylation and apoptosis in response to UV radiation. *EMBO J.* **1999**, *18*, 6845–6854. [CrossRef] [PubMed]

27. Huang, C.; Ma, W.Y.; Maxiner, A.; Sun, Y.; Dong, Z. p38 kinase mediates UV-induced phosphorylation of p53 protein at serine 389. *J. Biol. Chem.* **1999**, *274*, 12229–12235. [CrossRef] [PubMed]

28. Sanchez-Prieto, R.; Rojas, J.M.; Taya, Y.; Gutkind, J.S. A role for the p38 mitogen-activated protein kinase pathway in the transcriptional activation of p53 on genotoxic stress by chemotherapeutic agents. *Cancer Res.* **2000**, *60*, 2464–2472. [PubMed]

29. Lu, X.; Nguyen, T.A.; Moon, S.H.; Darlington, Y.; Sommer, M.; Donehower, L.A. The type 2C phosphatase Wip1: An oncogenic regulator of tumor suppressor and DNA damage response pathways. *Cancer Metastasis Rev.* **2008**, *27*, 123–135. [CrossRef] [PubMed]

30. Lahav, G.; Rosenfeld, N.; Sigal, A.; Geva-Zatorsky, N.; Levine, A.J.; Elowitz, M.B.; Alon, U. Dynamics of the p53-MDM2 feedback loop in individual cells. *Nat. Genet.* **2004**, *36*, 147–150. [CrossRef] [PubMed]

31. Batchelor, E.; Mock, C.S.; Bhan, I.; Loewer, A.; Lahav, G. Recurrent initiation: A mechanism for triggering p53 pulses in response to DNA damage. *Mol. Cell* **2008**, *9*, 277–289. [CrossRef] [PubMed]

32. Mirzayans, R.; Andrais, B.; Scott, A.; Murray, D. New insights into p53 signaling and cancer-cell response to DNA damage: Implications for cancer therapy. *J. Biomed. Biotechnol.* **2012**, *2012*, 170325. [CrossRef] [PubMed]

33. Yang, X.H.; Sladek, T.L.; Liu, X.; Butler, B.R.; Froelich, C.J.; Thor, A.D. Reconstitution of caspase 3 sensitizes MCF-7 breast cancer cells to doxorubicin- and etoposide-induced apoptosis. *Cancer Res.* **2001**, *61*, 348–354. [PubMed]

34. Jänicke, R.U. MCF-7 breast carcinoma cells do not express caspase-3. Breast. *Cancer Res. Treat.* **2009**, *117*, 219–221. [CrossRef] [PubMed]

35. Essmann, F.; Engels, I.H.; Totzke, G.; Schulze-Osthoff, K.; Jänicke, R.U. Apoptosis resistance of MCF-7 breast carcinoma cells to ionizing radiation is independent of p53 and cell cycle control but caused by the lack of caspase-3 and a caffeine-inhibitable event. *Cancer Res.* **2004**, *64*, 7065–7072. [CrossRef] [PubMed]

36. Purvis, J.E.; Karhohs, K.W.; Mock, C.; Batchelor, E.; Loewer, A.; Lahav, G. p53 dynamics control cell fate. *Science* **2012**, *336*, 1440–1444. [CrossRef] [PubMed]

37. Sun, T.; Cui, J. Dynamics of p53 in response to DNA damage: Mathematical modeling and perspective. *Prog. Biophys. Mol. Biol.* **2015**, *119*, 175–182. [CrossRef] [PubMed]

38. Jin, S.; Levine, A.J. The p53 functional circuit. *J. Cell Sci.* **2001**, *114*, 4139–4140. [PubMed]

39. Haupt, S.; Berger, M.; Goldberg, Z.; Haupt, Y. Apoptosis—The p53 network. *J. Cell Sci.* **2003**, *116*, 4077–4085. [CrossRef] [PubMed]

40. Oren, M. Decision making by p53: Life, death and cancer. *Cell Death Differ.* **2003**, *10*, 431–442. [CrossRef] [PubMed]

41. Kracikova, M.; Akiril, G.; George, A.; Sachidanandam, R.; Aaronson, S.A. A threshold mechanism mediates p53 cell fate decision between growth arrest and apoptosis. *Cell Death Differ.* **2013**, *20*, 576–588. [CrossRef] [PubMed]

42. Schug, T.T. Awakening p53 in senescent cells using nutlin-3. *Aging* **2009**, *1*, 842–844. [CrossRef] [PubMed]

43. Shen, H.; Maki, C.G. Persistent p21 expression after nutlin-3a removal is associated with senescence-like arrest in 4N cells. *J. Biol. Chem.* **2010**, *285*, 23105–23114. [CrossRef] [PubMed]

44. Arya, A.K.; El-Fert, A.; Devling, T.; Eccles, R.M.; Aslam, M.A.; Rubbi, C.P.; Vlatković, N.; Fenwick, J.; Lloyd, B.H.; Sibson, D.R.; et al. Nutlin-3, the small-molecule inhibitor of MDM2, promotes senescence and radiosensitises laryngeal carcinoma cells harbouring wild-type p53. *Br. J. Cancer* **2010**, *103*, 186–195. [CrossRef] [PubMed]

45. Polański, R.; Noon, A.P.; Blaydes, J.; Phillips, A.; Rubbi, C.P.; Parsons, K.; Vlatković, N.; Boyd, M.T. Senescence induction in renal carcinoma cells by Nutlin-3: A potential therapeutic strategy based on MDM2 antagonism. *Cancer Lett.* **2014**, *353*, 211–219. [CrossRef] [PubMed]

46. Jänicke, R.U.; Sohn, D.; Schulze-Osthoff, K. The dark side of a tumor suppressor: Anti-apoptotic p53. *Cell Death Differ.* **2008**, *15*, 959–976. [CrossRef] [PubMed]

47. Lee, H.J.; Kim, J.M.; Kim, K.H.; Heo, J.I.; Kwak, S.J.; Han, J.A. Genotoxic stress/p53-induced DNAJB9 inhibits the pro-apoptotic function of p53. *Cell Death Differ.* **2015**, *22*, 86–95. [CrossRef] [PubMed]

48. Mirzayans, R.; Murray, D. *Cellular Senescence: Implications for Cancer Therap*, 1st ed.; Nova Science Publishers: New York, NY, USA, 2009; pp. 1–130.

49. Warfel, N.A.; El-Deiry, W.S. p21[WAF1] and tumourigenesis: 20 years after. *Curr. Opin. Oncol.* **2013**, *25*, 52–58. [CrossRef] [PubMed]

50. Javelaud, D.; Besançon, F. Inactivation of p21[WAF1] sensitizes cells to apoptosis via an increase of both p14[ARF] and p53 levels and an alteration of the Bax/BCL-2 ratio. *J. Biol. Chem.* **2002**, *277*, 37849–37954. [CrossRef] [PubMed]

51. Zhang, Y.; Xiong, Y.; Yarbrough, W.G. ARF promotes MDM2 degradation and stabilizes p53: *ARF-INK4a* locus deletion impairs both the Rb and p53 tumor suppression pathways. *Cell* **1998**, *92*, 725–734. [CrossRef]

52. Honda, R.; Yasuda, H. Association of p19 (ARF) with MDM2 inhibits ubiquitin ligase activity of MDM2 for tumor suppressor p53. *EMBO J.* **1999**, *18*, 22–27. [CrossRef] [PubMed]

53. Mirzayans, R.; Andrais, B.; Scott, A.; Wang, Y.W.; Weiss, R.H.; Murray, D. Spontaneous γH2AX foci in human solid tumor-derived cell lines in relation to p21[WAF1] and WIP1 expression. *Int. J. Mol. Sci.* **2015**, *16*, 11609–11628. [CrossRef] [PubMed]

54. Sohn, D.; Essmann, F.; Schulze-Osthoff, K.; Jänicke, R.U. p21 blocks irradiation-induced apoptosis downstream of mitochondria by inhibition of cyclin-dependent kinase-mediated caspase-9 activation. *Cancer Res.* **2006**, *66*, 11254–11262. [CrossRef] [PubMed]

55. Roninson, I.B. Oncogenic functions of tumour suppressor p21 (Waf1/Cip1/Sdi1): Association with cell senescence and tumour-promoting activities of stromal fibroblasts. *Cancer Lett.* **2002**, *179*, 1–14. [CrossRef]

56. Dotto, G.P. p21$^{WAF1/Cip1}$: More than a break to the cell cycle? *Biochim. Biophys. Acta* **2000**, *1471*, M43–M56. [CrossRef]

57. Pang, L.Y.; Scott, M.; Hayward, R.L.; Mohammed, H.; Whitelaw, C.B.; Smith, G.C.; Hupp, T.R. p21^{WAF1} is component of a positive feedback loop that maintains the p53 transcriptional program. *Cell Cycle* **2011**, *10*, 932–950. [CrossRef] [PubMed]

58. Hill, R.; Leidal, A.M.; Madureira, P.A.; Gillis, L.D.; Waisman, D.M.; Chiu, A.; Lee, P.W. Chromium-mediated apoptosis: Involvement of DNA-dependent protein kinase (DNA-PK) and differential induction of p53 target genes. *DNA Repair* **2008**, *7*, 1484–1499. [CrossRef] [PubMed]

59. Broude, E.V.; Demidenko, Z.N.; Vivo, C.; Swift, M.E.; Davis, B.M.; Blagosklonny, M.V.; Roninson, I.B. p21 (CDKN1A) is a negative regulator of p53 stability. *Cell Cycle* **2007**, *6*, 1468–1471. [CrossRef] [PubMed]

60. Murray, D.; Mirzayans, R. Role of therapy-induced cellular senescence in tumor cells and its modification in radiotherapy; the good, the bad and the ugly. *J. Nucl. Med. Radiat. Ther.* **2013**, *6*, 018.

61. Crescenzi, E.; Palumbo, G.; de Boer, J.; Brady, H.J. Ataxia telangiectasia mutated and p21^{CIP1} modulate cell survival of drug-induced senescent tumor cells: Implications for chemotherapy. *Clin. Cancer Res.* **2008**, *14*, 1877–1887. [CrossRef] [PubMed]

62. Samuel, T.; Weber, H.O.; Rauch, P.; Verdoodt, B.; Eppe, J.T.; McShea, A.; Hermeking, H.; Funk, J.O. The G2/M regulator 14-3-3δ prevents apoptosis through sequestration of Bax. *J. Biol. Chem.* **2001**, *276*, 45201–45206. [CrossRef] [PubMed]

63. Nomura, M.; Shimizu, S.; Sugiyama, T.; Narita, M.; Ito, T.; Matsuda, H.; Tsujimoto, Y. 14-3-3δ interacts directly with and negatively regulates pro-apoptotic Bax. *J. Biol. Chem.* **2003**, *278*, 2058–2065. [CrossRef] [PubMed]

64. Datta, S.R.; Katsov, A.; Hu, L.; Petros, A.; Fesik, S.W.; Yaffe, M.B.; Greenberg, M.E. 14-3-3 proteins and survival kinases cooperate to inactivate BAD by BH3 domain phosphorylation. *Mol. Cell* **2000**, *6*, 41–51. [CrossRef]

65. Chiang, C.W.; Harris, G.; Ellig, C.; Masters, S.C.; Subramanian, R.; Shenolikar, S.; Wadzinski, B.E.; Yang, E. Protein phosphatase 2A activates the proapoptotic function of BAD in interleukin-3-dependent lymphoid cells by a mechanism requiring 14-3-3 dissociation. *Blood* **2001**, *97*, 1289–1297. [CrossRef] [PubMed]

66. Subramanian, R.R.; Masters, S.C.; Zhang, H.; Fu, H. Functional conservation of 14-3-3 isoforms in inhibiting bad-induced apoptosis. *Exp. Cell Res.* **2001**, *271*, 142–151. [CrossRef] [PubMed]

67. Ichim, G.; Tait, S.W.G. A fate worse than death: Apoptosis as an oncogenic process. *Nat. Rev. Cancer* **2016**, *16*, 539–548. [CrossRef] [PubMed]

68. Li, F.; He, Z.; Shen, J.; Huang, Q.; Li, W.; Liu, X.; He, Y.; Wolf, F.; Li, C.Y. Apoptotic caspases regulate induction of iPSCs from human fibroblasts. *Cell Stem Cell* **2010**, *7*, 508–520. [CrossRef] [PubMed]

69. Li, F.; Huang, Q.; Chen, J.; Peng, Y.; Roop, D.R.; Bedford, J.S.; Li, C.Y. Apoptotic cells activate the "phoenix rising" pathway to promote wound healing and tissue regeneration. *Sci. Signal.* **2010**, *3*, ra13. [CrossRef] [PubMed]

70. Huang, Q.; Li, F.; Liu, X.; Li, W.; Shi, W.; Liu, F.F.; O'Sullivan, B.; He, Z.; Peng, Y.; Tan, A.C.; et al. Caspase 3-mediated stimulation of tumor cell repopulation during cancer radiotherapy. *Nat. Med.* **2011**, *17*, 860–866. [CrossRef] [PubMed]

71. Boland, K.; Flanagan, L.; Prehn, J.H. Paracrine control of tissue regeneration and cell proliferation by caspase-3. *Cell Death Dis.* **2013**, *4*, e725. [CrossRef] [PubMed]

72. Mao, P.; Smith, L.; Xie, W.; Wang, M. Dying endothelial cells stimulate proliferation of malignant glioma cells via a caspase 3-mediated pathway. *Oncol. Lett.* **2013**, *5*, 1615–1620. [PubMed]

73. Liu, Y.R.; Sun, B.; Zhao, X.L.; Gu, Q.; Liu, Z.Y.; Dong, X.Y.; Che, N.; Mo, J. Basal caspase-3 activity promotes migration, invasion, and vasculogenic mimicry formation of melanoma cells. *Melanoma Res.* **2013**, *23*, 243–253. [CrossRef] [PubMed]

74. Donato, A.L.; Huang, Q.; Liu, X.; Li, F.; Zimmerman, M.A.; Li, C.Y. Caspase 3 promotes surviving melanoma tumor cell growth after cytotoxic therapy. *J. Invest. Dermatol.* **2014**, *134*, 1686–1692. [CrossRef] [PubMed]

75. Cheng, J.; Tian, L.; Ma, J.; Gong, Y.; Zhang, Z.; Chen, Z.; Xu, B.; Xiong, H.; Li, H.; Huang, Q. Dying tumor cells stimulate proliferation of living tumor cells via caspase-dependent protein kinase C-δ activation in pancreatic ductal adenocarcinoma. *Mol. Oncol.* **2015**, *9*, 105–114. [CrossRef] [PubMed]

76. Liu, X.; He, Y.; Li, F.; Huang, Q.; Kato, T.A.; Hall, R.P.; Li, C.Y. Caspase-3 promotes genetic instability and carcinogenesis. *Mol. Cell* **2015**, *58*, 284–296. [CrossRef] [PubMed]

77. Feng, X.; Yu, Y.; He, S.; Cheng, J.; Gong, Y.; Zhang, Z.; Yang, X.; Xu, B.; Liu, X.; Li, C.Y.; et al. Dying glioma cells establish a proangiogenic microenvironment through a caspase 3 dependent mechanism. *Cancer Lett.* **2017**, *385*, 12–20. [CrossRef] [PubMed]

78. Kroemer, G.; Galluzzi, L.; Vandenabeele, P.; Abrams, J.; Alnemri, E.S.; Baehrecke, E.H.; Blagosklonny, M.V.; El-Deiry, W.S.; Golstein, P.; Green, D.R.; et al. Classification of cell death: Recommendations of the Nomenclature Committee on Cell Death 2009. *Cell Death Differ.* **2009**, *16*, 3–11. [CrossRef] [PubMed]

79. Berndtsson, M.; Hägg, M.; Panaretakis, T.; Havelka, A.M.; Shoshan, M.C.; Linder, S. Acute apoptosis by cisplatin requires induction of reactive oxygen species but is not associated with damage to nuclear DNA. *Int. J. Cancer* **2007**, *120*, 175–180. [CrossRef] [PubMed]

80. Lee, S.L.; Hong, S.W.; Shin, J.S.; Kim, J.S.; Ko, S.G.; Hong, N.J.; Kim, D.J.; Lee, W.J.; Jin, D.H.; Lee, M.S. p34SEI-1 inhibits doxorubicin-induced senescence through a pathway mediated by protein kinase C-δ and c-Jun-NH2-kinase 1 activation in human breast cancer MCF7 cells. *Mol. Cancer Res.* **2009**, *7*, 1845–1853. [CrossRef] [PubMed]

81. Sliwinska, M.A.; Mosieniak, G.; Wolanin, K.; Babik, A.; Piwocka, K.; Magalska, A.; Szczepanowska, J.; Fronk, J.; Sikora, E. Induction of senescence with doxorubicin leads to increased genomic instability of HCT116 cells. *Mech. Ageing Dev.* **2009**, *130*, 24–32. [CrossRef] [PubMed]

82. Puig, P.E.; Guilly, M.N.; Bouchot, A.; Droin, N.; Cathelin, D.; Bouyer, F.; Favier, L.; Ghiringhelli, F.; Kroemer, G.; Solary, E.; et al. Tumor cells can escape DNA-damaging cisplatin through DNA endoreduplication and reversible polyploidy. *Cell Biol. Int.* **2008**, *32*, 1031–1043. [CrossRef] [PubMed]

83. Zhang, S.; Zhang, D.; Yang, Z.; Zhang, X. Tumor budding, micropapillary pattern, and polyploidy giant cancer cells in colorectal cancer: Current status and future prospects. *Stem Cells Int.* **2016**, *2016*, 4810734. [CrossRef] [PubMed]

84. Niu, N.; Zhang, J.; Zhang, N.; Mercado-Uribe, I.; Tao, F.; Han, Z.; Pathak, S.; Multani, A.S.; Kuang, J.; Yao, J.; et al. Linking genomic reorganization to tumor initiation via the giant cell cycle. *Oncogenesis* **2016**, *5*, e281. [CrossRef] [PubMed]

85. Mo, J.; Sun, B.; Zhao, X.; Gu, Q.; Dong, X.; Liu, Z.; Ma, Y.; Zhao, N.; Tang, R.; Liu, Y.; et al. Hypoxia-induced senescence contributes to the regulation of microenvironment in melanomas. *Pathol. Res. Pract.* **2013**, *209*, 640–647. [CrossRef] [PubMed]

86. Wang, W.; Wang, D.; Li, H. Initiation of premature senescence by BCL-2 in hypoxic condition. *Int. J. Clin. Exp. Pathol.* **2014**, *7*, 2446–2453. [PubMed]

87. Fei, F.; Zhang, D.; Yang, Z.; Wang, S.; Wang, X.; Wu, Z.; Wu, Q.; Zhang, S. The number of polyploid giant cancer cells and epithelial-mesenchymal transition-related proteins are associated with invasion and metastasis in human breast cancer. *J. Exp. Clin. Cancer Res.* **2015**, *34*, 158. [CrossRef] [PubMed]

88. Lv, H.; Shi, Y.; Zhang, L.; Zhang, D.; Liu, G.; Yang, Z.; Li, Y.; Fei, F.; Zhang, S. Polyploid giant cancer cells with budding and the expression of cyclin E, S-phase kinase-associated protein 2, stathmin associated with the grading and metastasis in serous ovarian tumor. *BMC Cancer* **2014**, *14*, 576. [CrossRef] [PubMed]

89. Zhang, S.; Mercado-Uribe, I.; Hanash, S.; Liu, J. iTRAQ-based proteomic analysis of polyploid giant cancer cells and budding progeny cells reveals several distinct pathways for ovarian cancer development. *PLoS ONE* **2013**, *8*, e80120. [CrossRef] [PubMed]

90. Zhivotovsky, B.; Kroemer, G. Apoptosis and genomic instability. *Nat. Rev. Mol. Cell Biol.* **2004**, *5*, 752–762. [CrossRef] [PubMed]

91. Coppé, J.P.; Desprez, P.Y.; Krtolica, A.; Campisi, J. The senescence-associated secretory phenotype: The dark side of tumor suppression. *Annu. Rev. Pathol.* **2010**, *5*, 99–118. [CrossRef] [PubMed]

92. Davalos, A.R.; Coppé, J.P.; Campisi, J.; Desprez, P.Y. Senescent cells as a source of inflammatory factors for tumor progression. *Cancer Metastasis Rev.* **2010**, *29*, 273–283. [CrossRef] [PubMed]

93. Sikora, E.; Mosieniak, G.; Sliwinska, M.A. Morphological and functional characteristic of senescent cancer cells. *Curr. Drug Targets* **2016**, *17*, 377–387. [CrossRef] [PubMed]

94. Castro-Vega, L.J.; Jouravleva, K.; Ortiz-Montero, P.; Liu, W.Y.; Galeano, J.L.; Romero, M.; Popova, T.; Bacchetti, S.; Vernot, J.P.; Londoño-Vallejo, A. The senescent microenvironment promotes the emergence of heterogeneous cancer stem-like cells. *Carcinogenesis* **2015**, *36*, 1180–1192. [CrossRef] [PubMed]

95. Cantor, D.J.; David, G. SIN3B, the SASP, and pancreatic cancer. *Mol. Cell. Oncol.* **2014**, *1*, e969167. [CrossRef] [PubMed]

96. Suzuki, M.; Boothman, D.A. Stress-induced premature senescence (SIPS)—Influence of SIPS on radiotherapy. *J. Radiat. Res.* **2008**, *49*, 105–112. [CrossRef] [PubMed]

97. Muscat, A.; Popovski, D.; Jayasekara, W.S.; Rossello, F.J.; Ferguson, M.; Marini, K.D.; Alamgeer, M.; Algar, E.M.; Downie, P.; Watkins, D.N.; et al. Low-dose histone deacetylase inhibitor treatment leads to tumor growth arrest and multi-lineage differentiation of malignant rhabdoid tumors. *Clin. Cancer Res.* **2016**, *22*, 3560–3570. [CrossRef] [PubMed]

98. Foerster, F.; Chen, T.; Altmann, K.H.; Vollmar, A.M. Actin-binding doliculide causes premature senescence in p53 wild type cells. *Bioorg. Med. Chem.* **2016**, *24*, 123–129. [CrossRef] [PubMed]

99. Chen, W.S.; Yu, Y.C.; Lee, Y.J.; Chen, J.H.; Hsu, H.Y.; Chiu, S.J. Depletion of securin induces senescence after irradiation and enhances radiosensitivity in human cancer cells regardless of functional p53 expression. *Int. J. Radiat. Oncol. Biol. Phys.* **2010**, *77*, 566–574. [CrossRef] [PubMed]

100. Fitzgerald, A.L.; Osman, A.A.; Xie, T.X.; Patel, A.; Skinner, H.; Sandulache, V.; Myers, J.N. Reactive oxygen species and p21$^{Waf1/Cip1}$ are both essential for p53-mediated senescence of head and neck cancer cells. *Cell Death Dis.* **2015**, *6*, e1678. [CrossRef] [PubMed]

101. Maier, P.; Hartmann, L.; Wenz, F.; Herskind, C. Cellular pathways in response to ionizing radiation and their targetability for tumor radiosensitization. *Int. J. Mol. Sci.* **2016**, *17*, 102. [CrossRef] [PubMed]

102. Leikam, C.; Hufnagel, A.L.; Otto, C.; Murphy, D.J.; Mühling, B.; Kneitz, S.; Nanda, I.; Schmid, M.; Wagner, T.U.; Haferkamp, S.; et al. In vitro evidence for senescent multinucleated melanocytes as a source for tumor-initiating cells. *Cell Death Dis.* **2015**, *6*, e1711. [CrossRef] [PubMed]

103. Wang, Q.; Wu, P.C.; Dong, D.Z.; Ivanova, I.; Chu, E.; Zeliadt, S.; Vesselle, H.; Wu, D.Y. Polyploidy road to therapy-induced cellular senescence and escape. *Int. J. Cancer* **2013**, *132*, 1505–1515. [CrossRef] [PubMed]

104. Puck, T.T.; Marcus, P.I. Action of X-rays on mammalian cells. *J. Exp. Med.* **1956**, *103*, 653–666. [CrossRef] [PubMed]

105. Puck, T.T.; Marcus, P.I. A rapid method for viable cell titration and clonal production with HeLa cells in tissue culture: The use of X-irradiated cells to supply conditioning factors. *Proc. Natl. Acad. Sci. USA* **1955**, *41*, 432–437. [CrossRef] [PubMed]

106. Puck, T.T.; Marcus, P.I.; Cieciura, S.J. Clonal growth of mammalian cells in vitro; growth characteristics of colonies from single HeLa cells with and without a feeder layer. *J. Exp. Med.* **1956**, *103*, 273–283. [CrossRef] [PubMed]

107. Narisawa-Saito, M.; Kiyono, T. Basic mechanisms of high-risk human papillomavirus-induced carcinogenesis: Roles of E6 and E7 proteins. *Cancer Sci.* **2007**, *98*, 1505–1511. [CrossRef] [PubMed]

108. Mirzayans, R.; Andrais, B.; Scott, A.; Wang, Y.W.; Kumar, P.; Murray, D. Multinucleated giant cancer cells produced in response to ionizing radiation retain viability and replicate their genome. *Int. J. Mol. Sci.* **2017**, *18*, 360. [CrossRef] [PubMed]

109. Kizaka-Kondoh, S.; Inoue, M.; Harada, H.; Hiraoka, M. Tumor hypoxia: A target for selective cancer therapy. *Cancer Sci.* **2003**, *94*, 1021–1028. [CrossRef] [PubMed]

110. Muz, B.; de la Puente, P.; Azab, F.; Azab, A.K. The role of hypoxia in cancer progression, angiogenesis, metastasis, and resistance to therapy. *Hypoxia* **2015**, *3*, 83–92. [CrossRef] [PubMed]

111. Aragonés, J.; Fraisl, P.; Baes, M.; Carmeliet, P. Oxygen sensors at the crossroad of metabolism. *Cell Metab.* **2009**, *9*, 11–22. [CrossRef] [PubMed]

112. Ho, V.T.; Bunn, H.F. Effects of transition metals on the expression of the erythropoietin gene: Further evidence that the oxygen sensor is a heme protein. *Biochem. Biophys. Res. Commun.* **1996**, *223*, 175–180. [CrossRef] [PubMed]

113. Piret, J.P.; Mottet, D.; Raes, M.; Michiels, C. CoCl$_2$, a chemical inducer of hypoxia-inducible factor-1, and hypoxia reduce apoptotic cell death in hepatoma cell line HepG2. *Ann. N. Y. Acad. Sci.* **2002**, *973*, 443–447. [CrossRef] [PubMed]

114. Piret, J.P.; Lecocq, C.; Toffoli, S.; Ninane, N.; Raes, M.; Michiels, C. Hypoxia and CoCl$_2$ protect HepG2 cells against serum deprivation- and t-BHP-induced apoptosis: A possible anti-apoptotic role for HIF-1. *Exp. Cell Res.* **2004**, *295*, 340–349. [CrossRef] [PubMed]

115. Weihua, Z.; Lin, Q.; Ramoth, A.J.; Fan, D.; Fidler, I.J. Formation of solid tumors by a single multinucleated cancer cell. *Cancer* **2011**, *117*, 4092–4099. [CrossRef] [PubMed]

116. Zhang, S.; Mercado-Uribe, I.; Xing, Z.; Sun, B.; Kuang, J.; Liu, J. Generation of cancer stem-like cells through the formation of polyploid giant cancer cells. *Oncogene* **2014**, *33*, 116–128. [CrossRef] [PubMed]

117. Erenpreisa, J.A.; Cragg, M.S.; Fringes, B.; Sharakhov, I.; Illidge, T.M. Release of mitotic descendants by giant cells from irradiated Burkitt's lymphoma cell lines. *Cell Biol. Int.* **2000**, *24*, 635–648. [CrossRef] [PubMed]

118. Illidge, T.M.; Cragg, M.S.; Fringes, B.; Olive, P.; Erenpresia, J.A. Polyploid giant cells provide a survival mechanism of p53 mutant cells after DNA damage. *Cell Biol. Int.* **2000**, *24*, 621–633. [CrossRef] [PubMed]

119. Lagadec, C.; Vlashi, E.; Della Donna, L.; Dekmezian, C.; Pajonk, F. Radiation-induced reprogramming of breast cancer cells. *Stem Cells* **2012**, *30*, 833–844. [CrossRef] [PubMed]

120. Erenpreisa, J.; Cragg, M.S. MOS, aneuploidy and the ploidy cycle of cancer cells. *Oncogene* **2010**, *29*, 5447–5451. [CrossRef] [PubMed]

121. Sundaram, M.; Guernsey, D.L.; Rajaraman, M.M.; Rajaraman, R. Neosis: A novel type of cell division in cancer. *Cancer Biol. Ther.* **2004**, *3*, 207–218. [CrossRef] [PubMed]

122. Walen, K.H. Spontaneous cell transformation: Karyoplasts derived from multinucleated cells produce new cell growth in senescent human epithelial cell cultures. *In Vitro Cell Dev. Biol. Anim.* **2004**, *40*, 150–158. [CrossRef]

123. Navolanic, P.M.; Akula, S.M.; McCubrey, J.A. Neosis and its potential role in cancer development and chemoresistance. *Cancer Biol. Ther.* **2004**, *3*, 219–320. [CrossRef] [PubMed]

124. Rajaraman, R.; Rajaraman, M.M.; Rajaraman, S.R.; Guernsey, R.L. Neosis—A paradigm of self-renewal in cancer. *Cell Biol. Int.* **2005**, *29*, 1084–1097. [CrossRef] [PubMed]

125. Walen, K.H. Budded karyoplasts from multinucleated fibroblast cells contain centrosomes and change their morphology to mitotic cells. *Cell Biol. Int.* **2005**, *29*, 1057–1065. [CrossRef] [PubMed]

126. Rajaraman, R.; Guernsey, D.L.; Rajaraman, M.M.; Rajaraman, S.R. Stem cells, senescence, neosis and self-renewal in cancer. *Cell Biol. Int.* **2006**, *29*, 1084–1097. [CrossRef] [PubMed]

127. Jiang, Q.; Zhang, Q.; Wang, S.; Xie, S.; Fang, W.; Liu, Z.; Liu, J.; Yao, K. A fraction of CD133+ CNE2 cells is made of giant cancer cells with morphological evidence of asymmetric mitosis. *J. Cancer* **2015**, *6*, 1236–1244. [CrossRef] [PubMed]

128. Esmatabadi, M.J.; Bakhshinejad, B.; Motlagh, F.M.; Babashah, S.; Sadeghizadeh, M. Therapeutic resistance and cancer recurrence mechanisms: Unfolding the story of tumour coming back. *J. Biosci.* **2016**, *41*, 497–506. [CrossRef] [PubMed]

129. Mosieniak, G.; Sliwinska, M.A.; Alster, O.; Strzeszewska, A.; Sunderland, P.; Piechota, M.; Was, H.; Sikora, E. Polyploidy formation in doxorubicin-treated cancer cells can favor escape from senescence. *Neoplasia* **2015**, *17*, 882–893. [CrossRef] [PubMed]

130. Shah, O.J.; Lin, X.; Li, L.; Huang, X.; Li, J.; Anderson, M.G.; Tang, H.; Rodriguez, L.E.; Warder, S.E.; McLoughlin, S.; et al. BCL-XL represents a druggable molecular vulnerability during aurora B inhibitor-mediated polyploidization. *Proc. Natl. Acad. Sci. USA* **2010**, *107*, 12634–12639. [CrossRef] [PubMed]

131. Martin, S.K.; Pu, H.; Penticuff, J.C.; Cao, Z.; Horbinski, C.; Kyprianou, N. Multinucleation and mesenchymal-to-epithelial transition alleviate resistance to combined cabazitaxel and antiandrogen therapy in advanced prostate cancer. *Cancer Res.* **2016**, *76*, 912–926. [CrossRef] [PubMed]

132. Nik-Zainal, S.; Alexandrov, L.B.; Wedge, D.C.; van Loo, P.; Greenman, C.D.; Raine, K.; Jones, D.; Hinton, J.; Marshall, J.; Stebbings, L.A.; et al. Mutational processes molding the genomes of 21 breast cancers. *Cell* **2012**, *149*, 979–993. [CrossRef] [PubMed]

133. Alexandrov, L.B.; Nik-Zainal, S.; Wedge, D.C.; Aparicio, S.A.; Behjati, S.; Biankin, A.V.; Bignell, G.R.; Bolli, N.; Borg, A.; Børresen-Dale, A.L.; et al. Signatures of mutational processes in human cancer. *Nature* **2013**, *500*, 415–421. [CrossRef] [PubMed]

134. Alexandrov, L.B.; Nik-Zainal, S.; Wedge, D.C.; Campbell, P.J.; Stratton, M.R. Deciphering signatures of mutational processes operative in human cancer. *Cell Rep.* **2013**, *3*, 246–259. [CrossRef] [PubMed]

135. Helleday, T.; Eshtad, S.; Nik-Zainal, S. Mechanisms underlying mutational signatures in human cancers. *Nat. Rev. Genet.* **2014**, *15*, 585–598. [CrossRef] [PubMed]

136. Alexandrov, L.B.; Jones, P.H.; Wedge, D.C.; Sale, J.E.; Campbell, P.J.; Nik-Zainal, S.; Stratton, M.R. Clock-like mutational processes in human somatic cells. *Nat. Genet.* **2015**, *47*, 1402–1407. [CrossRef] [PubMed]

137. Fox, E.J.; Salk, J.J.; Loeb, L.A. Exploring the implications of distinct mutational signatures and mutation rates in aging and cancer. *Genome Med.* **2016**, *8*, 30. [CrossRef] [PubMed]

138. Behjati, S.; Gundem, G.; Wedge, D.C.; Roberts, N.D.; Tarpey, P.S.; Cooke, S.L.; van Loo, P.; Alexandrov, L.B.; Ramakrishna, M.; Davies, H.; et al. Mutational signatures of ionizing radiation in second malignancies. *Nat. Commun.* **2016**, *7*, 12605. [CrossRef] [PubMed]

International Journal of
Molecular Sciences

MDPI

Article

A Case-Control Study of the Genetic Variability in Reactive Oxygen Species—Metabolizing Enzymes in Melanoma Risk

Tze-An Yuan [1] , Vandy Yourk [2], Ali Farhat [3], Argyrios Ziogas [4], Frank L. Meyskens [1,4,5],
Hoda Anton-Culver [4] and Feng Liu-Smith [4,5,*]

[1] Program in Public Health, University of California Irvine, Irvine, CA 92697, USA; tzeany@uci.edu (T.-A.Y.);
    flmeyske@uci.edu (F.L.M.)
[2] Department of Neurobiology and Behavior, School of Biological Sciences, University of California Irvine,
    Irvine, CA 92697, USA; vandyyourk@gmail.com
[3] Department of Biomedical Engineering, The Henry Samueli School of Engineering,
    University of California Irvine, Irvine, CA 92697, USA; farhatam@uci.edu
[4] Department of Epidemiology, School of Medicine, University of California, Irvine, CA 92697, USA;
    aziogas@uci.edu (A.Z.); hantoncu@uci.edu (H.A.-C.)
[5] Chao Family Comprehensive Cancer Center, Irvine, CA 92697, USA
*   Correspondence: liufe@uci.edu; Tel.: +1-949-824-2778

Received: 18 December 2017; Accepted: 12 January 2018; Published: 14 January 2018

Abstract: Recent studies have shown that ultraviolet (UV)-induced chemiexcitation of melanin fragments leads to DNA damage; and chemiexcitation of melanin fragments requires reactive oxygen species (ROS), as ROS excite an electron in the melanin fragments. In addition, ROS also cause DNA damages on their own. We hypothesized that ROS producing and metabolizing enzymes were major contributors in UV-driven melanomas. In this case-control study of 349 participants, we genotyped 23 prioritized single nucleotide polymorphisms (SNPs) in nicotinamide adenine dinucleotide phosphate (NADPH) oxidases 1 and 4 (*NOX1* and *NOX4*, respectively), *CYBA*, *RAC1*, superoxide dismutases (*SOD1*, *SOD2*, and *SOD3*) and catalase (*CAT*), and analyzed their associated melanoma risk. Five SNPs, namely rs1049255 (*CYBA*), rs4673 (*CYBA*), rs10951982 (*RAC1*), rs8031 (*SOD2*), and rs2536512 (*SOD3*), exhibited significant genotypic frequency differences between melanoma cases and healthy controls. In simple logistic regression, *RAC1* rs10951982 (odds ratio (OR) 8.98, 95% confidence interval (CI): 5.08 to 16.44; $p < 0.001$) reached universal significance ($p = 0.002$) and the minor alleles were associated with increased risk of melanoma. In contrast, minor alleles in *SOD2* rs8031 (OR 0.16, 95% CI: 0.06 to 0.39; $p < 0.001$) and *SOD3* rs2536512 (OR 0.08, 95% CI: 0.01 to 0.31; $p = 0.001$) were associated with reduced risk of melanoma. In multivariate logistic regression, *RAC1* rs10951982 (OR 6.15, 95% CI: 2.98 to 13.41; $p < 0.001$) remained significantly associated with increased risk of melanoma. Our results highlighted the importance of *RAC1*, *SOD2*, and *SOD3* variants in the risk of melanoma.

Keywords: melanoma; reactive oxygen species; ROS; NADPH oxidase; single nucleotide polymorphisms; SNP; superoxide dismutase; SOD2; SOD3; RAC1

---

1. Introduction

Ultraviolet (UV) rays are capable of inducing melanin production in melanocytes and promoting melanin transportation to the outermost layer of the skin—the keratinocytes. These melanins form a cap over the nucleus of both cell types and protect DNA from direct energy destruction [1,2]. On the other hand, UV rays are also able to initiate nicotinamide adenine dinucleotide phosphate (NADPH) oxidase (NOX) dominated reactive oxygen species (ROS) production and chemiexcitation of melanin fragments that affect DNA stability in melanocytes [3–5]. The oncogenic characteristics of UV-induced ROS signaling have not yet been fully elucidated, particularly in the transformation of melanocytes to melanomas.

Recent understanding of melanoma photobiology has implied the etiological role of NOX enzymes, particularly NOX1 and NOX4 [6–8]. NOX enzymes produce superoxide and/or hydrogen peroxide when coupled with CYBA (p22phox) membrane protein [9]. RAC1, a newly defined melanoma oncogene [10], is shown to enhance NOX1 activity [11]. The downstream ROS metabolizing enzymes, e.g., copper-zinc superoxide dismutase (Cu-ZnSOD, *SOD1*), manganese superoxide dismutase (MnSOD, *SOD2*), and extracellular superoxide dismutase (ECSOD, *SOD3*), convert superoxide to hydrogen peroxide. Catalase then transforms hydrogen peroxide to water molecules (Figure 1). The cellular locations of NOX1, RAC1, NOX4, CYBA, and SOD enzymes, and their functions in ROS production and metabolism are illustrated in Figure 1. Little is known about the comprehensive role of this entire pathway in melanoma formation. However, risk associated with these genes has been reported in various health conditions. For example, V16A variant in *SOD2* (rs4880) showed an impaired mitochondrial importing function and was associated with prostate cancer risk [12]. The rs7277748 and rs4998557 variants in *SOD1* were found to be associated with amyotrophic lateral sclerosis [13]. Variants rs2536512 and rs699473 in *SOD3* were linked to cerebral infarction [14] and brain tumor [15].

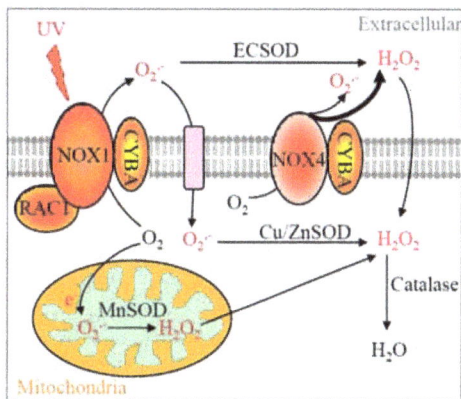

Figure 1. Diagram of the relevant reactive oxygen species (ROS) production pathway. NOX1, NOX4, CYBA, RAC1, SOD enzymes, catalase, their subcellular locations, and their functions in ROS production and metabolism are depicted in this diagram. NOX1 enzyme complex utilizes CYBA as one of its subunits and is activated by RAC1-GTPase to produce superoxide. On the other hand, NOX4 only couples with CYBA to generate hydrogen peroxide and superoxide. Of particular note, only plasma membrane NOX4 is shown in this diagram but mitochondrial or nuclear NOX4 has also been reported [16]. NOX1 is activated by UV to enhance its superoxide production, which requires the GTPase activity of RAC1. Superoxide is further metabolized into hydrogen peroxide at various subcellular locations by different SOD isozymes. Hydrogen peroxide is then converted into water molecules by catalase. Other additional redox enzymes (e.g., glutathione peroxidases, which also convert hydrogen peroxide into water) are not the focus in this study and therefore not included. Black arrows indicate the cellular movement of oxygen, ROS, and enzymatic metabolisms. A bold arrow represents a greater relative amount of ROS produced.

Although the causal network of melanoma has not yet been fully elucidated [17], UV exposure is the most tangible environmental risk factor that can be readily modified by behavioral precautions [18]. Therefore, the purpose of this study was to explore the relationship between the hypothesized photobiological pathway and risk of melanoma. Specifically, our aim was to use the candidate gene approach to discover the association of variations in the genetic profile of the redox enzymes with melanoma (Figure 1). Building upon this rationale, functional genetic variants, namely single nucleotide polymorphisms (SNPs), were identified in this study with a priori chance of being associated with the risk of melanoma based on the following criteria: (1) not a well-known somatic mutation found in tumors with an established causality; (2) presented strong associations with many other health conditions in humans; and (3) with a potential to alter normal protein function based on the nucleotide substitution. For instance, variant rs8031 in *SOD2* was found to be associated with kidney complications in subjects with Type 1 diabetes [19]. Variant rs10951982 in *RAC1* has been implied in the increased risk of hypertension [20]. Even though rs10951982 in *RAC1* has not yet been reported in ROS-related malignancies, somatic mutations of *RAC1* (e.g., *RAC1*^{P29S}) were found in 9.2% of sun-exposed melanoma tumors [21,22].

With this genetic profiling information in hand, we hope to lay a foundation to identify those individuals predisposed to UV exposure and risk of melanoma. This in turn will contribute to a better primary prevention strategy, such as earlier-life behavioral precautions. To the best of our knowledge, our work was the first to use a hypothesis-driven and pathway-based approach to study the association between genetic variations in the ROS pathway and risk of melanoma.

2. Results

2.1. Study Participants

Gender and age distributions of melanoma patients and healthy controls are listed in Table 1. In total, 177 retrieved cases and 172 recruited controls were approximately matched for age groups and gender. Overall, there are higher percentages of female patients aged 19–39 (55.4%) and 40–59 (26.5%), while, at age 60 and older, there is a higher percentage of male patients (47.9%). This may reflect the actual sex ratios of melanoma incidence at different age groups [23]. Of particular note, cases were retrieved from the international Genes, Environment, and Melanoma (GEM) study, which may not be strictly generalizable to a broader melanoma patient population.

Table 1. Characteristics of the study participants.

Study Participant	Gender		
	Male *n* (%) [1]	Female *n* (%)	Total *n* (%)
Patients (*n* = 177)			
Age (years)			
19–39	5 (5.32%)	15 (18.1%)	20 (11.3%)
40–59	44 (46.8%)	46 (55.4%)	90 (50.8%)
≥60	45 (47.9%)	22 (26.5%)	67 (37.9%)
Controls (*n* = 172)			
Age (years)			
19–39	7 (7.1%)	15 (20.3%)	22 (12.8%)
40–59	45 (45.9%)	41 (55.4%)	86 (50.0%)
≥60	46 (46.9%)	18 (24.3%)	64 (37.2%)

[1] Percentage may not add up to 100% due to rounding.

SNP candidates and their currently known disease associations are listed in Table 2. Whole genome DNA amplification was successfully carried out in 322 study participants including 170 (96%) melanoma patients and 152 (88.4%) healthy controls (Figure 2). However, for each SNP, there were different number of failed genotyping samples due to poor PCR reaction, and the overall successful genotyping rates were between 66.4% and 98.7% in the controls, and between 78.8% and 99.4% in the cases. SNPs with genotyping rate less than 75% on either arm (case or control group) of the participants were thus excluded from further analyses (SNPs rs13306296 and rs585197 were excluded, Table 3). Ultimately, 161–169 melanoma patients, and 116–150 healthy controls remained to be further analyzed (Figure 2).

Table 2. Twenty-three SNP candidates.

Gene	SNP	location	dbSNP ID	Disease Association	Reference
NOX1	944G>A	R315H	rs2071756	Diabetes	[24]
	1284G>A	D360N	rs34688635	Severe pancolitis	[25]
NOX4	T>C	Intron	rs11018628	Increased plasma homocysteine level (risk in cardiovascular diseases)	[26]
	−114 C>T	5′UTR	rs585197	Decreased risk of hepatic-pulmonary syndrome	[27]
	C>T	Intron	rs2164521	Decreased risk of hepatic-pulmonary syndrome	[27]
CYBA^{p22phox}	−930A>G	Promoter	rs9932581	Modulates CYBA promoter activity	[28,29]
	242C>T	Y72H	rs4673	Decreased NOX activity; protective role in coronary heart disease	[30–32]
	−675A>T	Promoter	rs13306296	Related to hypertension	[33]
	C>G	Intron	rs3180279	Associated with non-Hodgkin lymphoma prognosis	[13]
	640A>G	3′UTR	rs1049255	Associated with coronary heart disease	[34]
RAC1	G>A	Intron	rs10951982	Risks in ulcerative colitis, hypertension, inflammatory bowel disease, end-stage renal disease	[20]
	T>C	Exon	rs4720672	Risks in inflammatory bowel disease, ulcerative colitis	[35,36]
	C>T	Intron	rs836478	Hypertension risk factor	[20]
SOD1	A>G	5′UTR	rs7277748	Familial amyotrophic lateral sclerosis	[37]
	7958G>A	Intron	rs4998557	Caused amyotrophic lateral sclerosis	[37–39]
SOD2	399T>C	Ile58Thr	rs1141718	Reduced enzyme activity	[40]
	T>C,A,G	V16A,D,G	rs4880	Mitochondrial importing, diabetes and prostate cancer	[12,41–43]
	T>A	Intron	rs8031	Oxidative stress	[44]
	C>A	Intron	rs2758330	Protective role in prostate cancer	[45]
SOD3	C>T	Promoter	rs699473	Brain tumor	[15]
	G>A	A377T	rs2536512	Cerebral infarction	[14]
CAT	−262C>T	5′UTR	rs1049982	Down-regulated transcription upon oxidative stimulation	[46,47]
	C>T	5′UTR	rs1001179	Brain tumor	[15]

Table 3. Descriptive statistics of the 23 SNP candidates.

SNP[1]	Gene	Genotyping Rate[2]		Minor Allele Frequency (MAF)		Association (p-Value)[4]				HWE[5] (p-Value)	dbSNP MAF[6]
		Cases (n = 170)[3]	Controls (n = 152)[3]	Cases	Controls	Genotypic	Allelic	Recessive	Dominant		
rs10951982	RAC1	96.5%	83.0%	47.3%	23.6%	<0.001	<0.001	0.333	<0.001	0.459	16.6%
rs11018628	NOX4	99.4%	94.1%	50.0%	33.9%	<0.001	<0.001	0.458	<0.001	<0.001	16.7%
rs8031	SOD2	95.9%	87.5%	38.3%	49.6%	<0.001	0.008	<0.001	0.576	0.605	36.7%
rs2536512	SOD3	97.1%	75.8%	27.6%	37.5%	<0.001	0.016	<0.001	0.168	0.431	40.1%
rs4720672	RAC1	96.5%	92.2%	23.8%	17.6%	0.009	0.076	0.582	0.014	0.043	12.5%
rs4673	CYBA	98.8%	93.5%	36.6%	30.6%	0.014	0.132	0.561	0.013	0.238	33.6%
rs3180279	CYBA	98.8%	94.7%	48.2%	48.6%	0.022	0.985	0.154	0.127	0.030	44.5%
rs1049255	CYBA	97.1%	90.3%	48.2%	37.1%	0.027	0.007	0.034	0.041	0.719	46.9%
rs1001179	CAT	96.5%	95.4%	16.8%	24.1%	0.062	0.030	0.610	0.025	1.000	12.6%
rs4880	SOD2	98.2%	93.5%	53.3%	55.2%	0.074	0.685	0.133	0.402	0.175	41.1%
rs9932581	CYBA	99.4%	90.9%	45.9%	42.5%	0.357	0.450	1.000	0.212	0.168	41.7%
rs699473	SOD3	97.1%	90.9%	63.0%	59.3%	0.461	0.388	0.745	0.280	0.861	44.1%
rs7277748	SOD1	95.9%	85.6%	4.0%	2.7%	0.486	0.518	N/A	0.511	1.000	3.9%
rs2164521	NOX4	98.2%	86.8%	9.3%	10.6%	0.646	0.688	0.442	0.792	1.000	26.2%
rs836478	RAC1	97.1%	94.8%	51.8%	50.0%	0.740	0.710	0.551	1.000	1.000	30.9%
rs4998557	SOD1	98.8%	94.1%	10.4%	11.5%	0.749	0.774	1.000	0.675	0.695	32.9%
rs1049982	CAT	94.7%	88.8%	35.7%	36.3%	0.962	0.951	1.000	0.904	0.710	47.1%
rs34688635	NOX1	97.6%	90.2%	1.5%	1.1%	1.000	0.734	1.000	1.000	1.000	0.5%
rs1141718	SOD2	96.5%	98.7%	49.4%	49.0%	1.000	0.988	1.000	1.000	<0.001	N/A
rs2758330	SOD2	96.5%	96.1%	20.7%	20.6%	1.000	1.000	1.000	1.000	0.010	26.5%
rs2071756	NOX1	98.2%	92.2%	0%	0%	N/A	1.000	N/A	N/A	1.000	0.1%
rs585197	NOX4	73.5%	70.4%	Excluded from further analysis due to low genotyping rate (≤75.0%)							
rs13306296	CYBA	78.8%	66.4%	Excluded from further analysis due to low genotyping rate (≤75.0%)							

[1] Ordered according to smallest to largest genotypic p-values; [2] Percentage of participants with SNP genotyping success; [3] Participant number = n * %; [4] Chi-square or Fisher's exact test of independence between SNP models and melanoma status (case and control); [5] Exact test for Hardy–Weinberg equilibrium (HWE) on the controls, p < 0.05 counts as evidence against HWE. HWE is a test of genotype balance in a given population; [6] Reference minor allele frequencies documented in dbSNP database. N/A: not available.

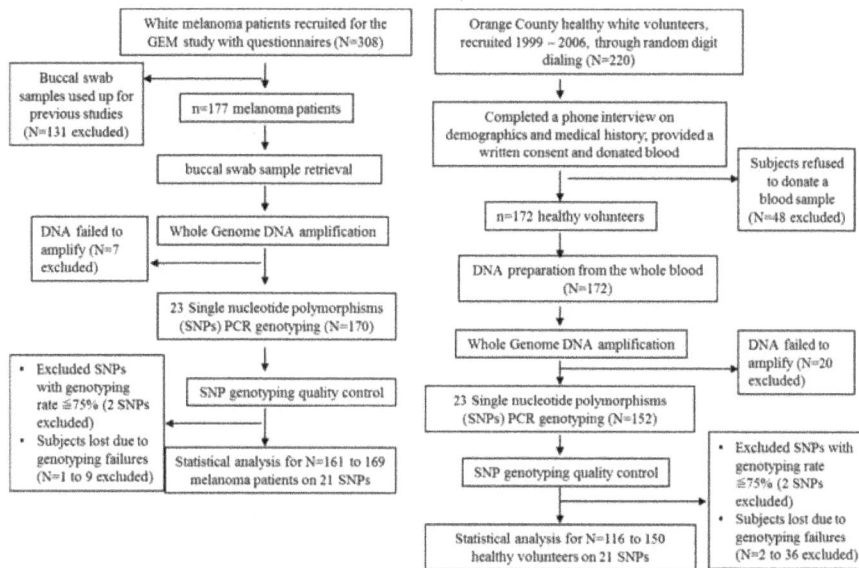

Figure 2. The inclusion and exclusion criteria of the participants in this study.

2.2. SNP Associations

Chi-square or Fisher's exact test of independence was performed to identify SNP frequency differences between melanoma patients and healthy controls under genotypic, allelic, recessive, and dominant SNP models (Table 3). An exact test of genotype counts on Hardy–Weinberg equilibrium (HWE) was conducted to identify and exclude SNPs not in genotype balance in our study sample. Under the genotypic model, five SNPs exhibited statistically significant ($p < 0.05$) frequency differences between cases and controls: rs10951982 (*RAC1*), rs8031 (*SOD2*), rs2536512 (*SOD3*), rs4673 (*CYBA*), and rs1049255 (*CYBA*) (Table 3). The allelic model only determined three of them as being significant: rs10951982 (*RAC1*), rs8031 (*SOD2*), and rs2536512 (*SOD3*). These three alleles exhibited significance in the recessive model as well. In the dominant model, rs10951982 (*RAC1*), rs4673 (*CYBA*), and rs1049255 (*CYBA*) showed significance. The rs1001179 (*CAT*) showed a significant difference between cases and controls in the dominant and recessive models but the significance disappeared in the other two models.

2.3. Bivariate Logistic Regression Analyses

The top five SNPs identified from the genotypic model without HWE violations were fitted into bivariate logistic regressions with additive, recessive, and dominant allele models, respectively. The odds ratios of melanoma risk were calculated using the homozygous major allele genotype as the reference (Table 4). Odds ratios derived from the regression models were compared to a corrected significance level at 0.00238 (0.05/21) to justify for multiple comparisons among the remaining 21 SNP candidates. Odds ratios with p-values < 0.00238 were considered having statistical significance in the results.

In the additive allele model, carrying one copy of minor allele A in rs10951982 (*RAC1*) was significantly associated with a higher risk of melanoma (OR 8.98, 95% CI: 5.08, 16.44, $p < 0.001$), as compared to those who carried homozygous minor alleles AA (OR 8.23, 95% CI: 2.73, 28.39, $p < 0.001$). Dominant allele model further showed that combined minor allele copies (GA+AA) as compared to homozygous major alleles GG exhibited the highest risk of melanoma (OR 8.91, 95% CI: 5.09, 16.19,

$p < 0.001$. A similar result was observed in rs4673 (*CYBA*), with one copy of the minor allele A exhibiting a higher risk of melanoma (OR 1.96, 95% CI: 1.23, 3.15, $p = 0.005$), and further confirmed in a dominant allele model (OR 1.84, 95% CI: 1.16, 2.92, $p = 0.010$). However, the p-values did not reach the corrected significance level of 0.00238.

Table 4. Crude associations between the top five SNPs and melanoma risk.

SNP/Model	Allele	Cases (*n* = 170) *n* (%) [1]	Controls (*n* = 152) *n* (%)	OR (95% CI)	*p*-Value [2]
rs10951982 (*RAC1*)					
	GG	21 (12.4%)	72 (47.4%)	Reference	–
Additive	GA	131 (77.1%)	50 (32.9%)	8.98 (5.08, 16.44)	<0.001
	AA	12 (7.1%)	5 (3.3%)	8.23 (2.73, 28.39)	<0.001
Recessive	GG+GA	152 (89.4%)	122 (80.3%)	Reference	–
	AA	12 (7.1%)	5 (3.3%)	1.93 (0.69, 6.19)	0.230
Dominant	GG	21 (12.4%)	72 (47.4%)	Reference	–
	GA+AA	143 (84.1%)	55 (36.2%)	8.91 (5.09, 16.19)	<0.001
rs1049255 (*CYBA*)					
	CC	47 (27.6%)	56 (36.8%)	Reference	–
Additive	CT	77 (45.3%)	63 (41.4%)	1.46 (0.88, 2.44)	0.149
	TT	41 (24.1%)	20 (13.2%)	2.44 (1.27, 4.79)	0.008
Recessive	CC+CT	124 (72.9%)	119 (78.3%)	Reference	–
	TT	41 (24.1%)	20 (13.2%)	1.97 (1.10, 3.61)	0.022
Dominant	CC	47 (27.6%)	56 (36.8%)	Reference	–
	CT+TT	118 (69.4%)	83 (54.6%)	1.69 (1.05, 2.74)	0.031
rs4673 (*CYBA*)					
	GG	53 (31.2%)	66 (43.4%)	Reference	–
Additive	GA	107 (62.9%)	68 (44.7%)	1.96 (1.23, 3.15)	0.005
	AA	8 (4.7%)	10 (6.6%)	1.00 (0.36, 2.70)	0.994
Recessive	GG+GA	160 (94.1%)	134 (88.2%)	Reference	–
	AA	8 (4.7%)	10 (6.6%)	0.67 (0.25, 1.75)	0.412
Dominant	GG	53 (31.2%)	66 (43.4%)	Reference	–
	GA+AA	115 (67.6%)	78 (51.3%)	1.84 (1.16, 2.92)	0.010
rs8031 (*SOD2*)					
	AA	45 (26.5%)	32 (21.1%)	Reference	–
Additive	AT	111 (65.3%)	70 (46.1%)	1.13 (0.65, 1.94)	0.665
	TT	7 (4.1%)	31 (20.4%)	0.16 (0.06, 0.39)	<0.001
Recessive	AA+AT	156 (91.8%)	102 (67.1%)	Reference	–
	TT	7 (4.1%)	31 (20.4%)	0.15 (0.06, 0.33)	<0.001
Dominant	AA	45 (26.5%)	32 (21.1%)	Reference	–
	AT+TT	118 (69.4%)	101 (66.4%)	0.83 (0.49, 1.40)	0.489
rs2536512 (*SOD3*)					
	GG	76 (44.7%)	43 (28.3%)	Reference	–
Additive	GA	87 (51.2%)	59 (38.8%)	0.83 (0.51, 1.37)	0.477
	AA	2 (1.2%)	14 (9.2%)	0.08 (0.01, 0.31)	0.001
Recessive	GG+GA	163 (95.9%)	102 (67.1%)	Reference	–
	AA	2 (1.2%)	14 (9.2%)	0.09 (0.01, 0.33)	0.002
Dominant	GG	76 (44.7%)	43 (28.3%)	Reference	–
	GA+AA	89 (52.4%)	73 (48.0%)	0.69 (0.42, 1.12)	0.134

[1] Participants lost due to genotyping failure; [2] *p*-value of the coefficient from the regression model. *p*-value was compared to a Bonferroni corrected significance level at 0.05/21 = 0.00238 to determine statistical significance. –: no *p*-value in the reference group.

The unadjusted odds of melanoma increased with homozygous minor allele T in rs1049255 (*CYBA*). TT exhibited an OR of 2.44 (95% CI: 1.27, 4.79, $p = 0.008$) in the additive model and an OR of 1.97 (95% CI: 1.10, 3.61, $p = 0.022$) in the recessive model. In both scenarios, *p*-values were greater than 0.00238, thus were non-significant because of the stringent Bonferroni correction for multiple comparison.

In contrast, homozygous minor allele genotypes at both rs8031 (*SOD2*) and rs2536512 (*SOD3*) exhibited significant association with a reduced risk of melanoma in the additive allele model, with

84% reduction in odds of melanoma (OR 0.16, 95% CI: 0.06, 0.39, $p < 0.001$) for rs8031 (*SOD2*), and 92% reduction in odds of melanoma (OR 0.08, 95% CI: 0.01, 0.31, $p = 0.001$) for rs2536512 (*SOD3*). Similar results were also observed in the recessive model, where an 85% reduction in the odds of melanoma (OR 0.15, 95% CI: 0.06, 0.33, $p < 0.001$) was observed for rs8031 (*SOD2*) with TT minor alleles, and a 91% reduction (OR 0.09, 95% CI: 0.01, 0.33, $p = 0.002$ with marginal significance) for rs2536512 (*SOD3*) with AA minor alleles.

2.4. Multivariate Logistic Regression Analyses

We continued to fit these top five SNPs into multivariate logistic regression under the three SNP models, controlling for major melanoma risk factors including gender, age at diagnosis, family history of melanoma, and lifetime ever-sunburned (Table 5). After adjusting for these risk factors, rs1049255 (*CYBA*), rs4673 (*CYBA*), rs8031 (*SOD2*), and rs2536512 (*SOD3*) were no longer associated with melanoma risk in all three models ($p > 0.00238$).

Table 5. Adjusted [1] associations between the top five SNPs and melanoma risk.

SNP/Model	Allele	Cases (*n* = 170) *n* (%) [2]	Controls (*n* = 152) *n* (%)	OR (95% CI)	*p*-Value [3]
rs10951982 (*RAC1*)					
Additive	GG	21 (12.4%)	72 (47.4%)	Reference	–
	GA	131 (77.1%)	50 (32.9%)	6.15 (2.98, 13.41)	<0.001
	AA	12 (7.1%)	5 (3.3%)	2.88 (0.68, 12.56)	0.149
Recessive	GG+GA	152 (89.4%)	122 (80.3%)	Reference	–
	AA	12 (7.1%)	5 (3.3%)	0.79 (0.21, 3.03)	0.719
Dominant	GG	21 (12.4%)	72 (47.4%)	Reference	–
	GA+AA	143 (84.1%)	55 (36.2%)	5.79 (2.84, 12.51)	<0.001
rs1049255 (*CYBA*)					
Additive	CC	47 (27.6%)	56 (36.8%)	Reference	–
	CT	77 (45.3%)	63 (41.4%)	1.20 (0.63, 2.30)	0.574
	TT	41 (24.1%)	20 (13.2%)	1.42 (0.61, 3.38)	0.420
Recessive	CC+CT	124 (72.9%)	119 (78.3%)	Reference	–
	TT	41 (24.1%)	20 (13.2%)	1.28 (0.59, 2.83)	0.531
Dominant	CC	47 (27.6%)	56 (36.8%)	Reference	–
	CT+TT	118 (69.4%)	83 (54.6%)	1.26 (0.69, 2.31)	0.456
rs4673 (*CYBA*)					
Additive	GG	53 (31.2%)	66 (43.4%)	Reference	–
	GA	107 (62.9%)	68 (44.7%)	2.17 (1.17, 4.07)	0.015
	AA	8 (4.7%)	10 (6.6%)	0.50 (0.11, 1.82)	0.315
Recessive	GG+GA	160 (94.1%)	134 (88.2%)	Reference	–
	AA	8 (4.7%)	10 (6.6%)	0.31 (0.07, 1.07)	0.080
Dominant	GG	53 (31.2%)	66 (43.4%)	Reference	–
	GA+AA	115 (67.6%)	78 (51.3%)	1.88 (1.03, 3.47)	0.042
rs8031 (*SOD2*)					
Additive	AA	45 (26.5%)	32 (21.1%)	Reference	–
	AT	111 (65.3%)	70 (46.1%)	1.33 (0.66, 2.65)	0.421
	TT	7 (4.1%)	31 (20.4%)	0.32 (0.09, 0.94)	0.047
Recessive	AA+AT	156 (91.8%)	102 (67.1%)	Reference	–
	TT	7 (4.1%)	31 (20.4%)	0.26 (0.08, 0.70)	0.011
Dominant	AA	45 (26.5%)	32 (21.1%)	Reference	–
	AT+TT	118 (69.4%)	101 (66.4%)	1.06 (0.54, 2.08)	0.864
rs2536512 (*SOD3*)					
Additive	GG	76 (44.7%)	43 (28.3%)	Reference	–
	GA	87 (51.2%)	59 (38.8%)	0.68 (0.35, 1.28)	0.232
	AA	2 (1.2%)	14 (9.2%)	0.26 (0.03, 1.50)	0.144
Recessive	GG+GA	163 (95.9%)	102 (67.1%)	Reference	–
	AA	2 (1.2%)	14 (9.2%)	0.33 (0.04, 1.83)	0.218
Dominant	GG	76 (44.7%)	43 (28.3%)	Reference	–
	GA+AA	89 (52.4%)	73 (48.0%)	0.65 (0.34, 1.21)	0.175

[1] Adjusted for gender, age at diagnosis/interview, family history of melanoma, and ever sunburned; [2] Participants lost due to genotyping failure; [3] *p*-value of the coefficient from the regression model. *p*-value was compared to a Bonferroni corrected significance level at 0.05/21 = 0.00238 to determine statistical significance. –: no *p*-value in the reference group.

Consistent with what we have found in Table 4, the most significant genotype was heterozygous GA genotype in rs10951982 (*RAC1*), which exhibited an OR of 6.15 (95% CI: 2.98, 13.44, $p < 0.001$) after controlling for other risk factors. This minor allele also showed a significant association with melanoma risk in the dominant model (OR 5.79, 95% CI: 2.84, 12.51, $p < 0.001$). Similar results were also found for rs4673 (*CYBA*) but with only marginal significance. Heterozygous GA genotype was associated with an increased risk of melanoma (OR 2.17, 95% CI: 1.17, 4.07, $p = 0.015$), which was further confirmed in the dominant allele model (OR 1.88, 95% CI: 1.03, 3.47, $p = 0.042$), although the p-values did not reach the corrected significance level of 0.00238.

The homozygous minor allele TT genotype in rs8031 (*SOD2*) was found associated with a decreased risk of melanoma, with an OR of 0.32 (95% CI: 0.09, 0.94, $p = 0.047$) in the additive model, and an OR of 0.26 (95% CI: 0.08, 0.70, $p = 0.011$) in the recessive allele model, which indicated that homozygous minor alleles TT reduced the odds of melanoma by 74%, but neither of these results reached the universal significance level of 0.00238.

3. Discussion

After removal of SNP markers with high error rates during the assessment of genotyping quality, 21 SNP candidates remained to be eligible for the genetic association analysis. Eight SNPs showed significant association with melanoma but three of them were not in Hardy–Weinberg equilibrium, which may suggest that there are multiple alleles in the same locus, and we missed genotyping of other alleles. Therefore, only five SNP candidates showed genotypic significance and were further analyzed in regression models, including rs10951982 (*RAC1*), rs1049255 (*CYBA*), rs4673 (*CYBA*), rs8031 (*SOD2*), and rs2536512 (*SOD3*). We corrected the universal p-value to be compared with at 0.00238 (0.05/21, 21 SNPs being tested) to justify the multiple comparison issue in genetic association studies, using a Bonferroni approach [48,49]. The rs10951982 (*RAC1*) and rs4673 (*CYBA*) exhibited the highest increased risk of melanoma when presenting one copy of the minor allele in the unadjusted regression model, but rs4673 did not reach the universal significance level at 0.00238 in the multivariate regression model with adjustments for melanoma risk factors including age, sex, family history of melanoma, and lifetime ever-sunburned. Of particular note, a homozygous minor allele TT genotype in rs8031 (*SOD2*) was found to be associated with reduced risk of melanoma in the bivariate regression, however significance was lost in the multivariate regression analyses.

SOD2 is known to be a major superoxide detoxifying enzyme of cells, and therefore an altered function or expression of this enzyme may lead to unbalanced redox homeostasis and thus potentially increase or decrease the risk of melanoma [40]. Since SOD2 converts superoxide to hydrogen peroxide (Figure 1), which belongs to a type of ROS, the function of SOD2 is thus double-edged. Our multivariate analysis indicated that homozygous TT allele in rs8031 reduced the risk of melanoma, but little is currently known about the molecular function of this variant. We suggest a lab-based functional molecular biology study to unravel the discrepancy between zygote expression and enzymatic activity in this particular SNP.

SNPs rs1049255 and rs4673 in *CYBA* showed genotypic frequency differences between cases and controls in the unadjusted model (Table 4), with more patients carrying higher copies of minor alleles in rs1049255. Variant rs4673 changes the amino acid at position 72 from a tyrosine to a histidine (Y72H) of the CYBA (p22phox) protein, which is frequently referred to a C242T variant in the literature [50]. The T allele exhibited decreased dimerization with NOX and therefore may potentially reduce NOX activity and cellular ROS level [32]. In fact, the CT and TT genotype showed lower NADPH oxidase activity in hypertensive patients as compared with CC genotype [51]. However, opposite observation was also reported, where the CT genotype and T allele are associated with higher risk of coronary artery disease [52]. In our study, the CT and TT (GA and AA) showed higher risk for melanoma as compared to CC (GG) allele (the dominant model in Tables 4 and 5). This observation needs further validation. Variant rs1049255 is located in the 3′ untranslated region (3′ UTR) of the *CYBA* gene. Although the molecular function of this SNP is unknown, current understanding of 3′ UTR is an important miRNA binding site, and SNPs located in this region might have the potential to regulate mRNA stability and translation efficiency [53,54].

RAC1-GTPase is an NOX1 activator which promotes binding of NOX1 with its subunits and forms the complete enzyme complex [55–57]. NOX1 was one of the first cellular molecules found to be directly regulated by RAC1 in the phagocytic process [58–60]. However, SNP rs10951982 in *RAC1* alone has not been reported in any ROS-related activities thus far. Information on the function of this locus and its association with any malignancy is limited in the current literature. Nevertheless, this variant has been reported to be associated with over-reactive immune diseases and an increased risk of hypertension [20,35,36,61]. Considering that *CYBA* variants have been widely studied in cardiovascular diseases, including coronary heart disease [34] and hypertension [33], which are tightly associated with increased levels of ROS, *RAC1* rs10951982 may also play a part in inducing oxidative stress. Since rs10951982 is the most significant variant in our current study, and in lieu of its function in immune diseases as well as a potential role in NOX1-induced oxidative stress, our discovery might not only suggest an inflammatory microenvironment created by RAC1 that is in favor of melanoma progression [62], but also indicate an elevation of ROS level via RAC1 in melanoma etiology. In addition, RAC1 is also a crucial kinase in the NRAS and PI3K pathway [63], both of which are key melanoma oncogenic pathways. Therefore, it is possible that RAC1 plays a non-ROS role and impacts these other oncogenic pathways.

Overall, of the three significant SNPs after adjustment against age, sex, family history and life time sun burn history, the minor allele of *RAC1* rs10951982 (the A allele) showed a consistent role with an increase ROS and thus increased melanoma risk. The minor allele of rs4673 (the A allele) was reported controversial role in ROS association [51,64], it may exhibit certain cell-specific effects. In our study, the minor allele showed higher risk for melanoma in a dominant model. The minor allele of rs8031 (the T allele) exhibited a protective role against melanoma risk in a recessive model. It is unclear how this allele modifies ROS levels. Based on our results, the T allele can be associated with either increased or decreased SOD activities as SOD2 is double-edged and can play dual roles in ROS metabolism.

Of particular note, in our regression models, we applied the most common ways of disease transmission, namely additive, recessive, and dominant modes, in our analyses. This was because we did not want to make any assumptions of the disease transmission modes. According to Sham and Purcell [49], a test that assumed additive effects would have greater power than a test that also allowed dominance, if the true effects at the locus were indeed additive and did not show dominance. Conversely, if the underlying causal variant was recessive, then power would be lost by carrying out an analysis that assumed additively. If there was uncertainty regarding the true pattern of effects at a locus, then it might be appropriate to use several statistical tests to ensure adequate statistical power for all possible scenarios. We therefore included results from these additional models that may provide more information and maintain statistical power as well. Although the covariates were not presented as part of the results in our tables, family history of melanoma and lifetime ever-sunburned controlled in the multivariate models consistently showed statistical significance, whereas sex and age did not. Family history of melanoma [23], along with fair skin, light hair and eye color are known melanoma genetic risk factors, whereas the levels of sun exposure including sunburns and moles or freckles are important environmental risk factors for melanoma [65]. The statistical significance of the covariates might indicate a mediating role in our primary study interest, from the susceptible familial genetic makeup of these participants, as well as the behavior or attitude towards sun exposure that resulted in getting sunburns or freckles.

Our study had a few limitations. First, the small sample size does not always provide sufficient power [66]. Second, by the experimental design, we could only genotype two alleles. Therefore, loci with multiple alleles may not show HWE and must be excluded for analysis. Third, our study participants included only those white individuals from the southern California area, and therefore a loss of generalizability to the broader white population might be expected. Last, a common limitation of case-control studies is that the results provide only an association with risk, but they are not necessarily connected to causality. Replicating findings from another dataset is a common strategy to validate the results identified in our current study. However, even with the most stringent statistical design, SNP findings are usually hard to replicate [48,49]. Multiple reasons are considered, such as there are still unknown and uncontrolled confounders, multiple comparisons only lead to chance findings, the gene and environment interaction is not easy to account for, and the target allele is in linkage disequilibrium with the identified allele but the chance finding failed to locate the target allele and thus make replication difficult to achieve. Nevertheless, we will still validate our findings in a separate dataset in our next study, as our ultimate goal is to develop useful markers in prevention.

To conclude, our initial analyses revealed an increased risk of melanoma associated with rs10951982 (*RAC1*), and a decreased risk associated with rs8031 (*SOD2*). Multivariate analyses further confirmed the association of an increased risk of melanoma with rs10951982 (*RAC1*). Our results highlighted the importance of RAC1 enzyme and cellular oxidation-metabolizing efficiency controlled by SOD2 in association with ROS-mediated risk of melanoma. We suggest that these results shall be further validated with the goal of designing novel screening targets to identify highly UV-susceptible individuals, particularly in the *RAC1* and *SOD2* genes, in order to take the melanoma primary prevention strategy to a precision level.

4. Materials and Methods

4.1. Ethics Statement

We obtained approval from the Institutional Review Board of the University of California Irvine Office of Research (protocol number 2011-8238, approved 27 June 2011).

4.2. Study Population

Our study subjects were adopted from a previously designed case-control study (the international Genes, Environment, and Melanoma study, the GEM study), although we made considerable modifications. The original GEM case-control study compared white multiple melanomas patients to primary melanoma patients [67]. In total, 177 patients were recruited between 1998 and 2003 in the southern California area as part of the GEM study, and consent forms were obtained accordingly [67]. In our study, we used both of these patients as our cases and we recruited additional healthy participants as controls. Healthy white volunteers from Orange County were recruited through random-digit-dialing by trained interviewers during 1999 to 2006.

Demographic information regarding age, sex, family history of melanoma, and lifetime sun exposure were recorded via in-person questionnaires and phone interviews, with written consents from the patients and their physicians [67–73]. Random-digit-dialing healthy respondents completed eligibility screening questions over the phone, including being Orange County residents and having no personal history of melanoma or any other types of cancer. Eligible respondents were asked for their verbal informed consents for a 20 min standardized phone interview [67], in which they were asked questions about basic demographics, personal medical history, and family cancer history. In total, 172 participants further agreed to donate a blood sample. A phlebotomist obtained written consents from these participants while performing the blood draw [67]. Participation rate after phone eligibility screening was approximately 78%. Population-based controls were frequency-matched to cases with respect to sex and age (Table 1).

4.3. DNA Extraction

Buccal cells from melanoma patients and whole blood cells from healthy participants were re-suspended in a phosphate-buffered saline system. Ten microliters of the cell suspension were used directly as a template for whole genome amplification (WGA). The WGA procedure was conducted following the manufacturer's instruction from Sigma. In brief, a cell suspension (10 μL each) was heated to 95 °C for 5 min in a PCR machine in a strip of PCR tubes and cooled down on ice. One microliter of 10× Fragmentation Buffer was added to each tube. Tubes were then heated again in a PCR machine at 95 °C for exactly 4 min. Samples were cooled down on ice immediately and then centrifuged briefly to consolidate the contents. Out of 70 μL of the amplified sample, 6 μL was mixed with 1 μL of 6× loading buffer and directly used to load on an agarose DNA gel containing ethidium bromide. DNA was visualized under a UV lamp and water was used as a non-DNA negative control to compare with the presence of the visualized DNA product. Participants with little to no whole genome amplified DNA product were excluded from SNP genotyping (7 patients and 20 healthy controls were excluded, Figure 2).

4.4. SNP Candidates

Functional SNPs were selected from a publicly available SNP database (dbSNP, NCBI) that have been found correlated with other diseases, based on the three criteria listed in the introduction (Table 2). In brief, 6 SNPs in the coding region of *NOX1* appeared in dbSNP. We were interested in D360N (rs34688635) and R315H (rs2071756) variants for the following reasons: (1) D360 is shared in *NOX1*, -2, -3, and -4 [25], and conserved in various species including fish, mouse, bird, amphibian, and man [9]; and (2) 315H allele was found associated with diabetic patients, suggesting that this is a functional allele and may be associated with other disease risks [24]. SNPs rs585197 and rs2164521 in *NOX4* have been linked to a protective effect on Hepatopulmonary Syndrome [27]; and rs11018628 has a possible effect on plasma homocysteine level [26]. −930A > G in *CYBA* promoter region (rs9932581) affects gene transcription activity and has been found to be associated with coronary heart disease due to ROS involvement in the pathogenesis of atherosclerosis [28]. Similarly, increased or decreased risks of hypertension [33] and coronary heart disease [34], respectively, have been found in *CYBA* alleles rs4673, rs13306296, and rs1049255. *CYBA* rs3180279 has been related to non-Hodgkin lymphoma prognosis [13]. Three SNPs, rs10951982, rs4720672, and rs836478 in *RAC1*, have been associated with risks in hypertension, inflammatory bowel disease, and end-stage renal disease [20,35,36,61]. Although these loci in *RAC1* have not yet been discussed in ROS-related malignancies, *RAC1* is a well-known melanoma oncogene with constantly activated mutations in some melanoma tumors [74,75].

SNPs in the three subtypes of *SOD* and *CAT* genes have been widely studied with various disease associations. For instance, rs7277748 and rs4998557 variants in *SOD1* (Cu-ZnSOD) were found to cause amyotrophic lateral sclerosis. Ile58Thr (rs1141718) in *SOD2* (MnSOD) severally impaired SOD2 enzymatic activity [40], while a variant of rs8031 increased oxidative stress [44]. V16A variant rs4880 in *SOD2* impaired mitochondrial importing and was found to be a risk factor for prostate cancer [12], whereas rs2758330 showed a protective effect on prostate cancer [45]. Variants rs2536512 and rs699473 in *SOD3* were associated with brain diseases, including cerebral infarction [14] and brain tumor [15]. The rs1001179 in *CAT* was also correlated to brain malignancy [15]. Additionally, −262C > T (rs1049982) variant in *CAT* showed a decreased interaction with HIF1α upon oxidative stress stimulation [46,47] (Table 2).

4.5. SNP Genotyping

SNP genotyping polymerase chain reaction (PCR) assay kit was purchased from Life Technologies™ (Carlsbad, CA, USA). Allele-specific primers and probe sets for each SNP were also purchased from Life Technologies™, either custom-designed or from the library. DNA sample per participant was genotyped for every SNP in duplicates to ensure accuracy. About 97% of the SNPs were replicable. By definition, if one allele was amplified during PCR reaction, the call for that SNP assay was homozygous alleles (inherited the same alleles from both parents); if both alleles were amplified, the call for that SNP assay was heterozygous alleles (inherited different alleles from the parents). However, if no significant PCR amplification for either allele was observed, then the SNP assay was defined as N/A (genotyping failure) due to no reaction to the designed allele primers and probe. SNPs with genotyping rate < 75% were excluded from statistical analysis (SNPs rs13306296 and rs585197 were excluded from further analysis, Table 3). SNPs with inconsistent duplicated results were validated manually by reading the raw real-time PCR amplification plots, or through additional genotyping reactions.

4.6. SNP Quality Control

The raw PCR amplification data was analyzed by QuantStudio™ (Thermo Fisher Scientific Inc., Huntington Beach, CA, USA) Real-Time PCR software (v1.2). Those duplicated samples presenting identical calls were automatically determined by the software. However, if the calls were made differently between duplicates, or, in some rare cases, if the calls were "undetermined" by the software, then the individual PCR amplification plots were read manually and subjectively. Any amplification curve appearing after 20 cycles of PCR, and being at least two-fold elevated from the threshold was determined as presenting a positive PCR amplification curve. Genotyping failure was assigned as N/A if no clear PCR amplification curve was observed.

4.7. Statistics

Allele frequency was determined by making counts of the participants based on different SNP conditions: genotypic, allelic, recessive, and dominant models. Chi-square or Fisher's exact test of independence was performed to examine the associations between SNP conditions and melanoma case-control status. Two-sided statistical significance level by default was set to be 0.05 (5%), and, to justify for multiple comparison among the SNP candidates, universal significance level was further adjusted to 0.05 divided by the number of final SNP candidates being tested, which was $0.05/21 = 0.00238$, applying the most stringent Bonferroni approach [48,49]. Participant numbers varied among SNPs due to different genotyping rates, and only complete data was used for statistical analysis (participants with N/A data were excluded per SNP analysis). Bivariate simple logistic regression models showing the unadjusted associations between the binary response variable (melanoma cases vs. controls) and primary study variables of interest (SNPs) were conducted separately based on additive, recessive, and dominant allele models. Dummy variables of the SNPs in the three allele models were created by default, making genotype with homozygous major alleles as the reference to compare with. Odds ratios and 95% confidence intervals were calculated accordingly in RStudio (v0.99.893). Adjusted associations between SNPs and melanoma status were analyzed by fitting multivariate logistic regression models with the three allele models separately, controlling for known melanoma risk factors, including gender [76], age at diagnosis [77], family history of melanoma [23], and ever sunburned [78]. Genotypic Hardy–Weinberg equilibrium exact test, which examines the expected frequencies of genotypes if mating is non-assortative and there are no mutations from one allele to another, was carried out by using R package HardyWeinberg. In brief, a two-sided test was performed on genotype counts, whether an excess or a dearth of heterozygotes counts as evidence ($p < 0.05$) against Hardy–Weinberg equilibrium.

Acknowledgments: This study was supported by NCI/NIH K07 grant to Feng Liu-Smith (CA 160756). Tze-An Yuan was supported by the UCI Public Health Graduate Program.

Author Contributions: Feng Liu-Smith conceived and designed the experiments; Vandy Yourk and Ali Farhat performed the experiments; Tze-An Yuan and Argyrios Ziogas analyzed the data; Hoda Anton-Culver and Frank L. Meyskens contributed materials; and Tze-An Yuan and Feng Liu-Smith wrote the paper.

Conflicts of Interest: The authors declare no conflict of interest. The founding sponsors had no role in the design of the study; in the collection, analyses, or interpretation of data; in the writing of the manuscript, and in the decision to publish the results.

Abbreviations

UV	Ultraviolet
ROS	Reactive oxygen species
SNP	Single nucleotide polymorphism
NOX	NADPH oxidase
SOD	Superoxide dismutase
CAT	Catalase
OR	Odds ratio
HWE	Hardy–Weinberg equilibrium

References

1. D'Orazio, J.; Jarrett, S.; Amaro-Ortiz, A.; Scott, T. UV radiation and the skin. *Int. J. Mol. Sci.* **2013**, *14*, 12222–12248. [CrossRef] [PubMed]
2. Kobayashi, N.; Nakagawa, A.; Muramatsu, T.; Yamashina, Y.; Shirai, T.; Hashimoto, M.W.; Ishigaki, Y.; Ohnishi, T.; Mori, T. Supranuclear melanin caps reduce ultraviolet induced DNA photoproducts in human epidermis. *J. Investig. Dermatol.* **1998**, *110*, 806–810. [CrossRef] [PubMed]
3. Raad, H.; Serrano-Sanchez, M.; Harfouche, G.; Mahfouf, W.; Bortolotto, D.; Bergeron, V.; Kasraian, Z.; Dousset, L.; Hosseini, M.; Taieb, A.; et al. NADPH oxidase-1 plays a key role in keratinocyte responses to ultraviolet radiation and UVB-induced skin carcinogenesis. *J. Investig. Dermatol.* **2017**. [CrossRef] [PubMed]
4. Premi, S.; Wallisch, S.; Mano, C.M.; Weiner, A.B.; Bacchiocchi, A.; Wakamatsu, K.; Bechara, E.J.; Halaban, R.; Douki, T.; Brash, D.E. Photochemistry. Chemiexcitation of melanin derivatives induces DNA photoproducts long after UV exposure. *Science* **2015**, *347*, 842–847. [CrossRef] [PubMed]
5. Liu-Smith, F.; Poe, C.; Farmer, P.J.; Meyskens, F.L., Jr. Amyloids, melanins and oxidative stress in melanomagenesis. *Exp. Dermatol.* **2015**, *24*, 171–174. [CrossRef] [PubMed]
6. Liu-Smith, F. Reactive Oxygen Species in Melanoma Etiology. In *Reactive Oxygen Species in Biology and Human Health*; Shamim, A., Ed.; CRC Press: Boca Raton, FL, USA, 2016; pp. 259–275.
7. Liu-Smith, F.; Dellinger, R.; Meyskens, F.L., Jr. Updates of reactive oxygen species in melanoma etiology and progression. *Arch. Biochem. Biophys.* **2014**, *563*, 51–55. [CrossRef] [PubMed]
8. Liu, F.; Garcia, A.M.G.; Meyskens, F.L. NADPH Oxidase 1 Overexpression Enhances Invasion via Matrix Metalloproteinase-2 and Epithelial-Mesenchymal Transition in Melanoma Cells. *J. Investig. Dermatol.* **2012**, *132*, 2033–2041. [CrossRef] [PubMed]
9. Kawahara, T.; Quinn, M.T.; Lambeth, J.D. Molecular evolution of the reactive oxygen-generating NADPH oxidase (Nox/Duox) family of enzymes. *BMC Evol. Biol.* **2007**, *7*, 109. [CrossRef] [PubMed]
10. Hodis, E.; Watson, I.R.; Kryukov, G.V.; Arold, S.T.; Imielinski, M.; Theurillat, J.P.; Nickerson, E.; Auclair, D.; Li, L.; Place, C.; et al. A landscape of driver mutations in melanoma. *Cell* **2012**, *150*, 251–263. [CrossRef] [PubMed]
11. Cheng, G.; Diebold, B.A.; Hughes, Y.; Lambeth, J.D. Nox1-dependent reactive oxygen generation is regulated by Rac1. *J. Biol. Chem.* **2006**, *281*, 17718–17726. [CrossRef] [PubMed]
12. Kang, D.; Lee, K.M.; Park, S.K.; Berndt, S.I.; Peters, U.; Reding, D.; Chatterjee, N.; Welch, R.; Chanock, S.; Huang, W.Y.; et al. Functional variant of manganese superoxide dismutase (SOD2 V16A) polymorphism is associated with prostate cancer risk in the prostate, lung, colorectal, and ovarian cancer study. *Cancer Epidemiol. Biomark. Prev.* **2007**, *16*, 1581–1586. [CrossRef] [PubMed]
13. Hoffmann, M.; Schirmer, M.A.; Tzvetkov, M.V.; Kreuz, M.; Ziepert, M.; Wojnowski, L.; Kube, D.; Pfreundschuh, M.; Trumper, L.; Loeffler, M.; et al. A Functional Polymorphism in the NAD(P)H Oxidase Subunit CYBA Is Related to Gene Expression, Enzyme Activity, and Outcome in Non-Hodgkin Lymphoma. *Cancer Res.* **2010**, *70*, 2328–2338. [CrossRef] [PubMed]
14. Naganuma, T.; Nakayama, T.; Sato, N.; Fu, Z.; Soma, M.; Aoi, N.; Hinohara, S.; Doba, N.; Usami, R. Association of extracellular superoxide dismutase gene with cerebral infarction in women: A haplotype-based case-control study. *Hereditas* **2008**, *145*, 283–292. [CrossRef] [PubMed]
15. Rajaraman, P.; Hutchinson, A.; Rothman, N.; Black, P.M.; Fine, H.A.; Loeffler, J.S.; Selker, R.G.; Shapiro, W.R.; Linet, M.S.; Inskip, P.D. Oxidative response gene polymorphisms and risk of adult brain tumors. *Neuro Oncol.* **2008**, *10*, 709–715. [CrossRef] [PubMed]

16. Case, A.J.; Li, S.; Basu, U.; Tian, J.; Zimmerman, M.C. Mitochondrial-localized NADPH oxidase 4 is a source of superoxide in angiotensin II-stimulated neurons. *Am. J. Physiol. Heart Circ. Physiol.* **2013**, *305*, H19–H28. [CrossRef] [PubMed]

17. Pal, A.; Alam, S.; Mittal, S.; Arjaria, N.; Shankar, J.; Kumar, M.; Singh, D.; Pandey, A.K.; Ansari, K.M. UVB irradiation-enhanced zinc oxide nanoparticles-induced DNA damage and cell death in mouse skin. *Mutat. Res. Genet. Toxicol. Environ. Mutagen.* **2016**, *807*, 15–24. [CrossRef] [PubMed]

18. Dellinger, R.W.; Liu-Smith, F.; Meyskens, F.L., Jr. Continuing to illuminate the mechanisms underlying UV-mediated melanomagenesis. *J. Photochem. Photobiol. B Biol.* **2014**, *138*, 317–323. [CrossRef] [PubMed]

19. Mohammedi, K.; Bellili-Munoz, N.; Driss, F.; Roussel, R.; Seta, N.; Fumeron, F.; Hadjadj, S.; Marre, M.; Velho, G. Manganese superoxide dismutase (SOD2) polymorphisms, plasma advanced oxidation protein products (AOPP) concentration and risk of kidney complications in subjects with type 1 diabetes. *PLoS ONE* **2014**, *9*, e96916. [CrossRef] [PubMed]

20. Tapia-Castillo, A.; Carvajal, C.A.; Campino, C.; Vecchiola, A.; Allende, F.; Solari, S.; Garcia, L.; Lavanderos, S.; Valdivia, C.; Fuentes, C.; et al. Polymorphisms in the RAC1 gene are associated with hypertension risk factors in a Chilean pediatric population. *Am. J. Hypertens.* **2014**, *27*, 299–307. [CrossRef] [PubMed]

21. Dulak, A.M.; Stojanov, P.; Peng, S.; Lawrence, M.S.; Fox, C.; Stewart, C.; Bandla, S.; Imamura, Y.; Schumacher, S.E.; Shefler, E.; et al. Exome and whole-genome sequencing of esophageal adenocarcinoma identifies recurrent driver events and mutational complexity. *Nat. Genet.* **2013**, *45*, 478–486. [CrossRef] [PubMed]

22. Krauthammer, M.; Kong, Y.; Ha, B.H.; Evans, P.; Bacchiocchi, A.; McCusker, J.P.; Cheng, E.; Davis, M.J.; Goh, G.; Choi, M.; et al. Exome sequencing identifies recurrent somatic RAC1 mutations in melanoma. *Nat. Genet.* **2012**, *44*, 1006–1014. [CrossRef] [PubMed]

23. Liu, F.; Bessonova, L.; Taylor, T.H.; Ziogas, A.; Meyskens, F.L., Jr.; Anton-Culver, H. A unique gender difference in early onset melanoma implies that in addition to ultraviolet light exposure other causative factors are important. *Pigment Cell Melanoma Res.* **2013**, *26*, 128–135. [CrossRef] [PubMed]

24. Lim, S.C.; Liu, J.J.; Low, H.Q.; Morgenthaler, N.G.; Li, Y.; Yeoh, L.Y.; Wu, Y.S.; Goh, S.K.; Chionh, C.Y.; Tan, S.H.; et al. Microarray analysis of multiple candidate genes and associated plasma proteins for nephropathy secondary to type 2 diabetes among Chinese individuals. *Diabetologia* **2009**, *52*, 1343–1351. [CrossRef] [PubMed]

25. O'Neill, S.; Brault, J.; Stasia, M.J.; Knaus, U.G. Genetic disorders coupled to ROS deficiency. *Redox Biol.* **2015**, *6*, 135–156. [CrossRef] [PubMed]

26. Pare, G.; Chasman, D.I.; Parker, A.N.; Zee, R.R.; Malarstig, A.; Seedorf, U.; Collins, R.; Watkins, H.; Hamsten, A.; Miletich, J.P.; et al. Novel associations of CPS1, MUT, NOX4, and DPEP1 with plasma homocysteine in a healthy population: A genome-wide evaluation of 13,974 participants in the Women's Genome Health Study. *Circ. Cardiovasc. Genet.* **2009**, *2*, 142–150. [CrossRef] [PubMed]

27. Roberts, K.E.; Kawut, S.M.; Krowka, M.J.; Brown, R.S., Jr.; Trotter, J.F.; Shah, V.; Peter, I.; Tighiouart, H.; Mitra, N.; Handorf, E.; et al. Genetic risk factors for hepatopulmonary syndrome in patients with advanced liver disease. *Gastroenterology* **2010**, *139*, 130–139. [CrossRef] [PubMed]

28. Niemiec, P.; Nowak, T.; Iwanicki, T.; Krauze, J.; Gorczynska-Kosiorz, S.; Grzeszczak, W.; Ochalska-Tyka, A.; Zak, I. The-930A > G polymorphism of the CYBA gene is associated with premature coronary artery disease. A case-control study and gene-risk factors interactions. *Mol. Biol. Rep.* **2014**, *41*, 3287–3294. [CrossRef] [PubMed]

29. San Jose, G.; Moreno, M.U.; Olivan, S.; Beloqui, O.; Fortuno, A.; Diez, J.; Zalba, G. Functional effect of the p22phox-930A/G polymorphism on p22phox expression and NADPH oxidase activity in hypertension. *Hypertension* **2004**, *44*, 163–169. [CrossRef] [PubMed]

30. Li, A.; Prasad, A.; Mincemoyer, R.; Satorius, C.; Epstein, N.; Finkel, T.; Quyyumi, A.A. Relationship of the C242T p22phox gene polymorphism to angiographic coronary artery disease and endothelial function. *Am. J. Med. Genet.* **1999**, *86*, 57–61. [CrossRef]

31. Wyche, K.E.; Wang, S.S.; Griendling, K.K.; Dikalov, S.I.; Austin, H.; Rao, S.; Fink, B.; Harrison, D.G.; Zafari, A.M. C242T CYBA polymorphism of the NADPH oxidase is associated with reduced respiratory burst in human neutrophils. *Hypertension* **2004**, *43*, 1246–1251. [CrossRef] [PubMed]

32. Guzik, T.J.; West, N.E.; Black, E.; McDonald, D.; Ratnatunga, C.; Pillai, R.; Channon, K.M. Functional effect of the C242T polymorphism in the NAD(P)H oxidase p22phox gene on vascular superoxide production in atherosclerosis. *Circulation* 2000, *102*, 1744–1747. [CrossRef] [PubMed]

33. Moreno, M.U.; San Jose, G.; Fortuno, A.; Beloqui, O.; Redon, J.; Chaves, F.J.; Corella, D.; Diez, J.; Zalba, G. A novel CYBA variant, the -675A/T polymorphism, is associated with essential hypertension. *J. Hypertens.* 2007, *25*, 1620–1626. [CrossRef] [PubMed]

34. Gardemann, A.; Mages, P.; Katz, N.; Tillmanns, H.; Haberbosch, W. The p22 phox A640G gene polymorphism but not the C242T gene variation is associated with coronary heart disease in younger individuals. *Atherosclerosis* 1999, *145*, 315–323. [CrossRef]

35. Muise, A.M.; Walters, T.; Xu, W.; Shen-Tu, G.; Guo, C.H.; Fattouh, R.; Lam, G.Y.; Wolters, V.M.; Bennitz, J.; van Limbergen, J.; et al. Single nucleotide polymorphisms that increase expression of the guanosine triphosphatase RAC1 are associated with ulcerative colitis. *Gastroenterology* 2011, *141*, 633–641. [CrossRef] [PubMed]

36. Lev-Tzion, R.; Renbaum, P.; Beeri, R.; Ledder, O.; Mevorach, R.; Karban, A.; Koifman, E.; Efrati, E.; Muise, A.M.; Chowers, Y.; et al. Rac1 Polymorphisms and Thiopurine Efficacy in Children With Inflammatory Bowel Disease. *J. Pediatr. Gastroenterol. Nutr.* 2015, *61*, 404–407. [CrossRef] [PubMed]

37. Saeed, M.; Yang, Y.; Deng, H.X.; Hung, W.Y.; Siddique, N.; Dellefave, L.; Gellera, C.; Andersen, P.M.; Siddique, T. Age and founder effect of SOD1 A4V mutation causing ALS. *Neurology* 2009, *72*, 1634–1639. [CrossRef] [PubMed]

38. Nyaga, S.G.; Lohani, A.; Jaruga, P.; Trzeciak, A.R.; Dizdaroglu, M.; Evans, M.K. Reduced repair of 8-hydroxyguanine in the human breast cancer cell line, HCC1937. *BMC Cancer* 2006, *6*, 297. [CrossRef] [PubMed]

39. Oestergaard, M.Z.; Tyrer, J.; Cebrian, A.; Shah, M.; Dunning, A.M.; Ponder, B.A.; Easton, D.F.; Pharoah, P.D. Interactions between genes involved in the antioxidant defence system and breast cancer risk. *Br. J. Cancer* 2006, *95*, 525–531. [CrossRef] [PubMed]

40. Holley, A.K.; Bakthavatchalu, V.; Velez-Roman, J.M.; St Clair, D.K. Manganese superoxide dismutase: Guardian of the powerhouse. *Int. J. Mol. Sci.* 2011, *12*, 7114–7162. [CrossRef] [PubMed]

41. Mollsten, A.; Marklund, S.L.; Wessman, M.; Svensson, M.; Forsblom, C.; Parkkonen, M.; Brismar, K.; Groop, P.H.; Dahlquist, G. A functional polymorphism in the manganese superoxide dismutase gene and diabetic nephropathy. *Diabetes* 2007, *56*, 265–269. [CrossRef] [PubMed]

42. Han, J.; Colditz, G.A.; Hunter, D.J. Manganese superoxide dismutase polymorphism and risk of skin cancer (United States). *Cancer Causes Control* 2007, *18*, 79–89. [CrossRef] [PubMed]

43. Rajaraman, P.; Wang, S.S.; Rothman, N.; Brown, M.M.; Black, P.M.; Fine, H.A.; Loeffler, J.S.; Selker, R.G.; Shapiro, W.R.; Chanock, S.J.; et al. Polymorphisms in apoptosis and cell cycle control genes and risk of brain tumors in adults. *Cancer Epidemiol. Biomark. Prev.* 2007, *16*, 1655–1661. [CrossRef] [PubMed]

44. Rodrigues, P.; de Marco, G.; Furriol, J.; Mansego, M.L.; Pineda-Alonso, M.; Gonzalez-Neira, A.; Martin-Escudero, J.C.; Benitez, J.; Lluch, A.; Chaves, F.J.; et al. Oxidative stress in susceptibility to breast cancer: Study in Spanish population. *BMC Cancer* 2014, *14*, 861. [CrossRef] [PubMed]

45. Abe, M.; Xie, W.; Regan, M.M.; King, I.B.; Stampfer, M.J.; Kantoff, P.W.; Oh, W.K.; Chan, J.M. Single-nucleotide polymorphisms within the antioxidant defence system and associations with aggressive prostate cancer. *BJU Int.* 2011, *107*, 126–134. [CrossRef] [PubMed]

46. Perianayagam, M.C.; Liangos, O.; Kolyada, A.Y.; Wald, R.; MacKinnon, R.W.; Li, L.; Rao, M.; Balakrishnan, V.S.; Bonventre, J.V.; Pereira, B.J.; et al. NADPH oxidase p22phox and catalase gene variants are associated with biomarkers of oxidative stress and adverse outcomes in acute renal failure. *J. Am. Soc. Nephrol.* 2007, *18*, 255–263. [CrossRef] [PubMed]

47. Nadif, R.; Mintz, M.; Jedlicka, A.; Bertrand, J.P.; Kleeberger, S.R.; Kauffmann, F. Association of CAT polymorphisms with catalase activity and exposure to environmental oxidative stimuli. *Free Radic. Res.* 2005, *39*, 1345–1350. [CrossRef] [PubMed]

48. Jorgensen, T.J.; Ruczinski, I.; Kessing, B.; Smith, M.W.; Shugart, Y.Y.; Alberg, A.J. Hypothesis-driven candidate gene association studies: Practical design and analytical considerations. *Am. J. Epidemiol.* 2009, *170*, 986–993. [CrossRef] [PubMed]

49. Sham, P.C.; Purcell, S.M. Statistical power and significance testing in large-scale genetic studies. *Nat. Rev. Genet.* 2014, *15*, 335–346. [CrossRef] [PubMed]

50. Taylor, R.M.; Dratz, E.A.; Jesaitis, A.J. Invariant local conformation in p22phox p.Y72H polymorphisms suggested by mass spectral analysis of crosslinked human neutrophil flavocytochrome b. *Biochimie* **2011**, *93*, 1502–1509. [CrossRef] [PubMed]

51. Moreno, M.U.; San Jose, G.; Fortuno, A.; Beloqui, O.; Diez, J.; Zalba, G. The C242T CYBA polymorphism of NADPH oxidase is associated with essential hypertension. *J. Hypertens.* **2006**, *24*, 1299–1306. [CrossRef] [PubMed]

52. Mazaheri, M.; Karimian, M.; Behjati, M.; Raygan, F.; Hosseinzadeh Colagar, A. Association analysis of rs1049255 and rs4673 transitions in p22phox gene with coronary artery disease: A case-control study and a computational analysis. *Ir. J. Med. Sci.* **2017**, *186*, 921–928. [CrossRef] [PubMed]

53. Tanguay, R.L.; Gallie, D.R. Translational efficiency is regulated by the length of the 3′ untranslated region. *Mol. Cell. Biol.* **1996**, *16*, 146–156. [CrossRef] [PubMed]

54. Wilkie, G.S.; Dickson, K.S.; Gray, N.K. Regulation of mRNA translation by 5′- and 3′-UTR-binding factors. *Trends Biochem. Sci.* **2003**, *28*, 182–188. [CrossRef]

55. Hordijk, P.L. Regulation of NADPH oxidases—The role of Rac proteins. *Circ. Res.* **2006**, *98*, 453–462. [CrossRef] [PubMed]

56. Raz, L.; Zhang, Q.G.; Zhou, C.F.; Han, D.; Gulati, P.; Yang, L.C.; Yang, F.; Wang, R.M.; Brann, D.W. Role of Rac1 GTPase in NADPH Oxidase Activation and Cognitive Impairment Following Cerebral Ischemia in the Rat. *PLoS ONE* **2010**, *5*, e12606. [CrossRef] [PubMed]

57. Nikolova, S.; Lee, Y.S.; Lee, Y.S.; Kim, J.A. Rac1-NADPH oxidase-regulated generation of reactive oxygen species mediates glutamate-induced apoptosis in SH-SY5Y human neuroblastoma cells. *Free Radic. Res.* **2005**, *39*, 1295–1304. [CrossRef] [PubMed]

58. Abo, A.; Boyhan, A.; West, I.; Thrasher, A.J.; Segal, A.W. Reconstitution of neutrophil NADPH oxidase activity in the cell-free system by four components: p67-phox, p47-phox, p21rac1, and cytochrome b-245. *J. Biol. Chem.* **1992**, *267*, 16767–16770. [PubMed]

59. Bokoch, G.M. Regulation of the phagocyte respiratory burst by small GTP-binding proteins. *Trends Cell Biol.* **1995**, *5*, 109–113. [CrossRef]

60. Bosco, E.E.; Mulloy, J.C.; Zheng, Y. Rac1 GTPase: A "Rac"of All Trades. *Cell. Mol. Life Sci.* **2009**, *66*, 370–374. [CrossRef] [PubMed]

61. Liu, Y.; Zhou, J.; Luo, X.; Yang, C.; Zhang, Y.; Shi, S. Association of RAC1 Gene Polymorphisms with Primary End-Stage Renal Disease in Chinese Renal Recipients. *PLoS ONE* **2016**, *11*, e0148270. [CrossRef] [PubMed]

62. Doma, V.; Gulya, E. Genetic diversity and immunological characteristics of malignant melanoma: The therapeutic spectrum. *Orv. Hetil.* **2015**, *156*, 583–591. [CrossRef] [PubMed]

63. Lissanu Deribe, Y. Interplay between PREX2 mutations and the PI3K pathway and its effect on epigenetic regulation of gene expression in NRAS-mutant melanoma. *Small GTPases* **2016**, *7*, 178–185. [CrossRef] [PubMed]

64. Genius, J.; Grau, A.J.; Lichy, C. The C242T polymorphism of the NAD(P)H oxidase p22phox subunit is associated with an enhanced risk for cerebrovascular disease at a young age. *Cerebrovasc. Dis.* **2008**, *26*, 430–433. [CrossRef] [PubMed]

65. Czene, K.; Lichtenstein, P.; Hemminki, K. Environmental and heritable causes of cancer among 9.6 million individuals in the Swedish family-cancer database. *Int. J. Cancer* **2002**, *99*, 260–266. [CrossRef] [PubMed]

66. Neil, A.; Campbell, J.B.R.; Urry, L.A.; Cain, M.L.; Wasserman, S.A.; Minorsky, P.V.; Jackson, R.B. *Campbell Biology*, 8th ed.; Pearson Benjamin Cummings: Boston, MA, USA, 2008; p. 1393.

67. Begg, C.B.; Hummer, A.J.; Mujumdar, U.; Armstrong, B.K.; Kricker, A.; Marrett, L.D.; Millikan, R.C.; Gruber, S.B.; Culver, H.A.; Zanetti, R.; et al. A design for cancer case-control studies using only incident cases: Experience with the GEM study of melanoma. *Int. J. Epidemiol.* **2006**, *35*, 756–764. [CrossRef] [PubMed]

68. Kricker, A.; Armstrong, B.K.; Goumas, C.; Litchfield, M.; Begg, C.B.; Hummer, A.J.; Marrett, L.D.; Theis, B.; Millikan, R.C.; Thomas, N.; et al. Ambient UV, personal sun exposure and risk of multiple primary melanomas. *Cancer Causes Control* **2007**, *18*, 295–304. [CrossRef] [PubMed]

69. Orlow, I.; Begg, C.B.; Cotignola, J.; Roy, P.; Hummer, A.J.; Clas, B.A.; Mujumdar, U.; Canchola, R.; Armstrong, B.K.; Kricker, A.; et al. CDKN2A germline mutations in individuals with cutaneous malignant melanoma. *J. Investig. Dermatol.* **2007**, *127*, 1234–1243. [CrossRef] [PubMed]

70. Kanetsky, P.A.; Rebbeck, T.R.; Hummer, A.J.; Panossian, S.; Armstrong, B.K.; Kricker, A.; Marrett, L.D.; Millikan, R.C.; Gruber, S.B.; Culver, H.A.; et al. Population-based study of natural variation in the melanocortin-1 receptor gene and melanoma. *Cancer Res.* **2006**, *66*, 9330–9337. [CrossRef] [PubMed]

71. Berwick, M.; Orlow, I.; Hummer, A.J.; Armstrong, B.K.; Kricker, A.; Marrett, L.D.; Millikan, R.C.; Gruber, S.B.; Anton-Culver, H.; Zanetti, R.; et al. The prevalence of CDKN2A germ-line mutations and relative risk for cutaneous malignant melanoma: An international population-based study. *Cancer Epidemiol. Biomark. Prev.* **2006**, *15*, 1520–1525. [CrossRef] [PubMed]

72. Millikan, R.C.; Hummer, A.; Begg, C.; Player, J.; de Cotret, A.R.; Winkel, S.; Mohrenweiser, H.; Thomas, N.; Armstrong, B.; Kricker, A.; et al. Polymorphisms in nucleotide excision repair genes and risk of multiple primary melanoma: The Genes Environment and Melanoma Study. *Carcinogenesis* **2006**, *27*, 610–618. [CrossRef] [PubMed]

73. Begg, C.B.; Orlow, I.; Hummer, A.J.; Armstrong, B.K.; Kricker, A.; Marrett, L.D.; Millikan, R.C.; Gruber, S.B.; Anton-Culver, H.; Zanetti, R.; et al. Lifetime risk of melanoma in CDKN2A mutation carriers in a population-based sample. *J. Natl. Cancer Inst.* **2005**, *97*, 1507–1515. [CrossRef] [PubMed]

74. Vu, H.L.; Rosenbaum, S.; Purwin, T.J.; Davies, M.A.; Aplin, A.E. RAC1 P29S regulates PD-L1 expression in melanoma. *Pigment Cell Melanoma Res.* **2015**, *28*, 590–598. [CrossRef] [PubMed]

75. Halaban, R. RAC1 and melanoma. *Clin. Ther.* **2015**, *37*, 682–685. [CrossRef] [PubMed]

76. Arce, P.M.; Camilon, P.R.; Stokes, W.A.; Nguyen, S.A.; Lentsch, E.J. Is Sex an Independent Prognostic Factor in Cutaneous Head and Neck Melanoma? *Laryngoscope* **2014**, *124*, 1363–1367. [CrossRef] [PubMed]

77. CDC. Melanoma Incidence Rates and Death Rates by Race and Ethnicity. Available online: http://www.cdc.gov/cancer/skin/statistics/race.htm (accessed on 18 December 2017).

78. Gandini, S.; Sera, F.; Cattaruzza, M.S.; Pasquini, P.; Picconi, O.; Boyle, P.; Melchi, C.F. Meta-analysis of risk factors for cutaneous melanoma: II. Sun exposure. *Eur. J. Cancer* **2005**, *41*, 45–60. [CrossRef] [PubMed]

International Journal of
Molecular Sciences

MDPI

Article

CoQ10 Deficiency May Indicate Mitochondrial Dysfunction in Cr(VI) Toxicity

Xiali Zhong [1,2], Xing Yi [1], Rita de Cássia da Silveira e Sá [3], Yujing Zhang [1], Kaihua Liu [1], Fang Xiao [1] and Caigao Zhong [1,*]

[1] Department of Health Toxicology, School of Public Health, Central South University, Changsha 410008, China; xializhong87@sina.com (X.Z.); yistar@163.com (X.Y.); zhangyujing24@163.com (Y.Z.); lecapher@gmail.com (K.L.); fangxiao@csu.edu.cn (F.X.)

[2] Department of Environmental Health Science, Bloomberg School of Public Health, Johns Hopkins University, Baltimore, MD 21205, USA

[3] Department of Physiology and Pathology, Health Sciences Center, Federal University of Paraíba, 58059-900 João Pessoa, Brazil; ritacassia.sa@bol.com.br

* Correspondence: caigaozhong@gmail.com; Tel.: +86-731-8480-5461

Academic Editors: Ashis Basu and Takehiko Nohmi

Received: 11 February 2017; Accepted: 7 April 2017; Published: 24 April 2017

Abstract: To investigate the toxic mechanism of hexavalent chromium Cr(VI) and search for an antidote for Cr(VI)-induced cytotoxicity, a study of mitochondrial dysfunction induced by Cr(VI) and cell survival by recovering mitochondrial function was performed. In the present study, we found that the gene expression of electron transfer flavoprotein dehydrogenase (ETFDH) was strongly downregulated by Cr(VI) exposure. The levels of coenzyme 10 (CoQ10) and mitochondrial biogenesis presented by mitochondrial mass and mitochondrial DNA copy number were also significantly reduced after Cr(VI) exposure. The subsequent, Cr(VI)-induced mitochondrial damage and apoptosis were characterized by reactive oxygen species (ROS) accumulation, caspase-3 and caspase-9 activation, decreased superoxide dismutase (SOD) and ATP production, increased methane dicarboxylic aldehyde (MDA) content, mitochondrial membrane depolarization and mitochondrial permeability transition pore (MPTP) opening, increased Ca^{2+} levels, Cyt c release, decreased Bcl-2 expression, and significantly elevated Bax expression. The Cr(VI)-induced deleterious changes were attenuated by pretreatment with CoQ10 in L-02 hepatocytes. These data suggest that Cr(VI) induces CoQ10 deficiency in L-02 hepatocytes, indicating that this deficiency may be a biomarker of mitochondrial dysfunction in Cr(VI) poisoning and that exogenous administration of CoQ10 may restore mitochondrial function and protect the liver from Cr(VI) exposure.

Keywords: hexavalent chromium Cr(VI); coenzyme Q10; reactive oxygen species (ROS); mitochondrial membrane potential (MMP); L-02 hepatocytes; apoptosis

---

1. Introduction

Chromium (Cr) and its compounds have become a serious public health issue, causing environmental pollution [1,2] and threatening human health. The health hazards associated with exposure to Cr are dependent on its oxidation state [3,4], with hexavalent chromium Cr(VI) being the most toxic component. Throughout the world, human exposure occurs mainly via industrial uses, such as leather tanning and steel manufacturing, as well as in food additives and tobacco [5,6]. Another source of contact is drinking water contaminated with Cr(VI). In vivo and in vitro studies have demonstrated that Cr(VI) can cause a wide range of toxic effects, including hepatotoxicity, in animals and humans [1–4]. It has been reported that the liver shows the highest accumulation following oral exposure to Cr(VI) [7–9]. The liver is the primary organ involved in xenobiotic metabolism and, for this reason, is particularly susceptible to injury. However, the liver as a target organ for Cr(VI) after oral exposure in humans remains controversial. For instance, Proctor et al. reported that Cr(VI) is not carcinogenic to humans via the oral route of exposure at permissible drinking water concentrations [10]. Several other studies have suggested that Cr(VI) could induce liver injury [1,3,11] and may cause primary cancer or increase the risk of liver cancer [12–14]. The increasing incidence of Cr(VI)-induced hepatotoxicity has emphasized the importance of elucidating the intoxication mechanism and identifying useful antidotes for Cr(VI) toxic effects on the liver.

Cr(VI) can easily enter cells through anion channels, and, once inside, it is reduced to its intermediate metabolites, Cr(IV), Cr(V), and the more stable form Cr(III), by enzymatic and non-enzymatic reductants [15]. Reactive oxygen species (ROS) are generated in the oxidation–reduction process and play a critical role in the mechanism of Cr(VI)-induced cytotoxicity. ROS accumulation, for instance, is known to cause the collapse of mitochondrial membrane potential and the opening of the mitochondrial permeability transition pore (MPTP) [16]. Cr(VI) induces cell apoptosis through intrinsic and extrinsic pathways involving the release of cytochrome c (Cyt c) from the mitochondrial intermembrane space. The release of Cyt c from the mitochondrial intermembrane space is regulated by B-cell lymphoma-2 (Bcl-2) family proteins, including anti-apoptotic (such as Bcl-2 and Bcl-xl) and pro-apoptotic proteins (such as Bcl-xs, Bax, and Bid) [17]. It has been postulated that the ratio of anti- and pro-apoptotic Bcl proteins regulates the function of MPTP within the mitochondria. Cr(VI)-induced ROS accumulation might cause an imbalance in Bcl-2 family proteins, swelling of the mitochondrial membrane, opening of the MPTP, release of Cyt c into the cytoplasm, and activation of caspase-9 and -3 [17,18], which would eventually trigger cell apoptosis.

Coenzyme Q (CoQ10) is an essential endogenous molecule in cell respiration and metabolism. It functions as a mitochondrial antioxidant, inhibiting lipid peroxidation and scavenging free radicals, as well as maintaining genome stability [19]. Moreover, CoQ10 possesses an independent anti-apoptosis function that regulates MPTP [20]. CoQ10 biosynthesis occurs in the mitochondrial matrix through the mevalonate pathway [21]. Evidence suggests that mutations in the genes involved in the biosynthesis of CoQ10 could cause primary and secondary CoQ10 deficiencies. They have also been linked to various clinical mitochondrial diseases [22,23]. CoQ10 deficiency could disturb mitochondrial bioenergetics and oxidative stress, as demonstrated by decreased ATP generation, increased ROS production, and cell death [24,25]. In general, secondary CoQ10 deficiency may be induced by dietary insufficiency or exposure to certain xenobiotics [26,27]. Several studies have shown that CoQ10 is susceptible to environmental toxins, which cause CoQ10 deficiency at both the cellular and in vivo levels [28,29]. Considering the pivotal role played by CoQ10 in mitochondrial function, this study aimed to investigate whether Cr(VI) can induce changes in the level of CoQ10, and whether CoQ10 treatment is effective against Cr(VI)-induced hepatotoxicity.

2. Results

2.1. Effect of Coenzyme Q10 and Cr(VI) on L-02 Hepatocyte Viability

To assay the changes in cell viability after exposure to Cr(VI), we evaluated the dose effects of Cr(VI) on cultured L-02 hepatocytes with or without CoQ10 pretreatment. It was observed that increasing Cr(VI) concentrations of 0.5, 1, 2, 4, and 8 μM significantly decreased cell viability ($p < 0.05$), as shown in Figure 1A. Treatment with CoQ10 at 0–5 μM increased cell viability, although not at a statistically significant level ($p > 0.05$). In contrast, treatment with concentrations of 5–20 μM CoQ10 significantly decreased cell viability ($p < 0.05$), as shown in Figure 1B.

2.2. Cr(VI) Decreases CoQ10 Content in L-02 Hepatocytes

The assays of CoQ10 level showed that Cr(VI) decreased the CoQ10 concentration in the mitochondria of L-02 hepatocytes when compared with the control group ($p < 0.05$). Pretreatment with CoQ10 restored the level of CoQ10 compared with the 2 μM Cr(VI) treatment group (*p* < 0.05), as shown in Figure 1C.

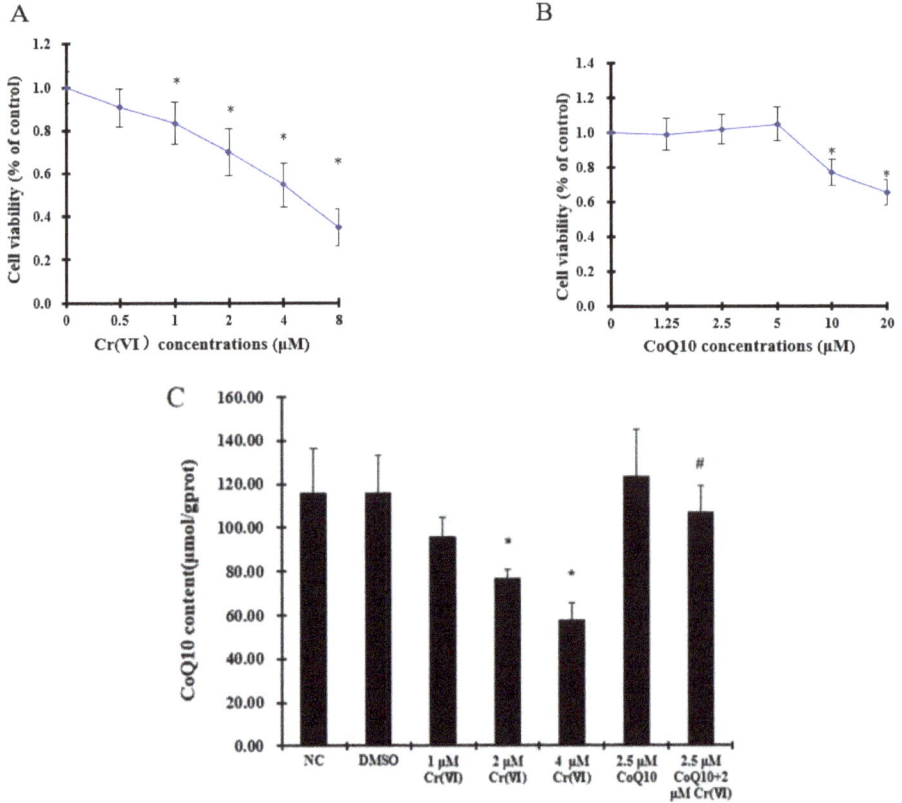

Figure 1. (**A**) Effect of different doses of Cr(VI) exposure on L-02 hepatocyte viability. Cells were cultured with different concentrations of Cr(VI), and the cell viability was determined by the 3-(4,5-dimethylthiazol-2-yl)-2,5-diphenyltetrazolium bromide (MTT) assay as described previously; (**B**) Effect of different doses of CoQ10 exposure on L-02 hepatocyte viability; (**C**) The CoQ10 content in the mitochondria of L-02 hepatocytes treated with CoQ10 and Cr(VI). The data were presented as mean ± SD (*n* = 6). * *p* < 0.05 compared with the control group; # *p* < 0.05 compared with the 2 μM Cr(VI) treatment group.

2.3. Effect of Cr(VI) on the Expression of Genes Involved in the CoQ10 Synthesis Pathway

Because Cr(VI) treatment decreased the level of CoQ10, and to understand the mechanism of CoQ10 deficiency, we analyzed the expression of genes involved directly or indirectly in the CoQ10 synthesis pathway. As shown in Table 1, electron transfer flavoprotein dehydrogenase (*ETFDH*) was the gene most strongly downregulated by Cr(VI) (log2(Ratio) = −1.41). *PDSS2*, *COQ5*, and *COQ9* were also downregulated by Cr(VI), but the fold changes were less than 2 (log2(Ratio) = −0.92, −0.99, and −0.94, respectively). Interestingly, aarF domain-containing kinase 3 (*ADCK3*) was upregulated after exposure to Cr(VI) (log2(Ratio) = 0.97). No significant changes were observed in the other genes. These results indicated that Cr(VI) affected the expression of genes directly or indirectly involved in the CoQ10 synthesis pathway to cause CoQ10 deficiency.

Table 1. Effect of Cr(VI) on the expression of genes involved in the CoQ10 biosynthesis pathway in L-02 hepatocytes.

Gene Name	Description	log2(Ratio)	*p*-Values
PDSS1	prenyl (decaprenyl) diphosphate synthase, subunit 1	0.079217	0.728622
PDSS2	prenyl (decaprenyl) diphosphate synthase, subunit 2	−0.917213	2.86×10^{-5}
COQ2	coenzyme Q2 homolog, prenyltransferase	−0.136864	0.434046
COQ3	coenzyme Q3 homolog, methyltransferase	−0.443767	0.002707
COQ4	coenzyme Q4 homolog	−0.646236	0.004524
COQ5	coenzyme Q5 homolog, methyltransferase	−0.986167	4.4×10^{-8}
COQ6	coenzyme Q6 homolog, monooxygenase	−0.185303	0.267227
COQ7	coenzyme Q7 homolog, ubiquinone	−0.308896	0.011926
COQ7	coenzyme Q7 homolog, ubiquinone	−0.528129	0.115161
ADCK3	aarF domain containing kinase 3	0.970573	6.12×10^{-9}
ADCK4	aarF domain containing kinase 4	NA	NA
COQ9	coenzyme Q9 homolog	−0.936477	3.1×10^{-18}
COQ10A	coenzyme Q10 homolog A	0.239029	0.045092
COQ10B	coenzyme Q10 homolog B	0.59875	0.000958
APTX	aprataxin	−0.648958	0.003698
ETFDH	electron-transferring-flavoprotein dehydrogenase	−1.412843	4.43×10^{-7}
BRAF	B-Raf proto-oncogene, serine/threonine kinase	−0.837738	0.008798
PMVK	phosphomevalonate kinase	−0.146285	0.455768
MVD	mevalonate (diphospho) decarboxylase	0.842443	0.01009
MVK	mevalonate kinase	0.612375	0.000892

Two-fold change indicates log2(Ratio) ≥ 1.0 or log2(Ratio) ≤ -1; log2(Ratio) = "NA" indicates that the difference in intensity between the two samples was ≥ 1000.0, $n = 3$.

2.4. Oxidative Damage Induced by Cr(VI) Is Reduced by CoQ10

We measured ROS production using dihydroethidine (DHE) and CellROX® Green Reagent, which detect superoxide and ROS, respectively. Cr(VI) significantly enhanced ROS and O_2^- generation compared with the control group ($p < 0.05$). Pretreatment with 2.5 μM CoQ10 prevented Cr(VI)-induced ROS accumulation and excessive O_2^-, as shown in Figure 2A,B. The levels of ROS in the CoQ10-treated group did not significantly change in comparison with those of the normal control group ($p > 0.05$). SOD is an enzyme that plays an important role in the protection of the cell membrane against oxidative stress. It can catalyze the dismutation of O_2^- to O_2 and to the less reactive species H_2O_2. Treatment with Cr(VI) caused a significant decrease in SOD levels when compared to control values. No significant difference in SOD activity was observed between the Cr(VI) group pretreated with CoQ10 and the control group ($p < 0.05$), as shown in Figure 2C. In addition, CoQ10 effectively restored the SOD level to protect hepatocytes against oxidative damage induced by Cr(VI). MDA was evaluated as an indicator of hepatocyte lipid peroxidation. Cr(VI) significantly increased MDA levels compared with the control group ($p < 0.05$), and pretreatment with CoQ10 markedly reduced the level of Cr(VI)-induced MDA ($p < 0.05$) (Figure 2C).

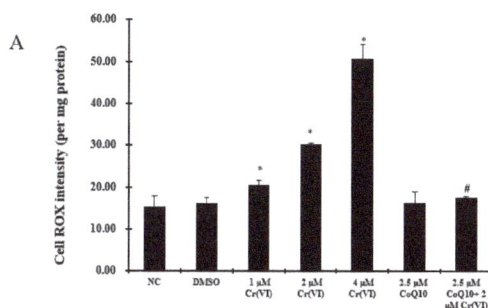

Figure 2. *Cont.*

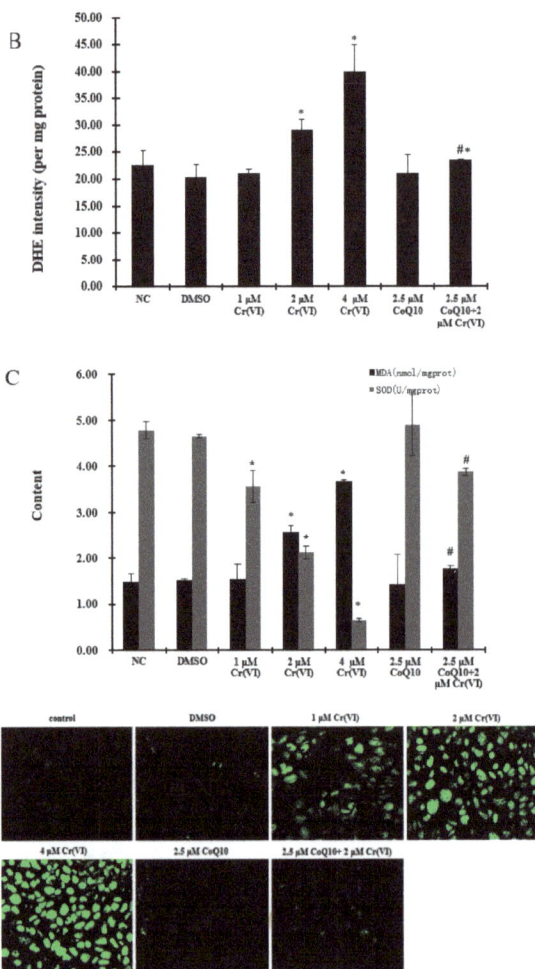

Figure 2. CoQ10 attenuates oxidative damage induced by Cr(VI). (**A**) Quantification of ROS levels. Effect of CoQ10 on Cr(VI)-induced ROS accumulation in L-02 hepatocytes and quantitation by fluorescence spectrophotometry; (**B**) After L-02 hepatocytes were treated with Cr(VI) (0~4 µM) for 24 h, with or without CoQ10 pretreatment for 2 h, O_2^- generation was detected with dihydroethidium; (**C**) CoQ10 reduced the oxidative damage induced by Cr(VI). The cells were treated with Cr(VI) (0~4 µM) for 24 h, with or without CoQ10 pretreatment, and MDA was detected by the MDA detection kit as the end product of lipid oxidation. SOD was measured using the total superoxide dismutase activity assay, which involves the inhibition of superoxide-induced chromogen chemiluminescence by SOD; (**D**) Effect of CoQ10 on Cr(VI)-induced ROS accumulation. The cells were incubated with 5 µM of CellROX® Green Reagent for 30 min and observed under a confocal microscope using a 40× objective. Brighter green fluorescence indicated greater ROS accumulation. The data are presented as mean ± SD ($n = 6$). * $p < 0.05$ compared with the control group; # $p < 0.05$ compared with the 2 µM Cr(VI) treatment group.

2.5. Induction of Mitochondrial Loss by Cr(VI) Can Be Counteracted by Supplementation with CoQ10

Mitochondrial loss was reflected by a decrease in the mitochondrial mass and mtDNA. We examined the mitochondrial mass using the 10-*N*-nonyl acridine orange (NAO) fluorescence intensity, which was lower in the different Cr(VI) concentration groups than in the control group. CoQ10 maintained the mitochondrial mass at a normal level against Cr(VI) exposure (Figure 3A). In addition, Cr(VI) treatment markedly reduced the mtDNA copy number, and CoQ10 preserved the mtDNA copy number (Figure 3B).

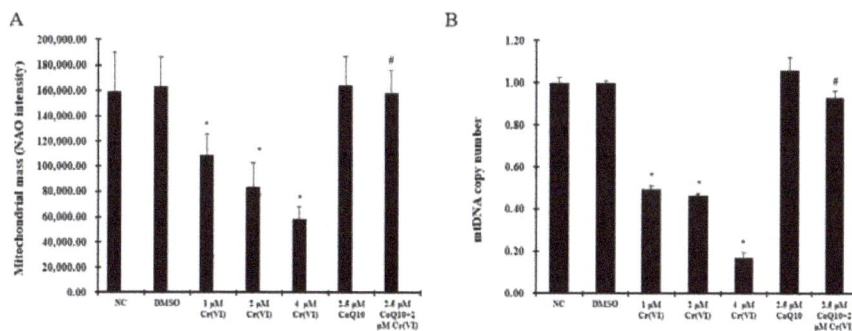

Figure 3. Cr(VI) triggers significant mitochondrial biogenesis loss. (**A**) NAO staining was used to analyze the mitochondrial mass using a microplate reader; (**B**) quantitative real-time PCR analysis was applied to detect the mtDNA copy number. The data are presented as mean $\pm$ SD ($n = 6$). * $p < 0.05$ compared with the control group; # $p < 0.05$ compared with the 2 μM Cr(VI) treatment group.

2.6. Cr(VI) Induces Mitochondrial Depolarization, MPTP Opening, Ca²⁺ Overload, and Decreased ATP Levels, and These Outcomes Are Attenuated by CoQ10

The effect of Cr(VI) on the mitochondrial membrane potential (MMP) was quantified by the uptake of JC-1, as illustrated in Figure 4A. The shift in the membrane potential was observed as the disappearance of fluorescent red/green-stained mitochondria, showing a large negative MMP, and as the increase in fluorescent green-stained mitochondria, indicating the loss of MMP. With increasing Cr(VI) concentrations, MMP significantly decreased as compared to the normal controls ($p < 0.05$), indicating mitochondrial membrane depolarization. Pretreatment with CoQ10 relieved Cr(VI)-induced mitochondrial membrane depolarization. To directly assess MPTP opening, the calcein-AM-cobalt assay was performed in L-02 hepatocytes. The inner mitochondrial membrane permeability was significantly increased in response to Cr(VI) stimuli in a concentration-dependent manner. The degree of MPTP opening in cells after co-incubation with 2 μM Cr(VI) and 2.5 μM CoQ10 was significantly decreased compared with cells incubated with 2 μM Cr(VI) alone (Figure 4B). MPTP opening is a Ca^{2+}-dependent event; hence, to assess the Ca^{2+} concentration in L-02 hepatocytes, as shown in Figure 4C, the cells were incubated with the Flo-3M probe to detect the intercellular Ca^{2+} concentration. Cr(VI) caused a statistically significant Ca^{2+} concentration-dependent increase ($p < 0.05$), while pretreatment with CoQ10 significantly attenuated Ca^{2+} overload when compared with treatment with 2 μM Cr(VI) alone ($p < 0.05$). These results suggest that CoQ10 suppressed Ca^{2+} overload and maintained a suitable degree of MPTP opening. Apoptosis induced by toxicants is an energy-consuming process and thus is accompanied by a massive decrease in cellular ATP production. To examine the effects of Cr(VI) on mitochondrial ATP production, L-02 hepatocytes were treated with 1, 2, or 4 μM Cr(VI) for 24 h or pretreated with 2.5 μM CoQ10 for 30 min, followed by the addition of 2 μM Cr(VI). ATP levels were then measured colorimetrically. As shown in Figure 4D, Cr(VI) induced a statistically significant and concentration-dependent decrease in ATP in L-02 hepatocytes ($p < 0.05$). Pretreatment with CoQ10 significantly decreased the Cr(VI)-induced decrease in ATP ($p < 0.05$).

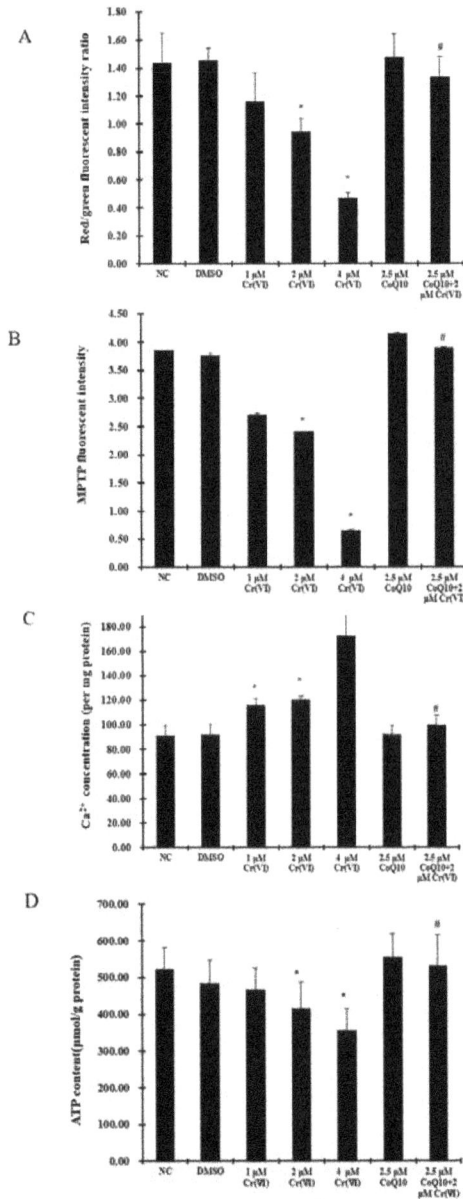

Figure 4. Cr(VI) induces mitochondrial depolarization, MPTP opening, Ca^{2+} overload, and ATP level decrease, and these outcomes are attenuated by CoQ10. (**A**) Effect of CoQ10 on Cr(VI)-increased mitochondrial membrane potential in L-02 hepatocytes. The mitochondrial membrane potential was examined by JC-1 staining; (**B**) The activity of MPTP was detected using the calcein-AM-cobalt assay; (**C**) the Ca^{2+} concentration was measured with Flo-3M by fluorescence spectrophotometry; (**D**) cells were treated with Cr(VI) (0–4 µM) for 24 h, with or without CoQ10 pretreatment for 2 h, and the ATP levels in L-02 hepatocytes were assessed. The data are presented as mean $\pm$ SD ($n = 6$). * $p < 0.05$ compared with the control group; # $p < 0.05$ compared with the 2 µM Cr(VI) treatment group.

2.7. Cr(VI) Induces Cyt c Release, Caspase-3 and Caspase-9 Activation, and Unbalanced Bcl-2/Bax Expression in Response to Apoptotic Stimuli, and CoQ10 Counteracts These Outcomes

Cyt c, caspase-3 and caspase-9 activities were analyzed as indexes of apoptosis execution via the intrinsic (mitochondrion-dependent) pathway. Figure 5 shows the release of Cyt c into the cytoplasm. At 24 h after application of Cr(VI), cytoplasmic Cyt c levels were markedly increased but remained substantially unaffected if treatment was preceded by CoQ10 administration. Similarly, Figure 6A,B show that caspase-3 and caspase-9 activities were enhanced at 24 h after Cr(VI) exposure. The enhancement was dramatically lower when Cr(VI) exposure was preceded by CoQ10 administration. CoQ10 had the ability to prevent Cyt c release, and caspase-3 and caspase-9 activation in response to Cr(VI) exposure, three events that are triggered by MPTP opening. CoQ10 inhibited apoptosis by directly maintaining MPTP in the closed conformation. Additionally, the process of apoptosis is regulated by the Bcl-2 family of proteins, which includes anti-apoptotic and pro-apoptotic proteins. As shown in Figure 6C,D, 24-h exposure to different concentrations of Cr(VI) induced significant concentration-dependent inhibition of Bcl-2 and induction of Bax. Pretreatment with 2.5 µM CoQ10 restored Bcl-2 expression and decreased Bax expression compared with treatment with 2 µM Cr(VI) alone.

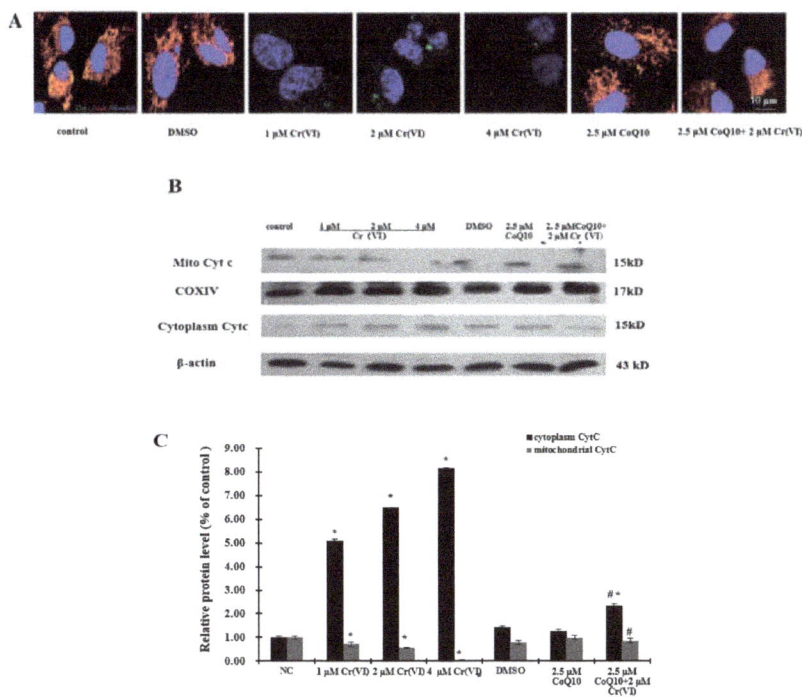

Figure 5. Cr(VI) induces cytochrome c release from the mitochondria to the cytoplasm. (**A**) Merged images of the mitochondria (red), Cyt c (green) and nucleus (blue) after exposure to Cr(VI) for 24 h in L-02 hepatocytes. Cyt c (green) and mitochondria (red) localization (yellow) indicates that Cyt c is still inside mitochondria. The separation of Cyt c and mitochondria suggests that Cyt c is no longer within the mitochondria and has been released into the cytoplasm, scale bar: 10 µm; (**B**) CoQ10 prevents Cyt c release to the cytoplasm; Cyt c protein expression was measured by Western blotting. COXIV and β-actin were used as loading controls; (**C**) The relative protein levels were calculated by Image J software. Experiments were repeated three times and showed similar results.

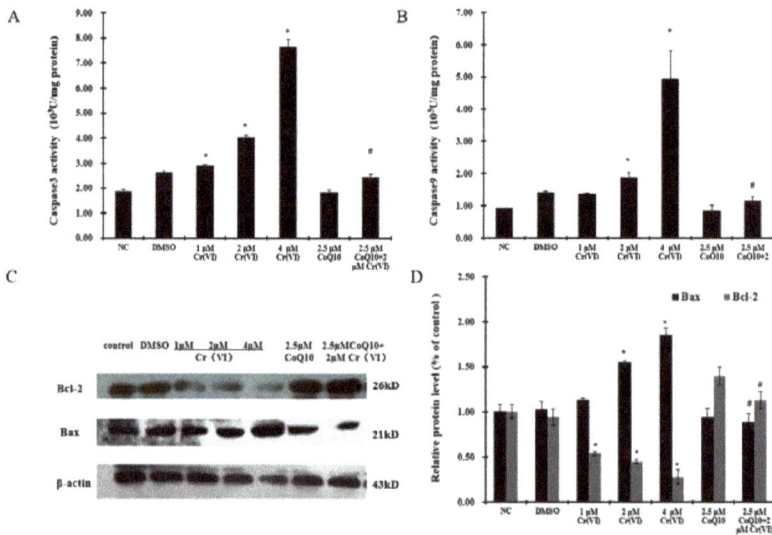

Figure 6. Cr(VI) induces caspase-3 and caspase-9 activation and unbalanced Bcl-2/Bax expression in response to apoptotic stimuli, and CoQ10 counteracts these outcomes. Cells were treated with Cr(VI) (0~4 μM) for 24 h, with or without CoQ10 pretreatment for 2 h. Caspase-3 (**A**) and -9 (**B**) activities were detected using a microplate reader; (**C**) The expression of Bcl-2 and Bax was measured by Western blotting and the relative protein levels were calculated by Image J software (**D**). The data are expressed as mean ± SD ($n = 6$). * $p < 0.05$ compared with the control group; # $p < 0.05$ compared with the 2 μM Cr(VI) treatment group.

2.8. Cr(VI) Induces L-02 Hepatocyte Apoptosis in a Concentration-Dependent Manner, and CoQ10 Might Reduce the Rate of Apoptosis

To measure the effects of CoQ10 on programmed cell death after Cr(VI) exposure, we analyzed cell apoptosis using the Annexin V-FITC and propidium iodide (PI) double staining methods after incubation for 30 min. As shown in Figure 7A,B, 24 h exposure of Cr(VI) increased the early and late apoptotic populations in L-02 hepatocytes. Approximately 5.95%–48.46% of the cell population expressed high FITC and low PI signals, which are indicative of apoptotic cells, following treatment with up to 4 μM Cr(VI). Pretreatment with CoQ10 attenuated the Cr(VI)-induced increase in Annexin V-positively stained cells. The protective effect of CoQ10 on Cr(VI)-induced apoptosis was 15.28% ($p < 0.05$), indicating that CoQ10 can attenuate Cr(VI)-induced apoptosis.

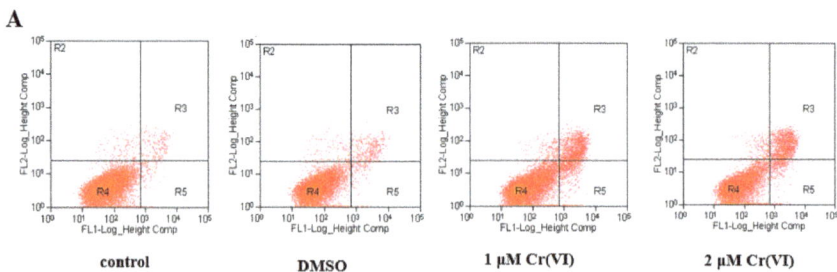

Figure 7. *Cont.*

B

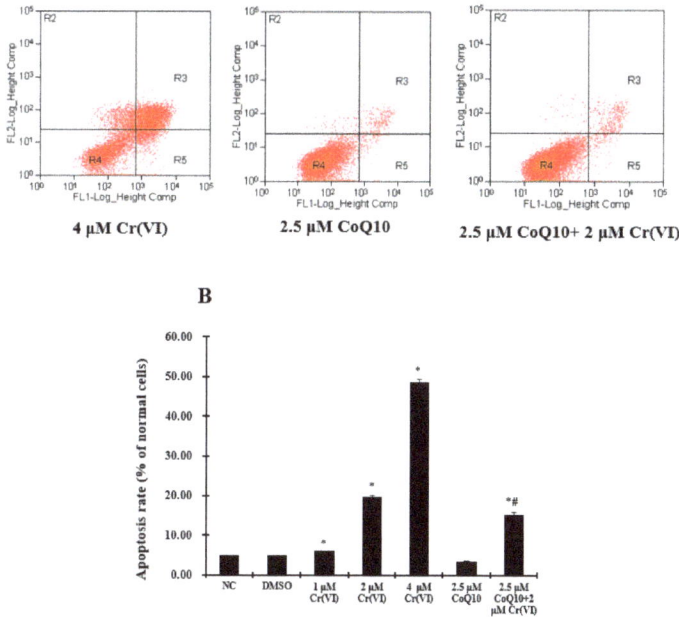

Figure 7. Cr(VI) induces apoptosis in L-02 hepatocytes, and CoQ10 antagonizes apoptosis. (**A**) Cells were stained with Annexin V-FITC/PI and analyzed by flow cytometry. Both early apoptotic and late apoptotic cells were assessed in the cell death determinations. The experiments were repeated three times; (**B**) Quantification of apoptotic cells. Data were obtained from flow cytometry assays and were expressed as mean $\pm$ SD ($n = 6$). * $p < 0.05$ compared with the control group; # $p < 0.05$ compared with the 2 μM Cr(VI) treatment group.

3. Discussion

In the present study, we demonstrated that Cr(VI)decreased the level of endogenous CoQ10 by disturbing the CoQ10 synthesis pathway, and that the pretreatment with CoQ10 maintained the level of endogenous CoQ10. These findings led us to investigate the role of CoQ10 in the mechanism of Cr(VI)-induced hepatotoxicity and its possible role as a hepatoprotective agent against Cr(VI)-induced hepatocyte damage.

In order to achieve these goals, we examined the genes involved both directly and indirectly in the CoQ10 biosynthetic pathway. Cr(VI) changed the expression of many genes, including *ETFDH*, which exhibited the strongest downregulation. *ETFDH* is indirectly involved in the biosynthesis of CoQ10 and encodes a component of the electron-transfer system in mitochondria. Gempel et al. reported that mutations in the *ETFDH* gene cause pure myopathy, as evidenced in seven patients from five different families with severely decreased activities of respiratory chain complexes I and II + III and CoQ10 deficiency [30]. We also previously demonstrated that Cr(VI) induces mitochondrial dysfunction by disturbing electron transport and inhibiting the respiratory chain complexes. As a link to our present study, we tentatively propose that Cr(VI) may cause secondary deficiency of CoQ10 in L-02 hepatocytes, resulting in mitochondrial dysfunction and cell apoptosis. This result provides a new perspective on the mechanism of Cr(VI)-induced hepatotoxicity.

In addition, Cr(VI) downregulated the expression of *PDSS2*, *COQ5*, and *COQ9*, which are directly involved in the CoQ10 biosynthesis pathway. Defects in these genes are a cause of CoQ10 deficiency [31]. *COQ5* catalyzes the only C-methylation step in the CoQ10 biosynthesis pathway in yeast. Chen et al. demonstrated that an uncoupling chemical in CoQ10 dose deficiency downregulates the mature form of *COQ5*. They also showed that knockdown of the *COQ5* gene reduces CoQ10 levels, indicating that *COQ5* plays a critical role in the biosynthesis of CoQ10 [32]. In the present study, Cr(VI) suppressed *COQ5* gene expression, but not strongly. It is possible that Cr(VI) induced *ETFDH* inhibition, causing mitochondrial dysfunction to disturb the expression of genes in CoQ10 biosynthesis. This is supported by Hsiu-Chuan's study, which reported the suppressive effect of FCCP on *COQ5* levels in association with decreased mitochondrial membrane potential, mitochondrial ATP production, and CoQ10 levels [33]. Interestingly, *ADCK3* was upregulated after exposure to Cr(VI), although not significantly. *ADCK3* is required for the biosynthesis of CoQ10, and mutation of *ADCK3* has been associated with CoQ10 deficiency in humans [34]. *ADCK3* also functions in an electron-transferring membrane protein complex in the respiratory chain. Tumor suppressor p53 can induce *ADCK3* expression, and in response to DNA damage, inhibition of *ADCK3* expression partially suppresses p53-induced apoptosis [35]. We believe that CoQ10 maintains homeostasis only when gene expression is maintained at normal levels. However, little is currently known about the regulation of CoQ10 gene expression. Here, we present preliminary data that mitochondrial CoQ10 deficiency may represent a potential biomarker of Cr(VI) toxicity. However, to better understand the mechanism of Cr(VI)-induced CoQ10 deficiency, more robust evidence is needed.

Mitochondrial loss has been indicated to play a prominent role in mitochondrial dysfunction. Our results showed that Cr(VI) exposure led to mitochondrial loss in hepatocytes, as reflected by a decrease in the mitochondrial mass, mtDNA copy number, and inhibition of expression of components of the mitochondrial respiratory chain [36]. The observed Cr(VI)-induced decrease in the level of mtDNA copy number supports a link between CoQ10 deficiency and mtDNA depletion [37,38]. Moreover, mtDNA is vulnerable to ROS due to the lack of protection from histones and a self-repair mechanism. Our results also sowed Cr(VI)-induced CoQ10 deficiency and ROS accumulation. It is reported that CoQ10 deficiency may cause mitochondrial dysfunction, thus triggering ROS generation. The capability of ROS scavenging is weakened, possibly further aggravating ROS accumulation due to CoQ10 deficiency [39]. Supplementation with CoQ10 could increase the mitochondrial mass, mtDNA copy number, and mitochondrial electron transport chain activity, as demonstrated by our study and Duberley's research [40]. Previous investigations have also shown that CoQ10 protects against neuron apoptosis induced by iron by reducing ROS accumulation and inhibiting lipid peroxidation [41]. Consistent with previous studies, we found that pretreatment with CoQ10 eliminated excessive ROS and O_2^-, reduced lipid peroxidation, and maintained SOD content. SOD plays an important role in the protection of cell membranes against oxidative stress by catalyzing the dismutation of O_2^- to O_2 and to the less reactive species H_2O_2 [42], corroborating the findings that CoQ10 plays a critical role in Cr(VI)-induced mitochondrial dysfunction.

Mitochondria are the factory of ATP production via oxidative phosphorylation [43]. In return, the primary function of mitochondria and homeostasis are maintained by adequate ATP levels. CoQ10 is an essential cofactor of oxidative phosphorylation to permit ATP biosynthesis [44]. Our data indicated that Cr(VI) causes CoQ10 deficiency and markedly reduces cellular ATP. ATP production is also related to Ca^{2+} and ROS generation. Calcium overload has been proposed to play a crucial role in ROS generation [45], which could aggravate mitochondrial damage in hepatocytes. Furthermore, Ca^{2+} induces Cyt c release from mitochondria by enhancing Cyt c dislocation, competing for cardiolipin-binding sites in the mitochondrial inner membrane, or activating MPTP opening, resulting in the disturbance of electron transfer and the Q cycle. As a consequence, there is an upsurge in ROS accumulation [46]. Therefore, eliminating Ca^{2+} overload and ROS accumulation are conducive to promoting ATP synthesis and repairing damaged mitochondria. In this study, we have demonstrated that CoQ10 could attenuate Cr(VI)-induced adverse effects by preventing Ca^{2+} overload and scavenging excessive ROS.

Apoptosis is an energy-consuming process that is regulated by the activation of caspases [47]. Cr(VI) induced the release of Cyt c from mitochondria, accompanied by activation of caspase-3 and 9, which are associated with significantly reduced Bcl-2 expression and enhanced Bax expression.

Previous studies have confirmed that, after Cyt c is released from the mitochondrial intermembrane space into the cytosol, it forms a complex with Apaf-1 and pro-caspase, thereby triggering caspase-9 activation and the consequent initiation of the caspase cascade and induction of cell apoptosis [48]. In addition, cell apoptosis is regulated by the balance of pro- and anti- apoptotic proteins. It is generally believed that the anti-apoptotic protein Bcl-2 counterbalances oxidative damage and maintains the structural and functional integrity of the mitochondrial membrane by preventing Cyt c release. The Bax protein exerts an important effect in cell apoptosis. High Bax expression levels and the formation of homo- or heterodimers with Bcl-2 lead to cell death [49]. We also saw that pretreatment with CoQ10 could block the release of Cyt c, inhibit the activation of caspase-3 and caspase-9, and regulate the expression of Bcl-2/Bax to reach equilibrium and prevent subsequent apoptosis in L-02 hepatocytes. These results are compatible with the findings of other studies using CoQ10 against iron-induced neuronal toxicity [28], statin toxicity in hepatocytes [50], or ethanol-induced apoptosis in corneal fibroblasts [51].

As mentioned above, CoQ10 could alleviate Cr(VI)-induced hepatotoxicity. However, the mechanism of the protective effects of CoQ10 is not fully understood. Possible reasons for the protective effects of CoQ10 in Cr(VI)-induced toxicity include the fact that Cr(VI) is a water-soluble compound that can be easily reduced by water-soluble antioxidants [52]. Therefore, Cr(VI) might be reduced to Cr(III) before entering the cell when co-incubated with CoQ10. However, considering that Cr(VI) can dissolve in the medium, CoQ10, being a lipid-soluble quinone, may fail to interact with Cr(VI) outside the cell. On the other hand, since the cells were pretreated with CoQ10 for 2 h prior to exposure to Cr(VI), it is possible that CoQ10 may have entered the cell, and CoQ10 was distributed on membranes to block the opening of anion channels to prevent the entrance of Cr(VI) into the cell. This is supported by several studies showing that quinones modulate MPTP through a common binding site rather than through redox reactions [53,54]. This study also demonstrated that CoQ10 could decrease the MPTP opening degree to antagonize the mitochondrial toxicity induced by Cr(VI). Moreover, considering that exogenous CoQ10 enhanced the level of CoQ10 in mitochondria, we propose that the protective effect of CoQ10 might be associated with its role as a mobile electron transporter. CoQ10 can correct the disorder of electron transfer and improve the Q cycle, thus attenuating Ca^{2+} overload and Cyt c release. We have shown that CoQ10 can prevent cell apoptosis, but the results did not indicate how CoQ10 plays a protective role against Cr(VI)-induced hepatotoxicity. The detoxification role of CoQ10 against Cr(VI)-induced hepatotoxicity should be more comprehensively investigated.

4. Materials and Methods

4.1. Materials

Potassium dichromate ($K_2Cr_2O_7$), coenzyme Q10, 3-(4,5-dimethylthiazol-2-yl)-2,5-diphenyl-tetrazolium bromide (MTT), and dimethyl sulfoxide (DMSO) were purchased from Sigma (St. Louis, MO, USA). RPMI-1640 culture medium, fetal bovine serum (FBS), trypsin, and penicillin-streptomycin were provided by Dingguo Changsheng Biotechnology Co. LTD (Beijing, China). All solvents and chemicals were analytical grade.

4.2. Cell Culture

The normal liver L-02 cell line (Type Culture Collection of the Chinese Academy of Sciences, Shanghai, China) was derived from adult human normal liver, immortalized by stable transfection with the hTERT [55], and was reported to be liver-specific [56]. L-02 hepatocytes were maintained in 1640 RPMI medium containing 10% fetal bovine serum and a 1% mixture of penicillin and streptomycin in a 5% CO_2 humidified atmosphere at 37 °C. The medium was changed every two days, and the cells were passaged using trypsin.

4.3. Treatment of Cells with Cr(VI) and CoQ10

Cells were treated with a final concentration of 1–4 µM $K_2Cr_2O_7$ for 24 h in a complete medium. Cells were pretreated for 2 h prior to $K_2Cr_2O_7$ exposure and were co-treated for the 24-h $K_2Cr_2O_7$ exposure period at a final concentration of 2.5 µM CoQ10. Control samples were exposed to an equivalent concentration of DMSO as the solvent control.

4.4. Cell Viability Assay

MTT was used to evaluate cell viability. Cells were seeded in 96-well plates (1×10^4/well) and cultured in the presence of 0, 0.5, 1, 2, 4, or 8 µM Cr(VI) or 0, 1.25, 2.5, 5, 10, or 20 µM CoQ10 for 24 h. After 24 h of incubation, 10 µL of MTT solution (stock solution of 5 mg/mL in PBS) was added to each well of the 96-well plates, and the plates were incubated for an additional 4 h at 37 °C. The MTT-reducing activity of the cells was measured by treatment with DMSO prior to reading at 490 nm with an automatic microplate reader.

4.5. Preparation of Mitochondria

The treated cells were washed with PBS and then centrifuged at 600 g for 5 min. The supernatant was discarded, and the pellets were suspended in ice-cold lysis buffer (250 mM sucrose, 20 mM *N*-(2-hydroxyethy)piperazine-*N'*-(2-ethanesulfonic acid (HEPES), pH 7.4, 10 mM KCl, 1.5 mM $MgCl_2$, 1 mM each of EGTA, EDTA, DTT, and PMSF, and 10 µg/mL each of leupeptin, aprotinin, and pepstatin A) and incubated for 20 min. The cells were homogenized up and down 20 times, on ice, at 1000 r.p.m., transferred to a new tube, and centrifuged at 600× *g* for 10 min at 4 °C The supernatant was then centrifuged at 12,000× *g* for 10 min at 4 °C. Afterwards, the supernatants were collected and centrifuged at 12,000× *g* for 15 min at 4 °C for preparation of the cytosolic fraction. The precipitated pellets were resuspended in the lysis buffer and were used as the mitochondrial fraction after centrifugation at 12,000× *g* for 10 min [57].

4.6. Extraction and Quantification of CoQ10

Extraction was performed as described previously. Briefly, 250 µL of samples and 750 µL of hexane:ethanol (5:2, *v/v*) were mixed together for 1 min using a vortex mixer. The mixture was centrifuged for 3 min at 4000× *g*, and 450 µL of the hexane layer was collected, dried under a stream of nitrogen, and dissolved in 100 µL of ethanol (1:1, *v/v*) [58]. Quantification of CoQ10 was performed by HPLC according to Lass and Sohal [59]. A 10-µL aliquot of the extract was chromatographed on a reverse-phase C180 HPLC column (150 mm × 4.6 mm, 5 µM; Thermo Hypersil), using a mobile phase consisting of ethanol: methanol (1:1, *v/v*) at a flow rate of 0.8 mL/min. The eluent was monitored with a UV detector at 275 nm.

4.7. Gene Chip ANALYSIS

Total RNA was isolated using Trigo (Sigma), and first-strand cDNA was synthesized using RevertAid M-MuL V Reverse Transcriptase (Thermo, Waltham, MA, USA). A Genechip 30 IVT Express Kit was used to synthesize double-stranded cDNA for in vitro transcription (IVT, standard Affymetrix procedure, Santa Clara, CA, USA). During the synthesis of the amplified RNA (aRNA), a biotinylated nucleotide analog was incorporated as a label for the message, and the aRNA was purified with magnetic beads. A 15-µg quantity of aRNA was fragmented with a fragmentation buffer according to the manufacturer's instructions. Next, 15 µg of fragmented aRNA were hybridized with Affymetrix Human Genome U133 plus 2.0 arrays, according to the manufacturer's instructions. The chips were heated in a GeneChip Hybridization Oven-645 for 16 h at 60 rpm and 45 °C. The chips were washed and stained using a Genechip Fluidics Station-450 with the Affymetrix HWS kit. Chip scanning was performed with an Affymetrix Gene-Chip Scanner-3000-7G, and the normalized data were extracted using Affymetrix GCOS software (1.0, Santa Clara, CA, USA). The normalized spot intensities were transformed to gene expression log 2 ratios, and comparisons between control and treated groups were conducted using a *t*-test. A *p*-value ≤ 0.05 and a fold change value ≥ 2 indicated statistically significant regulation. A two-fold change was indicated at log2(Ratio) ≥ 1.0 or log2(Ratio) ≤ -1.

4.8. Real-Time PCR

Total RNA was isolated using Trigo (Sigma), and the first-strand cDNA was synthesized using RevertAid M-MuL V Reverse Transcriptase (Thermo). The cDNA was amplified in 20-μL reactions using SYBR premix Dye I (TAKARA BIO, Shiga, Japan) in a Thermal Cycler Dice Real-Time System (TAKARA BIO). The mRNA expression was normalized to the expression of the standard reference gene GAPDH. The primer sequences are shown in the Table 2.

Table 2. Primer sequences.

Target	Forward Primer (5′→3′)	Reverse Primer (5′→3′)
mtDNA	CAAACCTACGCCAAAATCCA	GAAATGAATGAGCCTACAGA
GAPDH	TGACAACAGCCTCAAGAT	GAGTCCTTCCACGATACC

4.9. Determination of Reactive Oxygen Species (ROS)

Cells were washed three times with a serum-free medium, and thereafter CellROX® Green Reagent (Thermo Fisher, Waltham, MA, USA) was added to the cells at 5 μM final concentration, followed by incubation for 30 min at 37 °C. The cells were then washed twice with PBS and harvested. One hundred microliters of resuspended sample were added to 96-well plates, and ROS levels were assessed by fluorescence spectrophotometry with excitation at 485 nm and emission at 520 nm. Cells on the coverslip were observed under a confocal microscope.

4.10. Measurement of Superoxide Anion (O_2^-)

The treated cells were incubated with the cellular O_2^--sensitive fluorescent indicator dihydroethidium (DHE) at a final concentration of 5 μM for 30 min at 37 °C, and protected from light. They were washed three times with PBS, collected, resuspended in 2 mL of PBS, and finally examined by fluorescence spectrophotometry with excitation at 535 nm and emission at 610 nm.

4.11. Evaluation of Methane Dicarboxylic Aldehyde (MDA) and Superoxide Dismutase (SOD) Levels

MDA levels were determined using the trace MDA detection kit and a microplate reader at 530 nm. SOD levels were measured using the total superoxide dismutase activity assay, which involves the inhibition of superoxide-induced chromogen chemiluminescence by SOD, according to the manufacturer's instructions. The absorbance of the wells was read using a microplate reader at a primary wavelength of 550 nm.

4.12. Measurement of the Mitochondrial Mass

The mitochondrial mass was evaluated using the fluorescent probe 10-*N*-nonyl acridine orange (NAO) as previously described [30]. Treated cells were incubated in a medium containing 5 μM NAO for 30 min at 37 °C and protected from light. The NAO fluorescence intensity was determined using a microplate reader (Gemini EM, Molecular Devices, Sunnyvale, CA, USA). The emission and excitation wavelengths were 530 and 485 nm, respectively.

4.13. Measurement of the Mitochondrial Transmembrane Potential (MMP, Δψm) in Cells

The cells were loaded with JC-1 for 20 min at 37 °C. Depolarization of Δψm was assessed by measuring the fluorescence intensities at excitation and emission wavelengths of 490 and 539 nm, respectively, to measure JC-1 monomers. An excitation wavelength of 525 nm and an emission wavelength of 590 nm were used to measure JC-1 aggregates using a fluorescence microplate reader. During the measurements, the cells were maintained at 4 °C and protected from light. All fluorescence measurements were corrected by autofluorescence, which was determined using cells not loaded with JC-1.

4.14. Measurement of the MPTP Opening Degree

The cells were washed three times with PBS, and calcein-AM was added, followed by incubation for 20 min at 37 °C. Next, the cells were washed twice with GENMED cleaning liquid, harvested, resuspended in a cleaning liquid, and finally examined by fluorescence spectrophotometry with excitation at 488 nm and emission at 505 nm.

4.15. Measurement of Intracellular ATP Levels

The level of ATP was examined using the ATP assay kit (S0026, Beyotime, Shanghai, China), which utilized the catalysis of firefly luciferase to generate fluorescence requiring ATP to develop an ATP quantified method. The measurement was performed following the manufacturer's instructions. The absorbance was detected using a Luminometer.

4.16. Measurement of the Cellular Calcium Concentration (Ca^{2+})

Cultured cells were incubated with the cellular Ca^{2+}-sensitive fluorescence indicator Fluo-3AM at a final concentration of 2.5 μM for 30 min at 37 °C and protected from light. The cells were subsequently washed three times with PBS, collected, and resuspended in 2 mL of PBS for posterior examination by fluorescence spectrophotometry with excitation at 488 nm and emission at 525 nm.

4.17. Caspase Activity Assay

Caspase levels were measured with caspase-3 and caspase-9 activity assay kits according to the manufacturer's instructions (Beyotime). The absorbance was measured on a microplate reader at 405 nm, and the caspase activities were subsequently calculated based on the absorbance.

4.18. Western Blot Analysis

For Western blot analysis, cytosolic and mitochondrial fractions were prepared as reported previously. Samples containing an equal amount of concentrated proteins were separated on 10% gradient SDS-polyacrylamide gel and transferred to a polyvinylidene difluoride membrane by electroblotting for 90 min at 100 V and 4 °C. Non-specific membrane binding sites were blocked with blocking solution (PBS, 0.5% Tween-20, pH 7.4), containing 5% non-fat dry milk for 1 h at 4 °C. The membrane was incubated with primary mouse anti-Cyt c monoclonal antibody (Abcam, Cambridge, UK) diluted 1:1500 or goat anti-rabbit Bcl-2 (1:500, Santa Cruz, Santa Cruz, CA, USA) and Bax antibody (1:500, Santa Cruz) in blocking solution overnight at 4 °C. The membrane was washed thoroughly with PBS-T and then incubated for 1 h with horseradish peroxidase-conjugated anti-mouse (1:4000; Santa Cruz) or anti-rabbit IgG antibody (1:6000; Santa Cruz) in blocking solution, detected by chemiluminescence reagent plus, and exposed to film.

4.19. Immunofluorescence

Cyt c translocation from mitochondria to cytoplasm was analyzed by immunofluorescence. Treated cells were incubated with Mito-Tracker Red (Thermo Fisher) at 500 nM for 45 min, washed twice with PBS, and with 2% paraformaldehyde for 20 min at room temperature. The cells were blocked with blocking solution (1% BSA, 0.15% saponin and 10% goat serum in PBS) for 30 min at room temperature, incubated with primary antibody specific for Cyt c (1:200) overnight at 4 °C, and then incubated with secondary antibody for 1 h at room temperature. The cells were washed three times with PBS, and then the nuclei were stained with Hoechst for 5 min. Images were captured with a confocal microscope.

4.20. FITC Annexin V/propidium Iodide (PI) Staining for Apoptotic Cells

Cells were washed three times with PBS before suspension in binding buffer, and 10 μL of Annexin V-FITC were mixed with 200 μL of cell suspension containing 10^6 cells. The cells were incubated at room temperature for 30 min and shielded from light. Then, 10 μL of PI solution were added to the cells and they were incubated for 10 min on ice. The scatter parameters of the cells were analyzed using a flow cytometer. Usually, four cell populations are identified by the flow cytometer. The viable population is displayed in the lower-left quadrant; the early apoptotic population and late apoptotic population are presented in the lower-right quadrant and in the upper-right quadrant, respectively. Signals in the upper-left quadrant present a necrotic population.

4.21. Protein Assay

All protein assays in this study were measured using a Q5000 UV-Vis spectrophotometer (Quawell, Sunnyvale, CA, USA).

4.22. Statistical Analysis

All data were expressed as the group mean $\pm$ SD. Comparisons between control and treated groups were conducted using one-way ANOVA, as appropriate, followed by LSD. A value of $p < 0.05$ was considered to indicate statistical significance. Statistical analysis of the data was performed using SPSS18.0. All experiments were performed three times.

5. Conclusions

In conclusion, we have demonstrated that Cr(VI) may induce mitochondrial CoQ10 deficiency by inhibiting the expression of genes involved in the CoQ10 biosynthesis pathway, and by subsequently causing oxidative stress, altering the mitochondrial network, and reducing mitochondrial biogenesis. CoQ10 exerts a potent protective effect against Cr(VI)-induced apoptosis, reduces intracellular oxidative stress, inhibits mitochondrial depolarization, increases mitochondrial mass and the mtDNA copy number, decreases the degree of MPTP opening, ameliorates caspase activity, equilibrates Bcl-2/Bax expression, and antagonizes subsequent apoptosis. We conclude that Cr(VI)-induced CoQ10 deficiency might be corrected by supplementation with CoQ10, which appears to stimulate mitochondrial biogenesis and prevent apoptosis. Thus, new perspectives on mitochondrial toxicity related to Cr(VI) exposure are suggested, and we provide new insights into therapeutic potentials and strategies for protecting hepatocytes against Cr(VI)-induced oxidative stress and apoptosis.

Acknowledgments: This study was supported, in part, by a grant from the Chinese National Natural Science Foundation (No. 81172701), a Chinese Council Scholarship (201506370074) and the Graduate Student Innovation Project in Hunan Province of China (CX2016B057). The authors appreciate the assistance of their excellent colleagues with this study.

Author Contributions: Xiali Zhong and Caigao Zhong designed the experiments; Xiali Zhong, Xing Yi, Yujing Zhang, and Kaihua Liu performed the experiments; Xiali Zhong and Fang Xiao analyzed the data; Xiali Zhong wrote the manuscript; Rita de Cássia da Silveira e Sá and Caigao Zhong revised the manuscript.

Conflicts of Interest: The authors declare no conflict of interest.

Abbreviations

Cr(VI)	hexavalent chromium
CoQ10	coenzyme Q10
ROS	reactive oxygen species
MMP	mitochondrial membrane potential
Cyt c	cytochrome c

MPTP	mitochondrial permeability transition pore
MDA	methane dicarboxylic aldehyde
SOD	superoxide dismutase
PDSS1	prenyl(decaprenyl) diphosphate synthase, subunit 1
PDSS2	prenyl (decaprenyl) diphosphate synthase, subunit 2
COQ2	4-hydroxybenzoate polyprenyltransferase
COQ3	coenzyme Q3 methyltransferase
COQ4	coenzyme Q4
COQ5	coenzyme Q5 methyltransferase
COQ6	coenzyme Q6 monooxygenase
COQ7	coenzyme Q7 homolog
ADCK3	aarF domain-containing kinase 3
ADCK4	aarF domain-containing kinase 4
COQ9	coenzyme Q9
COQ10A	coenzyme Q10A
COA10B	coenzyme Q10B
APTX	ataxin
ETFDH	electron transfer flavoprotein dehydrogenase
BRAF	B-Raf proto-oncogene, serine/threonine kinase
PMVK	phosphomevalonate kinase
MVD	mevalonate diphosphate decarboxylase
MVK	mevalonate kinase

References

1. Krumschnabel, G.; Nawaz, M. Acute toxicity of hexavalent chromium in isolated teleost hepatocytes. *Aquat. Toxicol.* **2004**, *70*, 159–167. [CrossRef] [PubMed]
2. Thompson, C.M.; Kirman, C.R.; Proctor, D.M.; Haws, L.C.; Suh, M.; Hays, S.M.; Hixon, J.G.; Harris, M.A. A chronic oral reference dose for hexavalent chromium-induced intestinal cancer. *J. Appl. Toxicol.* **2014**, *34*, 525–536. [CrossRef] [PubMed]
3. Stout, M.D.; Herbert, R.A.; Kissling, G.E.; Collins, B.J.; Travlos, G.S.; Witt, K.L.; Melnick, R.L.; Abdo, K.M.; Malarkey, D.E.; Hooth, M.J. Hexavalent chromium is carcinogenic to F344/N rats and B6C3F1 mice after chronic oral exposure. *Environ. Health Perspect.* **2009**, *117*, 716–722. [CrossRef] [PubMed]
4. Costa, M. Toxicity and carcinogenicity of Cr(VI) in animal models and humans. *Crit. Rev. Toxicol.* **1997**, *27*, 431–442. [CrossRef] [PubMed]
5. Felter, S.P.; Dourson, M.L. Hexavalent chromium-contaminated soils: Options for risk assessment and risk management. *Regul. Toxicol. Pharmacol.* **1997**, *25*, 43–59. [CrossRef] [PubMed]
6. Megharaj, M.; Avudainayagam, S.; Naidu, R. Toxicity of hexavalent chromium and its reduction by bacteria isolated from soil contaminated with tannery waste. *Curr. Microbial.* **2003**, *47*, 51–54. [CrossRef] [PubMed]
7. Nudler, S.I.; Quinteros, F.A.; Miler, E.A.; Cabilla, J.P.; Ronchetti, S.A.; Duvilanski, B.H. Chromium VI administration induces oxidative stress in hypothalamus and anterior pituitary gland from male rats. *Toxicol. Lett.* **2009**, *185*, 187–192. [CrossRef] [PubMed]
8. Michie, C.A.; Hayhurst, M.; Knobel, G.J.; Stokol, J.M.; Hensley, B. Poisoning with a traditional remedy containing potassium dichromate. *Hum. Exp. Toxicol.* **1991**, *10*, 129–131. [CrossRef] [PubMed]
9. Collins, B.J.; Stout, M.D.; Levine, K.E.; Kissling, G.E.; Melnick, R.L.; Fennell, T.R.; Walden, R.; Abdo, K.; Pritchard, J.B.; Fernando, R.A.; et al. Exposure to hexavalent chromium resulted in significantly higher tissue chromium burden compared with trivalent chromium following similar oral doses to male F344/N rats and female B6C3F1 mice. *Toxicol. Sci.* **2010**, *118*, 368–379. [CrossRef] [PubMed]
10. Proctor, D.M.; Otani, J.M.; Finley, B.L.; Paustenbach, D.J.; Bland, J.A.; Speizer, N.; Sargent, E.V. Is hexavalent chromium carcinogenic via ingestion? A weight-of-evidence review. *J. Toxicol. Environ. Health Part A* **2002**, *65*, 701–746. [CrossRef] [PubMed]

11. Xiao, F.; Feng, X.; Zeng, M.; Guan, L.; Hu, Q.; Zhong, C. Hexavalent chromium induces energy metabolism disturbance and p53-dependent cell cycle arrest via reactive oxygen species in L-02 hepatocytes. *Mol. Cell. Biochem.* **2012**, *371*, 65–76. [CrossRef] [PubMed]

12. Linos, A.; Petralias, A.; Christophi, C.A.; Christoforidou, E.; Kouroutou, P.; Stoltidis, M.; Veloudaki, A.; Tzala, E.; Makris, K.C.; Karagas, M.R. Oral ingestion of hexavalent chromium through drinking water and cancer mortality in an industrial area of Greece—An ecological study. *Environ. Health* **2011**, *10*, 50. [CrossRef] [PubMed]

13. Beaumont, J.J.; Sedman, R.M.; Reynolds, S.D.; Sherman, C.D.; Li, L.H.; Howd, R.A.; Sandy, M.S.; Zeise, L.; Alexeeff, G.V. Cancer mortality in a chinese population exposed to hexavalent chromium in drinking water. *Epidemiology* **2008**, *19*, 12–23. [CrossRef] [PubMed]

14. Yang, Y.; Liu, H.; Xiang, X.H.; Liu, F.Y. Outline of occupational chromium poisoning in china. *Bull. Environ. Contam. Toxicol.* **2013**, *90*, 742–749. [CrossRef] [PubMed]

15. Myers, J.M.; Antholine, W.E.; Myers, C.R. The intracellular redox stress caused by hexavalent chromium is selective for proteins that have key roles in cell survival and thiol redox control. *Toxicology* **2011**, *281*, 37–47. [CrossRef] [PubMed]

16. Wang, C.C.; Fang, K.M.; Yang, C.S.; Tzeng, S.F. Reactive oxygen species-induced cell death of rat primary astrocytes through mitochondria-mediated mechanism. *J. Cell. Biochem.* **2009**, *107*, 933–943. [CrossRef] [PubMed]

17. Banu, S.K.; Stanley, J.A.; Lee, J.; Stephen, S.D.; Arosh, J.A.; Hoyer, P.B.; Burghardt, R.C. Hexavalent chromium-induced apoptosis of granulosa cells involves selective sub-cellular translocation of Bcl-2 members, ERK1/2 and p53. *Toxicol. Appl. Pharmacol.* **2011**, *251*, 253–266. [CrossRef] [PubMed]

18. Marouani, N.; Tebourbi, O.; Mokni, M.; Yacoubi, M.T.; Sakly, M.; Benkhalifa, M.; Rhouma, K.B. Hexavalent chromium-induced apoptosis in rat uterus: Involvement of oxidative stress. *Arch. Environ. Occup. Health* **2015**, *70*, 189–195. [CrossRef] [PubMed]

19. McCarthy, S.; Somayajulu, M.; Sikorska, M.; Borowy-Borowski, H.; Pandey, S. Paraquat induces oxidative stress and neuronal cell death; neuroprotection by water-soluble coenzyme Q10. *Toxicol. Appl. Pharmacol.* **2004**, *201*, 21–31. [CrossRef] [PubMed]

20. Papucci, L.; Schiavone, N.; Witort, E.; Donnini, M.; Lapucci, A.; Tempestini, A.; Formigli, L.; Zecchi-Orlandini, S.; Orlandini, G.; Carella, G.; et al. Coenzyme Q10 prevents apoptosis by inhibiting mitochondrial depolarization independently of its free radical scavenging property. *J. Biol. Chem.* **2003**, *278*, 28220–28228. [CrossRef] [PubMed]

21. Doimo, M.; Desbats, M.A.; Cerqua, C.; Cassina, M.; Trevisson, E.; Salviati, L. Genetics of coenzyme Q10 deficiency. *Mol. Syndromol.* **2014**, *5*, 156–162. [CrossRef] [PubMed]

22. Kawamukai, M. Biosynthesis of coenzyme Q in eukaryotes. *Biosci. Biotechnol. Biochem.* **2015**, *80*, 23–33. [CrossRef] [PubMed]

23. Quinzii, C.M.; Lopez, L.C.; Naini, A.; DiMauro, S.; Hirano, M. Human COQ10 deficiencies. *BioFactors* **2008**, *32*, 113–118. [CrossRef] [PubMed]

24. Ben-Meir, A.; Burstein, E.; Borrego-Alvarez, A.; Chong, J.; Wong, E.; Yavorska, T.; Naranian, T.; Chi, M.; Wang, Y.; Bentov, Y.; et al. Coenzyme Q10 restores oocyte mitochondrial function and fertility during reproductive aging. *Aging Cell* **2015**, *14*, 887–895. [CrossRef] [PubMed]

25. Quinzii, C.M.; Lopez, L.C.; Gilkerson, R.W.; Dorado, B.; Coku, J.; Naini, A.B.; Lagier-Tourenne, C.; Schuelke, M.; Salviati, L.; Carrozzo, R.; et al. Reactive oxygen species, oxidative stress, and cell death correlate with level of COQ10 deficiency. *FASEB J.* **2010**, *24*, 3733–3743. [CrossRef] [PubMed]

26. Quinzii, C.M.; Emmanuele, V.; Hirano, M. Clinical presentations of coenzyme Q10 deficiency syndrome. *Mol. Syndromol.* **2014**, *5*, 141–146. [CrossRef] [PubMed]

27. Potgieter, M.; Pretorius, E.; Pepper, M.S. Primary and secondary coenzyme Q10 deficiency: The role of therapeutic supplementation. *Nutr. Rev.* **2013**, *71*, 180–188. [CrossRef] [PubMed]

28. Kooncumchoo, P.; Sharma, S.; Porter, J.; Govitrapong, P.; Ebadi, M. Coenzyme Q10 provides neuroprotection in iron-induced apoptosis in dopaminergic neurons. *J. Mol. Neurosci.* **2006**, *28*, 125–141. [CrossRef]

29. Abdallah, G.M.; El-Sayed el, S.M.; Abo-Salem, O.M. Effect of lead toxicity on coenzyme Q levels in rat tissues. *Food Chem. Toxicol.* **2010**, *48*, 1753–1756. [CrossRef] [PubMed]

30. Gempel, K.; Topaloglu, H.; Talim, B.; Schneiderat, P.; Schoser, B.G.H.; Hans, V.H.; Palmafy, B.; Kale, G.; Tokatli, A.; Quinzii, C.; et al. The myopathic form of coenzyme Q10 deficiency is caused by mutations in the electron-transferring-flavoprotein dehydrogenase (*ETFDH*) gene. *Brain* **2007**, *130*, 2037–2044. [CrossRef] [PubMed]

31. Acosta, M.J.; Fonseca, L.V.; Desbats, M.A.; Cerqua, C.; Zordan, R.; Trevisson, E.; Salviati, L. Coenzyme Q biosynthesis in health and disease. *Biochim. Biophys. Acta.* **2016**, *1857*, 1079–1085. [CrossRef] [PubMed]

32. Chen, S.W.; Liu, C.C.; Yen, H.C. Detection of suppressed maturation of the human COQ5 protein in the mitochondria following mitochondrial uncoupling by an antibody recognizing both precursor and mature forms of COQ5. *Mitochondrion* **2013**, *13*, 143–152. [CrossRef] [PubMed]

33. Yen, H.C.; Liu, Y.C.; Kan, C.C.; Wei, H.J.; Lee, S.H.; Wei, Y.H.; Feng, Y.H.; Chen, C.W.; Huang, C.C. Disruption of the human COQ5-containing protein complex is associated with diminished coenzyme Q10 levels under two different conditions of mitochondrial energy deficiency. *Biochim. Biophys. Acta* **2016**, *1860*, 1864–1876. [CrossRef] [PubMed]

34. Wheeler, B.; Jia, Z.C. Preparation and characterization of human ADCK3, a putative atypical kinase. *Protein Expr. Purif.* **2015**, *108*, 13–17. [CrossRef] [PubMed]

35. Liu, X.; Yang, J.; Zhang, Y.; Fang, Y.; Wang, F.; Wang, J.; Zheng, X.; Yang, J. A systematic study on drug-response associated genes using baseline gene expressions of the cancer cell line encyclopedia. *Sci. Rep.* **2016**, *6*, 22811. [CrossRef] [PubMed]

36. Xiao, F.; Li, Y.; Luo, L.; Xie, Y.; Zeng, M.; Wang, A.; Chen, H.; Zhong, C. Role of mitochondrial electron transport chain dysfunction in Cr(VI)-induced cytotoxicity in L-02 hepatocytes. *Cell. Physiol. Biochem.* **2014**, *33*, 1013–1025. [CrossRef] [PubMed]

37. Montero, R.; Pineda, M.; Aracil, A.; Vilaseca, M.A.; Briones, P.; Sanchez-Alcazar, J.A.; Navas, P.; Artuch, R. Clinical, biochemical and molecular aspects of cerebellar ataxia and coenzyme Q10 deficiency. *Cerebellum* **2007**, *6*, 118–122. [CrossRef] [PubMed]

38. Lopez-Martin, J.M.; Salviati, L.; Trevisson, E.; Montini, G.; DiMauro, S.; Quinzii, C.; Hirano, M.; Rodriguez-Hernandez, A.; Cordero, M.D.; Sanchez-Alcazar, J.A.; et al. Missense mutation of the *COQ2* gene causes defects of bioenergetics and de novo pyrimidine synthesis. *Hum. Mol. Genet.* **2007**, *16*, 1091–1097. [CrossRef] [PubMed]

39. Maguire, J.J.; Kagan, V.; Ackrell, B.A.; Serbinova, E.; Packer, L. Succinate-ubiquinone reductase linked recycling of α-tocopherol in reconstituted systems and mitochondria: Requirement for reduced ubiquinone. *Arch. Biochem. Biophys.* **1992**, *292*, 47–53. [CrossRef]

40. Duberley, K.E.; Heales, S.J.R.; Abramov, A.Y.; Chalasani, A.; Land, J.M.; Rahman, S.; Hargreaves, I.P. Effect of coenzyme Q10 supplementation on mitochondrial electron transport chain activity and mitochondrial oxidative stress in coenzyme Q10 deficient human neuronal cells. *Int. J. Biochem. Cell Biol.* **2014**, *50*, 60–63. [CrossRef] [PubMed]

41. Garrido-Maraver, J.; Cordero, M.D.; Oropesa-Avila, M.; Fernandez Vega, A.; de la Mata, M.; Delgado Pavon, A.; de Miguel, M.; Perez Calero, C.; Villanueva Paz, M.; Cotan, D.; et al. Coenzyme Q10 therapy. *Mol. Syndromol.* **2014**, *5*, 187–197. [CrossRef] [PubMed]

42. Garcia-Sevillano, M.A.; Garcia-Barrera, T.; Navarro, F.; Gomez-Ariza, J.L. Absolute quantification of superoxide dismutase in cytosol and mitochondria of mice hepatic cells exposed to mercury by a novel metallomic approach. *Anal. Chim. Acta* **2014**, *842*, 42–50. [CrossRef] [PubMed]

43. Gong, L.L.; Wang, Z.H.; Li, G.R.; Liu, L.H. Protective effects of akebia saponin D against rotenone-induced hepatic mitochondria dysfunction. *J. Pharmacol. Sci.* **2014**, *126*, 243–252. [CrossRef] [PubMed]

44. Sohal, R.S.; Forster, M.J. Coenzyme Q, oxidative stress and aging. *Mitochondrion* **2007**, *7*, S103–S111. [CrossRef] [PubMed]

45. Kovac, S.; Domijan, A.M.; Walker, M.C.; Abramov, A.Y. Seizure activity results in calcium- and mitochondria-independent ROS production via NADPH and xanthine oxidase activation. *Cell Death Dis.* **2014**, *5*, e1442. [CrossRef] [PubMed]

46. Peng, T.I.; Jou, M.J. Oxidative stress caused by mitochondrial calcium overload. *Ann. N. Y. Acad. Sci.* **2010**, *1201*, 183–188. [CrossRef] [PubMed]

47. Richter, C.; Schweizer, M.; Cossarizza, A.; Franceschi, C. Control of apoptosis by the cellular ATP level. *FEBS Lett.* **1996**, *378*, 107–110. [CrossRef]

48. Petronilli, V.; Penzo, D.; Scorrano, L.; Bernardi, P.; di Lisa, F. The mitochondrial permeability transition, release of cytochrome C and cell death. Correlation with the duration of pore openings in situ. *J. Biol. Chem.* **2001**, *276*, 12030–12034. [CrossRef] [PubMed]

49. Liu, G.; Wang, T.; Wang, T.; Song, J.; Zhou, Z. Effects of apoptosis-related proteins caspase-3, Bax and Bcl-2 on cerebral ischemia rats. *Biomed. Rep.* **2013**, *1*, 861–867. [CrossRef] [PubMed]

50. Eghbal, M.A.; Abdoli, N.; Azarmi, Y. Efficiency of hepatocyte pretreatment with coenzyme Q10 against statin toxicity. *Arh. Hig. Rada Toksikol.* **2014**, *65*, 101–108. [CrossRef] [PubMed]

51. Chen, C.C.; Liou, S.W.; Chen, C.C.; Chen, W.C.; Hu, F.R.; Wang, I.J.; Lin, S.J. Coenzyme Q10 reduces ethanol-induced apoptosis in corneal fibroblasts. *PLoS ONE* **2011**, *6*, e19111. [CrossRef] [PubMed]

52. Xu, X.R.; Li, H.B.; Gu, J.D.; Li, X.Y. Kinetics of the reduction of chromium VI by vitamin C. *Environ. Toxicol. Chem.* **2005**, *24*, 1310–1314. [CrossRef] [PubMed]

53. Turunen, M.; Olsson, J.; Dallner, G. Metabolism and function of coenzyme Q. *Biochim. Biophys. Acta Biomembr.* **2004**, *1660*, 171–199. [CrossRef]

54. Fontaine, E.; Bernardi, P. Progress on the mitochondrial permeability transition pore: Regulation by complex I and ubiquinone analogs. *J. Bioenerg. Biomembr.* **1999**, *31*, 335–345. [CrossRef] [PubMed]

55. Zhang, W.Y.; Cai, N.; Ye, L.H.; Zhang, X.D. Transformation of human liver L-02 cells mediated by stable HBX transfection. *Acta Pharmacol. Sin.* **2009**, *30*, 1153–1161. [CrossRef] [PubMed]

56. Ding, Z.B.; Shi, Y.H.; Zhou, J.; Qiu, S.J.; Xu, Y.; Dai, Z.; Shi, G.M.; Wang, X.Y.; Ke, A.W.; Wu, B.; et al. Association of autophagy defect with a malignant phenotype and poor prognosis of hepatocellular carcinoma. *Cancer Res.* **2008**, *68*, 9167–9175. [CrossRef] [PubMed]

57. Frezza, C.; Cipolat, S.; Scorrano, L. Organelle isolation: Functional mitochondria from mouse liver, muscle and cultured fibroblasts. *Nat. Protoc.* **2007**, *2*, 287–295. [CrossRef] [PubMed]

58. Turkowicz, M.J.; Karpinska, J. Analytical problems with the determination of coenzyme Q10 in biological samples. *BioFactors* **2013**, *39*, 176–185. [CrossRef] [PubMed]

59. Lass, A.; Sohal, R.S. Electron transport-linked ubiquinone-dependent recycling of α-tocopherol inhibits autooxidation of mitochondrial membranes. *Arch. Biochem. Biophys.* **1998**, *352*, 229–236. [CrossRef] [PubMed]

International Journal of
Molecular Sciences

MDPI

Article

The Effect of VPA on Increasing Radiosensitivity in Osteosarcoma Cells and Primary-Culture Cells from Chemical Carcinogen-Induced Breast Cancer in Rats

Guochao Liu [1,†], Hui Wang [1,†], Fengmei Zhang [1,†], Youjia Tian [1], Zhujun Tian [1], Zuchao Cai [1], David Lim [2] and Zhihui Feng [1,*]

[1] Department of Occupational Health and Occupational Medicine, School of Public Health, Shandong University, Jinan 250012, China; 201514107@mail.sdu.edu (G.L.); sduwanghui03@163.com (H.W.); fengmeizhang2003@sdu.edu.cn (F.Z.); 201514110@mail.sdu.edu (Y.T.); tianzhujun_mercy@163.com (Z.T.); 201614266@mail.sdu.edu (Z.C.)
[2] Flinders Rural Health South Australia, Victor Harbor, SA 5211, Australia; david.lim@flinders.edu.au or c113.lim@qut.edu.au
* Correspondence: fengzhihui@sdu.edu.cn; Tel.: +86-135-7314-6295
† These authors contributed equally to this work.

Academic Editors: Ashis Basu and Takehiko Nohmi
Received: 30 March 2017; Accepted: 5 May 2017; Published: 10 May 2017

Abstract: This study explored whether valproic acid (VPA, a histone deacetylase inhibitor) could radiosensitize osteosarcoma and primary-culture tumor cells, and determined the mechanism of VPA-induced radiosensitization. The working system included osteosarcoma cells (U2OS) and primary-culture cells from chemical carcinogen (DMBA)-induced breast cancer in rats; and clonogenic survival, immunofluorescence, fluorescent in situ hybridization (FISH) for chromosome aberrations, and comet assays were used in this study. It was found that VPA at the safe or critical safe concentration of 0.5 or 1.0 mM VPA could result in the accumulation of more ionizing radiation (IR)-induced DNA double strand breaks, and increase the cell radiosensitivity. VPA-induced radiosensitivity was associated with the inhibition of DNA repair activity in the working systems. In addition, the chromosome aberrations including chromosome breaks, chromatid breaks, and radial structures significantly increased after the combination treatment of VPA and IR. Importantly, the results obtained by primary-culture cells from the tissue of chemical carcinogen-induced breast cancer in rats further confirmed our findings. The data in this study demonstrated that VPA at a safe dose was a radiosensitizer for osteosarcoma and primary-culture tumor cells through suppressing DNA-double strand breaks repair function.

Keywords: VPA; DNA double-strand breaks; radiosensitivity; DNA repair; U2OS; chemical carcinogen (DMBA)-induced tumor

---

1. Introduction

Among other methods, chemotherapy and radiotherapy are generally prescribed for the treatment of cancers, and such DNA-damaging cytotoxic therapies remain the main treatment of cancers such as osteosarcoma and breast cancer. However, with time, tumor cells develop mechanisms of resistance to such treatments. Recently, considerable attention has been on researching effective strategies to understand and develop means of decreasing tumor cellular sensitization and resistance to DNA-damaging agents. Histone deacetylase (HDAC) was identified as a promising therapeutic target for cancer treatment as it plays a central role in chromosome structural remodeling and gene-transcriptional regulation, with altered expression and mutation of HDAC linked to tumor development and occurrence [1]. Eighteen mammalian HDACs have been identified so far [2], and have been subdivided into four different classes based on their homology with yeast HDACs. HDAC inhibitors (HDACi), such as the anticonvulsant drug valproic acid (VPA), have been identified as neoadjuvant to chemotherapy and radiotherapy [3]. The VPA-induced sensitization of tumor cells has been attributed to its effect on HDAC-dependent transcriptional repression and hyperacetylation of histones, which resulted in the differentiation of tumor cells and increased both apoptotic and non-apoptotic cell death [4,5]. Previous studies have demonstrated that VPA downregulated key proteins such as BRAC1, RAD51, Ku70, Ku80, and prolonged radiation-induced repair protein foci such as γH2AX and 53BP1 in tumor cells [1,6–9]. This is important as DNA-damaging cytotoxic therapies are intended to induce DNA-double strand breaks (DSBs). We have previously demonstrated that VPA increased the radiosensitivity of breast cancer cells through the disruption of both BRAC1-Rad51-mediated homologous recombination and Ku80-mediated non-homologous end-jointing [7]. However, some results indicated that radiotherapy was largely ineffective in osteosarcoma [10,11], so it would be very interesting to investigate whether VPA could enhance the radiosensitivity of osteosarcoma cells, which may be helpful for osteosarcoma treatment in medical clinics.

The safe blood concentration of VPA for the treatment of epilepsy in clinic is 50–100 μg/mL, which is equal to 0.3–0.8 mM. Based on this information, our study selected 0.5 mM and 1.0 mM as a safe dose and a critical safe dose, respectively, for the treatment of epilepsy in clinic to explore the effect of VPA on radiosensitivity and its mechanism in osteosarcoma cells (U2OS cell line) and primary-culture cells from the tissue of chemical carcinogen (DMBA)-induced breast cancer in rats. Our results clearly suggest that a safe dose of VPA could induce more DSBs in both working systems in response to DNA damage induced by IR, and increase radiosensitivity and genetic instability in the cells by disrupting DNA repair function.

2. Results

2.1. Effects of VPA on DNA-Double Strand Breaks (DSB) in Osteosarcoma Cells

To quantify the effects of VPA on the DSB using neutral comet assay, a U2OS cell line was pretreated with 0.5 mM VPA and subjected to 4 Gy ionizing radiotherapy (IR). With and without IR, there was no statistically significant difference between the DNA-tail of VPA versus untreated-control; however, visually it does appear that cell-lines pretreated with VPA exhibited a longer DNA tail (Figure 1A upper). VPA + IR had statistically more relative DSB compared to post-IR at both 30- and 120-min (Figure 1A lower right, $p < 0.05$). The findings inferred that VPA caused the accumulation of more IR-induced DSB in osteosarcoma cells, and a slower recovery of DSB in a time-dependent manner.

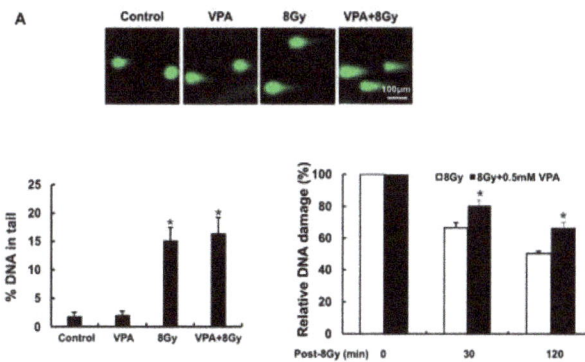

Figure 1. *Cont.*

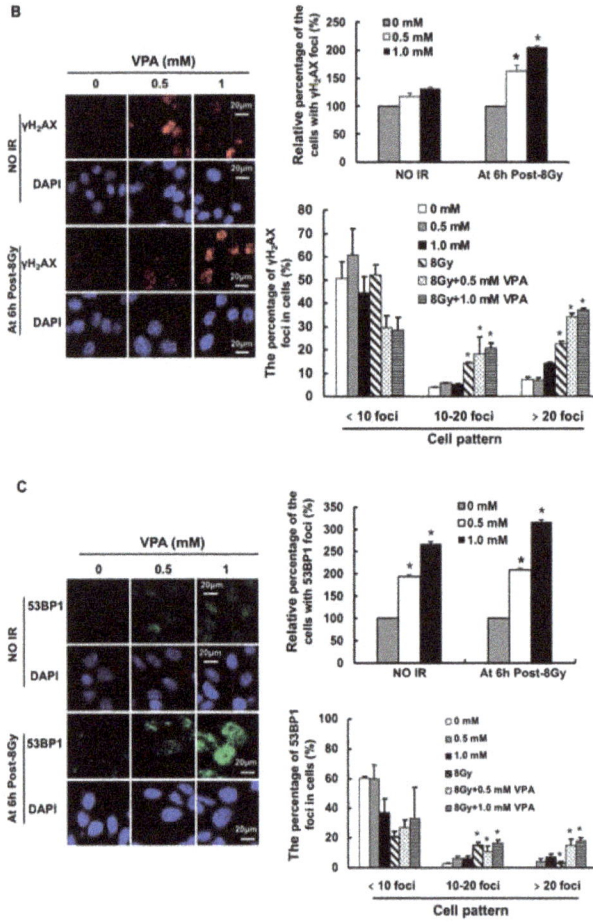

Figure 1. VPA can cause the accumulation of more IR-induced DNA DSBs in U2OS cells. (**A**) 0.5 mM VPA-treated and untreated cells before and after 8 Gy treatment are presented in the images from comet assay (**upper**), and the olive moment was further analyzed (**lower left**); after correcting the data, the relative olive moment at 0, 30, and 120 min post-IR was exhibited in the cells (**lower right**); (**B**) The images represent the γH2AX foci formation in the cells treated with 0.5 or 1.0 mM VPA before and after IR (8 Gy) treatment (**left**), and the percentage of cells with γH2AX foci formation in each group was calculated (**right upper panel**, the cell with >10 foci was called positive and counted), also the percentage of cells with different patterns divided by the number of foci per nucleus in each group estimated (**right lower panel**); (**C**) The images represent 53BP1 foci formation in the cells treated with 0.5 or 1.0 mM VPA before and after IR (8 Gy) treatment (**left**), and the percentage of cells with 53BP1 foci formation in each group was calculated (**right upper panel**, the cell with >10 foci was called positive and counted), the percentage of cells with different patterns divided by the number of 53BP1 foci per nucleus in each group estimated (**right lower panel**). 4′,6-diamidino-2-phenylindole (DAPI) was used for nuclear staining. 2.5×10^4 U2OS cells were seeded on the chamber in immunofluorescence assays. Each data point in the graphs was from three independent experiments (mean ± SD). *p*-Values were calculated by Student's *t*-test (* *p* < 0.05).

To triangulate the above-mentioned findings, DSB-induced histone H2AX phosphorylation on serine 139 (γH2AX) formation [12–14], and p53 binding protein 1 (53BP1)—the markers for DSB [15–18]

—were analyzed. The U2O2 cell line was pretreated with 0.5 mM and 1 mM VPA for 24 h before being subjected to 8 Gy radiation. At 6 h post-IR, the immunofluorescence staining showed that safe doses of VPA at 0.5 and 1.0 mM induced an increased γH2AX foci formation when compared to the control group (Figure 1B). The foci's size and density in the VPA-pretreated group appeared larger and brighter than the IR-alone group (Figure 1B left). The relative percentage of cells with γH2AX foci increased by 1.6 and 2.1-fold, respectively (Figure 1B right upper, $p < 0.05$). We further analyzed the above data in another way by categorizing the cells containing γH2AX foci into three patterns according to the number of foci in each cell: <10, 10–20, and >20 (Figure 1B right lower). There was a positive association between the number of γH2AX foci and VPA-treatment ($p < 0.05$). Likewise, the relative ratio of U2O2 cells with 53BP1 foci pretreated with 0.5 mM and 1.0 mM VPA increased by 2.1 and 3.2-fold, respectively (Figure 1C left and right upper, $p < 0.05$). Similarly, we observed appositive association between number of 53BP1 foci and VPA-treatment (Figure 1C right lower, $p < 0.05$).

To confirm the radiosensitization effect of VPA on the tumor cells, a model of chemical carcinogen (DMBA)-induced breast cancer in SD rats was established to obtain primary-culture tumor cells. Rats at 50 days old were gavaged DMBA to induce tumor formation around the rats' nipples, which was detached from skin around 90 days after DMBA administration (Figure 2(A1–A3)). The morphological structure of the tissue was observed by hematoxylin-eosin (HE) staining. Figure 2(A4) shows that the structure of breast tissue in normal rats in contrast with a large number of hyperplasia cells in the DMBA-induced breast cancer tissue (Figure 2(A5)), indicating that breast cancer in rats was successfully induced by this chemical carcinogen. Primary-culture tumor cells were then obtained from this breast cancer tissue (Figure 2(A6)). Two methods of neutral comet assay and γH2AX foci were used to test the radiosensitivity effect of VPA on the cells. At 0 min post-IR, the combination of 0.5 mM VPA and 8 Gy significantly increased the olive moments in the cells when compared with IR alone (Figure 2B, $p < 0.05$), suggesting that VPA could induce more IR-caused DSBs. Additionally, similar results were found via the γH2AX foci formation assay. For the combined treatment group, at 6 h post-IR treatment, the percentage of primary-culture tumor cells containing γH2AX foci was obviously higher than that of the IR alone group (Figure 2C, $p < 0.05$), confirming that VPA can lead to more DSBs damage in response to IR treatment. The above-mentioned results clearly revealed that VPA was a radiosensitizer not only for the tumor cell line, but also for the primary-culture tumor cells.

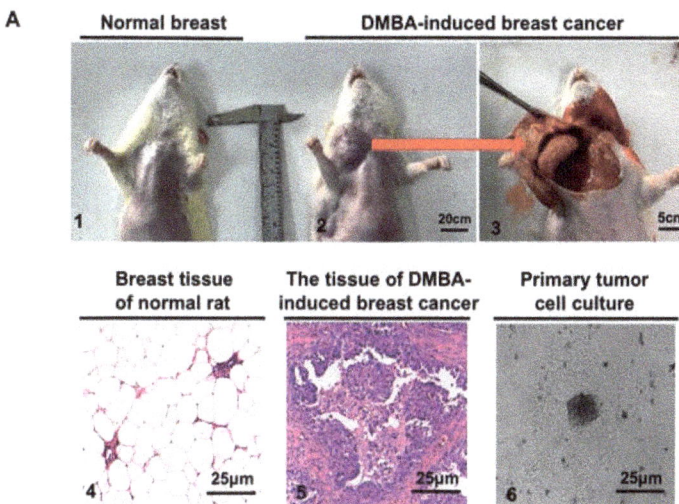

Figure 2. *Cont.*

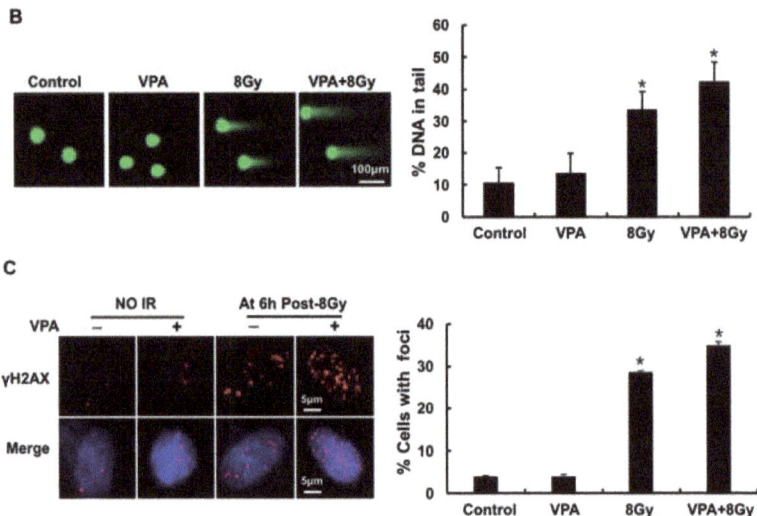

Figure 2. The effect of the combination of VPA and IR on primary-culture cells of breast cancer tissue in rats. (**A**) Normal breast (**1**), and DMBA-induced breast cancer (**2** and **3**) of rats under gross observation; HE staining for the morphology of normal tissue (**4**) and DMBA-induced breast cancer (**5**); the primary cell culture of breast cancer tissue (**6**); (**B**) The untreated and 0.5 mM VPA-treated cells are presented in the images from comet assay before and after 8 Gy treatment (**left**), and the olive moment was further analyzed (**right**); (**C**) The images represent the γH2AX foci formation in the cells treated with 0.5 mM VPA at 6 h post-IR, "+" and "−" indicated whether VPA was added in the groups (**left**); the percentage of cells with γH2AX foci formation in each group was calculated (**right**, the cell with >10 foci were called positive and counted). DAPI was used for nuclear staining. Each data point in the graphs was from three independent experiments (mean ± SD). *p*-Values were calculated by Student's *t*-test (* $p < 0.05$).

2.2. Effects of VPA on Radiosensitivity of Tumor Cells

To understand whether the above-observed VPA effects on DSB may be associated with cellular radiosensitivity, a clonogenic survival assay was employed. The U2O2 cell line was pretreated with 0.5 mM VPA and then exposed to 0, 2, 4, and 6 Gy radiation, respectively. The cells were cultured for a further 14 days, and the clonogenic colonies were stained and counted (Figure 3 upper). The VPA-treated group showed a decreased survival fraction when compared to the control group (Figure 3 lower left, $p < 0.05$). After the survival fraction of IR and the combination of VPA with IR was corrected by the corresponding control group, the data showed that there was a significant decrease in all combinations of VPA and IR groups when compared with relative IR groups (Figure 3, bottom right, $p < 0.05$). Additionally, the size of the colonies in all groups of the combination of VPA and IR were smaller in appearance than those of IR alone (Figure 3, top). The findings suggest that VPA increased the radio-sensitivity of tumor cells and suppressed tumor cell growth in response to DNA damage.

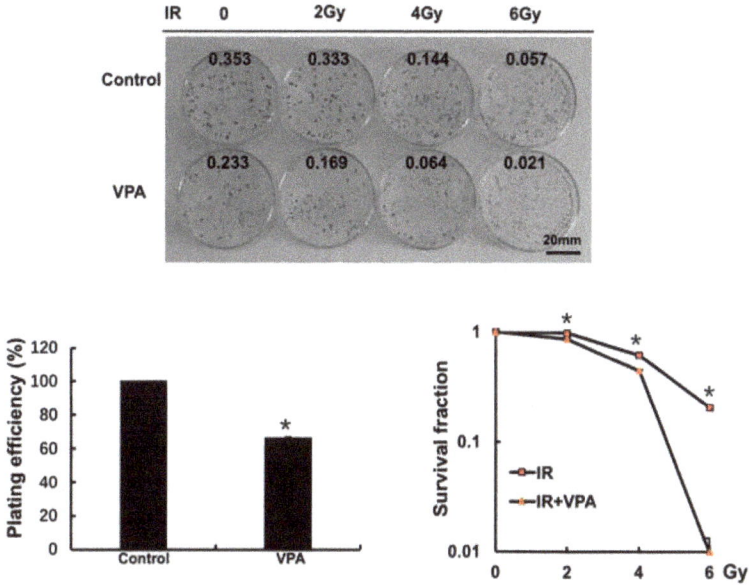

Figure 3. VPA can increase the radiosensitivity in U2OS cells. Plating Efficiency (PE) was presented in untreated and VPA-treated cells (**lower left**) A clonogenic survival assay was used to detect survival in the cells treated with different doses of IR (0, 2, 4, or 6 Gy) and the combination of 0.5 mM VPA with different doses of IR (**lower right**). Each data point in the graphs was from three independent experiments (mean ± SD). *p*-Values were calculated by Student's *t*-test (* $p < 0.05$).

2.3. VPA at Safe Dose Can Arise the Dysfunction of DNA Repair

There is a biphasic DNA repair mechanism post-IR: the early (0–6 h post-IR), and late (6–24 h post-IR) phases. Congruent with previous study [7], immunofluorescence staining of γH2AX and 53BP1 foci was utilized to study the mechanisms of VPA-induced radio-sensitivity in U2O2 cells 24 h post-6 Gy IR treatment. There was a statistically significant increase in the percentage of cells with γH2AX (0.5 mM: 1.27-fold, $p < 0.05$; 1 mM: 1.62-fold, $p < 0.05$) and 53BP1 (0.5 mM: 2.24-fold, $p < 0.05$; 1 mM: 3.43-fold, $p < 0.05$) foci with VPA-treatment (Figure 4A). These findings indicated that VPA may impact the repair ability of DNA in the late phase.

Next, primary-culture tumor cells from breast cancer tissue were used to test DNA repair activity. After the data was corrected with the corresponding control group at 120 min post-IR, the olive moments in the cells treated with a combination of VPA and 8 Gy were significantly higher when compared with IR alone (Figure 4B upper, $p < 0.05$). Similar results were also found via a γH2AX foci formation assay, at both 6 h and 24 h post-IR treatment, the percentage of primary-culture tumor cells containing γ-H2AX foci in the combined treatment group was obviously higher than that of the IR alone group (Figure 4B lower, $p < 0.05$). Thus, the results suggested that DNA repair activity was suppressed by VPA as a late response to IR treatment.

Synthetic lethality (SL) was first defined as a genetic combination of mutations in two or more genes that leads to cell death, whereas a mutation in any one of the genes does not [19,20]. The theory of SL in the DNA damage repair field has recently grown in popularity with the finding that poly (adenosine diphosphate (ADP)-ribose) polymerase inhibitors (PARPi) are specifically toxic to BRCA1 or BRCA2-associated homologues recombination (HR)-defective cells [21,22]. We speculated that the combination of VPA with PARPi would cause cell death if the VPA could inhibit HR function; thus, a clonogenic survival assay was used to study the effect of VPA and a typical poly ADP-ribose polymerase inhibitor, ABT888, on cell survival. Figure 4C demonstrates that 10 μM ABT888 alone, VPA alone, and VPA + ABT888 significantly reduced the relative survival fraction. The combination of 1 mM VPA and ABT888 had the lowest relative survival fraction (33.51%). The results indicated that the actions of VPA on suppressing tumor cell growth may be through its effect on DNA repair functions.

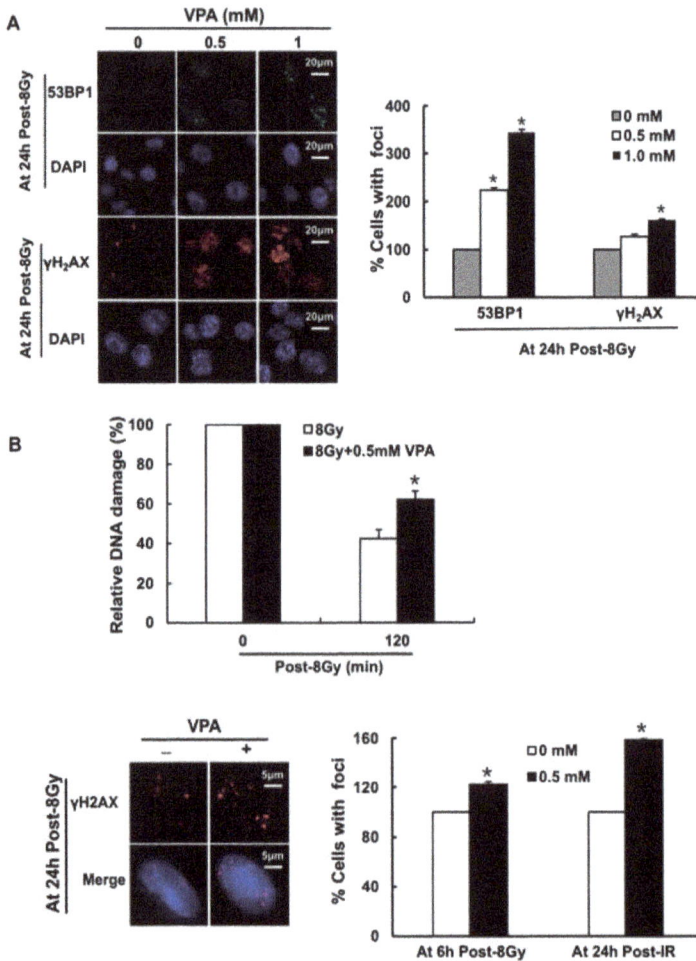

Figure 4. *Cont.*

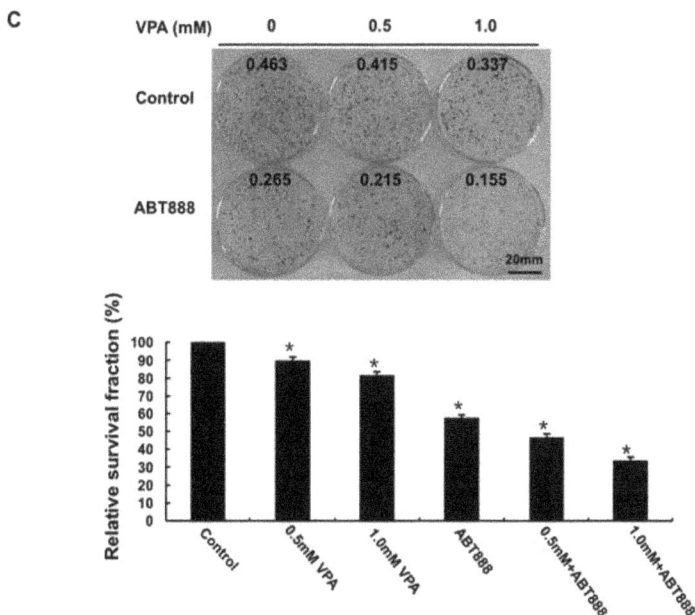

Figure 4. VPA at a safe dose can lead to the dysfunction of DNA repair function. (**A**) The images represent γH2AX and 53BP1 foci formation in U2OS cells treated with 0.5 or 1.0 mM VPA at 24 h post-IR treatment (**left panel**), and the percentage of cells with γH2AX foci or 53BP1 foci formation in each group was calculated (**right panel**, the cell with >10 foci was called positive and counted). DAPI was used for nuclear staining; (**B**) The relative DNA damage in primary-culture tumor cells at 120 min post-IR was analyzed by comet assay (**upper**), and γH2AX foci formation in primary-culture tumor cells at 24 h post-IR was presented, "+" and "−" indicated whether VPA was added in the groups (**lower left**) and calculated (**lower right**); (**C**) The clonogenic survival assay was used to detect survival in the U2OS cells treated with the combination of 0.5 or 1.0 mM VPA with 10 μM ABT888 (PARPi). Each data point in the graphs was from three independent experiments (mean ± SD). *p*-Values were calculated by Student's *t*-test (* *p* < 0.05).

2.4. Effects of VPA on Chromosome Aberrations

To test the effects of VPA on genomic stability, Q-FISH was utilized for the analysis of chromosome aberrations. Figure 5 showed no statistical difference in the number of chromatid and chromosome breaks between control and IR-treatment groups, whilst IR increased the number of radical structure from 1.59 per 1000 chromosomes to 7.34 (*p* < 0.05). The pre-treatment with 0.5 mM VPA significantly increased the number of chromatid breaks (4.57 per 1000 chromosomes to 17.24, *p* < 0.01), chromosome breaks (18.27 per 1000 chromosomes to 43.10, *p* < 0.01), and radical structure (4.57 per 1000 chromosomes to 12.93, *p* < 0.01). The findings demonstrated that VPA could lead to genomic instability through its effects on chromosome aberrations in response to IR.

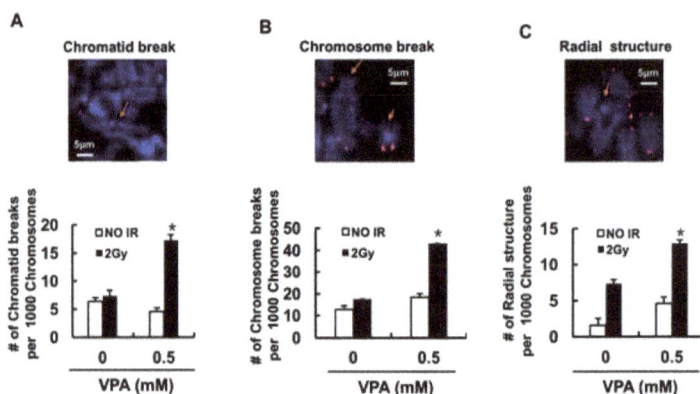

Figure 5. The effect of VPA on chromosome aberrations in U2OS cells. The untreated or VPA-treated cells were irradiated by 2 Gy, and the frequencies of IR-induced chromosome aberrations were analyzed. Fluorescence in situ hybridization using a telomeric probe is indicated in pink, and chromosomes were stained with DAPI in blue. Fifty metaphases for each sample were scored. Chromatid break (**A**), chromosome break (**B**), or radial structure (**C**) were presented and pointed by the arrow. "#"in each graph was indicated the number of the break. Each data point in the graphs was from three independent experiments (mean ± SD). *p*-Values were calculated by Student's *t*-test (* *p* < 0.05).

3. Discussion

It has been increasingly proposed that the effect of HDAC inhibitors in the radiosensitization of tumor cells occurred via their effects on the DNA repair pathway [7,23]. Our previous results demonstrated that safe doses of VPA can radiosensitize the breast cancer cells by affecting both DNA DSB repair pathways, as well as decrease the frequency of homologous and non-homologous end joining [7,24,25]. In this study, we investigated whether safe concentrations of VPA could induce more IR-induced DSBs and inhibit cell survival in vivo using osteosarcoma cells and chemical-induced breast cancer cells. The use of primary tissue culture was important as this model mimicked the development of human primary tumors in situ. Our findings demonstrated that VPA did induce the radiosensitization of tumor cells and the effects of this HDAC inhibitor operates through suppressed DNA repair and associated genomic instability. Together with previous reports of VPA augmented radiation-induced apoptosis through targeted activity on BRCA1, Rad51 and Ku80 proteins [7,26,27], this study advanced the proposal for the use of VPA as a neoadjuvant to radiotherapy for cancer treatment.

As DNA repair functions, such as HR and Non-Homologous End Joining (NHEJ), are an important mechanism for HDACi-radiosensitization in tumor cells, some results indicated that the effect of HDACi on them was inconsistent, which may be relative to the HDACi used as in the study as our data demonstrated that VPA could decrease the frequency of HR and NHEJ in breast cancer cells [7,24]. Other reports also found that both NHEJ and HR rates decreased in the presence of butyrate, it was estimated that NHEJ decreased by 40% and HR decreased by 60% [28]. However, it was observed that 5 mM or 10 mM VPA could enhance HR after treatment for 24 h in Chinese hamster ovary (CHO) 3–6 cells [25], and Suberoylanilide hydroxamic acid (SAHA) and Trichostatin A (TSA) could only increase NHEJ activity but did not change the HR frequency in HeLa cells [7,28]. HDACi can affect several key proteins in DSBs repair such as p53, BRCA1, RAD51, and Ku80. Our previous report pointed out that VPA could disrupt HR and NHEJ through targeting the activity of BRCA1, Rad51, and Ku80 [7], may enhance radiation-induced apoptosis and serve as a radiosensitizer in a p53-dependent manner in colorectal cancer cells [26], and downregulate both protein expression and foci accumulation of BRCA1 and RAD51 in LNCaP and DU-145 cells [27,29]. Vorinostat and TSA could also attenuate upregulation of Ku80 and DNA-PKcs in prostate and colon cancer cells [29]. SAHA attenuated radiation-induced Rad51 and Ku80 protein expression in two sarcoma cell lines (KHOS-24OS, SAOS2) [30].

In this study, we also used primary-culture cells of chemical-induced breast cancer model to detect VPA-induced radiosensitivity. This model successfully mimicked the development of human primary tumor by a chemical carcinogen, DMBA. The results indicated that VPA can radiosensitize tumor cells through inhibiting DNA repair function, which provides strong evidence in support of the effects of VPA on radiosensitivity and exhibited a worthy implication for the study of its clinical trial and preclinical study. However, we still need to further explore how VPA influences tumor radiosensitivity in vivo in this primary tumor model.

Therefore, sensitization of tumor cells via inhibition of the DNA damage repair response may contribute a broader and more meaningful strategy to improve radio-therapy efficacy for tumor patients.

4. Materials and Methods

4.1. Materials

VPA was purchased from Sigma (St. Louis, MO, USA). The concentration of 0.5 mM and 1.0 mM were chosen as a safe and critical dose, respectively, as informed by previous work [7].

4.2. Cell Line

The U2OS osteosarcoma cell line was obtained from Maria Jesin's Lab in Developmental Biology Program, Memorial Sloan-Kettering Cancer Centre, New York, NY, USA. The U2OS cell line was cultured in DMEM medium (Gibco, Carlsbad, CA, USA) with 10% fetal bovine serum (Gibco, Carlsbad, CA, USA), 100 µg/mL streptomycin and 100 units/mL penicillin (Sigma). The cell line was grown at 37 °C with a humidified environment of 5% carbon dioxide. The cell line was treated with VPA and 10 µM ABT888 (poly ADP-ribose polymerase inhibitor, Active Biochemicals, Hong Kong, China) for in vitro clonogenic assay.

4.3. Tissue Culture and Animal Husbandry

Female Sprague-Dawley (SD) rats were purchased from Pengyue Laboratory Animal Co. Ltd. Jinan, China. The studies of animal tissue were performed in accordance with the requirements of the Shandong University Human and Animal Ethics Research Committee (The project identification code is 81472800, the date of approval was 3 March 2014 issued by the ethics committee review board of prevention medicine in Shandong University of China). All rats were housed in a specific-pathogen-free environment, at a temperature of 23 ± 1 °C. The lights were at a daily rhythm of 12 h and the SD rats were fed fresh food and water ad libitum throughout the experiment. The care of the animals was in accordance with the relevant Chinese laws and guidelines used for experimentation and scientific purposes. Breast tumors were induced in 50 day old female SD rats (weighted 150 ± 15 g; $n = x$) by a single administration of 20 mg/mL DMBA (7,12-dimethylbenzanthracene, Sigma, St. Louis, MO, USA) dissolved in sesame oil by oral gavage. The rats were palpated twice weekly for tumors. The rats where tumor burden was approximately 10% of total body weight were killed on day 90. All other rats were euthanized 20 weeks after the administration of DMBA.

Rats were injected intraperitoneal with 1 mL chloral hydrate (Sigma, St. Louis, MO, USA) for anesthesia, then sodium sulfide (Sigma, St. Louis, MO, USA) was used for unhairing. Breast tumor induced by DMBA in rats was sterilely isolated and mechanically dissociated into approximately 2 mm^3 of tissue was utilized. The tumor specimens were put onto P60 dishes and incubated with 20% fetal bovine serum and cultured at 37 °C with a humidified environment of 5% carbon dioxide for primary cell culture. Around 10 days, the cells grew from tissue and the cells were used for relative study.

The morphological structure of the tissue was observed by HE staining. Figure 2(A4) showed that the structure of breast tissue in normal rats had a few duct and acinus; In contrast, a large number of hyperplasia cells were found in the breast cancer tissue and the cell arrangement in the tumor tissue was also part of the disorder (Figure 2(A5)), indicating that the breast cancer in these rats was successfully induced by this chemical carcinogen. The primary culture tumor cells were obtained from the breast cancer tissue (Figure 2(A6)).

4.4. Clonogenic Survival Assay

The clonogenic survival assay was described in our previous publications [7,31]. In brief, the U2OS cells and primary culture cells from breast cancer tissue in rats were treated with 2, 4, or 6 Gy of IR using a Siemens Stabilipan 2 X-ray generator (Qilu Hospital, Jinan, China) operating at 250 kVp 12 mA at a dose rate of 2.08 Gy/min. For the combination group, the cells were pretreated with 0.5 mM for 24 h, then further irradiated with different doses. The number of cell colonies ($\geq$50 cells per clone) was counted and cell survival was presented by the survival fraction (SF): SF = (the number of clones/seeded cells)/plating efficiency (PE).

For clonogenic survival assay in the cells treated with both VPA and ABT888, the cells were pretreated with 0.5 or 1.0 mM VPA for 24 h, and then 10 μM ABT888 was added for a further 24 h incubation. The SF in each group was also analyzed.

4.5. Quantitative Fluorescence In Situ Hybridization (Q-FISH) for Chromosomal Aberration Analysis

As described in Reference [31–33], the 2 Gy treated cells with the pretreatment by VPA for 24 h and culture were incubated for 20 h before 0.05 g/mL colcemid (Gibco, Carlsbad, CA, USA) was added for a further 4 h incubation to obtain metaphase cells.

4.6. Comet Assay for DNA DSBs

The neutral comet assay was performed using the Trevigen Comet Assay kit and was described in our recent publication [7]. Simply speaking, the comet tail in VPA-treated, or untreated cells at 0, 60, and 120 min post-8 Gy were analyzed. For the comet assay used to examine comet tail in the cells at 0 min post-IR, VPA-treated or untreated cells were on ice during the whole irradiation process to allow the cell have minimum chance to repair damaged DNA. At this time point, whole DNA DSBs in the cells were presented. However, for the comet assay of cells at 60 and 120 min post-IR, VPA-treated or untreated cells did not require ice. Other steps of the comet assay were done in accordance with the standard procedures provided by the manufacturer (Trevigen Company, Gaithersburg, Montgomery County, MD, USA).

4.7. Immunofluorescence Assay of γH2AX and 53BP1

The cells were pretreated by VPA for 24 h and further irradiated. Then treated and untreated cells were rinsed with phosphate buffer saline (PBS) and fixed with paraformaldehyde. Cells were washed with PBST buffer (PBS + 0.2% Triton X-100), then blocked with 10% serum for 1 h and incubated with a primary antibody of γH2AX (Ser139, clone JBW301, Millipore, Darmstadt, Germany), or 53BP1 (NB100-304, NOVUS) overnight at 4 °C. The cells were further incubated with a secondary antibody of AlexaFluor 594-labeled goat anti-mouse I gG, or AlexaFluor 488-labeled chicken anti-rabbit (Thermo Fisher, Waltham, MA, USA) at a 1:300 dilution for 1 h in the dark after washing with PBST buffer, then stained with DAPI for nucleus [7,31].

4.8. Statistical Analysis

Results are expressed as means $\pm$ standard deviation for the groups. Data were analyzed by independent sample *t*-test. $p < 0.05$ indicated a statistically significant difference.

Acknowledgments: This work was supported by the Natural Science Foundation of China (81472800, 81172527), and Shandong University of Science and Technology of Shandong Province (2013GGE27052, 2014GGH218010).

Author Contributions: Zhihui Feng conceived and designed the experiments; Guochao Liu, Hui Wang, Fengmei Zhang, Youjia Tian, Zhujun Tian, and Zuchao Cai performed the experiments; David Lim and Zhihui Feng analyzed the data; Guochao Liu, Hui Wang, David Lim, and Zhihui Feng wrote the paper.

Conflicts of Interest: The authors declare no conflict of interest.

References

1. Groselj, B.; Sharma, N.L.; Hamdy, F.C.; Kerr, M.; Kiltie, A.E. Histone deacetylase inhibitors as radiosensitisers: Effects on DNA damage signalling and repair. *Br. J. Cancer* **2013**, *108*, 748–754. [CrossRef] [PubMed]

2. Mawatari, T.; Ninomiya, I.; Inokuchi, M.; Harada, S.; Hayashi, H.; Oyama, K.; Makino, I.; Nakagawara, H.; Miyashita, T.; Tajima, H.; et al. Valproic acid inhibits proliferation of HER2-expressing breast cancer cells by inducing cell cycle arrest and apoptosis through Hsp70 acetylation. *Int. J. Oncol.* **2015**, *47*, 2073–2081. [CrossRef] [PubMed]

3. Elbadawi, M.A.A.; Awadalla, M.K.A.; Hamid, M.M.A.; Mohamed, M.A.; Awad, T.A. Valproic Acid as a Potential Inhibitor of Plasmodium falciparum Histone Deacetylase 1 (PfHDAC1): An in Silico Approach. *Int. J. Mol. Sci.* **2015**, *16*, 3915–3931. [CrossRef] [PubMed]

4. Munster, P.; Marchion, D.; Bicaku, E.; Schmitt, M.; Lee, J.H.; DeConti, R.; Simon, G.; Fishman, M.; Minton, S.; Garrett, C.; et al. Phase I trial of histone deacetylase inhibition by valproic acid followed by the topoisomerase II inhibitor epirubicin in advanced solid tumors: A clinical and translational study. *J. Clin. Oncol.* **2007**, *25*, 1979–1985. [CrossRef] [PubMed]

5. Marchion, D.C.; Bicaku, E.; Daud, A.I.; Sullivan, D.M.; Munster, P.N. Valproic acid alters chromatin structure by regulation of chromatin modulation proteins. *Cancer Res.* **2005**, *65*, 3815–3822. [CrossRef] [PubMed]

6. Makita, N.; Ninomiya, I.; Tsukada, T.; Okamoto, K.; Harada, S.; Nakanuma, S.; Sakai, S.; Makino, I.; Kinoshita, J.; Hayashi, H.; et al. Inhibitory effects of valproic acid in DNA double-strand break repair after irradiation in esophageal squamous carcinoma cells. *Oncol. Rep.* **2015**, *34*, 1185–1192. [CrossRef] [PubMed]

7. Luo, Y.; Wang, H.; Zhao, X.P.; Dong, C.; Zhang, F.M.; Guo, G.; Guo, G.S.; Wang, X.W.; Powell, S.N.; Feng, Z.H. Valproic acid causes radiosensitivity of breast cancer cells via disrupting the DNA repair pathway. *Toxicol. Res. UK* **2016**, *5*, 859–870. [CrossRef]

8. Van Oorschot, B.; Granata, G.; Di Franco, S.; ten Cate, R.; Rodermond, H.M.; Todaro, M.; Medema, J.P.; Franken, N.A.P. Targeting DNA double strand break repair with hyperthermia and DNA-PKCS inhibition to enhance the effect of radiation treatment. *Oncotarget* **2016**, *7*, 65504–65513. [CrossRef] [PubMed]

9. Chinnaiyan, P.; Cerna, D.; Burgan, W.E.; Beam, K.; Williams, E.S.; Camphausen, K.; Tofilon, P.J. Postradiation sensitization of the histone deacetylase inhibitor valproic acid. *Clin. Cancer Res.* **2008**, *14*, 5410–5415. [CrossRef] [PubMed]

10. Mamo, T.; Mladek, A.C.; Shogren, K.L.; Gustafson, C.; Gupta, S.K.; Riester, S.M.; Maran, A.; Galindo, M.; van Wijnen, A.J.; Sarkaria, J.N.; et al. Inhibiting DNA-PKCS radiosensitizes human osteosarcoma cells. *Biochem. Biophys. Res. Commun.* **2017**, *486*, 307–313. [CrossRef] [PubMed]

11. Zuch, D.; Giang, A.H.; Shapovalov, Y.; Schwarz, E.; Rosier, R.; O'Keefe, R.; Eliseev, R.A. Targeting radioresistant osteosarcoma cells with parthenolide. *J. Cell. Biochem.* **2012**, *113*, 1282–1291. [CrossRef] [PubMed]

12. Rogakou, E.P.; Pilch, D.R.; Orr, A.H.; Ivanova, V.S.; Bonner, W.M. DNA double-stranded breaks induce histone H2AX phosphorylation on serine 139. *J. Biol. Chem.* **1998**, *273*, 5858–5868. [CrossRef] [PubMed]

13. Li, Y.H.; Wang, X.; Pan, Y.; Lee, D.H.; Chowdhury, D.; Kimmelman, A.C. Inhibition of non-homologous end joining repair impairs pancreatic cancer growth and enhances radiation response. *PLoS ONE* **2012**, *7*, e39588. [CrossRef] [PubMed]

14. Lobrich, M.; Shibata, A.; Beucher, A.; Fisher, A.; Ensminger, M.; Goodarzi, A.A.; Barton, O.; Jeggo, P.A. gamma H2AX foci analysis for monitoring DNA double-strand break repair Strengths, limitations and optimization. *Cell Cycle* **2010**, *9*, 662–669. [CrossRef] [PubMed]

15. Malewicz, M. The role of 53BP1 protein in homology-directed DNA repair: Things get a bit complicated. *Cell Death Differ.* **2016**, *23*, 1902–1903. [CrossRef] [PubMed]

16. Maes, K.; De Smedt, E.; Lemaire, M.; De Raeve, H.; Menu, E.; Van Valckenborgh, E.; McClue, S.; Vanderkerken, K.; De Bruyne, E. The role of DNA damage and repair in decitabine-mediated apoptosis in multiple myeloma. *Oncotarget* **2014**, *5*, 3115–3129. [CrossRef] [PubMed]

17. Lassmann, M.; Hanscheid, H.; Gassen, D.; Biko, J.; Meineke, V.; Reiners, C.; Scherthan, H. In Vivo Formation of gamma-H2AX and 53BP1 DNA Repair Foci in Blood Cells After Radioiodine Therapy of Differentiated Thyroid Cancer. *J. Nucl. Med.* **2010**, *51*, 1318–1325. [CrossRef] [PubMed]

18. Croco, E.; Marchionni, S.; Bocchini, M.; Angeloni, C.; Stamato, T.; Stefanelli, C.; Hrelia, S.; Sell, C.; Lorenzini, A. DNA Damage Detection by 53BP1: Relationship to Species Longevity. *J. Gerontol. A Biol. Sci. Med. Sci.* **2016**. [CrossRef] [PubMed]

19. Nijman, S.M.B. Synthetic lethality: General principles, utility and detection using genetic screens in human cells. *FEBS Lett.* **2011**, *585*, 1–6. [CrossRef] [PubMed]

20. Brunen, D.; Bernards, R. Drug therapy: Exploiting synthetic lethality to improve cancer therapy. *Nat. Rev. Clin. Oncol.* **2017**. [CrossRef] [PubMed]

21. Bryant, H.E.; Schultz, N.; Thomas, H.D.; Parker, K.M.; Flower, D.; Lopez, E.; Kyle, S.; Meuth, M.; Curtin, N.J.; Helleday, T. Specific killing of BRCA2-deficient tumours with inhibitors of poly(ADP-ribose) polymerase. *Nature* **2005**, *434*, 913–917. [CrossRef] [PubMed]

22. Guo, G.S.; Zhang, F.M.; Gao, R.J.; Delsite, R.; Feng, Z.H.; Powell, S.N. DNA repair and synthetic lethality. *Int. J. Oral Sci.* **2011**, *3*, 176–179. [CrossRef] [PubMed]

23. Moynahan, M.E.; Jasin, M. Mitotic homologous recombination maintains genomic stability and suppresses tumorigenesis. *Nat. Rev. Mol. Cell Biol.* **2010**, *11*, 196–207. [CrossRef] [PubMed]

24. Shoji, M.; Ninomiya, I.; Makino, I.; Kinoshita, J.; Nakamura, K.; Oyama, K.; Nakagawara, H.; Fujita, H.; Tajima, H.; Takamura, H.; et al. Valproic acid, a histone deacetylase inhibitor, enhances radiosensitivity in esophageal squamous cell carcinoma. *Int. J. Oncol.* **2012**, *40*, 2140–2146. [CrossRef] [PubMed]

25. Defoort, E.N.; Kim, P.M.; Winn, L.M. Valproic acid increases conservative homologous recombination frequency and reactive oxygen species formation: A potential mechanism for valproic acid-induced neural tube defects. *Mol. Pharmacol.* **2006**, *69*, 1304–1310. [CrossRef] [PubMed]

26. Chen, X.; Wong, P.; Radany, E.; Wong, J.Y. HDAC inhibitor, valproic acid, induces p53-dependent radiosensitization of colon cancer cells. *Cancer Biother. Radiopharm.* **2009**, *24*, 689–699. [CrossRef] [PubMed]

27. Adimoolam, S.; Sirisawad, M.; Chen, J.; Thiemann, P.; Ford, J.M.; Buggy, J.J. HDAC inhibitor PCI-24781 decreases RAD51 expression and inhibits homologous recombination. *Proc. Natl. Acad. Sci. USA* **2007**, *104*, 19482–19487. [CrossRef] [PubMed]

28. Koprinarova, M.; Botev, P.; Russev, G. Histone deacetylase inhibitor sodium butyrate enhances cellular radiosensitivity by inhibiting both DNA nonhomologous end joining and homologous recombination. *DNA Repair* **2011**, *10*, 970–977. [CrossRef] [PubMed]

29. Kachhap, S.K.; Rosmus, N.; Collis, S.J.; Kortenhorst, M.S.; Wissing, M.D.; Hedayati, M.; Shabbeer, S.; Mendonca, J.; Deangelis, J.; Marchionni, L.; et al. Downregulation of homologous recombination DNA repair genes by HDAC inhibition in prostate cancer is mediated through the E2F1 transcription factor. *PLoS ONE* **2010**, *5*, e11208. [CrossRef] [PubMed]

30. Blattmann, C.; Oertel, S.; Ehemann, V.; Thiemann, M.; Huber, P.E.; Bischof, M.; Witt, O.; Deubzer, H.E.; Kulozik, A.E.; Debus, J.; et al. Enhancement of radiation response in osteosarcoma and rhabdomyosarcoma cell lines by histone deacetylase inhibition. *Int. J. Radiat. Oncol. Biol. Phys.* **2010**, *78*, 237–245. [CrossRef] [PubMed]

31. Dong, C.; Zhang, F.; Luo, Y.; Wang, H.; Zhao, X.; Guo, G.; Powell, S.N.; Feng, Z. p53 suppresses hyper-recombination by modulating BRCA1 function. *DNA Repair* **2015**, *33*, 60–69. [CrossRef] [PubMed]

32. Feng, Z.H.; Scott, S.P.; Bussen, W.; Sharma, G.G.; Guo, G.S.; Pandita, T.K.; Powell, S.N. Rad52 inactivation is synthetically lethal with BRCA2 deficiency. *Proc. Natl. Acad. Sci. USA* **2011**, *108*, 686–691. [CrossRef] [PubMed]

33. Feng, Z.H.; Zhang, J.R. A dual role of BRCA1 in two distinct homologous recombination mediated repair in response to replication arrest. *Nucleic Acids Res.* **2012**, *40*, 726–738. [CrossRef] [PubMed]

International Journal of
Molecular Sciences

MDPI

Article

High NOTCH1 mRNA Expression Is Associated with Better Survival in HNSCC

Markus Wirth [1,*,†], **Daniel Jira** [1,†], **Armin Ott** [2] , **Guido Piontek** [1] and **Anja Pickhard** [1]

1   Department of Otolaryngology-Head and Neck Surgery, Technical University of Munich,
    Ismaninger Straße 22, 81675 Muenchen, Germany; daniel.jira@web.de (D.J.);
    piontek@lrz.tu-muenchen.de (G.P.); a.pickhard@lrz.tu-muenchen.de (A.P.)
2   Institute of Medical Informatics Statistics and Epidemiology, Technical University of Munich,
    Ismaninger Straße 22, 81675 Muenchen, Germany; armin.ott@tum.de
*   Correspondence: markus.wirth@tum.de; Tel.: +49-89-4140-9416
†   These authors contributed equally to this work.

Received: 25 February 2018; Accepted: 6 March 2018; Published: 13 March 2018

Abstract: The clinical impact of the expression of NOTCH1 signaling components in squamous cell carcinoma of the pharynx and larynx has only been evaluated in subgroups. The aim of this study was therefore to evaluate NOTCH1 expression in head and neck squamous cell cancer (HNSCC) patient tissue and cell lines. We analyzed tissue from 195 HNSCCs and tissue from 30 normal patients for mRNA expression of NOTCH1, NOTCH3, HES1, HEY1, and JAG1 using quantitative real-time PCR. Association of expression results and clinical orpathological factors was examined with multivariate Cox regression. NOTCH1 expression was determined in three Human Papilloma Virus (HPV)-positive and nine HPV-negative HNSCC cell lines. High expression of NOTCH1 was associated with better overall survival ($p = 0.013$) and disease-free survival ($p = 0.040$). Multivariate Cox regression confirmed the significant influence of NOTCH1 expression on overall survival ($p = 0.033$) and disease-free survival ($p = 0.029$). A significant correlation was found between p16 staining and NOTCH1 mRNA expression (correlation coefficient 0.28; $p = 0.01$). NOTCH1 was expressed at higher levels in HPV-positive HNSCC cell lines compared with HPV-negative cell lines, which was not statistically significant ($p = 0.068$). We conclude that NOTCH1 expression is associated with overall survival, and that inhibition of NOTCH1 therefore seems less promising.

Keywords: NOTCH signaling; HNSCC; overall survival

1. Introduction

Head and neck cancer is ranked as the seventh most frequent cause of cancer death in the world and squamous cell carcinoma comprises the most common subgroup [1]. Despite ongoing advances in surgery and in radio- and chemotherapy, five-year survival rates for head and neck squamous cell carcinoma (HNSCC) remain still in the order of 50% to 60% [1,2]. Besides Human Papilloma Virus (HPV) status, prediction of clinical outcome and therapy are still based on histopathological and clinical parameters [3,4]. Novel therapeutic targets and markers to stratify patients are therefore urgently needed.

The NOTCH signaling pathway is becoming increasingly relevant in diverse tumor entities including HNSCC [5,6]. NOTCH signaling plays an integral part in cell fate and development by controlling proliferation, differentiation, angiogenesis, and apoptosis [7]. Initiation of signaling is mediated through binding of the ligands Jagged or Delta-like resulting in the cleavage and release of NOTCH intracellular fragments (NOTCH-IC) [5,7–9]. Subsequently, NOTCH-IC are translocated to the nucleus and interact with RBPJ, a DNA-binding protein [5,7]. This leads to transcription of targets such as the MYC transcription factor and HES and HEY family proteins [5,7].

Alterations in NOTCH signaling have been described in several cancers and result in tumor promotion or suppression depending on the cancer entity and context [5]. The predominant function of NOTCH signaling remains controversial—it may act as an oncogene, tumor suppressor, or even have a bimodal role [10]. Frequent mutations of the NOTCH receptor family were detected in HNSCC and most likely result in loss of function of the receptors [10–12]. A high incidence of nonsynonymous mutations was identified in 43% of Chinese patients with oral squamous cell carcinoma (OSCC) and the occurrence of mutations was associated with poor overall survival [13,14]. The expression of NOTCH1 pathway genes, however, has only been studied in small patient cohorts or in subgroups of HNSCC with contradictory results.

Currently, the clinical relevance of transcriptional alterations in the NOTCH signaling pathway in HNSCC is not well understood. The aim of this study was therefore to evaluate the expression of key components of the NOTCH pathway with quantitative real-time PCR in a larger HNSCC collective and in normal tissue. Secondly, the association of the expression with clinical and pathological parameters was investigated. Moreover, NOTCH1 expression was also analyzed in HPV-positive and -negative HNSCC cell lines.

2. Results

The clinical and pathological characteristics of this cohort are depicted in Table 1. Significant associations between high NOTCH1 expression and nodal stage and p16 status were found.

Table 1. Depiction of the clinical and pathological characteristics of the 195 head and neck squamous cell carcinoma (HNSCC) patients included in this study and association with NOTCH1 expression.

Clinical Characteristics	Overall	NOTCH1 Expression		*p* Value (Fisher Exact)
		Intermediate–Low	High	
Overall	195	175 (89.7%)	20 (10.3%)	
Primary site				1
Oral cavity	32 (16.4%)	30 (17.1%)	2 (10.0%)	
Oropharynx	83 (42.6%)	70 (40.0%)	13 (65.0%)	
Larynx	42 (21.5%)	40 (22.9%)	2 (10%)	
Hypopharynx	38 (19.5%)	35 (20.0%)	3 (15.0%)	
Alcohol consumption				0.128
Daily	153 (78.5%)	142 (85.5%)	11 (61.1%)	
Rare/never	31 (15.9%)	24 (14.5%)	7 (38.9%)	
Unknown	11 (5.6%)			
Tobacco exposure				0.072
Smoker	166 (85.1%)	154 (90.6%)	12 (66.7%)	
Nonsmoker	22 (11.3%)	16 (9.4%)	6 (33.3%)	
Unknown	7 (3.6%)			
Staging and Grading				
Tumor stage (pathological)				0.096
T1	46 (23.6%)	38 (21.7%)	8 (40.0%)	
T2	53 (27.2%)	44 (25.1%)	9 (45.0%)	
T3	50 (25.6%)	48 (27.4%)	2 (10.0%)	
T4	46 (23.6%)	45 (25.7%)	1 (5%)	
Nodal stage (pathological)				0.032
N0	65 (33.3%)	64 (36.6%)	1 (5.3%)	
N1–3	129 (66.2%)	111 63.4%)	18 (94.7%)	
NX	1 (0.5%)			
Metastasis (initial stage)				1
M0	172 (88.2%)	155 (88.6%)	17 (85%)	
M1	8 (4.1%)	8 (4.6%)	0	
MX	15 (7.7%)	12 (6.9%)	3 (15.0%)	

Table 1. *Cont.*

Clinical Characteristics	Overall	NOTCH1 Expression		*p* Value (Fisher Exact)
		Intermediate–Low	High	
Overall	195	175 (89.7%)	20 (10.3%)	
Grading				1
G1/G2	104 (53.5%)	95 (54.3%)	9 (45%)	
G3/G4	91 (46.7%)	80 (45.7%)	11 (55%)	
p16 Status				0.048
p16 positive	50 (25.6%)	41 (26.6%)	9 (64.3%)	
p16 negative	118 (60.5%)	113 (73.4%)	5 (35.7%)	
p16 unknown	27 (13.8%)			

2.1. NOTCH1 and 3 and HES1 Significantly Lower in Tumor

Relative mRNA expression of NOTCH1 and 3 and HES1 mRNA was significantly lower in tumor compared with in normal tissue (Table 2 and Figure 1). HEY1 mRNA was increased in HNSCC, but this was not significant ($p = 0.254$, Table 2). No significant difference between tumor tissue and normal tissue was detected for JAG1 mRNA expression ($p = 0.270$, Table 2).

Table 2. Depiction of relative mRNA expression in tumor tissue vs. in normal tissue. Relative expression was compared with the $\Delta\Delta$Ct method and *p*-value calculated by Mann–Whitney Test.

Target	Normal Tissue			HNSCC			*p* Value
	Min	Max	Median	Min	Max	Median	
NOTCH1	0.29	2.48	1.15	0.10	5.94	0.69	0.003
NOTCH3	0.41	5.52	0.89	0.11	2.80	0.57	<0.001
HES1	0.33	9.97	0.85	0.16	3.21	0.69	0.049
HEY1	0.26	3.99	0.85	0.01	15.57	1.24	0.254
JAG1	0.18	3.13	1.10	0.19	3.63	0.92	0.270

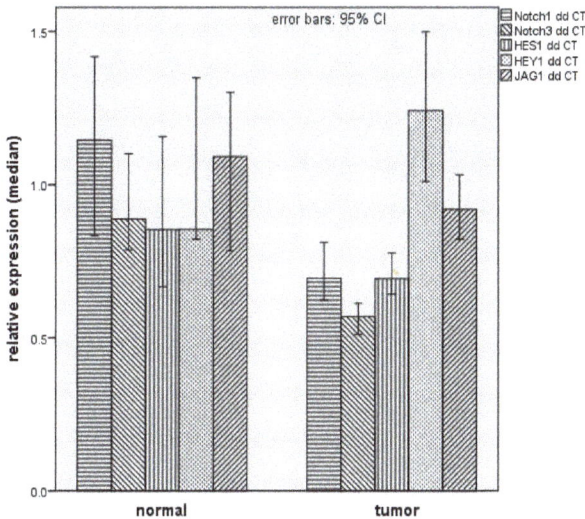

Figure 1. Depiction of median of relative mRNA expression in tumor tissue and normal tissue. Data were normalized to GAPDH per sample.

2.2. High NOTCH1 Expression with Significant Longer Overall Survival

In HNSCC patients, high mRNA expression of NOTCH1 was associated with better overall survival (OS, $p = 0.013$) and better disease-free survival (DFS, $p = 0.040$), as illustrated in Figure 2 and Table 3. Multivariate Cox regression confirmed the significant influence of high NOTCH1 expression on OS ($p = 0.033$) and DFS ($p = 0.029$) (Table 4).

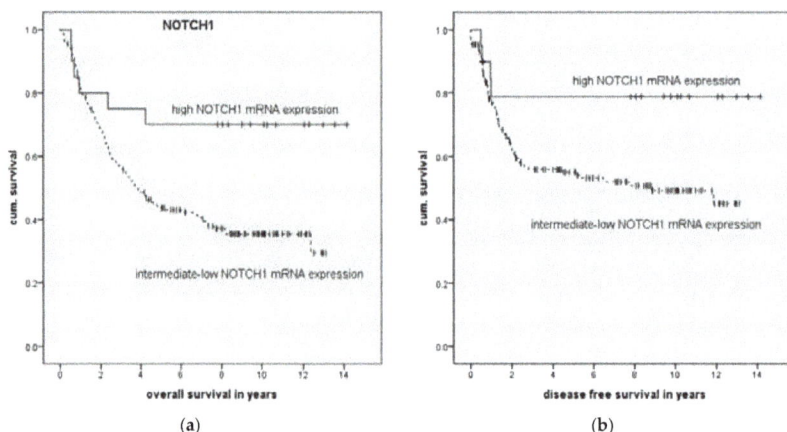

(a) (b)

Figure 2. Association of relative NOTCH1 mRNA expression with overall (OS) and disease-free (DFS) survival was analyzed using the Kaplan-Meier method as well as the log-rank test. Patients were stratified into a relative high expression (mean mRNA expression + 1 standard deviation (SD)) group and an intermediate–low mRNA expression group (remainder). Significantly longer overall survival (**a**) and disease-free survival (**b**) were found for patients with relatively high mRNA expression of NOTCH1 compared with those with intermediate or low expression ($p = 0.013$ (OS) and $p = 0.040$ (DFS)).

Table 3. Depiction of results of survival analysis with Kaplan–Meier method and log-rank test. Association of relative mRNA expression and overall and disease-free survival. For each target, two comparisons were made: (1) high mRNA expression (mean + 1 standard deviation) vs. intermediate and low expression (remainder); (2) relative low mRNA expression (mean − 1 standard deviation) vs. intermediate and high expression (remainder).

Comparison of Relative mRNA Expression	OS (*p* Value)	DFS (*p* Value)
Comparison of high vs. intermediate—low expression		
NOTCH1	0.013	0.040
NOTCH3	0.568	0.896
HEY1	0.419	0.077
HES1	0.268	0.240
JAG1	0.461	0.481
Comparison of low vs. intermediate—high expression		
NOTCH1	0.611	0.673
NOTCH3	0.633	0.082
HEY1	0.755	0.040
HES1	0.590	0.065
JAG1	0.322	0.529

Patients with low HEY1 mRNA expression demonstrated a significantly shortened disease-free survival ($p = 0.040$); this was also confirmed by multivariate Cox regression ($p = 0.027$, HR 1.69, Cox regression performed with clinical and pathological factors listed in Table 4 and low expression of

NOTCH1 and HEY1). Prolonged DFS was seen in patients with high mRNA expression of downstream HEY1 ($p = 0.077$). Patients with low expression of NOTCH3 and downstream HES1 also showed a shortened DFS ($p = 0.082$ and $p = 0.065$). No significant alteration was observed for OS or DFS by JAG1 mRNA expression.

Table 4. Forward stepwise Cox regression with age; sex; T, N, and M status; grading; p16 staining; and high NOTCH1 and high HEY1 expression (mean + 1 SD) as criteria was performed for OS (a) and DFS (b).

(a) Overall Survival			
Factor	*p* Value	HR	95% CI
T status	<0.01	1.95	(1.33–2.87)
N status	<0.01	3.29	(2.09–5.18)
Age	0.01	1.03	(1.01–1.05)
p16 positive	<0.01	0.42	(0.26–0.69)
High expression of NOTCH1 (mean + 1 SD)	0.03	0.38	(0.16–0.93)
(b) Disease-Free Survival—High NOTCH1 and HEY1			
Factor	*p* Value	HR	95% CI
N status	<0.01	2.60	(1.55–4.35)
M status	0.01	2.38	(1.29–4.38)
p16 positive	<0.01	0.39	(0.21–0.71)
High expression of HEY1 (mean + 1 SD)	0.06	0.41	(0.16–1.02)
High expression of NOTCH1 (mean + 1 SD)	0.03	0.31	(0.11–0.89)

2.3. NOTCH1 Expression and p16 Staining

Overall, 26% of the patients were p16 positive with most frequent staining in the oropharynx (43%) and least frequent staining in the hypopharynx (8%). Expression of p16 was associated with significantly prolonged overall and disease-free survival ($p = 0.018$ and $p = 0.008$), which was also confirmed in the Cox regression analysis ($p = 0.001$ and $p = 0.002$). The median NOTCH1 mRNA expression was significantly higher in p16-positive patients (median 0.91 vs. median 0.61, $p < 0.001$, Figure 3). A significant correlation was found between p16 staining and NOTCH1 mRNA expression (correlation coefficient: 0.280; $p = 0.01$). Patients with p16-positive tumors with high NOTCH1 expression had longer overall survival ($n = 50$, $p = 0.139$, Figure 4). No noticeable effect was seen in p16-negative patients ($n = 118$, $p = 0.967$, Figure 4).

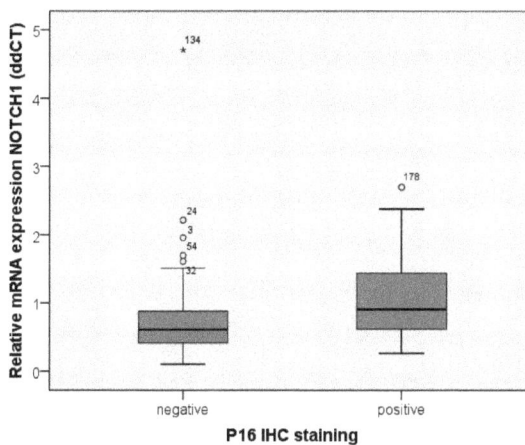

Figure 3. Depiction of median of relative mRNA expression of NOTCH1 in p16-positive and p16-negative tumor tissue. Expression of NOTCH1 was significantly higher in p16-positive tumor tissue ($p \leq 0.001$, Mann–Whitney U Test). * represents extreme outliers.

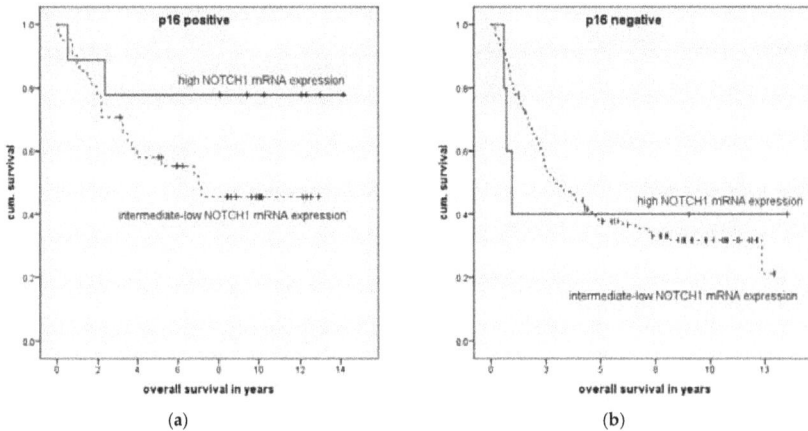

(a)　　　　　　　　　　　　　　　　　　　(b)

Figure 4. Association of relative NOTCH1 mRNA expression and overall survival in p16-positive (**a**) and -negative patients (**b**) was analyzed by the Kaplan-Meier method followed by log-rank test. A longer overall survival was found for p16-positive patients with relative high mRNA expression of NOTCH1 (**a**) ($n = 50$, $p = 0.139$) but not for p16-negative patients (**b**) ($n = 118$, $p = 0.967$).

2.4. NOTCH1 Expression in HNSCC Cell Lines

HPV16 status of cell lines was confirmed by measuring viral oncogenes E6 and E7 with RT-PCR (Figure S1). NOTCH1 signaling was characterized in three HPV-positive (UD-SCC-2, UP-SCC-154, 93VU) and nine HPV-negative (UD-SCC-3, -4, -5, -6, -7, UP-SCC-111, HN, Cal27, and SAS) HNSCC cell lines with Western blotting (Figure 5). All cell lines except for HN expressed NOTCH1. NOTCH1 (NTM) protein expression was higher in the HPV-positive cell lines compared with in the HPV-negative cell lines, but this was not statistically significant ($p = 0.068$). The NOTCH1, HES1, and HEY1 expression varied considerably in the HPV-negative cell lines.

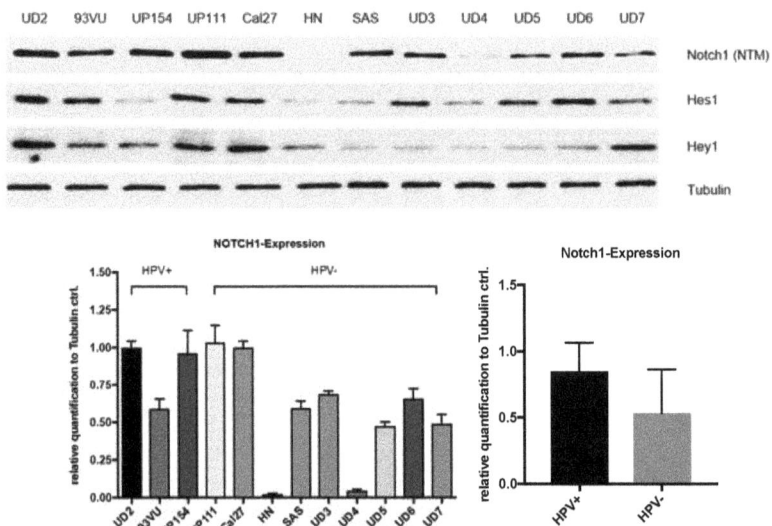

Figure 5. In Western blot analyses, NOTCH1 (NTM) could be detected in all cell lines except for HN (**top**). Quantification revealed higher NOTCH1 protein expression in HPV-positive cell lines compared with in HPV-negative cell lines ($p = 0.068$) (**bottom**). Error bars represent one standard deviation.

3. Discussion

3.1. Expression Analysis

A significant downregulation of NOTCH1, 3, and HES1 was found in comparison with normal tissue. Other researchers have found dissimilar results: in a cohort study of 44 patients with HNSCC, researchers found an upregulation of NOTCH3, HES1, HEY1, JAG1, and JAG2 on the mRNA level in comparison with noncancerous soft palate tissue [10]. The differing results could be due to the different location of normal tissue collected and also due to the smaller sample size analyzed. We analyzed normal mucosa from the oral cavity, oropharynx, hypopharynx, and larynx, which were excised during panendoscopy in proportion to the HNSCC tumor location in our cohort.

3.2. Overall Survival Analysis

Subsequently, the association of the differential mRNA expression with clinical outcome was examined. Patients were grouped into high and low expression cohorts (cut-off one standard deviation above or below mean). The subgroup with high NOTCH1 expression had significantly longer overall survival and disease-free survival. Corresponding to this result, patients with high expression of downstream effector HEY1 showed significantly longer DFS. Both results were confirmed in a multivariate Cox regression. As of yet, the role of the expression of key signaling components in HNSCC has only been examined in small patient cohorts or subgroups and differing results have been reported. Consistent with these results, in oropharyngeal squamous cell carcinoma patients, NOTCH1 staining correlated with improved survival [15]. Furthermore, negative staining for NOTCH1 intracellular domain in HNSCC tumors was associated with less differentiation [16]. However, in two reports, HNSCC patients with high NOTCH1 protein expression in immunohistochemistry showed poor prognosis [17,18]. Zhang et al. described higher expression of NOTCH1 and JAG1 in lymph node metastasis-positive tongue cancer [19]. Varying mutation rates and forms of NOTCH1 could explain the discrepancy between immunohistochemistry and quantitative real-time PCR. Mutation rates between 9% to 15% in Western cohorts and of 43% in a Chinese cohort were detected [6,11–13]. Predominantly inactivating mutations were reported in Western cohorts whereas a substantial proportion of probable oncogenic mutations were detected in the Chinese cohort [6,11–13]. Mutant receptors could, for example, be less degraded and therefore induce higher protein level measurements or truncated receptors to not be expressed at all, so that no protein can be detected. Moreover, NOTCH1 receptor mutation can alter downstream activation significantly, and downstream alterations can also change the effect of NOTCH signaling. Sun et al. reported that patients with mutant NOTCH1 receptor expressed downstream HES1/HEY1 similar to normal epithelium but a large subset of NOTCH1 wildtype patients showed an overexpression of HES1/HEY1 [10]. The proposed bimodal pattern of activation and suppression of NOTCH signaling in HNSCC could also contribute to seemingly contradictory results.

3.3. Association of NOTCH1 Expression, p16 Status, and Survival

Significantly higher NOTCH1 mRNA expression was found in p16-positive tumor probes. Additionally, a significant correlation between NOTCH1 mRNA expression and p16 staining was detected. Both high NOTCH1 expression and p16 staining were independent significant factors in the multiple stepwise Cox regression analysis for overall survival. A nonsignificant association was found between patients with high NOTCH1 expression and longer overall survival in the p16-positive patient group ($n = 50$, $p = 0.139$), but not in the p16-negative group ($n = 118$, $p = 0.967$). Unfortunately, p16 immunohistochemistry was only possible in 168 of 195 patient probes due to the consumption of formalin-fixed and paraffin-embedded (FFPE) material in the RNA isolation. The p16-positive group was therefore possibly too small to reach significance. High NOTCH1 expression is therefore probably an independent positive prognostic factor in p16-positive patients only. The differing expression and clinical relevance in p16-positive and -negative tumors could explain part of the above-described discrepancies on the role of NOTCH1 as a tumor suppressor or inhibitor. Since patients with p16-positive tumors are on average younger and healthier, long-term treatment side effects are increasingly becoming relevant. After validation in a bigger p16-positive patient cohort, high NOTCH1 expression could be further analyzed for use as a predictive marker and potentially used, e.g., in future de-escalation studies in p16-positive patients. In line with our results, the immunohistochemical examination of cleaved NOTCH1 expression revealed that negative staining HNSCC tumors were less likely to be HPV-positive [16]. Conversely, Troy et al. detected no correlation between NOTCH1 staining and HPV status in a small cohort of 27 HPV-positive and 40 HPV-negative HNSCCs [20]. This may be due to the small cohort size, disallowing the detection of a correlation. Additionally, protein staining could differ from RNA expression due to receptor degradation.

3.4. NOTCH1 Expression In Vitro

Since different NOTCH1 expression levels were detected in p16-positive and -negative HNSCC patients and high NOTCH1 expression was only a positive prognostic factor in p16-positive patients, NOTCH1 signaling was analyzed in three HPV-positive and nine HPV-negative HNSCC cell lines. Corresponding to the higher NOTCH1 expression in p16-positive patient tissues, we also found a trend for higher NOTCH1 expression in the HPV-positive cell lines. NOTCH1 expression greatly varied in the HPV-negative cell lines. There are only limited published data available on NOTCH1 expression in the examined HNSCC cells. Pickering also analyzed NOTCH1 expression in different HNSCC cell lines and detected an association between NOTCH1 mutational status and NOTCH1 expression [6]. Interestingly, NOTCH1 mutations were predominantly found in HPV-negative patient tissue in another investigation [12]. Higher NOTCH1 expression in HPV-positive patients could therefore be due to NOTCH1 wildtype expression. Moreover, it has been reported that cutaneous HPV E6 proteins inactivate NOTCH1 signaling downstream via interaction with mastermind-like (MAML) [21] and thus promotes dedifferentiation. HPV may also inactivate NOTCH1 signaling downstream in HNSCC, explaining the higher expression in p16-positive tumors.

In summary, this is to our knowledge the largest analysis towards determining the clinical relevance of the expression of NOTCH1 signaling components in HNSCC patients. Large alterations in the expression of NOTCH1, 3, and HES1 were detected in comparison with normal tissue. Patients with high expression of NOTCH1 showed significantly better prognosis. The favorable prognostic relevance of high NOTCH1 expression was only seen in p16-positive patients. High NOTCH1 expression should therefore be further evaluated as a prognostic marker in a larger cohort of p16-positive patients. There was also a trend seen for higher NOTCH1 expression in HPV-positive cell lines as compared to HPV-negative cells. Inhibitors of NOTCH signaling are already in clinical testing in other malignancies such as pancreatic and small-cell lung cancers [22,23]. However, NOTCH inhibition seems less promising in HNSCC, since patients with high NOTCH1 expression demonstrated better survival in our study. The findings in this study are therefore highly relevant for the development of NOTCH-targeting therapies in HNSCC.

4. Materials and Methods

4.1. Patient Tissue Samples

We obtained tissue samples from 195 HNSCC Patients (163 males, 32 females; median age 59 years, range 35 to 89 years) diagnosed between January 2002 and December 2005. As a control group we used mucosa obtained through panendoscopy or tonsillectomy ($n = 30$, 18 males, 12 females; median age 51 years, range 25 to 87 years) from oropharynx ($n = 7$), hypopharynx ($n = 17$), and larynx ($n = 6$). All patients were treated in the Department of Otorhinolaryngology at Klinikum rechts der Isar, Technical University of Munich. Sixty-five patients of the cohort have been used for a previous study [9]. All tissue samples were formalin fixed and paraffin embedded (FFPE). Only one patient received neoadjuvant therapy; all remaining specimens were retrieved before adjuvant or primary radiochemotherapy. The independent ethics committee of the Technical University of Munich approved the study, under project number 1420/05 (13 June 2014) and 107/15 (12 March 2015).

4.2. Clinical Data

Clinical data provided by the Munich Cancer Registry were verified with data gathered from filed and electronical medical records. Thirty-nine patients (20.0%) were treated solely with surgery; 95 patients (48.7%) underwent surgical resection combined with radio(-chemo)therapy. Primary radiation or radio-chemotherapy was used in 48 patients (24.6%); 5 patients (2.6%) received palliative radio- and/or chemotherapy. In 8 cases (4.1%), no treatment was applied or treatment could not be reproduced. The median survival calculated by Kaplan-Meier analysis was 4.24 years (min–max follow up period 0.09–14.14 years); the overall five-year survival rate was 46.6%.

4.2. RNA Isolation and cDNA Synthesis

FFPE tissue samples were deparaffinized and digested using 40 µL Proteinase K (Roche Diagnostics GmbH, Unterhaching, Germany) in 100 µL PK buffer (50 mM Tris, 1 mM EDTA, and 25% Tween 20 diluted in water) added to 16 µL 10% SDS (10 g Sodiumdodecylsulfat diluted in 100 mL water) at 55 °C. After 24 h, 10 µL Proteinase K was added again and samples were incubated for another day. Further processing was performed using an InviTrap® RNA Mini Kit (Stratec, Birkenfeld, Germany) according to the manufacturer's protocol. RNA from HNSCC cells was isolated with the RNeasy-Mini-Kit (Qiagen, Hilden, Germany). After isolation the RNA concentration was quantified using the NanoDrop 1000 system (PEQLAB, Erlangen, Germany). We used only probes with a minimal RNA concentration of 10 ng/µL. Afterwards, probes were diluted to a concentration of 10 or 25 ng/µL depending on the initial concentration of RNA and stored at −20 °C. cDNA synthesis was performed using Maxima® reverse transcriptase (Fermentas, Waltham, MA, USA) according to the manufacturer's protocol.

4.3. Quantitative Real-Time PCR

Quantitative real-time PCR (qPCR) was performed to quantify mRNA expression of NOTCH1, NOTCH3, HES1, HEY1, JAG1, HPV16 E6, and HPV16 E7. For normalization of expression levels, GAPDH was used per sample. For qPCR mix, 50 ng cDNA template was added to 12.5 µL KAPA-SYBR Fast Universal (PeqLab, Erlangen, Germany) and 0.5 µL of 20 pmol of each primer. Water was added to a final volume of 25 µL. Primer sequences and specific annealing temperatures are depicted in Table 5. NOTCH1, NOTCH3, HES1, and HEY1 primers were newly designed and GAPDH, JAG1, E6, and E7 primers were used as previously described [7,24–26]. After normalization, the ΔΔCt method was used to compare relative expression for NOTCH1 pathway components and gel electrophoresis for HPV E6 and E7.

Table 5. Primer sequences and specific annealing temperatures used for quantitative real-time PCR.

Primer	Sequence	Annealing Temperature
NOTCH1 forward	TGAATGGCGGGAAGTGTGAAG	62.0 °C
NOTCH1 reverse	GGTTGGGGTCCTGGCATCG	
NOTCH3 forward	ATGGTATCTGCACCAACCTGG	63.0 °C
NOTCH3 reverse	GATGTCCTGATCGCAGGAAGG	
HES1 forward	AAGAAAGATAGCTCGCGGCA	57.0 °C
HES1 reverse	CGGAGGTGCTTCACTGTCAT	
HEY1 forward	CCGACGAGACCGGATCAATA	64.0 °C
HEY1 reverse	GCTTAGCAGATCCCTGCTTCT	
JAG1 forward	ATCGTGCTGCCTTTCAGTTT	56.3 °C
JAG1 reverse	TCAGGTTGAACGGTGTCATT	
HPV16 E6 forward	CAAACCGTTGTGTGATTTGTTAATTA	61.0 °C
HPV16 E6 reverse	GCTTTTTGTCCAGATGTCTTTGC	
HPV16 E7 forward	TTTGCAACCAGAGACAACTGA	58.0 °C
HPV16 E7 reverse	GCCCATTAACAGGTCTTCCA	
GAPDH forward	AGCCACATCGCTCAGACA	56.0 °C
GAPDH reverse	GCCCAATACGACCAAATCC	

4.4. Immunohistochemical Study

To identify carcinomas associated with HPV infection, p16 expression was analyzed in tumor tissue. From previously identified FFPE blocks, 1.5 µm sections (cut with Microm HM 355 S (International GmbH, Walldorf, Germany)) were placed on glass slides, dewaxed, and rehydrated. Microwave oven heating in citrate-buffered saline was used for antigen retrieval, as recommended by the manufacturer. After cooling, the slides were incubated with the antibody. A CINtec® Histology Kit (Roche Diagnostics GmbH, Mannheim, Germany) containing mouse monoclonal anti-p16INK4a (E6H4) at concentration 1 µg/ml was used according to the manufacturer's protocol. Tissue with known expression of p16 was used as a positive control.

P16 expression level was described using a scoring system including staining intensity and percentage of stained tumor cells (Table 6). HPV positivity was considered at 3 or more points.

Table 6. Scoring system used for immunohistochemical identification of HPV-positive carcinomas. Points for staining intensity and staining proportion of tumor cells were summated; HPV positivity was considered at 3 or more points.

Staining Intensity	Points	Staining Proportion	Points
No staining	0	0%	0
Low	1	<10%	1
Moderate	2	10–29%	2
High	3	30–59%	3
		60–100%	4

4.5. Cell Culture

The Cal27, HN, and UP-SCC-154 cell lines were obtained from DSMZ (Braunschweig, Germany), the UD-SCC-2-7 cell lines were obtained from the University of Düsseldorf (Department of Otorhinolaryngology, Düsseldorf, Germany), the 93VU cell line from VU University Medical Center Amsterdam (Department of Clinical Genetics, Amsterdam, Netherlands), and SAS from JCRB cell bank (Osaka, Japan). All cell lines have been STR profiled and were routinely prophylactically treated against mycoplasma infection. The cells were cultured in Dulbecco's Modified Eagle Medium (DMEM) (Invitrogen, Darmstadt, Germany) containing 10% fetal bovine serum (FBS) (Biochrom,

Berlin, Germany), 2 mM glutamine, 100 µg/mL streptomycin, and 100 U/mL penicillin (Biochrom), maintained at 37 °C in an atmosphere of 5% CO_2, and grown to 70–90% confluence.

4.6. Western Blot Analysis

For protein analysis, cells were grown to 70% confluence in 10 cm tissue culture dishes. Cells were washed with ice-cold 1× DPBS and lysed with 500 µL of cell lysis buffer. The buffer contained 1× Cell Lysis Buffer (Cell Signaling, Danvers, MA, USA), 1 mM PMSF (Carl Roth, Karlsruhe, Germany), and 1× Protease Inhibitory Cocktail (Cell Signaling). The lysis buffer (10×) (Cell Signaling) included 20 mM Tris-HCl (pH 7.5), 150 mM NaCl, 1 mM Na_2EDTA, 1 mM EGTA, 1% Triton, 2.5 mM sodium pyrophosphate, 1 mM β-glycerophosphate, 1 mM Na_3VO_4, and 1 µg/mL leupeptin. Next, the cells were scraped off the culture dishes, pipetted into 1.5 mL microtubes, incubated on ice, and centrifuged at 4 °C and 10,000 rpm for 15 min to isolate the soluble protein fraction. The clarified lysate was frozen at −20 °C until use in the Bradford assay. The Bradford assay was used to verify that equal amounts were loaded per lane on an SDS-PAGE.

Equal protein concentrations (15 µg) were separated for 3 h at 120 V using an SDS-PAGE (Blotting System Mini-PROTEAN® Tetra System and PowerPac™ HC from Bio-Rad Laboratories, Munich, Germany) in a Tris-glycine running buffer. The densities of the running gels ranged from 7.5% to 12.5%, and the stacking gels possessed a density of 5%. The proteins were then transferred to a polyvinylidene fluoride (PVDF) membrane (Merck Millipore, Darmstadt, Germany) using a Trans-Blot® SD Semi Dry Transfer Cell (Bio-Rad Laboratories, Munich, Germany) at 225 mA for 80 min. A solution containing 5% nonfat dry milk in 1× TBS and 0.1% Tween-20 was used to block unspecific binding sites. The membranes were then incubated with the primary antibodies against NOTCH1 (NTM), HES1, HEY1, and Tubulin (all from Cell Signaling Technologies, Danvers, MA, USA) in 1× TBS + 0.1% Tween-20 for 12 h at 4 °C, washed, and incubated with an HRP-linked secondary antibody (Cell Signaling technologies) in 5% nonfat dry milk in 1× TBS and 0.1% Tween-20 for 1 h at room temperature. Next, the membranes were washed and incubated in Thermo Scientific™ Pierce™ ECL Western Blotting Substrate (Fisher Scientific, Waltham, MA, USA) for 1 minute. Immunoreactivity was visualized by ChemiDoc XRS+ with Image Lab™ Software (Bio-Rad Laboratories, Munich, Germany). Protein expression was quantified with scanning densitometry and values normalized to a tubulin control.

4.7. Statistical Analysis

All statistical tests were two-sided and significance was determined at a level of 5%. For comparison of mRNA expression in normal tissue versus tumor tissue and p16-positive and p16-negative tissue, Mann–Whitney U Test or Kruskal–Wallis Test was used. Correlation was calculated according to Spearman Rho. Expression in Western blots was compared with *t*-test.

To examine the impact of mRNA expression on clinical parameters we categorized patients into high and low expression groups and compared these to the remaining patients. High and low mRNA expression was defined as mean relative mRNA expression plus or minus one standard deviation.

Association of clinical parameters and high NOTCH1 expression was compared with Fisher's exact test with Bonferroni correction applied. The impact of expression levels on survival was analyzed with Kaplan–Meier curves, and significance was calculated using log-rank testing. To examine the association of expression levels with clinical data, multivariate forward stepwise Cox regression was performed. Statistical calculations were done in SPSS version 23 (IBM, Ehningen, Germany) or GraphPad Prism 6.0 (GraphPad Software, La Jolla, CA, USA).

Supplementary Materials: Supplementary materials can be found at www.mdpi.com/1422-0067/19/3/830/s1.

Acknowledgments: No grant or external financial support was received for this work.

Author Contributions: Markus Wirth, Anja Pickhard, Guido Piontek and Daniel Jira conceived and designed the experiments; Daniel Jira and Guido Piontek performed the experiments; Markus Wirth, Armin Ott and Daniel Jira analyzed the data; Markus Wirth wrote the paper.

Conflicts of Interest: The authors declare no conflict of interest.

Abbreviations

FFPE	Formalin fixed and paraffin embedded
HNSCC	Head and neck squamous cell carcinoma
HPV	Human papilloma virus
OSCC	Oral squamous cell carcinoma
PCR	Polymerase chain reaction
RBPJ	Recombining binding protein suppressor of hairless

References

1. Jemal, A.; Siegel, R.; Xu, J.; Ward, E. Cancer Statistics, 2010. *CA Cancer J. Clin.* **2010**, *60*, 277–300. [CrossRef]
2. Hoellein, A.; Pickhard, A.; von Keitz, F.; Schoeffmann, S.; Piontek, G.; Rudelius, M.; Baumgart, A.; Wagenpfeil, S.; Peschel, C.; Dechow, T.; et al. Aurora kinase inhibition overcomes cetuximab resistance in squamous cell cancer of the head and neck. *Oncotarget* **2011**, *2*, 599–609. [CrossRef]
3. Reiter, R.; Gais, P.; Jutting, U.; Steuer-Vogt, M.K.; Pickhard, A.; Bink, K.; Rauser, S.; Lassmann, S.; Hofler, H.; Werner, M.; et al. Aurora kinase A messenger RNA overexpression is correlated with tumor progression and shortened survival in head and neck squamous cell carcinoma. *Clin. Cancer Res.* **2006**, *12*, 5136–5141. [CrossRef]
4. Fakhry, C.; Westra, W.H.; Li, S.; Cmelak, A.; Ridge, J.A.; Pinto, H.; Forastiere, A.; Gillison, M.L. Improved Survival of Patients with Human Papillomavirus–Positive Head and Neck Squamous Cell Carcinoma in a Prospective Clinical Trial. *J. Natl. Cancer Inst.* **2008**, *100*, 261–269. [CrossRef]
5. Ntziachristos, P.; Lim, J.S.; Sage, J.; Aifantis, I. From Fly Wings to Targeted Cancer Therapies: A Centennial for Notch Signaling. *Cancer Cell* **2014**, *25*, 318–334. [CrossRef]
6. Pickering, C.R.; Zhang, J.; Yoo, S.Y.; Bengtsson, L.; Moorthy, S.; Neskey, D.M.; Zhao, M.; Ortega Alves, M.V.; Chang, K.; Drummond, J.; et al. Integrative Genomic Characterization of Oral Squamous Cell Carcinoma Identifies Frequent Somatic Drivers. *Cancer Discov.* **2013**, *3*, 770–781. [CrossRef]
7. Man, C.-H.; Wei-Man Lun, S.; Wai-Ying Hui, J.; To, K.-F.; Choy, K.-W.; Wing-Hung Chan, A.; Chow, C.; Tin-Yun Chung, G.; Tsao, S.-W.; Tak-Chun Yip, T.; et al. Inhibition of NOTCH3 signalling significantly enhances sensitivity to cisplatin in EBV-associated nasopharyngeal carcinoma. *J. Pathol.* **2012**, *226*, 471–481. [CrossRef]
8. Chillakuri, C.R.; Sheppard, D.; Lea, S.M.; Handford, P.A. Notch receptor–ligand binding and activation: Insights from molecular studies. *Semin. Cell Dev. Biol.* **2012**, *23*, 421–428. [CrossRef]
9. Wirth, M.; Doescher, J.; Jira, D.; Meier, M.A.; Piontek, G.; Reiter, R.; Schlegel, J.; Pickhard, A. HES1 mRNA expression is associated with survival in sinonasal squamous cell carcinoma. *Oral Surg. Oral Med. Oral Pathol. Oral Radiol.* **2016**, *122*, 491–499. [CrossRef]
10. Sun, W.; Gaykalova, D.A.; Ochs, M.F.; Mambo, E.; Arnaoutakis, D.; Liu, Y.; Loyo, M.; Agrawal, N.; Howard, J.; Li, R.; et al. Activation of the NOTCH Pathway in Head and Neck Cancer. *Cancer Res.* **2014**, *74*, 1091–1104. [CrossRef]
11. Agrawal, N.; Frederick, M.J.; Pickering, C.R.; Bettegowda, C.; Chang, K.; Li, R.J.; Fakhry, C.; Xie, T.X.; Zhang, J.; Wang, J.; et al. Exome sequencing of head and neck squamous cell carcinoma reveals inactivating mutations in NOTCH1. *Science* **2011**, *333*, 1154–1157. [CrossRef] [PubMed]
12. Stransky, N.; Egloff, A.M.; Tward, A.D.; Kostic, A.D.; Cibulskis, K.; Sivachenko, A.; Kryukov, G.V.; Lawrence, M.S.; Sougnez, C.; McKenna, A.; et al. The Mutational Landscape of Head and Neck Squamous Cell Carcinoma. *Science* **2011**, *333*, 1157–1160. [CrossRef] [PubMed]
13. Song, X.; Xia, R.; Li, J.; Long, Z.; Ren, H.; Chen, W.; Mao, L. Common and Complex Notch1 Mutations in Chinese Oral Squamous Cell Carcinoma. *Clin. Cancer Res.* **2014**, *20*, 701–710. [CrossRef] [PubMed]
14. Yap, L.F.; Lee, D.; Khairuddin, A.N.M.; Pairan, M.F.; Puspita, B.; Siar, C.H.; Paterson, I.C. The opposing roles of NOTCH signalling in head and neck cancer: A mini review. *Oral Dis.* **2015**, *21*, 850–857. [CrossRef] [PubMed]

15. Kaka, A.S.; Nowacki, N.B.; Kumar, B.; Zhao, S.; Old, M.O.; Agrawal, A.; Ozer, E.; Carrau, R.L.; Schuller, D.E.; Kumar, P.; et al. Notch1 Overexpression Correlates to Improved Survival in Cancer of the Oropharynx. *Otolaryngol. Head Neck Surg.* **2017**, *156*, 652–659. [CrossRef] [PubMed]

16. Rettig, E.M.; Chung, C.H.; Bishop, J.A.; Howard, J.D.; Sharma, R.; Li, R.J.; Douville, C.; Karchin, R.; Izumchenko, E.; Sidransky, D.; et al. Cleaved NOTCH1 expression pattern in head and neck squamous cell carcinoma is associated with NOTCH1 mutation, HPV status and high-risk features. *Cancer Prev. Res.* **2015**, *8*, 287–295. [CrossRef] [PubMed]

17. Lee, S.H.; Do, S.I.; Lee, H.J.; Kang, H.J.; Koo, B.S.; Lim, Y.C. Notch1 signaling contributes to stemness in head and neck squamous cell carcinoma. *Lab. Investig.* **2016**, *96*, 508–516. [CrossRef] [PubMed]

18. Lin, J.-T.; Chen, M.-K.; Yeh, K.-T.; Chang, C.-S.; Chang, T.-H.; Lin, C.-Y.; Wu, Y.-C.; Su, B.-W.; Lee, K.-D.; Chang, P.-J. Association of High Levels of Jagged-1 and Notch-1 Expression with Poor Prognosis in Head and Neck Cancer. *Ann. Surg. Oncol.* **2010**, *17*, 2976–2983. [CrossRef] [PubMed]

19. Zhang, T.-H.; Liu, H.-C.; Zhu, L.-J.; Chu, M.; Liang, Y.-J.; Liang, L.-Z.; Liao, G.-Q. Activation of Notch signaling in human tongue carcinoma. *J. Oral Pathol. Med.* **2011**, *40*, 37–45. [CrossRef] [PubMed]

20. Troy, J.D.; Weissfeld, J.L.; Youk, A.O.; Thomas, S.; Wang, L.; Grandis, J.R. Expression of EGFR, VEGF, and NOTCH1 suggest differences in tumor angiogenesis in HPV-positive and HPV-negative head and neck squamous cell carcinoma. *Head Neck Pathol.* **2013**, *7*, 344–355. [CrossRef] [PubMed]

21. Brimer, N.; Lyons, C.; Wallberg, A.E.; Vande Pol, S.B. Cutaneous Papillomavirus E6 oncoproteins associate with MAML1 to repress transactivation and NOTCH signaling. *Oncogene* **2012**, *31*, 4639–4646. [PubMed]

22. Yen, W.-C.; Fischer, M.M.; Axelrod, F.; Bond, C.; Cain, J.; Cancilla, B.; Henner, W.R.; Meisner, R.; Sato, A.; Shah, J.; et al. Targeting Notch Signaling with a Notch2/Notch3 Antagonist (Tarextumab) Inhibits Tumor Growth and Decreases Tumor-Initiating Cell Frequency. *Clin. Cancer Res.* **2015**, *21*, 2084–2095. [PubMed]

23. Takebe, N.; Miele, L.; Harris, P.J.; Jeong, W.; Bando, H.; Kahn, M.; Yang, S.X.; Ivy, S.P. Targeting Notch, Hedgehog, and Wnt pathways in cancer stem cells: Clinical update. *Nat. Rev. Clin. Oncol.* **2015**, *12*, 445–464. [PubMed]

24. Kwon, M.J.; Oh, E.; Lee, S.; Roh, M.R.; Kim, S.E.; Lee, Y.; Choi, Y.-L.; In, Y.-H.; Park, T.; Koh, S.S.; et al. Identification of Novel Reference Genes Using Multiplatform Expression Data and Their Validation for Quantitative Gene Expression Analysis. *PLoS ONE* **2009**, *4*, e6162. [CrossRef]

25. Caicedo-Granados, E.; Lin, R.; Clements-Green, C.; Yueh, B.; Sangwan, V.; Saluja, A. Wild-type p53 reactivation by small-molecule Minnelide™ in human papillomavirus (HPV)-positive head and neck squamous cell carcinoma. *Oral Oncol.* **2014**, *50*, 1149–1156. [PubMed]

26. Taguchi, A.; Kawana, K.; Tomio, K.; Yamashita, A.; Isobe, Y.; Nagasaka, K.; Koga, K.; Inoue, T.; Nishida, H.; Kojima, S.; et al. Matrix Metalloproteinase (MMP)-9 in Cancer-Associated Fibroblasts (CAFs) Is Suppressed by Omega-3 Polyunsaturated Fatty Acids In Vitro and In Vivo. *PLoS ONE* **2014**, *9*, e89605. [CrossRef]

International Journal of
Molecular Sciences

MDPI

Review

Dynamic Reorganization of the Cytoskeleton during Apoptosis: The Two Coffins Hypothesis

Suleva Povea-Cabello [1], Manuel Oropesa-Ávila [1], Patricia de la Cruz-Ojeda [1],
Marina Villanueva-Paz [1], Mario de la Mata [1], Juan Miguel Suárez-Rivero [1],
Mónica Álvarez-Córdoba [1], Irene Villalón-García [1], David Cotán [1], Patricia Ybot-González [2] and
José A. Sánchez-Alcázar [1,*]

[1]   Centro Andaluz de Biología del Desarrollo (CABD), and Centro de Investigación Biomédica en Red:
      Enfermedades Raras, Instituto de Salud Carlos III, Consejo Superior de Investigaciones Científicas,
      Universidad Pablo de, Carretera de Utrera Km 1, 41013 Sevilla, Spain; sulevapovea@gmail.com (S.P.-C.);
      manueloropesa@hotmail.com (M.O.-Á.); patricia_dlcruz_ojeda@hotmail.com (P.d.l.C.-O.);
      marvp75@gmail.com (M.V.-P.); mrdelamata@gmail.com (M.d.l.M.); juasuariv@gmail.com (J.M.S.-R.);
      monikalvarez11@hotmail.com (M.Á.-C.); villalon.irene@gmail.com (I.V.-G.); lobolivares@hotmail.com (D.C.)
[2]   Grupo de Neurodesarrollo, Unidad de Gestión de Pediatría, Instituto de Biomedicina de Sevilla (IBIS),
      Hospital Universitario Virgen del Rocío, 41013 Sevilla, Spain; pachybot@yahoo.co.uk
*   Correspondence: jasanalc@upo.es; Tel.: +34-954-978071; Fax: +34-954-349376

Received: 14 October 2017; Accepted: 9 November 2017; Published: 11 November 2017

Abstract: During apoptosis, cells undergo characteristic morphological changes in which the cytoskeleton plays an active role. The cytoskeleton rearrangements have been mainly attributed to actinomyosin ring contraction, while microtubule and intermediate filaments are depolymerized at early stages of apoptosis. However, recent results have shown that microtubules are reorganized during the execution phase of apoptosis forming an apoptotic microtubule network (AMN). Evidence suggests that AMN is required to maintain plasma membrane integrity and cell morphology during the execution phase of apoptosis. The new "two coffins" hypothesis proposes that both AMN and apoptotic cells can adopt two morphological patterns, round or irregular, which result from different cytoskeleton kinetic reorganization during the execution phase of apoptosis induced by genotoxic agents. In addition, round and irregular-shaped apoptosis showed different biological properties with respect to AMN maintenance, plasma membrane integrity and phagocyte responses. These findings suggest that knowing the type of apoptosis may be important to predict how fast apoptotic cells undergo secondary necrosis and the subsequent immune response. From a pathological point of view, round-shaped apoptosis can be seen as a physiological and controlled type of apoptosis, while irregular-shaped apoptosis can be considered as a pathological type of cell death closer to necrosis.

Keywords: apoptosis; apoptotic microtubule network; microtubules; actin filaments; genotoxic drugs

---

1. Introduction: An Overview of Apoptosis

The term "apoptosis" was coined by Kerr et al. in the early 1970s to describe the ultrastructural features of dying cells seen during development of hepatocytes [1]. This has given rise to the concept that these cells have an intrinsic suicide program that predetermines their fate. Nowadays, apoptosis is conceived as the major type of programmed cell death (PCD) in multicellular organisms, being distinct but connected to other types of PCD like autophagy or programmed necrosis [2]. It is characterized by several morphological and biochemical features which take place while membrane integrity is maintained [3]. The typical hallmarks of apoptosis include cell shrinkage, convolution of the nuclear and cellular outlines, cell membrane blebbing, formation of apoptotic bodies, chromatin condensation, caspase activation and DNA fragmentation [4].

Animal development and tissue homeostasis depend on the appropriate regulation of apoptosis. It is involved in sculpting and deleting structures, morphogenesis, regulation of cell number and elimination of aberrant cells [5]. Paradoxically, cell death turns out to be essential in the proliferative environment of development through the release of factors that influence cell division and survival of adjacent tissues. It has been shown to be necessary during different stages of embryonic growth, including the mammalian blastocyst stage, when both the inner cell mass and the trophectoderm undergo apoptosis. However, apoptosis levels must be strictly programmed in order to maintain embryonic homeostasis [6]. Apoptosis also contributes to the formation of vesicles and tubes (e.g., neural tube) when epithelial sheets invaginate and tissue inside has to be eliminated [7]. Neurons and oligodendrocytes which are overproduced during the development of the nervous system are also eliminated by apoptosis [8]. One example of the role of apoptosis can be seen during palate fusion. Here, unbalanced apoptosis has been shown to be a cause for cleft palate, one of the most common oral malformations [9].

However, most of the knowledge that we have today about the regulation of PCD comes from three model organisms: *Caenorhabditis elegans*, *Drosophila melanogaster* and the mouse. In *C. elegans*, 131 of the 1090 final somatic cells undergo programmed cell death during embryogenesis [10]. Programmed cell death is an intrinsic characteristic of somatic cells, strictly controlled by cell lineage. Ablation of cell death genes *ced-3* and *ced-4* prevents apoptosis in cells that normally die. Instead, these cells survive and differentiate. The concept of apoptosis extends in *Drosophila*, as it is controlled by both environmental and genetic factors. During the developmental stage of *Drosophila*, PCD is present from the embryo stage until oogenesis [11]. The regulation of apoptosis in vertebrates appears considerably more complex and vast numbers of cells undergo apoptosis throughout development and tissue homeostasis in adulthood [6]. In humans, perturbations of the signalling cascades regulating apoptosis can result in a wide variety of human diseases such as cancers, infectious diseases including AIDS (Acquired Immune Deficiency Syndrome), autoimmune diseases and neurodegenerative diseases [12].

Apoptotic cell death develops in three distinct phases: induction, execution and clearance of the dying cell. The fate of apoptotic cells in multicellular organisms is their immediate elimination by phagocytes. However, cells that perform apoptosis in in vitro cultures progress to secondary necrosis, a process which entails the loss of membrane integrity and the release of cellular content into the surrounding interstitial tissue [13]. In vivo, secondary necrosis is also likely to happen in case of extensive cell death or impaired phagocytosis and it has been hypothesized to participate in the genesis of many human diseases [14].

The phase of induction encompasses all the intrinsic or extrinsic environmental changes that lead to the activation of the apoptotic signalling. Following induction, the execution phase takes place thanks to the activation of a caspase-dependent proteolytic cascade [15]. Caspases are aspartic acid-specific proteases responsible for cellular component degradation. Some of them, like caspase-8 and -9, act as initiators of the apoptotic signalling pathway, while other caspases like caspases-3, -6 and -7, operate as executor caspases which actively participate in the degradation of cell substrates [16]. Caspase activation can be initiated by two main apoptotic pathways, the extrinsic or death receptor pathway and the intrinsic or mitochondrial pathway. However, there is evidence that the two pathways are interconnected and that molecules in one pathway can influence the other [17]. Eventually, the dying cell is engulfed by professional phagocytes or by neighbouring cells. Efficient apoptotic cell removal is driven by the interaction with phagocytes through the expression of "eat-me" signals and the release of "find-me" signals, which facilitate the engulfment of the dying cell and its eventual digestion in their phagolysosomes. This process of apoptotic cell clearance is essential for tissue turnover and homeostasis [18]. In fact, this interaction prevents undesired immune reactions by contributing to the development of an immunomodulatory environment [19].

2. Genotoxic Cell Response and Cytoskeleton

In the context of human disease, cancer is one of the most outstanding pathologies, in which apoptosis plays a major role. Evading apoptosis has been shown to be a hallmark of cancer as tumour progression is linked, not only to cell proliferation, but also to death insensitivity [20]. Despite being one of its main causes, apoptosis has been traditionally used as a target for cancer treatment [21]. Typical therapies involve genotoxic drugs (chemotherapy or ionising radiation) with the aim of targeting cell proliferation [22]. Many of the cytotoxic agents commonly used to treat cancer patients such as alkylating agents, platinum drugs, antimetabolites, topoisomerase poisons and ionising radiation cause high levels of DNA damage [23]. However, to prevent the transmission of damaged DNA during cell division, cells activate the DNA damage response (DDR) which depends on DNA damage repair pathways as well as cell cycle checkpoint activation to arrest the cell cycle [24]. If DNA damage is irreparable cells may signal for senescence (growth arrest), apoptosis or other pathways leading to cell death [25]. The DDR enables cells to detect damage, recruit multi-protein complexes at these foci and activate downstream signalling [26]. Depending on the extent of DNA damage, the DDR distinguishes between repairable and non-repairable DNA damage, and controls different cellular responses such as transient cell cycle arrest and DNA repair, senescence or cell death [27].

Central components of the DDR machinery are the phosphoinositide 3-kinase related kinases ATM (Ataxia-telangiectasia-mutated) and ATR (ataxia telangiectasia and Rad3-related). ATM responds mainly to double-strand break (DSBs), whereas ATR is activated by single-strand break (ssDNA) and stalled replication forks [28]. When ATM and ATR are recruited to sites of damage, they target many substrates, including downstream kinases such as checkpoint kinases Chk2 and Chk1, regulatory proteins such as p53, and scaffolding proteins such as BRCA1 and BRCA2 (breast cancer 1 and 2). Once these proteins are activated, they regulate the function of downstream effector proteins such as p21, Cdc25A and cyclin-dependent kinases (CDKs). Phosphorylation of p53 at serine 46 serves as a pro-apoptotic mark that induces the transcription of apoptotic genes such as BAX (BCL2 Associated X), PUMA (p53 upregulated modulator of apoptosis), NOXA and p53AIP1 (p53-regulated apoptosis-inducing protein 1) that finally activate the cell death pathway via the mitochondrial, intrinsic pathway [29]. However, p53 can also act in a transcription-independent mode targeting mitochondria and inducing BAX activation and mitochondrial outer membrane permeabilization (MOMP) [30].

The DDR response also includes cytoskeleton reorganizations. Thus, following DNA damage RhoA (Ras homolog gene family, member A) specific guanine nucleotide exchange factor (GEFs) such as neuroepithelioma transforming gene 1 (Net1) get activated [31], leading to a Fen1 dependent activation of the RhoA/ROCK (Rho-associated protein kinase) axis [32], which controls the organization of the actin cytoskeleton [33]. Net1 is a RhoA specific GEF that is frequently overexpressed in human cancer [34]. It has been reported that DNA damage activates Net1 to control RhoA- and p38 MAPK-mediated cell survival pathways in response to DNA damage [35]. In adherent cells, the cellular response to DNA damage involves Net1 dephosphorylation and translocation from the nucleus to the cytosol where it activates RhoA GTPase [36]. In turn, Rho A activation controls actin filaments reorganization through the activation of ROCK and MLC (myosin light chain) phosphorylation and actinomyosin contractility [37]. Knock down of Net1 by RNAi prevents RhoA activation, inhibits the formation of stress fibres and enhances cell death [36]. This indicates that Net1 activation is required for RhoA mediated response to genotoxic stress and that cytoskeleton reorganization may play an important role in DDR. The Net1 and the RhoA dependent signals also converge in the activation of mitogen-activated protein kinase p38 (p38 MAPK) and its downstream target MAPK-activated protein kinase 2 (MK2) [36].

The importance of cytoskeleton reorganization during DDR and its role in genotoxic resistance or apoptosis induction is not completely understood and needs further research.

3. Cytoskeleton Rearrangements during the Execution Phase of Apoptosis

The execution phase of apoptosis is denoted by cell contraction, plasma membrane blebbing, chromatin condensation and DNA fragmentation [38]. To achieve such dramatic morphological changes, apoptotic cells make profound cytoskeleton reorganizations, and caspase-mediated digestion of cytoskeleton proteins ensures the proper dismantlement of the dying cell during this process [39].

The eukaryotic cytoskeleton is mainly composed of actin filaments, microtubules and intermediate filaments. These three constituents act coordinately to increase tensile strength, allow cell motility, maintain plasma integrity, participate in cell division, contribute to cell morphology and provide a network for cellular transport [40]. Classically, it has been accepted that microtubules and intermediate filaments are disorganized at the onset of the execution phase [38], while the actin cytoskeleton is responsible for cell remodelling during this phase [41]. At later stages, it has been observed that microtubules are reorganized [42,43], giving rise to the apoptotic microtubule network (AMN), a structure that sustains apoptotic cell morphology and maintains plasma membrane integrity [44,45] and participates in the dispersion of cellular and nuclear fragments [41,46]. However, this model has recently been expanded by new evidence that supports the hypothesis that genotoxic drugs induce two dose-dependent types of apoptosis characterized by different cytoskeleton rearrangement kinetics depending on caspase activation timing (Figure 1) [47]. Thus, "slow" or round-shaped apoptosis is characterized by late caspase activation, slow actinomyosin ring contraction, plasma membrane blebbing, cell detachment, microtubules remodelling, and formation of a round-shaped AMN and apoptotic cell morphology. In contrast, "fast" or irregular–shaped apoptosis is characterized by early caspase activation, initial microtubules depolymerisation, fast actinomyosin ring contraction without cell detachment, and formation of an irregular-shaped AMN and apoptotic cell morphology which frequently shows apoptotic membrane protrusions or microtubule spikes. Both round and irregular AMN have been observed during apoptosis induced by a variety of genotoxic agents (camptothecin, doxorubicin, teniposide and cisplatin) in several cell lines (H460, HeLa, MCF7 and LLCPK-1α) [47].

How do cells undergo round or irregular-shaped apoptosis after the exposition to apoptosis inducers such as chemotherapeutic compounds? In part, genotoxic agent concentration and cell cycle phase determine the cell response [47]. First, round and irregular AMNs are dependent on the concentration of the apoptotic stimulus. At low concentrations of genotoxic agent, cells undergo slow apoptosis and display a rounded AMN, whereas at higher concentrations, cells undergo fast apoptosis and show an irregular AMN. It is reasonable to infer that treatment with low doses of chemotherapeutic agents induces a slower cell death than that produced by high doses which can rapidly activate caspases by the intrinsic pathway. Second, apoptotic and AMN morphology are also dependent on the cell cycle phase. Thus, cells in G1 undergo round-shaped apoptosis while cells in G2/M undergo irregular-shaped apoptosis, irrespective of the concentration of the genotoxic agent [47]. Induction of tumour cell death by chemotherapeutic modalities often occurs in a cell cycle-dependent manner. Thus, it has been observed that several regulatory proteins involved in tumour chemosensitivity and apoptosis are expressed periodically during cell cycle progression [48–50]. Experimental studies have previously shown that apoptotic cell death can occur either fast (~min) or very slow (~h) [51]. Results from the Monte Carlo study also showed two types of apoptosis that can switch from slow (~h) to fast (~min), as the strength of an apoptotic stimulus increases [52]. Traditionally, slow apoptosis can be initially considered as a caspase-independent cell death in which caspases may be activated at late stages [53]. However, more research is needed to clarify the mechanisms behind the cell cycle dependency of cytoskeleton reorganization during apoptosis.

A. Round-shaped apoptosis

B. Irregular-shaped apoptosis

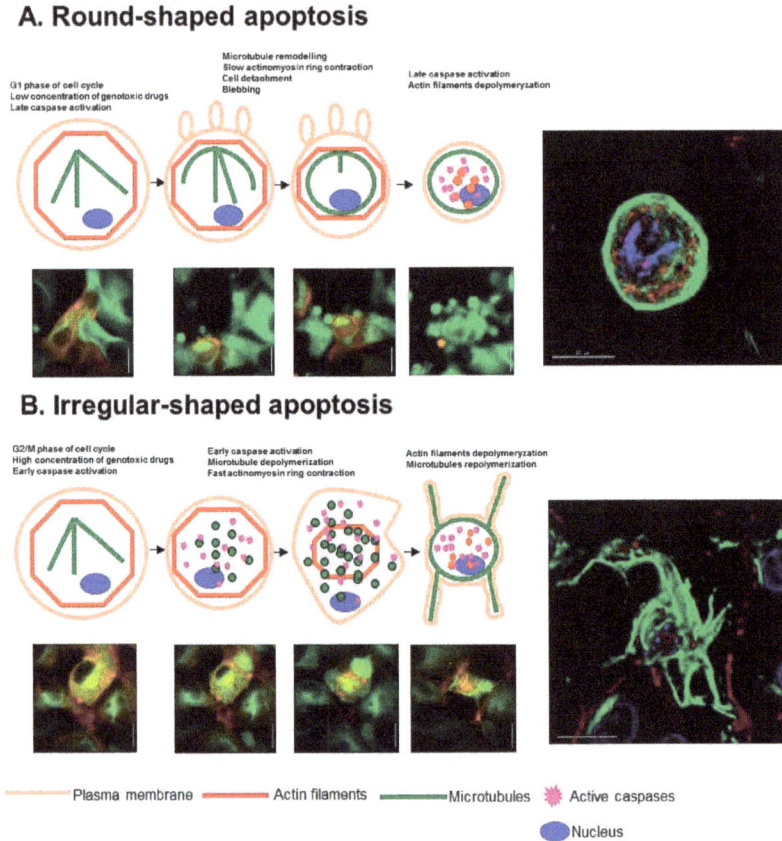

Figure 1. Schematic representation of the reorganization of actin filaments and microtubules during round (**A**) and irregular (**B**) -shaped apoptosis. Brown = plasma membrane; Blue = nucleus; Green = microtubules; Red = actin filaments; Pink = active caspases. Representative sequential images of round and irregular-shaped apoptosis in LLCPK-1α cells expressing GFP-αtubulin and pdsRed-monomer-actin are also included. Apoptosis was induced by camptothecin treatment. Right panels, immunofluorescence microscopy of round and irregular H460 apoptotic cells. Green = anti-α-tubulin: Red = anti-actin; Blue = Hoechst staining for nuclei. Scale bar= 15 μm.

3.1. Reorganization of Microtubules during Apoptosis

Microtubules are polar protofilaments made up of α and β tubulin which are involved in supporting and maintaining the shape of cells, cell polarity and migration, intracellular transport of vesicles and organelles and segregation of chromosomes in mitosis [54].

The recent hypothesis of two types of apoptosis proposes that microtubules undergo two kinetically different processes (Figure 1) [47]. In round-shaped apoptosis, microtubules are not depolymerized at early stages of apoptosis. Instead, they are remodelled and acquire a concentric organization forced by the actinomyosin ring contraction. In round-shaped apoptosis, microtubule nucleation depends on the γ-tubulin ring complexes (γTuRC) which are not disorganized by caspases that are activated at later stages. In contrast, when caspases are activated at early stages during irregular-shaped apoptosis, γTuRC are degraded and interphase microtubules are soon disorganized [55–57]. This disassembly of microtubules is followed by a fast and full actinomyosin ring contraction. Next, actin filaments are depolymerised. In this situation, apoptotic cells are devoid of the main cytoskeletal elements. Soon after, microtubules are repolymerized adopting an irregular disposition beneath the plasma membrane [58].

The molecular mechanisms involved in microtubule depolymerisation during the initial stages of irregular-shaped apoptosis are still unknown (Figure 2). One possibility is that early caspase activation can directly cleave γ-TURC provoking microtubule disassembly [47]. Alternatively, pericentriolar proteins could be targeted by caspases. Thus, GRASP65 (Golgi reassembly-stacking protein of 65 kDa), a pericentriolar protein related to the Golgi apparatus, has been described as a caspase target [59,60]. Another possible explanation relies on dynein, a microtubule motor protein, which is essential for the centrosomal localization of pericentrin and γ-tubulin in living cells [61]. Caspase cleavage of the dynein intermediate chain stops its motility and reduces the content of pericentrin and γ-tubulin at the centrosome, thereby impairing its capacity to nucleate microtubules [62]. Another hypothesis relies in the concept that microtubule dynamics are governed by several effectors such as motor proteins, gradients Ran-GTP, + ends proteins and microtubule-associated proteins (MAPs), which in turn are under the control of phosphatases and kinases [63–65]. One of the main kinases involved in phosphorylation is the cyclin-dependent kinase 1 (Cdk1), which associates with cyclin B as a key regulatory kinase that controls the entry into mitosis and regulates microtubule dynamics [66]. Cdk1 regulates some microtubule effectors by phosphorylation. For instance, MAP4 reduces its ability to stabilize microtubules after this modification [67]. In addition, Cdk1 is able to block protofilament growth during mitosis by phosphorylating β-tubulin [68]. Although Cdk1 activity is not apoptosis-specific, it has been observed during cell death. Therefore, it has been suggested that it may act as an essential regulator of microtubule reorganizations [57]. On the other hand, other authors have shown that microtubule depolymerisation at the onset of apoptosis is associated with activation of the PP2A-like phosphatase, dephosphorylation of the microtubule regulator Tau (τ) protein and tubulin deacetylation [69]. Both mechanisms can coexist in apoptosis as PP2A-mediated dephosphorylation of cdc25, a CdK1 regulator, precludes mitotic entry [70].

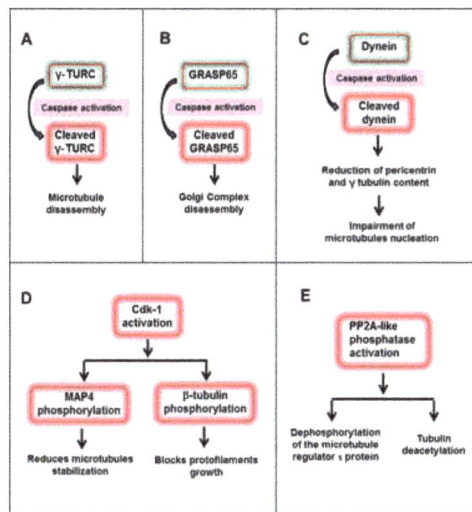

Figure 2. Molecular mechanisms involved in microtubules depolymerisation during the initial stages of irregular-shaped apoptosis: Cleavage of γ-TURC (**A**); cleavage of pericentriolar proteins such as GRASP65 (**B**); cleavage of dynein, a microtubule motor protein (**C**); activation of Cdk1, a kinase which regulates several microtubule effectors (**D**); and PP2A-like phosphatase activation which induces dephosphorylation of the microtubule regulator τ protein and tubulin deacetylation (**E**).

Once interphase microtubules are depolymerized and the actinomyosin ring contracts, apoptotic microtubules are reassembled during irregular-shaped apoptosis. The complete sequence of events that guide this repolymerisation are still undiscovered. However, it is known that irregular AMN is organized independently of γ-TURC, indicating that AMN formation is regulated by other mechanisms [47]. It has been hypothesized that active caspases may cleave the C-terminal regulatory region of tubulin during the execution phase, thereby increasing its ability to polymerize and thus facilitating the formation of apoptotic microtubules [60,71]. Another possible candidate for AMN nucleation could be core centrioles, which are not degraded during the execution phase. However, apoptotic microtubules are not displayed in the typical radial pattern of interphase microtubules, suggesting that centrioles are unlikely to guide the formation of AMN in irregular apoptosis.

Even though the mechanisms governing AMN arrangement in irregular apoptosis are unknown, this process seems to be tightly regulated. It has been proposed that the Ras-like small GTPase Ran could be responsible for this. Ran-GTP's best known role focuses on regulating microtubule nucleation and dynamics during mitosis and meiosis [72]. Interestingly, Ran-GTP controls microtubule dynamics and motor activity [64]. It has been described that active Ran-GTP is necessary for apoptotic microtubule polymerization, and that its release into the apoptotic cytoplasm triggers microtubule nucleation [73]. Furthermore, RanGTP-activated spindle-assembly factor, TPX2 (targeting protein for Xklp2), escapes from the nucleus during the execution phase and associates with apoptotic microtubule and promotes their assembly [74]. Therefore, it has been hypothesized that apoptotic microtubule polymerization shares several mutual features with mitotic and meiotic spindle assembly, with a particular dependence on Ran-GTP and TPX2 [75].

After AMN reorganization, there is another level of regulation since apoptotic microtubules remain dynamic as it has been demonstrated by time-lapse imaging of the EB1 protein, a plus-end tracking protein [56,76].

3.2. Reorganization of Actin Cytoskeleton during Apoptosis

Unlike microtubules, actin cytoskeleton has been traditionally considered as a highly dynamic cytoskeletal element at the onset of apoptosis [58]. According to recent findings, actin cytoskeleton also suffers two kinetically different reorganizations during apoptosis (Figure 1). While round-shaped apoptosis depends on a slow contraction of the actinomyosin ring, irregular-shaped apoptosis is the result of a faster and full contraction.

Slow or round-shaped apoptosis coincides with traditional cytoskeletal reorganizations described in apoptosis [39]. After apoptosis induction, adherent cells lose their focal adhesion sites and partially detach from their substrates. Then, actin filaments are reorganized beneath the plasma membrane into an actinomyosin cortical ring with contractile force (Figure 1). Characteristically, actinomyosin contraction produces plasma membrane blebbing which could travel away from the apoptotic cell and alert surrounding and immune cells before secondary necrotic membrane breakdown [77,78].

Actinomyosin contraction is activated via the RhoA/ROCK signalling pathway (Figure 3) [79]. RhoA GTPases are activated by interchanging GDP for GTP. This activation is controlled by Rho guanine-nucleotide-exchange factors (Rho GEFs), such as Net1, when cells are exposed to DNA damaging agents [31,75]. Net1 has been shown to localize preferentially within the nucleus at steady state [80]. Nuclear import of Net1 is mediated by two nuclear localization signals present in the N-terminus of the protein, and forced cytoplasmic localization of Net1 is sufficient to activate Rho. Net1 can move in and out of the nucleus, and the activation of RhoA by Net1 is controlled by changes in its subcellular localization. A logical prediction of this hypothesis is that in order for Net1 to be functionally active, it must be transported out of the nucleus into the cytosol, where it can activate RhoA. However, DNA damage signals such as ionizing radiation (IR), which has been previously shown to stimulate RhoA, specifically promoted the activation of the nuclear pool of RhoA in a Net1-dependent manner while the cytoplasmic activity was not affected [81].

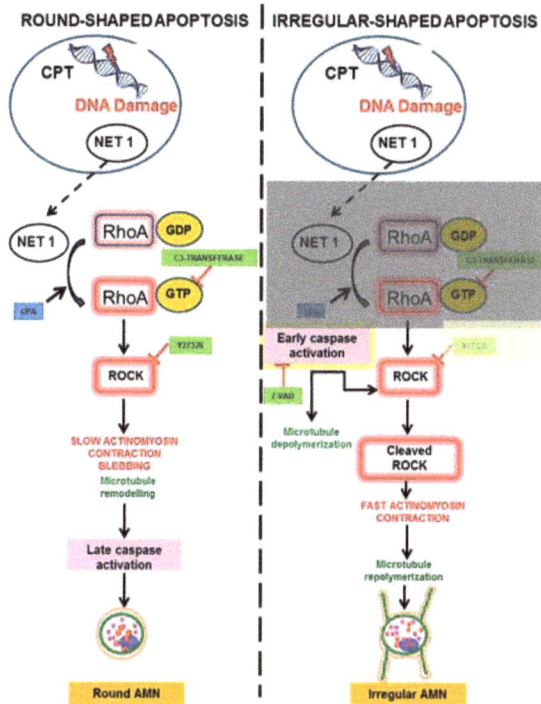

Figure 3. Schematic representation of signalling pathways involved in cytoskeleton reorganizations in round and irregular apoptosis. T arrow=inhibition; dashed arrow= translocation from nucleus to cytosol; solid arrow=activation; shaded area= disruption by caspases.

Irrespective of the localization of RhoA activation by Net1, it activates its downstream effector, the Rho-associated coiled-coil protein kinase 1 (ROCK-1). Next, ROCK phosphorylates myosin light chain phosphatase (MLCP), which in turns increases the phosphorylation levels of myosin light chain II (MLC). As a consequence, contraction of the actinomyosin ring is induced [82]. This sequence of events can be considered as "slow" (hours) in comparison to irregular-shaped apoptosis (minutes). In summary, when caspases are not activated at early stages of apoptosis, actinomyosin ring contracts via the NET1/RhoA/ROCK/MLC phosphorylation pathway, causing a slow round-shaped apoptosis. As microtubules are not depolymerized, they are adjusted to the actinomyosin ring acquiring a circular organization. Later, when caspases are activated, the actinomyosin ring disappears and a round AMN remains in apoptotic cells. The reasons of why in this situation active caspases do not cleave γ-TURC and induce AMN depolymerisation are not known, although we can speculate that the concentric remodelling of microtubules during round-shaped apoptosis can make the access of caspases to γ-TURC difficult.

In contrast, early caspase activation in irregular apoptosis completely changes the kinetics of actin reorganization during apoptosis. First, as mentioned above, microtubules are depolymerized presumably by degradation of γ-TURC. Second, caspase degradation of NET 1 interferes with NET1/RhoA/ROCK/MLC phosphorylation/actinomyosin ring contraction pathway. Furthermore, caspase activation generates a constitutively active fragment of ROCK1 [83] that causes a rapid and full contraction of the actinomyosin ring and, as a result, the whole cell became a "big bleb" that remained attached to the substrate. Once actinomyosin contraction is finished, apoptotic cells are devoid of the main cytoskeletal elements and an irregular AMN is formed independently of γ-TURC.

3.3. Reorganization of Intermediate Filaments during Apoptosis

Intermediate filaments that help maintain the integrity of tissues and cells are disrupted early at the onset of apoptosis by the action of caspases. The intermediate filament cleavage cause fragmentation and aggregation, and the breaking of the nuclear lamins facilitates nuclear disintegration [84–86].

The influence of intermediate filaments (IF) proteins on the cytoskeleton reorganization kinetics that lead to round of irregular-shaped apoptosis has not been investigated yet, so this review will only focus on the current knowledge of the role of microtubules and actin filaments during apoptosis.

4. Modulation of Round and Irregular-Shaped Apoptosis

As irregular-shaped apoptosis is dependent on early caspase activation, inhibition of caspases by z-VAD (benzyloxycarbonyl-valine-alanine-aspartate-fluoromethylketone) prevents both microtubules depolymerisation and ROCK1 and NET1 cleavage, allowing a slow actinomyosin ring contraction and, consequently, apoptotic cells adopt a round-shaped morphology [47].

On the other hand, inhibition of actinomyosin ring contraction by C3-transferase, a RhoA inhibitor, or Y27632, a ROCK inhibitor, induce early caspase activation and apoptotic cells adopt an irregular morphology [47]. On the contrary, activation of actinomyosin ring contraction by lysophosphatidic acid (LPA), a RhoA activator, induces apoptotic cells with round-shaped AMN [47]. These findings indicate that round- and irregular-shaped apoptosis can be modulated by specific inhibitors or activators of the NET1-RhoA-ROCK-MLC pathway.

5. Biological Implications of the "Two Coffins" Model in Apoptosis

In a metaphorical sense, AMN can be considered as an intracellular "coffin" that protects the plasma membrane and confines the degradative processes of apoptotic cells. According to the proposed hypothesis, cells undergoing apoptosis can actually exhibit two types of "coffins" with kinetically different cytoskeleton reorganizations and distinctive properties with respect to resistance of apoptotic cells to undergo secondary necrosis and the ability of being phagocytosed.

The evaluation of the resistance of apoptotic cells to undergo secondary necrosis revealed that round-shaped apoptotic cells were more resistant to secondary necrosis than irregular apoptotic cells, consistent with a more homogeneous organization of apoptotic microtubules in apoptotic cells with round-shaped AMN [47]. This property of round-shaped apoptosis can be interesting for design therapies with low inflammatory responses. On the contrary, irregular-shaped apoptotic cells undergo secondary necrosis more easily, and consequently, they are more prone to induce inflammation.

The last stage of apoptosis comprises the elimination of apoptotic cells by macrophages or neighbouring cells. In the former case, clearance of apoptotic cells by phagocytes during efferocytosis can be divided into four distinct processes: recruitment of phagocytes near apoptotic cells; recognition of dying cells thanks to surface bridge molecules or receptors; engulfment of apoptotic cells; and degradation [87]. Efficient phagocytosis of apoptotic cells by macrophages depends on the presence of apoptotic microtubules [88]. Indeed, it has been shown that apoptotic cells with AMN show high expression of phosphatidylserine on the cell surface and increased phagocytosis rate. However, both processes were markedly reduced when AMN was depolymerized by colchicine treatment and cells undergo secondary necrosis. The ability of apoptotic cells to stimulate their phagocytosis by macrophages before cell lysis is crucial to prevent adverse effects, such as tissue damage and inflammation, associated with secondary necrosis [13]. As a consequence, phagocytosis reduces the probability of inflammation by ensuring that apoptotic cells are eliminated before the release of intracellular contents into the extracellular medium.

The intensity of externalization of phosphatidylserine is similar in both round- and irregular-shaped apoptosis [47]. Despite this, round apoptotic cells are more efficiently engulfed by professional phagocytes rather than irregular apoptotic cells [47]. These differences could be due to the different pattern of cytoskeleton organization and final apoptotic cell morphology. In addition, the process of blebbing, which takes place during round-shaped apoptosis, can be also important for efficient phagocytosis of apoptotic cells [78]. Accordingly, less blebbing capacity correlates with less engulfment efficiency. Hence, the inability of irregular-shaped apoptotic cells to undergo blebbing would explain why these cells show less phagocytosis potential.

To evaluate whether rounded or irregular-shaped apoptosis became more pro-inflammatory, the levels of IL-1β in the medium from the phagocytosis assays were analysed [47]. IL-1β levels were significantly increased when macrophages were incubated with apoptotic cells with irregular morphology. However, IL-1β levels were not significantly increased in co-cultures with rounded apoptotic cells (which are more resistant to undergo secondary necrosis and more efficiently phagocytosed). These data suggest that irregular-shaped apoptosis may promote a higher production of pro-inflammatory cytokines by macrophages.

6. Conclusions and Future Perspectives

During the execution phase of apoptosis, the apoptotic microtubule network (AMN) adopts two different morphological patterns, round and irregular. Irrespective of different cytoskeletal rearrangements kinetics, AMN is required to maintain plasma membrane integrity and cell morphology during the execution phase of apoptosis.

The NET1/RhoA/ROCK1/MLC phosphorylation/actinomyosin contraction signalling pathway operates when apoptosis is induced by low concentrations of genotoxic drugs, promoting round-shaped apoptosis. In contrast, early caspase activation in response to high concentrations of genotoxic drugs (that induces NET 1 and ROCK1 cleavage) disrupts this signalling pathway and promotes irregular-shaped apoptosis.

Round- and irregular-shaped apoptosis are also dependent on cell cycle phase. Thus, cells in G1 undergo round-shaped apoptosis while cells in G2/M undergo irregular-shaped apoptosis, irrespective of the concentration of the genotoxic agent.

Round- and irregular-shaped apoptosis present different biological significance. Thus, round-shaped AMN makes apoptotic cells more resistant to secondary necrosis and less pro-inflammatory than irregular-shaped AMN.

It would be interesting to explore whether these two types of apoptotic cells are present in different physiological or pathological situations, and whether they have distinct signalling roles or produce different types of signalling molecules. Furthermore, the knowledge and modulation of round and irregular apoptosis may be important for deciding better therapeutic options and predicting the subsequent immune response.

Acknowledgments: This work was supported by PI16/00786 grant, Instituto de Salud Carlos III, Spain and Fondo Europeo de Desarrollo Regional (FEDER-Unión Europea), and by AEPMI (Asociación de Enfermos de Patología Mitocondrial) and ENACH (Asociación de Enfermos de Neurodegeneración con Acumulación Cerebral de Hierro).

Author Contributions: Suleva Povea-Cabello, Manuel Oropesa-Ávila and Patricia de la Cruz-Ojeda wrote the whole manuscript; Marina Villanueva-Paz, Mario de la Mata, Juan Miguel Suárez-Rivero, Mónica Álvarez-Córdoba, Irene Villalón-García, David Cotán contributed to the literature search; Patricia Ybot-González and José A. Sánchez-Alcázar revised and approved the manuscript.

Conflicts of Interest: The authors declare no conflict of interest.

References

1. Kerr, J.F.; Wyllie, A.H.; Currie, A.R. Apoptosis: A basic biological phenomenon with wide-ranging implications in tissue kinetics. *Br. J. Cancer* **1972**, *26*, 239–257. [CrossRef] [PubMed]
2. Ouyang, L.; Shi, Z.; Zhao, S.; Wang, F.T.; Zhou, T.T.; Liu, B.; Bao, J.K. Programmed cell death pathways in cancer: A review of apoptosis, autophagy and programmed necrosis. *Cell Prolif.* **2012**, *45*, 487–498. [CrossRef] [PubMed]

3. Elmore, S. Apoptosis: A review of programmed cell death. *Toxicol. Pathol.* **2007**, *35*, 495–516. [CrossRef] [PubMed]

4. Saraste, A.; Pulkki, K. Morphologic and biochemical hallmarks of apoptosis. *Cardiovasc. Res.* **2000**, *45*, 528–537. [CrossRef]

5. Meier, P.; Finch, A.; Evan, G. Apoptosis in development. *Nature* **2000**, *407*, 796–801. [CrossRef] [PubMed]

6. Fuchs, Y.; Steller, H. Programmed cell death in animal development and disease. *Cell* **2014**, *147*, 742–758. [CrossRef] [PubMed]

7. Jacobson, M.D.; Weil, M.; Raff, M.C. Programmed cell death in animal development. *Cell* **1997**, *88*, 347–354. [CrossRef]

8. Narayanan, V. Apoptosis in development and disease of the nervous system: 1. Naturally occurring cell death in the developing nervous system. *Pediatr. Neurol.* **1997**, *16*, 9–13. [CrossRef]

9. Burg, M.L.; Chai, Y.; Yao, C.A.; Magee, W., 3rd; Figueiredo, J.C. Epidemiology, etiology, and treatment of isolated cleft palate. *Front. Physiol.* **2016**, *7*, 67. [CrossRef] [PubMed]

10. Ellis, H.M.; Horvitz, H.R. Genetic control of programmed cell death in the nematode *C. elegans. Cell* **1986**, *44*, 817–829. [CrossRef]

11. Xu, D.; Woodfield, S.E.; Lee, T.V.; Fan, Y.; Antonio, C.; Bergmann, A. Genetic control of programmed cell death (apoptosis) in *drosophila. Fly* **2009**, *3*, 78–90. [CrossRef] [PubMed]

12. Rudin, C.M.; Thompson, C.B. Apoptosis and disease: Regulation and clinical relevance of programmed cell death. *Annu. Rev. Med.* **1997**, *48*, 267–281. [PubMed]

13. Silva, M.T. Secondary necrosis: The natural outcome of the complete apoptotic program. *FEBS Lett.* **2010**, *584*, 4491–4499. [CrossRef] [PubMed]

14. Silva, M.T.; do Vale, A.; dos Santos, N.M. Secondary necrosis in multicellular animals: An outcome of apoptosis with pathogenic implications. *Apoptosis* **2008**, *13*, 463–482. [CrossRef] [PubMed]

15. Logue, S.E.; Martin, S.J. Caspase activation cascades in apoptosis. *Biochem. Soc. Trans.* **2008**, *36*, 1–9. [CrossRef] [PubMed]

16. Wilson, M.R. Apoptosis: Unmasking the executioner. *Cell Death Differ.* **1998**, *5*, 646–652. [CrossRef] [PubMed]

17. Igney, F.H.; Krammer, P.H. Death and anti-death: Tumour resistance to apoptosis. *Nat. Rev. Cancer* **2002**, *2*, 277–288. [CrossRef] [PubMed]

18. Poon, I.K.; Hulett, M.D.; Parish, C.R. Molecular mechanisms of late apoptotic/necrotic cell clearance. *Cell Death Differ.* **2010**, *17*, 381–397. [CrossRef] [PubMed]

19. Krysko, D.V.; D'Herde, K.; Vandenabeele, P. Clearance of apoptotic and necrotic cells and its immunological consequences. *Apoptosis* **2006**, *11*, 1709–1726. [CrossRef] [PubMed]

20. Qi, S.; Calvi, B.R. Different cell cycle modifications repress apoptosis at different steps independent of developmental signaling in *drosophila. Mol. Biol. Cell* **2016**, *27*, 1885–1897. [CrossRef] [PubMed]

21. Hanahan, D.; Weinberg, R.A. The hallmarks of cancer. *Cell* **2000**, *100*, 57–70. [CrossRef]

22. Wong, R.S. Apoptosis in cancer: From pathogenesis to treatment. *J. Exp. Clin. Cancer Res. CR* **2011**, *30*, 87. [CrossRef] [PubMed]

23. Helleday, T.; Petermann, E.; Lundin, C.; Hodgson, B.; Sharma, R.A. DNA repair pathways as targets for cancer therapy. *Nat. Rev. Cancer* **2008**, *8*, 193–204. [CrossRef] [PubMed]

24. Woods, D.; Turchi, J.J. Chemotherapy induced DNA damage response: Convergence of drugs and pathways. *Cancer Biol. Ther.* **2013**, *14*, 379–389. [CrossRef] [PubMed]

25. Kastan, M.B.; Bartek, J. Cell-cycle checkpoints and cancer. *Nature* **2004**, *432*, 316–323. [CrossRef] [PubMed]

26. Marechal, A.; Zou, L. DNA damage sensing by the atm and atr kinases. *Cold Spring Harb. Perspect. Biol.* **2013**, *5*, a012716. [CrossRef] [PubMed]

27. Furgason, J.M.; Bahassi el, M. Targeting DNA repair mechanisms in cancer. *Pharmacol. Ther.* **2013**, *137*, 298–308. [CrossRef] [PubMed]

28. Matsuoka, S.; Ballif, B.A.; Smogorzewska, A.; McDonald, E.R., 3rd; Hurov, K.E.; Luo, J.; Bakalarski, C.E.; Zhao, Z.; Solimini, N.; Lerenthal, Y.; et al. Atm and atr substrate analysis reveals extensive protein networks responsive to DNA damage. *Science* **2007**, *316*, 1160–1166. [CrossRef] [PubMed]

29. Matt, S.; Hofmann, T.G. The DNA damage-induced cell death response: A roadmap to kill cancer cells. *Cell Mol. Life Sci.* **2016**, *73*, 2829–2850. [CrossRef] [PubMed]

30. Follis, A.V.; Llambi, F.; Merritt, P.; Chipuk, J.E.; Green, D.R.; Kriwacki, R.W. Pin1-induced proline isomerization in cytosolic p53 mediates bax activation and apoptosis. *Mol. Cell* **2015**, *59*, 677–684. [CrossRef] [PubMed]

31. Schwartz, M. Rho signalling at a glance. *J. Cell Sci.* **2004**, *117*, 5457–5458. [CrossRef] [PubMed]

32. Iden, S.; Collard, J.G. Crosstalk between small gtpases and polarity proteins in cell polarization. *Nat. Rev. Mol. Cell Biol.* **2008**, *9*, 846–859. [CrossRef] [PubMed]

33. Kjoller, L.; Hall, A. Signaling to rho gtpases. *Exp. Cell Res.* **1999**, *253*, 166–179. [CrossRef] [PubMed]

34. Chan, A.M.; Takai, S.; Yamada, K.; Miki, T. Isolation of a novel oncogene, net1, from neuroepithelioma cells by expression cDNA cloning. *Oncogene* **1996**, *12*, 1259–1266. [PubMed]

35. Oh, W.; Frost, J.A. Rho gtpase independent regulation of atm activation and cell survival by the rhogef net1a. *Cell Cycle* **2014**, *13*, 2765–2772. [CrossRef] [PubMed]

36. Guerra, L.; Carr, H.S.; Richter-Dahlfors, A.; Masucci, M.G.; Thelestam, M.; Frost, J.A.; Frisan, T. A bacterial cytotoxin identifies the RhoA exchange factor Net1 as a key effector in the response to DNA damage. *PLoS ONE* **2008**, *3*, e2254. [CrossRef] [PubMed]

37. Amano, M.; Nakayama, M.; Kaibuchi, K. Rho-kinase/rock: A key regulator of the cytoskeleton and cell polarity. *Cytoskeleton* **2010**, *67*, 545–554. [CrossRef] [PubMed]

38. Perruche, S.; Saae, P. L14. Immunomodulatory properties of apoptotic cells. *Presse Med.* **2013**, *42*, 537–543. [CrossRef] [PubMed]

39. Mills, J.C.; Stone, N.L.; Pittman, R.N. Extranuclear apoptosis. The role of the cytoplasm in the execution phase. *J. Cell Biol.* **1999**, *146*, 703–708. [CrossRef] [PubMed]

40. Ndozangue-Touriguine, O.; Hamelin, J.; Breard, J. Cytoskeleton and apoptosis. *Biochem. Pharmacol.* **2008**, *76*, 11–18. [CrossRef] [PubMed]

41. Moss, D.K.; Lane, J.D. Microtubules: Forgotten players in the apoptotic execution phase. *Trends Cell Biol.* **2006**, *16*, 330–338. [CrossRef] [PubMed]

42. Pittman, S.; Geyp, M.; Fraser, M.; Ellem, K.; Peaston, A.; Ireland, C. Multiple centrosomal microtubule organising centres and increased microtubule stability are early features of vp-16-induced apoptosis in ccrf-cem cells. *Leuk. Res.* **1997**, *21*, 491–499. [CrossRef]

43. Pittman, S.M.; Strickland, D.; Ireland, C.M. Polymerization of tubulin in apoptotic cells is not cell cycle dependent. *Exp. Cell Res.* **1994**, *215*, 263–272. [CrossRef] [PubMed]

44. Desouza, M.; Gunning, P.W.; Stehn, J.R. The actin cytoskeleton as a sensor and mediator of apoptosis. *Bioarchitecture* **2012**, *2*, 75–87. [CrossRef] [PubMed]

45. Van Engeland, M.; Kuijpers, H.J.; Ramaekers, F.C.; Reutelingsperger, C.P.; Schutte, B. Plasma membrane alterations and cytoskeletal changes in apoptosis. *Exp. Cell Res.* **1997**, *235*, 421–430. [CrossRef] [PubMed]

46. Moss, D.K.; Betin, V.M.; Malesinski, S.D.; Lane, J.D. A novel role for microtubules in apoptotic chromatin dynamics and cellular fragmentation. *J. Cell Sci.* **2006**, *119*, 2362–2374. [CrossRef] [PubMed]

47. Oropesa-Avila, M.; de la Cruz-Ojeda, P.; Porcuna, J.; Villanueva-Paz, M.; Fernandez-Vega, A.; de la Mata, M.; de Lavera, I.; Rivero, J.M.; Luzon-Hidalgo, R.; Alvarez-Cordoba, M.; et al. Two coffins and a funeral: Early or late caspase activation determines two types of apoptosis induced by DNA damaging agents. *Apoptosis* **2017**, *22*, 421–436. [CrossRef] [PubMed]

48. Brady, H.J.; Gil-Gomez, G.; Kirberg, J.; Berns, A.J. Bax α perturbs t cell development and affects cell cycle entry of t cells. *EMBO J.* **1996**, *15*, 6991–7001. [PubMed]

49. North, S.; Hainaut, P. P53 and cell-cycle control: A finger in every pie. *Pathol. Biol.* **2000**, *48*, 255–270. [PubMed]

50. Vairo, G.; Soos, T.J.; Upton, T.M.; Zalvide, J.; DeCaprio, J.A.; Ewen, M.E.; Koff, A.; Adams, J.M. Bcl-2 retards cell cycle entry through p27(kip1), prb relative p130, and altered e2f regulation. *Mol. Cell. Biol.* **2000**, *20*, 4745–4753. [CrossRef] [PubMed]

51. Blagosklonny, M.V. Cell death beyond apoptosis. *Leukemia* **2000**, *14*, 1502–1508. [CrossRef] [PubMed]

52. Raychaudhuri, S.; Willgohs, E.; Nguyen, T.N.; Khan, E.M.; Goldkorn, T. Monte carlo simulation of cell death signaling predicts large cell-to-cell stochastic fluctuations through the type 2 pathway of apoptosis. *Biophys. J.* **2008**, *95*, 3559–3562. [CrossRef] [PubMed]

53. Broker, L.E.; Huisman, C.; Ferreira, C.G.; Rodriguez, J.A.; Kruyt, F.A.; Giaccone, G. Late activation of apoptotic pathways plays a negligible role in mediating the cytotoxic effects of discodermolide and epothilone b in non-small cell lung cancer cells. *Cancer Res.* **2002**, *62*, 4081–4088. [PubMed]

54. Etienne-Manneville, S. From signaling pathways to microtubule dynamics: The key players. *Curr. Opin. Cell Biol.* **2010**, *22*, 104–111. [CrossRef] [PubMed]

55. Chang, J.; Xie, M.; Shah, V.R.; Schneider, M.D.; Entman, M.L.; Wei, L.; Schwartz, R.J. Activation of Rho-associated coiled-coil protein kinase 1 (rock-1) by caspase-3 cleavage plays an essential role in cardiac myocyte apoptosis. *Proc. Natl. Acad. Sci. USA* **2006**, *103*, 14495–14500. [CrossRef] [PubMed]

56. Bonfoco, E.; Zhivotovsky, B.; Orrenius, S.; Lipton, A.; Nicotera, P. Cytoskeletal breakdown and apoptosis elicited by no donors in cerebellar granule cells require nmda receptor activation. *J Neurochem.* **1996**, *67*, 2484–2493. [CrossRef] [PubMed]

57. Golsteyn, R.M. Cdk1 and Cdk2 complexes (cyclin dependent kinases) in apoptosis: A role beyond the cell cycle. *Cancer Lett.* **2005**, *217*, 129–138. [CrossRef] [PubMed]

58. Sanchez-Alcazar, J.A.; Rodriguez-Hernandez, A.; Cordero, M.D.; Fernandez-Ayala, D.J.; Brea-Calvo, G.; Garcia, K.; Navas, P. The apoptotic microtubule network preserves plasma membrane integrity during the execution phase of apoptosis. *Apoptosis* **2007**, *12*, 1195–1208. [CrossRef] [PubMed]

59. Zuo, D.; Jiang, X.; Han, M.; Shen, J.; Lang, B.; Guan, Q.; Bai, Z.; Han, C.; Li, Z.; Zhang, W.; et al. Methyl 5-[(1h-indol-3-yl)selanyl]-1h-benzoimidazol-2-ylcarbamate (m-24), a novel tubulin inhibitor, causes g2/m arrest and cell apoptosis by disrupting tubulin polymerization in human cervical and breast cancer cells. *Toxicol. In Vitro* **2017**, *42*, 139–149. [CrossRef] [PubMed]

60. Fischer, U.; Janicke, R.U.; Schulze-Osthoff, K. Many cuts to ruin: A comprehensive update of caspase substrates. *Cell Death Differ.* **2003**, *10*, 76–100. [CrossRef] [PubMed]

61. Gerner, C.; Frohwein, U.; Gotzmann, J.; Bayer, E.; Gelbmann, D.; Bursch, W.; Schulte-Hermann, R. The fas-induced apoptosis analyzed by high throughput proteome analysis. *J. Biol. Chem.* **2000**, *275*, 39018–39026. [CrossRef] [PubMed]

62. Young, A.; Dictenberg, J.B.; Purohit, A.; Tuft, R.; Doxsey, S.J. Cytoplasmic dynein-mediated assembly of pericentrin and γ tubulin onto centrosomes. *Mol. Biol. Cell* **2000**, *11*, 2047–2056. [CrossRef] [PubMed]

63. Andersen, S.S. Spindle assembly and the art of regulating microtubule dynamics by maps and stathmin/op18. *Trends Cell Biol.* **2000**, *10*, 261–267. [CrossRef]

64. Carazo-Salas, R.E.; Gruss, O.J.; Mattaj, I.W.; Karsenti, E. Ran-gtp coordinates regulation of microtubule nucleation and dynamics during mitotic-spindle assembly. *Nat. Cell Biol.* **2001**, *3*, 228–234. [CrossRef] [PubMed]

65. Galjart, N.; Perez, F. A plus-end raft to control microtubule dynamics and function. *Curr. Opin. Cell Biol.* **2003**, *15*, 48–53. [CrossRef]

66. Lamb, N.J.; Fernandez, A.; Watrin, A.; Labbe, J.C.; Cavadore, J.C. Microinjection of p34cdc2 kinase induces marked changes in cell shape, cytoskeletal organization, and chromatin structure in mammalian fibroblasts. *Cell* **1990**, *60*, 151–165. [CrossRef]

67. Sugawara, E.; Nikaido, H. Properties of adeabc and adeijk efflux systems of acinetobacter baumannii compared with those of the acrab-tolc system of *Escherichia coli*. *Antimicrob. Agents Chemother.* **2014**, *58*, 7250–7257. [CrossRef] [PubMed]

68. Fourest-Lieuvin, A.; Peris, L.; Gache, V.; Garcia-Saez, I.; Juillan-Binard, C.; Lantez, V.; Job, D. Microtubule regulation in mitosis: Tubulin phosphorylation by the cyclin-dependent kinase Cdk1. *Mol. Biol. Cell* **2006**, *17*, 1041–1050. [CrossRef] [PubMed]

69. Mills, J.C.; Lee, V.M.; Pittman, R.N. Activation of a PP2A-like phosphatase and dephosphorylation of tau protein characterize onset of the execution phase of apoptosis. *J. Cell Sci.* **1998**, *111*, 625–636. [PubMed]

70. Jiang, Y. Regulation of the cell cycle by protein phosphatase 2a in saccharomyces cerevisiae. *Microbiol. Mol. Biol. Rev.* **2006**, *70*, 440–449. [CrossRef] [PubMed]

71. Lane, J.D.; Lucocq, J.; Pryde, J.; Barr, F.A.; Woodman, P.G.; Allan, V.J.; Lowe, M. Caspase-mediated cleavage of the stacking protein grasp65 is required for golgi fragmentation during apoptosis. *J. Cell Biol.* **2002**, *156*, 495–509. [CrossRef] [PubMed]

72. Gromley, A.; Jurczyk, A.; Sillibourne, J.; Halilovic, E.; Mogensen, M.; Groisman, I.; Blomberg, M.; Doxsey, S. A novel human protein of the maternal centriole is required for the final stages of cytokinesis and entry into s phase. *J. Cell Biol.* **2003**, *161*, 535–545. [CrossRef] [PubMed]

73. Wilde, A.; Lizarraga, S.B.; Zhang, L.; Wiese, C.; Gliksman, N.R.; Walczak, C.E.; Zheng, Y. Ran stimulates spindle assembly by altering microtubule dynamics and the balance of motor activities. *Nat. Cell Biol.* **2001**, *3*, 221–227. [CrossRef] [PubMed]

74. Ridley, A.J.; Hall, A. The small GTP-binding protein rho regulates the assembly of focal adhesions and actin stress fibers in response to growth factors. *Cell* **1992**, *70*, 389–399. [CrossRef]

75. Coleman, M.L.; Olson, M.F. Rho gtpase signalling pathways in the morphological changes associated with apoptosis. *Cell Death Differ.* **2002**, *9*, 493–504. [CrossRef] [PubMed]

76. Wittmann, T.; Wilm, M.; Karsenti, E.; Vernos, I. Tpx2, a novel xenopus map involved in spindle pole organization. *J. Cell Biol.* **2000**, *149*, 1405–1418. [CrossRef] [PubMed]

77. Wickman, G.; Julian, L.; Olson, M.F. How apoptotic cells aid in the removal of their own cold dead bodies. *Cell Death Differ.* **2012**, *19*, 735–742. [CrossRef] [PubMed]

78. Wickman, G.R.; Julian, L.; Mardilovich, K.; Schumacher, S.; Munro, J.; Rath, N.; Zander, S.A.; Mleczak, A.; Sumpton, D.; Morrice, N.; et al. Blebs produced by actin-myosin contraction during apoptosis release damage-associated molecular pattern proteins before secondary necrosis occurs. *Cell Death Differ.* **2013**, *20*, 1293–1305. [CrossRef] [PubMed]

79. Gourlay, C.W.; Ayscough, K.R. The actin cytoskeleton in ageing and apoptosis. *FEMS Yeast Res.* **2005**, *5*, 1193–1198. [CrossRef] [PubMed]

80. Schmidt, A.; Hall, A. The rho exchange factor net1 is regulated by nuclear sequestration. *J. Biol. Chem.* **2002**, *277*, 14581–14588. [CrossRef] [PubMed]

81. Dubash, A.D.; Guilluy, C.; Srougi, M.C.; Boulter, E.; Burridge, K.; Garcia-Mata, R. The small gtpase rhoa localizes to the nucleus and is activated by Net1 and DNA damage signals. *PLoS ONE* **2011**, *6*, e17380. [CrossRef] [PubMed]

82. Srinivasan, S.; Ashok, V.; Mohanty, S.; Das, A.; Das, S.; Kumar, S.; Sen, S.; Purwar, R. Blockade of rho-associated protein kinase (rock) inhibits the contractility and invasion potential of cancer stem like cells. *Oncotarget* **2017**, *8*, 21418–21428. [CrossRef] [PubMed]

83. Somlyo, A.P.; Somlyo, A.V. Signal transduction by g-proteins, rho-kinase and protein phosphatase to smooth muscle and non-muscle myosin ii. *J. Physiol.* **2000**, *522 Pt 2*, 177–185. [CrossRef]

84. Byun, Y.; Chen, F.; Chang, R.; Trivedi, M.; Green, K.J.; Cryns, V.L. Caspase cleavage of vimentin disrupts intermediate filaments and promotes apoptosis. *Cell Death Differ.* **2001**, *8*, 443–450. [CrossRef] [PubMed]

85. Chen, F.; Chang, R.; Trivedi, M.; Capetanaki, Y.; Cryns, V.L. Caspase proteolysis of desmin produces a dominant-negative inhibitor of intermediate filaments and promotes apoptosis. *J. Biol. Chem.* **2003**, *278*, 6848–6853. [CrossRef] [PubMed]

86. Marceau, N.; Schutte, B.; Gilbert, S.; Loranger, A.; Henfling, M.E.; Broers, J.L.; Mathew, J.; Ramaekers, F.C. Dual roles of intermediate filaments in apoptosis. *Exp. Cell Res.* **2007**, *313*, 2265–2281. [CrossRef] [PubMed]

87. Erwig, L.P.; Henson, P.M. Clearance of apoptotic cells by phagocytes. *Cell Death Differ.* **2008**, *15*, 243–250. [CrossRef] [PubMed]

88. Oropesa-Avila, M.; Fernandez-Vega, A.; de la Mata, M.; Maraver, J.G.; Cordero, M.D.; Cotan, D.; de Miguel, M.; Calero, C.P.; Paz, M.V.; Pavon, A.D.; et al. Apoptotic microtubules delimit an active caspase free area in the cellular cortex during the execution phase of apoptosis. *Cell Death Dis.* **2013**, *4*, e527. [CrossRef] [PubMed]

MDPI

St. Alban-Anlage 66

4052 Basel

Switzerland

Tel. +41 61 683 77 34

Fax +41 61 302 89 18

www.mdpi.com

International Journal of Molecular Sciences Editorial Office

E-mail: ijms@mdpi.com

www.mdpi.com/journal/ijms

www.ingramcontent.com/pod-product-compliance
Lightning Source LLC
Chambersburg PA
CBHW051722210326
41597CB00032B/5576